ENVIRONMENTAL HEALTH

DADE W. MOELLER

ENVIRONMENTAL HEALTH

Third Edition

HARVARD UNIVERSITY PRESS
Cambridge, Massachusetts
London, England
2005

Printed in the United States of America

Library of Congress Cataloging-in-Publication Data

Moeller, D. W. (Dade W.)
Environmental health / Dade W. Moeller—3rd ed.
p. cm.
Includes bibliographical references and index.
ISBN 0–674–01494–4 (alk. paper)
1. Environmental health. I. Title.
RA565.M64 2004
616.9′8—dc22 2004047468

To Betty Jean, who for more than fifty years—until her death in 1998—was, and will continue to be, the joy of my life, and to Rad, Mark, Kehne, Matt, and Anne, their spouses, and our sixteen grandchildren, who never cease to make us proud

CONTENTS

PREFACE TO THE THIRD EDITION

The primary objectives in preparing this third edition were to incorporate new developments in the field and to add coverage of subject areas not previously included, such as environmental economics, terrorism, and ecosystems. As part of this effort, every chapter has been extensively revised and rewritten. In preparation for this effort, more than 1,000 articles and reports published during the 7 years since issuance of the second edition were reviewed and digested, and their salient features were incorporated into the text. These included articles published in scientific journals, as well as timely articles published in other well-respected publications. In some cases, such as the presentation of a preview of the revised simplified approach for explaining radiation protection standards to the public that is being developed by the International Commission on Radiological Protection, readers are being provided with information that will not be officially released until 2005. Another example is the inclusion of the guidelines that are anticipated to be provided in the revised "food pyramid" to be issued by the U.S. Department of Agriculture.

As part of this process, the text has been enriched by the addition of new tables and graphs and the insertion of brief case studies on a variety of topics. The chapters on "Air in the Home and Community" and "Drinking Water" now include standards for key contaminants, as well as discussions of key biological organisms and chemical contaminants that are of concern. As in the past, a concerted effort has been made to write a book that provides comprehensive coverage of the field. In the attempt to achieve this goal, many environmental problems are discussed in terms of both their local and global implications, both their short- and long-range impacts, and their importance to people who live in both the developed and less developed countries. In all cases, every effort has been made to ensure that the information being presented is based on "good science," and that the ensuing discussion offers a balanced assessment of current conditions.

Discussions of a number of emerging and controversial issues in environmental and public health are also incorporated into this new edition. These range from consideration of environmental justice, deforestation, and the protection of endangered species to topics such as multiple chemical sensitivity, the application of the threshold concept in evaluating the effects of toxic and radioactive materials, and assessments of the uncertainties in extrapolating laboratory data obtained through studies with small animals, such as mice, in estimating potential health effects in humans.

Care has also been taken to ensure that the reader understands the limitations associated with techniques, such as epidemiology and risk assessment, that are commonly applied in evaluating the impacts of various environmental stresses. At the same time, the potential for advancing the evidence that can be derived and the conclusions that can be reached through applications of newer techniques, such as molecular toxicology and epidemiology, is clearly enunciated.

Another feature of the third edition is the effort to ensure that the reader understands the differences among clinical medicine, public health, and environmental health. Explaining these relationships is important because previous editions of the book have been widely used as a text for teaching environmental health to MPH students, whose primary education is largely in the field of medicine. Also emphasized throughout the book is the need to adopt a systems approach in assessing environmental problems. Although almost everyone recognizes the need to manage and control various pollutants within individual components of the environment (air, water, soil, and food), there is a need to understand and account for interrelationships of these segments. Within this context, care has been exercised to ensure that the reader is aware of the need to protect both human and natural resources. This is exemplified by the concept of ecological risk assessment and the discussion of primary and secondary standards for airborne and waterborne contaminants—primary to protect the health of humans; secondary to protect the environment. The latter are emphasized through the discussion of acid precipitation, ozone depletion, and global warming.

As anyone who has undertaken the writing of a book of this magnitude would recognize, it would be next to impossible for one person to have all the knowledge and insight required to address the multitude of subjects, challenges, and issues involved. In this regard, multitudes of people willingly shared their talents and expertise in revising the book. Some did so by providing key references and data; others did so by volunteering to conduct independent reviews of individual chapters. These included former colleagues at the Harvard School of Public Health, faculty members at the School of Public Health, University of North Carolina, Chapel Hill, a host of employees in Dade Moeller & Associates, and many others who possessed expertise in specialized areas of environmental and public health. To all of these, whose names are listed below, I extend my gratitude and heartfelt appreciation.

Michelle Allen
Barbara D. Beck
William Burgess
Bill Craig
Theodore Daniell
Harold Denton
Douglas W. Dockery
John S. Evans
Robert M. Hallisey
Mickey Hunacek
Tracey Ikenberry
William E. Kennedy
Judson Kenoyer
Eric Krouse
Ralph Larsen
John B. Little
Steven E. Merwin
Ellen Messer-Wright
Carrie Moeller
Mark Moeller
Matthew Moeller
Peter Moeller
Thayer Moeller
Richard R. Monson
JoLynn Montgomery
Cynthia Palmer Olsen
Richard J. Pollack
Paul R. Portney
Ross Potter
Marc J. Roberts
Gene Rollins
Wendy Rosen
Michael T. Ryan
Jacob Shapiro
Cheryl Smith
Andrew Spielman
Casper Sun
Russ Treat
Robert Walker
Garret P. Westerhoff
Wesley Winne
Ellen Messer Wright
R. Craig Yoder

Finally, a special expression of gratitude is due to Charles Eberline for his editorial suggestions and to Michael Fisher and Sara Davis of Harvard University Press, who provided guidance throughout the preparation and completion of the new edition.

And God pronounced a blessing upon Noah and his sons and said to them, be fruitful and multiply and fill the earth.

And the fear of you and the dread and terror of you shall be upon every beast of the land, every bird of the air, all that creeps upon the ground, and upon all the fishes of the sea. Into your hands they are delivered.

Genesis 9:1–2

Hurt not the earth, neither the sea, nor the trees . . .

Revelation 7:3

ABBREVIATIONS

ACGIH	American Conference of Governmental Industrial Hygienists
ACSH	American Council on Science and Health
AEA	Atomic Energy Act
AEC	Atomic Energy Commission
AIDS	Acquired immune deficiency syndrome
AIHA	American Industrial Hygiene Association
AIHAJ	*American Industrial Hygiene Association Journal*
ALARA	As low as reasonably achievable
AMA	American Medical Association
APHA	American Public Health Association
AQI	Air-quality index
ASHRAE	American Society of Heating, Refrigeration, and Air-Conditioning Engineers
ASTM	American Society for Testing and Materials
ATSDR	Agency for Toxic Substances and Disease Registry, U.S. Department of Health and Human Services
AWWA	American Water Works Association
BACT	Best available control technology
BART	Best available retrofit technology
BAT	Best available technology
BEIs	Biological exposure indices
BEIR	Committee on the Biological Effects of Ionizing Radiation, National Research Council
BOD	Biochemical oxygen demand
Bt	*Bacillus thuringiensis*
Bti	*Bacillus thuringiensis israeliensis*
Btk	*Bacillus thuringiensis kurstaki*
BWR	Boiling-water reactor
C & C	Command and control

CAFE	Corporate average fuel economy
CAT	Computer-assisted tomography
CCA	Chromated copper arsenate
CDC	Centers for Disease Control and Prevention, U.S. Department of Health and Human Services
CEQ	Council on Environmental Quality
CERCLA	Comprehensive Environmental Response, Compensation, and Liability Act (Superfund Act)
CFC	Chlorofluorocarbon
CFR	Code of Federal Regulations
CIIT	Chemical Industry Institute of Toxicology
CO	Carbon monoxide
CO_2	Carbon dioxide
COD	Chemical oxygen demand
CRS	Congressional Research Service
CRT	Cathode-ray tube
DDD	1,1-dichloro-2,2-bis(*p*-chlorophenyl)ethane
DDE	1,1-dichloro-2,2-bis(*p*-chlorophenyl)ethylene
DDT	1,1,1-trichloro-2,2-bis(*p*-chlorophenyl)ethane
DEET	Diethyltoluamide
DMC	Dimethyl carbonate
DNA	Deoxyribonucleic acid
DO	Dissolved oxygen
DOE	U.S. Department of Energy
EIS	Environmental impact statement
EMAP	Environmental Monitoring and Assessment Program
EPA	Environmental Protection Agency
EPCRA	Emergency Planning and Community Right-to-Know Act
EPRI	Electric Power Research Institute
EU	European Union
eV	Electron volt
FDA	Food and Drug Administration, U.S. Department of Health and Human Services
FIFRA	Federal Insecticide, Fungicide, and Rodenticide Act
GI tract	Gastrointestinal tract
GM	Genetically modified
GRAS	Generally recognized as safe
HACCP	Hazard Analysis and Critical Control Points
HCs	Hydrocarbons
HHS	U.S. Department of Health and Human Services
HVAC	Heating, ventilating, and air conditioning
Hz	Hertz (cycles per second)
IARC	International Agency for Research on Cancer
ICNIRP	International Commission on Non-Ionizing Radiation Protection
ICRP	International Commission on Radiological Protection
IFT	Institute of Food Technologists

IIHS	Insurance Institute for Highway Safety
INPO	Institute of Nuclear Power Operations
ISM	Integrated safety management
IVHS	Intelligent vehicle highway systems
JAMA	*Journal of the American Medical Association*
LD_{50}	Lethal dose for 50 percent of the exposed population
LLRW	Low-level radioactive waste
LLRWPAA	Low-Level Radioactive Waste Policy Amendments Act
MADD	Mothers against Drunk Driving
MCL	Maximum contaminant level
MTBE	Methyl tertiary-butyl ether
NAAQS	National ambient air-quality standards
NADP	National Atmospheric Deposition Program
NAE	National Academy of Engineering
NAFTA	North American Free Trade Agreement
NASA	National Aeronautics and Space Administration
NCEA	National Center for Environmental Assessment
NCRP	National Council on Radiation Protection and Measurements
NEI	Nuclear Energy Institute
NEPA	National Environmental Policy Act
NEWWA	New England Water Works Association
NHEXAS	National Human Exposure Assessment Survey
NIOSH	National Institute for Occupational Safety and Health, U.S. Department of Health and Human Services
NIST	National Institute of Standards and Technology
NLVs	Norwalk-like viruses
NO_2	Nitrogen dioxide
NO_X	Nitrogen oxides
NPDES	National Pollution Discharge Elimination System
NPL	National Priorities List
NRC	National Research Council
NRPB	National Radiological Protection Board (United Kingdom)
NSC	National Safety Council
NSR	New Source Review
NTS	Nevada Test Site
O_3	Ozone
OECD	Organization for Economic Cooperation and Development
OSHA	Occupational Safety and Health Administration, U.S. Department of Labor
PAHO	Pan American Health Organization
PC	Personal computer
PCBs	Polychlorinated biphenyls
PET/CT	Positron emission tomography/computed tomography
$PM_{2.5}$	Particulate matter, 2.5 micrometers or smaller in size
PM_{10}	Particulate matter, 10 micrometers or smaller in size
PNNL	Pacific Northwest National Laboratories, U.S. Department of Energy

ppb	Parts per billion
ppm	Parts per million
PWR	Pressurized-water reactor
RACT	Reasonably available control technology
RCRA	Resource Conservation and Recovery Act
RFF	Resources for the Future
S & H	Safety and health
SARS	Severe acute respiratory syndrome
SO_2	Sulfur dioxide
TLVs	Threshold limit values
TSCA	Toxic Substances Control Act
UHF	Ultrahigh-frequency
UN	United Nations
USDA	U.S. Department of Agriculture
USNRC	U.S. Nuclear Regulatory Commission
USPHS	U.S. Public Health Service
UVR	Ultraviolet radiation
VOCs	Volatile organic compounds
VRE	Vancomycin-resistant enterococci
WANO	World Association of Nuclear Operators
WEF	Water Environment Federation
WHO	World Health Organization
WNV	West Nile virus

1

THE SCOPE

Many aspects of human well-being are influenced by the environment, and many diseases can be initiated, promoted, sustained, or stimulated by environmental factors. For this reason, the interactions of people with their environment are an important component of public health.

In its broadest sense, *environmental health* is the segment of public health that is concerned with assessing, understanding, and controlling the impacts of people on their environment and the impacts of the environment on them. Even so, this field is defined more by the problems it faces than by the approaches it uses. These problems include the treatment and disposal of liquid and airborne wastes, the elimination or reduction of stresses in the workplace, the purification of drinking-water supplies, the provision of food supplies that are adequate and safe, and the development and application of measures to protect hospital and medical workers from being infected with diseases such as acquired immune deficiency syndrome (AIDS) and severe acute respiratory syndrome (SARS). As this list implies, the basic source of our environmental problems is, in essence, the impact of an ever-increasing population (Figure 1.1).

Environmental health professionals also face long-range problems that include the effects of toxic chemicals and radioactive wastes, acidic deposition, depletion of the ozone layer, global warming, resource depletion, and the loss of forests and topsoil. The complexity of these issues requires multidisciplinary approaches. Thus a team that is coping with a major environmental health problem may include scientists, physicians, epidemiologists, engineers, economists, lawyers, mathematicians, and man-

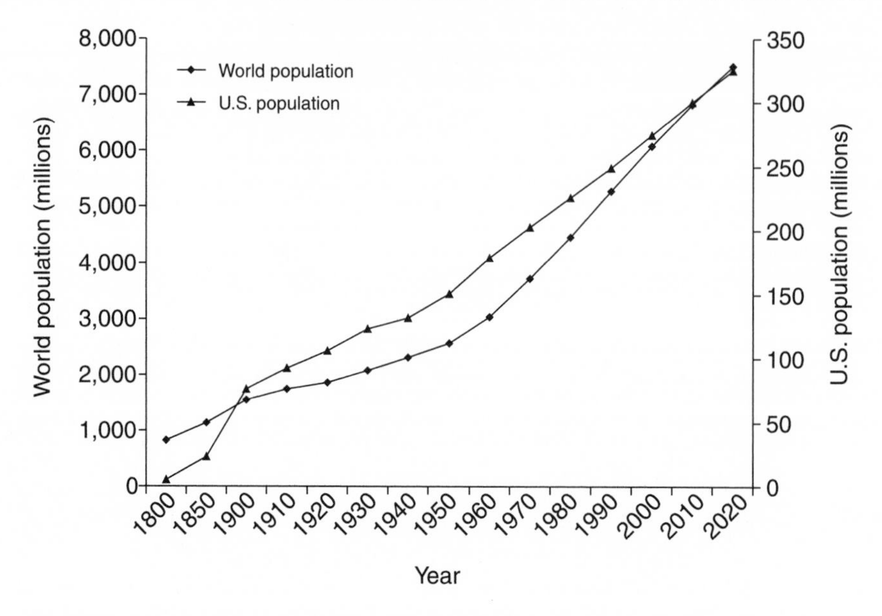

Figure 1.1 Trends in U.S. and world population from 1800 with projections to 2002

agers. Input from experts in these and related areas is essential for the development, application, and success of the control strategies necessary to encompass the full range of people's lifestyles and their environment.

Just as the field of public health involves more than disease (for example, health-care management, maternal and child health, epidemiology), the field of environmental health encompasses the effects of the environment on animals other than humans, as well as on trees and vegetation and on natural and historic landmarks. While many aspects of public health deal with the "here and now," many of the topics addressed within the subspecialty of environmental health are concerned with the previously cited impacts of a long-range nature.

Defining the Environment

To accomplish their goals effectively, environmental health professionals must keep in mind that there are many ways to define the environment. Some of the more prominent of these are described here.

THE INNER VERSUS OUTER ENVIRONMENT

From the standpoint of the human body, there are two environments: the one within the body and the one outside it. Separating them are three principal protective barriers: the skin, which protects the body from contaminants outside the body; the gastrointestinal (GI) tract, which protects the inner body from contaminants that have been ingested; and the membranes within the lungs, which protect the inner body from contaminants that have been inhaled (Figure 1.2, Table 1.1).

Although they may provide protection, each of these barriers is vulnerable under certain conditions. Contaminants can penetrate to the inner body through the skin by dissolving the layer of wax generated by the sebaceous glands. The GI tract, which has by far the largest surface area of any of the three barriers, is particularly vulnerable to compounds that are soluble and can be readily absorbed and taken into the body cells. Fortunately, the body has mechanisms that can protect the GI tract: un-

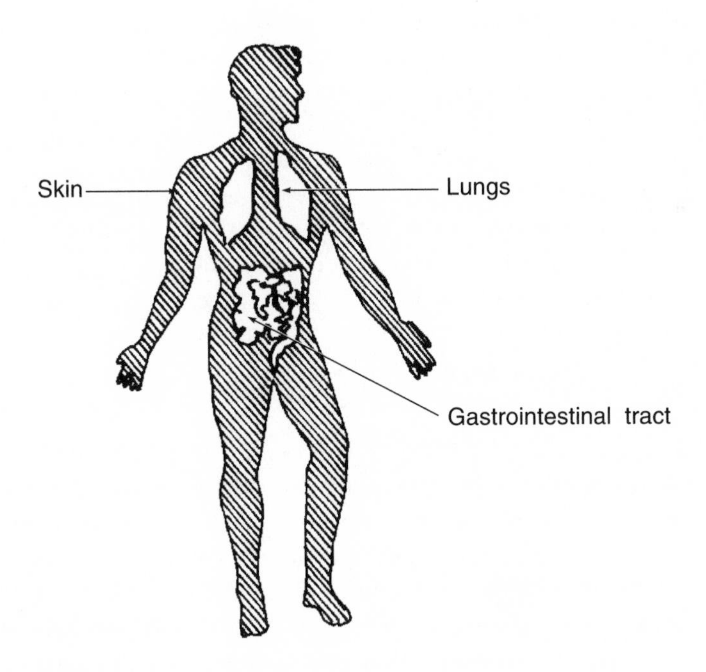

Figure 1.2 Barriers between the inner and outer environments

Table 1.1 Characteristics of the principal barriers between the outer and inner body

	Area		Thickness		Weight		Daily exposure	
Barrier	m^2	ft^2	μm	in	kg	lb	kg	lb
Skin	2	21	100	4×10^{-3}	12–16	30	Variable	
GI tract	200	2150	10–12	4×10^{-4}	7	15	3–4	6½–9
Lungs	140	1500	0.2–0.4	1×10^{-5}	0.8–0.9	2	24	50

wanted material can be vomited via the mouth or rapidly excreted through the bowels (as in the case of diarrhea). Airborne materials in the respirable size range may be deposited in the lungs and, if they are soluble, may be absorbed. Mechanisms for protecting the lungs range from simple coughing to cleansing by macrophages that engulf and promote the removal of foreign materials. Unless an environmental contaminant penetrates one of the three barriers, it will not gain access to the inner body, and even if a contaminant is successful in gaining access, the body still has mechanisms for controlling and/or removing it. For example, materials that enter the circulatory system can be detoxified in the liver or excreted through the kidneys.

Although an average adult ingests about 1.5 kilograms of food and 2 kilograms of water every day, he or she breathes roughly 20 cubic meters of air per day. This amount of air weighs more than 24 kilograms. Because people usually cannot be selective about what air is available, the lungs are the most important pathway for the intake of environmental contaminants into the body. The lungs are also by far the most fragile and susceptible of the three principal barriers.

THE PERSONAL VERSUS AMBIENT ENVIRONMENT

In another definition, people's "personal" environment, the one over which they have control, is contrasted with the working (Chapter 4) or ambient (outdoor) environment, over which they may have essentially no control. Although people commonly think of the working or outdoor environment as posing the higher threat, environmental health experts (as noted in the discussions that follow) estimate that the personal environment, influenced by hygiene, diet, sexual practices, exercise, use of tobacco, drugs, and alcohol, and frequency of medical checkups, often has much more, if not a dominating, influence on human well-being. This is illustrated in Table 1.2, which summarizes the estimated contributions of these various factors to cancer deaths in an industrialized society. As may be noted, the personal environment and the lifestyles followed by individuals account for about 70 percent or more of such deaths. For this reason, the influence of the personal environment on cancer will be discussed in more detail in one of the sections that follow.

THE GASEOUS, LIQUID, AND SOLID ENVIRONMENTS

The environment can also be considered as existing in one of three forms—gaseous, liquid, or solid. Each of these is subject to pollution, and people interact with all of them (Figure 1.3). Particulates and gases are often released into the atmosphere, sewage and liquid wastes are dis-

Table 1.2 Relative importance of various causes of cancer, United States

Risk factor	Estimated percentage of total cancer deaths attributable this factor
Tobacco	30
Adult diet/obesity	30
Sedentary lifestyle	5
Occupational factors	5
Family history of cancer	5
Viruses/other biologic agents	5
Perinatal factors/growth	5
Reproductive factors	3
Alcohol	3
Socioeconomic status	3
Environmental pollution	2
Ionizing/ultraviolet radiation	2
Prescription drugs/medical procedures	1
Salt/other food additives and contaminants	1

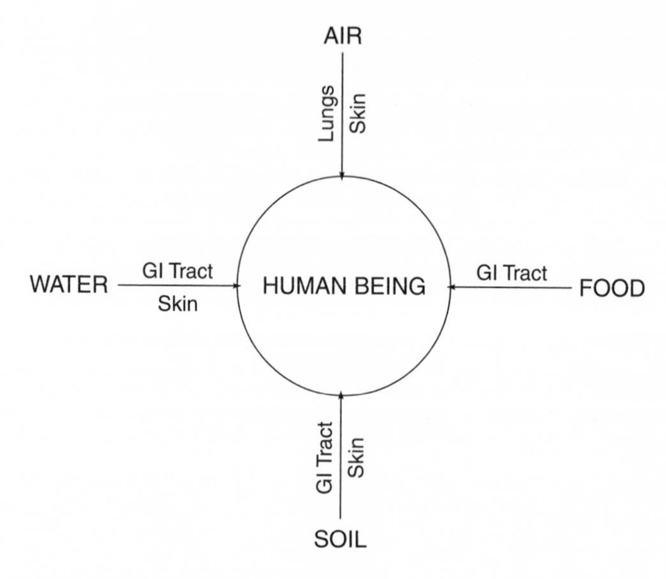

Figure 1.3 Routes of human exposure through the gaseous, liquid, and solid environment

charged into water (Chapter 8), and solid wastes, particularly plastics and toxic chemicals, are disposed of on land (Chapter 9).

THE CHEMICAL, BIOLOGICAL, PHYSICAL, AND SOCIOECONOMIC ENVIRONMENTS

Another perspective considers the environment in terms of the four avenues or mechanisms by which various factors affect people's health.

1. *Chemical* constituents and contaminants include toxic wastes and pesticides in the general environment, chemicals used in the home and in industrial operations (Chapter 4), and preservatives used in foods (Chapter 6).
2. *Biological* contaminants include various disease organisms that may be present in food and water (Chapters 6 and 7), those that can be transmitted by insects and animals (Chapter 10), and those that can be transmitted by person-to-person contact.
3. *Physical* factors that influence health and well-being range from injuries and deaths caused by accidents (Chapter 11) to excessive

noise, heat, and cold and to the harmful effects of ionizing and nonionizing radiation (Chapter 12).

4. *Socioeconomic* factors, though perhaps more difficult to measure and evaluate, significantly affect people's lives and health. Statistics demonstrate compelling relationships between morbidity and mortality and socioeconomic status. People who live in economically depressed neighborhoods are less healthy than those who live in more affluent areas.

Clearly, illness and well-being are the products of community, as well as of chemical, biological, and physical, forces. Factors that contribute to the differences range from the unavailability of jobs, inadequate nutrition, and lack of medical care to stressful social conditions, such as substandard housing and high crime rates. The contributing factors, however, extend far beyond socioeconomics. Studies have shown that people without political power, especially disadvantaged groups who live in lower-income neighborhoods, often bear a disproportionate share of the risks of environmental pollution. One common example is increased air and water pollution due to nearby industrial and toxic waste facilities. Disadvantaged groups also suffer more frequent exposure to lead paint in their homes and to pesticides and industrial chemicals in their work.

Taking action to correct such inequities, the U.S. President in 1994 signed an executive order on environmental justice. In implementing the accompanying requirements, the Environmental Protection Agency (EPA, 2001) defined the term as "the fair treatment of people of all races, cultures, and incomes with respect to the development, implementation, and enforcement of environmental laws and policies, and their meaningful involvement in the decision-making process of the government." Included among the objectives of this effort is a reaffirmation that all communities and individuals, regardless of economic status or racial makeup, are entitled to a safe and healthful environment and that, in the future, the risks associated with hazardous industrial facilities will be distributed equitably across population groups. As part of the process of selecting a site for any potentially hazardous operation, regulators are required to identify and critically examine all potentially adverse impacts on the health and environment of minority and low-income populations. As noted in the EPA definition, the order specifically requires that disadvantaged populations have an opportunity to participate fully in decisions that affect their health and environment.

One of the stimuli for this action was that all too often questions relating to environmental justice on past projects have not been discussed until the permitting or decision-making stage. At that point, any revisions of the plans or the development of alternatives were likely to be costly and difficult. As a result, it was often "too late" for changes to be made. Fortunately, in more recent years, federal and environmental regulators, strongly supported by community-based organizations, have worked together to ensure not only that issues relating to environmental justice are properly addressed but also that community involvement is undertaken sufficiently early in the decision-making process to have an impact. As a result, environmental managers are learning the advantages of being proactive on this issue. They are also learning to be prepared to treat this subject with the importance it deserves and to have procedures in place for addressing questions when they are raised (Targ and Bowen, 2002).

None of the preceding definitions of the environment is without its deficiencies, and, as noted in the section that follows, the list is by no means complete. Classification in terms of inner and outer environments or in terms of gaseous, liquid, and solid environments, for example, fails to take into account the significant socioeconomic factors cited earlier or physical factors such as noise and ionizing and nonionizing radiation. As a result, consideration of the full range of existing environments is essential for understanding the complexities involved and controlling the associated problems.

THE URBAN ENVIRONMENT

Another environment that is assuming increasing importance is that of large cities, the so-called urban environment. One of the primary reasons is that today about half of the world's population lives in urban centers (Figure 1.4). As noted later (Chapter 20), this is projected to increase to 60 percent within the next 20 years, with a major share of the change occurring within the less developed countries (Bugliarello, 2001). Unfortunately, the quality of life in cities throughout the world has been declining. As a result, many urban environments today are noisy, congested, frustrating, and unhealthy. Wildlife habitat is scarce, streams flow in artificial channels, wetlands are being filled, and aquifers are being depleted. Furthermore, the heat islands created by urban centers increase both the costs for cooling and the concentrations of air pollutants (DeKay and O'Brien, 2001).

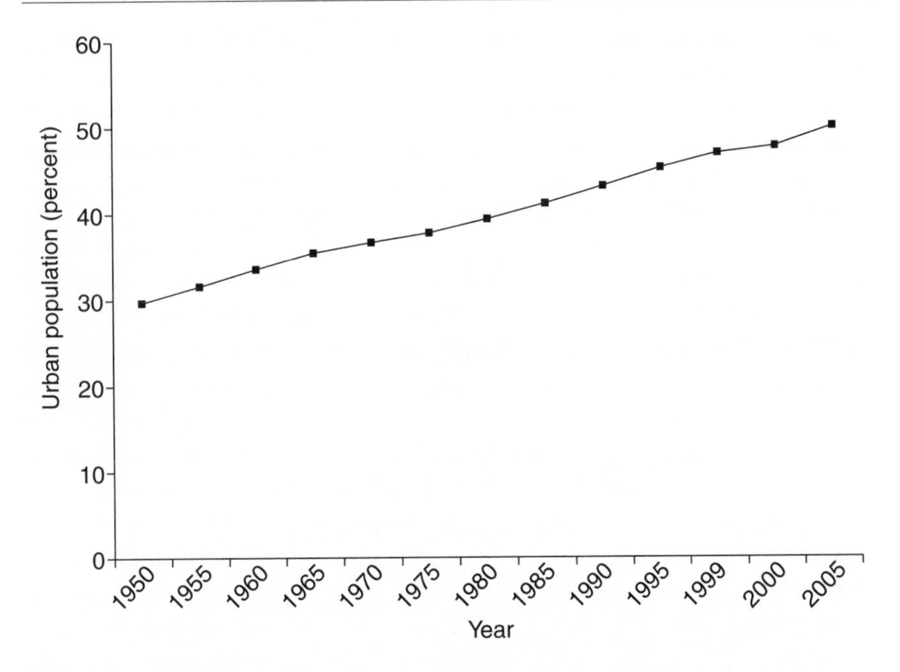

Figure 1.4 Increase in percentage of world population living in urban areas

If these problems are to be mitigated, methods must be found to make urban areas environmentally and socially sustainable. Recognition of this need is not new. Leonardo da Vinci proposed some 500 years ago that pedestrian and vehicular travel within cities be separated by placing them on two different levels (Bugliarello, 2001). This has been accomplished in the city of Boston, where, by moving a major vehicular transportation artery underground, officials have created large open spaces in the heart of the downtown metropolitan area. Another approach is to develop what might be called a hybrid city by making gardens an integral part of the urban area. This has been common practice in China for centuries, as has the reservation of areas surrounding its cities for agriculture. Extending the concept, the Chinese use city-generated waste to fertilize these areas. In a similar, but more modest effort, officials in New York City now promote the development of ad hoc urban gardens. Through this step, they have, in a sense, made urban agriculture an explicit element in city planning.

Other measures being used to revitalize cities are typified by officials in

Bogotá, Colombia, who have constructed riding pathways to encourage people to use bicycles for shorter trips. In Copenhagen, Denmark, officials have actually replaced curbside parking spaces for automobiles with bicycle lanes and walkways. As a result, that city has changed from being car oriented to being people oriented. The success of this latter effort is demonstrated by the fact that the total distance traveled by motor vehicles in Copenhagen is 10 percent less today than it was in 1970. Similar changes are taking place in Washington, DC, through the revitalization of portions of the downtown area that were left abandoned during the 1970s and 1980s (Sheehan, 2002). In multitudes of other U.S. cities, abandoned commercial properties are being revitalized through the brownfields program (Chapter 9).

Another approach is to incorporate a range of environmental features into the design, construction, and landscaping of city buildings. This approach, which is called "green architecture," includes the installation of systems for treating wastewater from toilets so that it can be recycled and the use of windows designed not only to open, but also to admit sunlight while concurrently reducing the addition or loss of heat through radiation. Still other revitalization steps include shifting to smaller, decentralized sources of energy, such as wind and solar power, while taking advantage of cogeneration and conservation (Chapter 18). The planting of trees is also being encouraged to provide both shade and sinks for stormwater runoff. Adding to the benefits is that the leaves of trees absorb airborne gases, such as sulfur dioxide, carbon monoxide, nitrogen dioxide, and ozone. The leaves also serve as sticky surfaces for the removal of airborne particles (Fields, 2002).

Cancer and the Personal Environment

One of the stated objectives of the National Cancer Institute is to translate the results of its research into ways of saving lives (Kaiser, 2002). Much the same approach has been adopted by other groups, including members of the Harvard Center for Cancer Prevention (Colditz et al., 2002). In fact, it is the judgment of the latter group that sufficient information is now available on the causes and prevention of cancer to enable the burden of this disease in the United States to be reduced by more than half during coming decades. One of the dominating supporting points is the information previously discussed (Table 1.2). At the same time, the Center staff is quick to acknowledge that this is not to imply that everything about cancer is known. Rather, it is to emphasize that the information that is available today can be effectively used to reduce the incidence of these types of diseases. What is needed is the initiation of programs that relay this information to members of the public and encourage them to change their lifestyles and personal habits in ways that will enable them to benefit. If long-term progress is to be achieved, however, the proposed programs must be supported by the enactment of a series of public policies that are designed to make risk-reduction behaviors easier for individuals to choose and maintain. Therefore, the recommendations of the Harvard Center focus on the following five major behavioral risk factors.

TOBACCO USE

It is estimated that the use of tobacco in the United States causes more than 450,000 deaths each year and is responsible (Table 1.2) for about 30 percent of all cancer-related deaths. This includes an estimated 35,000 deaths attributable to cancers in people who are subjected to secondhand smoke. On a global basis, tobacco use causes more than 4 million deaths each year, about 11,000 per day (Meister et al., 2003). While in the past, cigarette smoking was primarily linked to cancers of the lungs, oral cavity, gullet, larynx, pharynx, pancreas, and bladder, the latest reviews show that the list also includes cancers of the stomach, liver, cervix, uterus, kidney, and nasal sinuses, as well as myeloid leukemia. The years of life that are lost due to the early deaths of smokers are also significant. These range from 13 years for males to 14 years for females. The associated annual medical costs and productivity losses exceed \$50–\$70 billion, and \$50 billion, respectively. In fact, the medical costs represent about 8 percent of personal health-care expenditures (CDC, 2002a).

Even so, rates of tobacco use among high-school students are either increasing or remaining discouragingly high. Indeed, about 3,000 U.S. young people under the age of 18 years become regular smokers every day. At the same time, cigarette smoking is becoming an even more important health problem in some of the less developed countries of the world. The more than 300 million smokers in the People's Republic of China alone equal the total number in all the developed countries of the world combined. Adding to the concern is that the most recent epidemiological studies show that unless these people quit, upwards of half of them will eventually be killed by their habit (Hesketh, Ding, and Tomkins, 2001). If current smoking patterns continue, the annual number of people killed by tobacco will increase from a level of about 3 million per year in 1990 to about 10 million per year in 2030. This is truly an epidemic (Holden, 2001).

Given the pervasiveness of smoking in our society, effective control will require a multifaceted approach that addresses both prevention and cessation. In particular, attention needs to be directed to counteracting the multiple and somewhat devious avenues through which cigarettes are promoted. These include the glamorization of smoking in movies and television programs and the marketing of cigarettes through what otherwise would be prohibited avenues. The latter is accomplished through methods such as "brand stretching," wherein cigarette companies put their brand names on clothing lines and then advertise the clothing through various media outlets. This permits them to achieve brand recognition without violating marketing restrictions. Also essential are increased efforts to designate areas, such as restaurants, as smoke-free zones. An equivalent effort should be made with respect to workplaces, an action that reduced cigarette consumption in Australia between 1988 and 1995 by more than 20 percent. Similar programs have proved effective in universities through banning smoking in dormitories and student residence halls. Although the primary benefit is a smoke-free residential life, such bans may actually prevent the initiation of smoking and promote cessation (Fisher, 2001).

Increased attention also needs to be given to the design of and space allotted to warning labels on packages of cigarettes. In contrast to the approach required in the United States—a small label that is limited to the side of the package—the labels in Canada are required to cover half of the front and back of the package and to include pictures that graphically demonstrate the damage cigarettes can cause. In addition, manufacturers must include written guidance inside the package on methods for ceasing to smoke. Some 90 percent of the smokers surveyed in Canada said that they "noticed" the warning, and almost half said that the warnings increased their motivation to cease smoking. In contrast, the warnings required in the United States regularly go unnoticed. Stimulated by the success of the Canadian approach, the World Health Organization is leading a program to have similar requirements adopted throughout the world (Late, 2002).

PHYSICAL ACTIVITY

Physical activity has numerous health benefits, including reductions in the risk of colon and breast cancer and possibly a reduced risk of lung and prostate cancer. It also reduces premature mortality, cardiovascular disease, hypertension, diabetes, and osteoporosis. Even so, 75 percent of the U.S. population do not achieve the minimum daily recommended 30 minutes of walking or its equivalent (Shaw, 2004), and 40 percent of adults engage in no leisure-time physical activity at all (USPHS, 2001). That this is not surprising is illustrated by the fact that most U.S. communities are not structured to accommodate or encourage physically active lifestyles.

What is needed is the development of policies and environmental approaches that enhance and facilitate opportunities for young people to participate in physical activity (Colditz et al., 2002). A first step is the inclusion of physical education and fitness as part of the curriculum in public schools. As of 2002, only one U.S. state (Illinois) required daily physical education for kindergarten through twelfth grade, and even there, waivers were available to permit students to replace physical education with activities, such as band or choir. Nationwide, only 8 percent of elementary schools, 6.4 percent of middle schools, and 5.8 percent of high schools provide daily physical education for the entire school year. Twenty-five percent of schoolchildren are not provided an opportunity for any form of physical education whatsoever (Brink, 2002).

While exercise is important during childhood and adolescence, parents and school officials must also recognize that exposure of these age groups to ultraviolet radiation plays a role in the future development of skin cancer. The extent of this problem is illustrated by the fact that the number of new cases of melanoma, the most serious form of skin cancer, in this country has increased by 150 percent since 1975; at the same time, the number of related deaths has increased by more than 40 percent. Since more than half of a person's lifetime exposure to the sun occurs during the younger years, it is particularly important that these age groups be protected. Care must be taken not to exacerbate one problem while protecting people from other types of cancers and diseases (CDC, 2002b).

WEIGHT MAINTENANCE

Weight maintenance and the avoidance of obesity have long been a problem in the United States. Indeed, the latest surveys indicate that an estimated 61 percent of adults, 13 percent of children, and 14 percent of adolescents in this country are overweight, a threefold increase since 1980 (USPHS, 2001). Application of these numbers nationwide reveals that more than half of the population is either overweight or obese (Friedman, 2003). In terms of its public health impacts, obesity is estimated to be responsible for nearly 300,000 deaths in this country each year. Substantial portions of these deaths are attributable to diabetes, heart disease, hypertension, and cancer (Colditz et al., 2002).

The problem is compounded by the lack of a unified national strategy for the prevention of obesity and/or the promotion of programs to facilitate weight maintenance. Because obesity is closely intertwined with physical exercise (discussed earlier) and the maintenance of a healthy diet (discussed in the next section), the policies recommended for addressing these two problems are equally applicable to weight reduction and maintenance. This is one of the many reasons that the National Cancer Society strongly recommends that adults engage in leisure-time physical activities (Holden, 2001). Unless changes are made, public health officials predict that obesity could soon overtake tobacco use as the major source of preventable death in this country (USPHS, 2001). Various levels of the problem, however, exist in countries throughout the world. Indeed, obesity has been recognized by the World Health Organization as one of the top 10 global health problems (Kelmer and Helmuth, 2003).

HEALTHY DIETS

The previously cited epidemic of obesity in the United States is only the most conspicuous manifestation of the deplorable state of our diets. There is clear and convincing evidence, for example, that a diet rich in plant foods and moderate in animal products reduces the risk of cardiovascular disease and diabetes (Chapter 6). Although the evidence that links diet and cancer is not as robust, many studies suggest that such a diet might also modestly reduce this risk (Colditz et al., 2002). Yet the U.S. population continues to consume increasing amounts of refined starch, sugar, and hydrogenated fats (transfatty acids). The estimated annual costs of these and other impacts are of the order of $150 billion and are continuing to increase (Willett, 2002).

The collective evidence clearly shows that the U.S. public needs to change its dietary habits. As a beginning, policies should be developed and implemented that stimulate the provision of healthy meals and snacks in schools and workplaces and encourage nationwide consumption of the products of community gardens and farmers' markets (Colditz et al., 2002). Undoubtedly, similar problems need to be addressed in certain other parts of the world.

ALCOHOL

On the basis of epidemiological evidence, the International Agency for Research on Cancer concluded in 1988 that alcohol is a carcinogen and an independent risk factor for cancers of the liver and upper aerodigestive tract. Subsequent evidence has confirmed that alcohol consumption also increases the risk of breast cancer and possibly colon cancer. Although this evidence, as well as the role that alcohol plays in increased vehicular deaths and injuries, clearly calls for a reduction in its use, other data show that such a reduction would lead to a higher rate of cardiovascular disease. On the basis of this evidence, it would appear that the only ethical course of action is to conduct public educational campaigns that focus on reducing the abuse of alcohol and encouraging those who drink to do so moderately (Colditz et al., 2002). Such campaigns should include efforts to make people aware that the same benefits can be obtained through the consumption of certain types of grape juice.

The overriding message is that actions initiated by people on an individual basis represent an extremely effective method for controlling cancer. As noted, it is important that these actions be facilitated by government and private institutions through the development and implementation of policies that encourage and support the required behavioral changes. A good example of such support is the "Healthy People 2010" initiative of the U.S. Department of Health and Human Services (HHS, 2000). This initiative emphasizes the need for individuals to choose healthy lifestyles for themselves and their families. It also challenges communities and businesses to support health-promoting policies in schools, workplaces, and other settings. Ten leading health indicators have been designated for measuring success in achieving the goals of the initiative (Table 1.3). As may be noted, the first five of these involve individual choices. While the second five relate primarily to what might be called systemwide issues, at least two, namely, injury and violence and immunization, clearly depend on personal choices to some degree.

Specific Problems of Different Age Groups

While the environments in which people live and work are important, there are different factors that must be taken into consideration in assessing the problems of each specific age group. Examples of several of these are discussed here.

THE ELDERLY

For the elderly, one of the major sources of potential hazards is the home and the safety of the environment it provides. Specific problems include areas that are poorly lighted, combined with light switches that are either not clearly marked or cannot be seen in the dark; pathways that have

Table 1.3 Leading health indicators, *Healthy People 2010*

1.	Physical activity
2.	Overweight and obesity
3.	Tobacco use
4.	Substance abuse
5.	Responsible sexual behavior
6.	Mental health
7.	Injury and violence
8.	Environmental quality
9.	Immunization
10.	Access to health care

Note: The listing is not intended to represent the relative importance of the various indicators

obstructions, such as cords, loose throw rugs, or carpet edges that are curled; chairs and tables that are not sturdy and/or move easily; chairs and toilet seats that are low and difficult to get out or off of; areas that are slippery, particularly in bathtubs and showers; and tubs and showers that are not equipped with grab bars. Even though these hazards are well known, in many cases they are not addressed even in homes that are supposedly designed for the elderly. Surveys show, for example, that two or more such deficiencies exist in almost 60 percent of the bathrooms and in 23 to 42 percent of the other rooms in such homes. Nearly all homes in one survey had at least two potential hazards (Gill et al., 1999).

While these problems can readily be solved, others represent a more formidable challenge. One is to ensure that actions of the elderly do not cause them injury and/or death. Many approaches being applied in this case are based on sophisticated electronic systems, such as ones that will check whether ill people have taken their medication and recognize if they have become immobile or have fallen and injured themselves. Other systems can regulate the temperature of the water in the bathtub and even jog the memory of an occupant if a kettle on the stove has been left unwatched too long. A primary reason for the increasing importance of these types of problems is the escalating life expectancy. While during the period 1980–2000 the total population in the United States increased by 20 percent, that of people 65 years of age or older increased by more than 35 percent, and that of people 85 years of age or older increased by more than 90 percent (Martin, 2001). Worldwide, similar changes are taking place. For example, the number of people 65 years of age or older more than tripled between 1950 and 2000.

YOUNG PEOPLE

As has been emphasized in the preceding sections, data consistently demonstrate that lifestyles and personal habits have major influences on the health of individuals. These include their behavior in transportation vehicles, their choice of diets, and their decision on whether to smoke. Since patterns of adult behavior are largely established during youth, it is imperative that this age group be a primary audience for the receipt of information on these matters. As will be noted in the data presented here, the situation with respect to young people in the United States is particularly disturbing.

Recent surveys reveal, for example, that almost 20 percent of the young people in the age category 10–14 years in this country have rarely or never worn a safety belt while in a car. Furthermore, during the preceding 30 days more than a third of them had ridden in a car with a driver who had been drinking alcohol, more than half had consumed alcohol, and more than a third had smoked cigarettes. Also indicative of such lifestyle choices was that less than a third of them had eaten the recommended number of servings of fruits and vegetables during the preceding day (CDC, 1998). Although these revelations are due to a range of factors, one of the most significant is that they reflect an apparent lack of understanding on the part of parents and caregivers of the importance of helping children develop healthy living habits. Unfortunately, they may also reflect a lack of communication between parents and their children.

The types of problems being faced with young people, however, do not end here. As often is the case, unexpected and more subtle problems are discovered. One example is the harm caused by the backpacks used by young students to carry their books and other personal items to school. Noting that she was suffering back pain, a 14-year-old female student in Texas interviewed her classmates and found not only that a number of them were experiencing similar discomfort but also that some of them were suffering shoulder discomfort. On the basis of her study, she recommended that students carry no more than 10 percent of their body weight in such devices, that they carry them using both straps (not by slinging one strap over one shoulder), and that they place the heaviest items in the bottom of the pack so that they are close to the body (Guyer, 2001). Follow-up studies showed that backpack injuries are sufficiently painful to cause an estimated 3,000 to 4,000 U.S. school students to report to emergency rooms each year. Possible solutions include replacing hardbacks with paperbacks; printing slimmer, two-volume sets (one for each semester); or issuing smaller textbooks supplemented by CD-ROMs.

Other concerns include the quality of the air inside schools (Chapter 5). In many cases, poor air quality is caused by inadequate ventilation rates. In fact, studies show that the air in 30 to 40 percent of the nation's schools contains molds and other pollutants, such as volatile organic chemicals emitted from cleaning products, photocopiers, and classroom furnishings. Trailer units, which are used to provide additional space in overcrowded schools, have been found to have high airborne concentrations of formaldehyde and benzene. Noise is also a problem in teaching facilities located near airports. Such problems represent not only a risk to health but also a detriment to the learning process (Wakefield, 2002).

These types of problems are primarily those of children in the developed nations. Worldwide, there are many additional factors and activities to consider. According to the International Labour Office, for example, an estimated 100 million or more children aged 5 to 14 years work full-time. Many of them, the vast majority of whom live in the less developed countries, particularly those in Asia and Africa, are employed in tasks in which the accident rates are high. In addition, they are frequently subjected to harmful substances, physical agents, and psychosocial hazards (Forastieri, 1997).

CHILDREN

A close examination of the behavior and biological characteristics of infants and young children shows that there are multiple reasons that they are more susceptible to certain types of environmental stresses. For example, the metabolic pathways of the young, especially during the first few months of life, are immature; children are in the growing stage of life, a time during which their development processes are easily disrupted; and after exposure, they have more years in which to develop the range of chronic diseases that may be initiated. Another contributing factor is that infants and small children spend a considerable amount of time crawling either on the floor indoors or on the ground outdoors. This not only exposes them to higher levels of environmental toxicants, but also increases the possibility of absorbing toxic chemicals through the skin and ingesting them through hand- and object-to-mouth activities (Suk, 2002). Even if these factors are taken into account, additional, frequently surprising problems are often discovered. One was that the activation of a safety air bag in a motor vehicle could be fatal to infants and small children (Chapter 11).

Even so, the need to address the environmental health problems of children was late in being recognized, especially in the United States. Fortunately, this problem is now being corrected. The primary stimuli for these changes were (1) the issuance in 1993 of the National Research Council (NRC) report on "Pesticides in the Diets of Infants and Children" and (2) the convening in 1994 by the Children's Environmental Health Network of the first scientific conference on this subject. In rapid sequence thereafter, the U.S. Congress passed the Food Quality Protection Act of 1996, which incorporated the major recommendations of the NRC report, including a requirement that pesticide standards be set at levels that are protective of the health of children; the same year, the Environmental Protection Agency (EPA) established an Office of Children's Health Protection; the following year (1997), the President issued an executive order requiring that all federal agencies reduce environmental threats to children; and Congress followed with passage of the Children's Health Act in 2000.

In a similar manner, there has been increasing emphasis on the problem worldwide. For example, in 2002 the World Health Organization convened the International Conference the Environmental Health Threats to the Health of Children: Hazards and Vulnerability. One of the highlights of this conference was the issuance of "The Bangkok Statement," a pledge to protect children against environmental stresses (Table 14.3, Chapter 14). Also significant was the designation of "Healthy Environments for Children" as the theme of World Health Day for 2003 (Eskenazi and Landrigan, 2002).

Assessing Problems in the Ambient Environment

Among the many tasks that confront environmental health professionals is understanding the various ways in which humans interact with the ambient (indoor or outdoor) environment. In fulfilling this task, a primary step is to study the process or operation that leads to the generation of a problem and to determine how best to achieve control. Components of such an analysis include (1) determining the source and nature of each environmental contaminant or stress; (2) assessing how and in what form that contaminant or stress comes into contact with people; (3) measuring the resulting physical and economic (Chapter 13) impacts; and (4) applying controls when and where appropriate. In the case of air and water pollution, experience shows that instead of focusing on one or more individual sources within a given facility, every effort should be made to gather data on all the discharges from the facility, all the sources of each specific pollutant, and all the pollutants being deposited in the adjoining region, regardless of their nature, origin, or pathway (Chapter 16).

Even though tracing the source and pathways of each contaminant is important, an essential part of the process is to determine the effects on human health and the environment. When a pollutant is being evaluated for the first time, and exposure limits have not been established, such efforts may entail determining relationships between the exposure, the resulting dose, and the associated effects (Chapter 3). Armed with this information, appropriate governmental bodies, often in concert with various professional societies and organizations, can then move forward to establish standards for limiting exposures to the contaminant or stress (Chapter 15).

To assess the effects of exposures correctly, care must be taken to account not only for the fact that they can derive from multiple sources and enter the body by several routes, but also that elements in the environment are constantly interacting. In the course of transport or degradation, agents that were not originally toxic to people may become so, and vice versa. If the concentration of a contaminant in the environment (for example, a substance in the air) is relatively uniform, local or regional sampling may yield data adequate to estimate human exposure (Chapter 16). If concentrations vary considerably over space and time (as is true of certain indoor pollutants) and the people who are being exposed move about extensively, it may be necessary to measure exposure of individual workers or members of the public by providing them with small, lightweight, battery-operated portable monitoring units (Chapter 4). Development of such monitors and the specifications for their use requires the expertise of air-pollution engineers, industrial hygienists, chemists and chemical engineers, electronics experts, and quality-control personnel.

At the same time, advances in technology have produced highly sophisticated and sensitive analytical instruments that can measure many environmental contaminants at concentrations below those that have been demonstrated to cause harm to health or the environment. For example, techniques capable of measuring contaminants in parts per billion are common. The mere act of measuring and reporting the presence of certain contaminants in the environment often leads to concern on the part of the public, even though the reported levels may be well within the acceptable range. The accompanying fears, justified or not, can lead to expenditures on the control of environmental contaminants instead of on other, more urgent problems. Those responsible for protecting people's health must be wary of demands for "zero" pollution: it is neither realistic nor achievable as a goal in today's world. Rather, given the host of factors that are an integral part of our daily lives, the goal should be an optimal level of human and environmental well-being.

The Systems Approach

Attempts to control pollution in one segment of the environment can often result in the transfer of pollution to a different segments or the creation of a different form of pollution. Such interactions can be immediate or can take place over time; they can occur in the same general locality or at some distance. On a short-term basis, the incineration of solid wastes can cause atmospheric pollution; the application of scrubbers and other types of air-cleaning systems to airborne effluents can produce large amounts of solid wastes; and the chemical treatment of liquid wastes can produce large quantities of sludge. On a longer-term basis, the discharge of sulfur and nitrogen oxides into the atmosphere can result in acidic deposition at some distance from the point of release; the discharge of chlorofluorocarbons can lead to the destruction of the ozone layer in the upper atmosphere; and the discharge of carbon dioxide can lead to global warming (Chapter 20).

While the uses of chemicals have brought major benefits to humankind, in many cases they have had harmful effects. Once again, this emphasizes the need to consider all the ramifications of such uses. The chlorination of drinking water, for example, has led to significant reductions in the rates of many infectious diseases. As was the case with the discharge of chlorofluorocarbons, however, this process was later shown to have potentially harmful effects, namely, that through interactions with organic chemicals, chlorine can produce toxic compounds in drinking water (Chapter 7). While insecticides and pesticides have enabled farmers to achieve dramatic increases in the production of agricultural crops, the widespread and indiscriminant uses of such chemicals in this manner and in various types of industrial operations have led to a global legacy of enormous chemical contamination. Unless environmental health professionals recognize the severity and widespread nature of these problems, attempts to deal with them will be inadequate, piecemeal, and destined to fail.

Clearly, what is done to the environment in one place will almost certainly affect it elsewhere. A systems approach ensures that each problem is examined not in isolation, but in terms of how it interacts with and affects other segments of the environment and our daily lives.

Intervention and Control

Because the complexity of the problems in environmental health requires multidisciplinary approaches to their evaluation and control, the techniques for addressing them often differ from those applied in medical practice. Physicians traditionally deal with one patient at a time, whereas environmental health specialists must consider entire populations. To the extent possible, they must also try to anticipate problems to prevent them from developing. As depicted in the clinical intervention model (Figure 1.5a), the goal of the physician is to prevent a specific disease from leading to death (Morris and Hendee, 1992). The public health intervention model (Figure 1.5b), in contrast, calls for preventing the development of disease. Far superior to either is the environmental stewardship model (Figure 1.5c), in which the goal is to protect humans by preventing environmental degradation and its resulting impacts on health.

Even after a problem is understood, environmental health personnel need strong support from other groups if their goals are to be achieved. A prime necessity is the assurance of legislators that the requisite laws and regulations, as well as financial resources, are available (Chapter 14). Public health educators need to ensure that the public participates in the development of control programs, and that the associated regulations and requirements are fully understood by the industrial organizations and other groups who are expected to comply. The input of program planners and economists is also needed to assure that the available funds, invariably limited in quantity, are spent in the most effective manner. Far too often, decisions on where and how monies should be spent to improve the environment are based on emotions, not science. Since current programs on environmental protection in this country necessitate expenditures of more than $100 billion annually, it is imperative that the funds be directed to the most pressing situations.

Regardless of how competent they may be, environmental health pro-

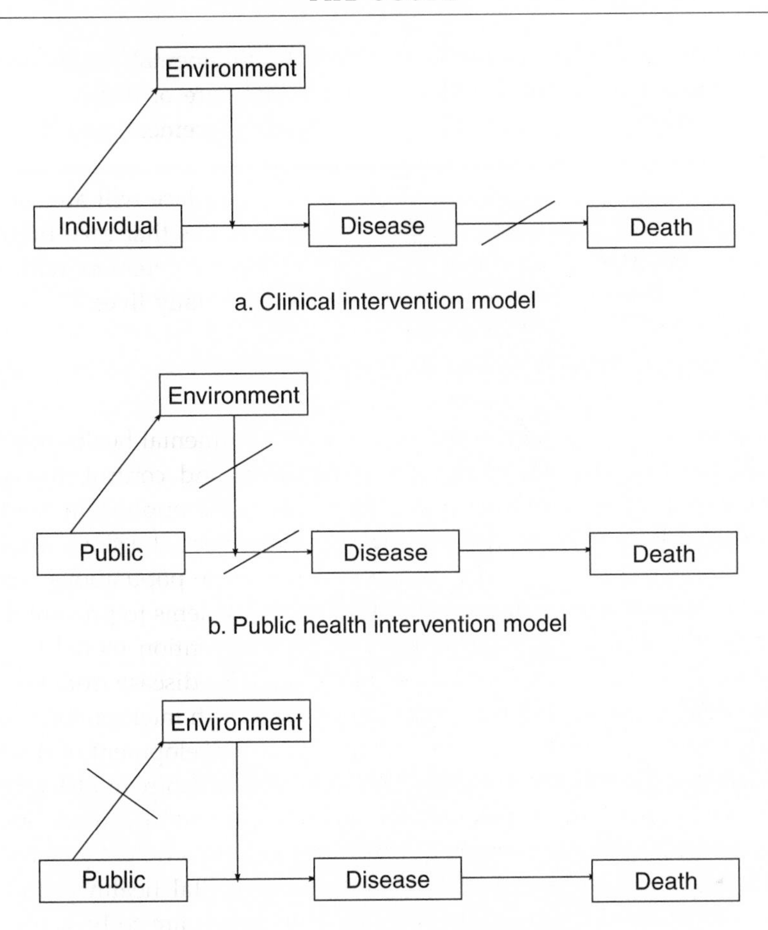

Figure 1.5 Various models for improving the state of human health and the environment

fessionals cannot be expected to solve these problems alone. As noted earlier, in the long run, the commitment and support of individual members of the public will be essential for success. Members of society must be constantly reminded that they can reduce the production of solid wastes by recycling newspapers, plastics, glass bottles, and metal (aluminum) cans. They can reduce the consumption of energy by car pooling, by minimizing home heating and cooling costs through the installation of storm windows and other weatherproofing measures, and by conserving water through the use of low-flow showerheads and the installation of low-water-consuming flush toilets (Chapter 18).

The General Outlook

In the course of their work, medical and public health personnel have achieved remarkable success in reducing human morbidity and mortality. One major benefit has been a significant increase in the average human life span. One important consequence has been a dramatic growth in the world's population and an accompanying heavier burden on the environment. In fact, a large share of the social, economic, and environmental decline in many parts of the world today results from the increased production of materials and wastes and higher consumption of resources in order to meet the expanding expectations of an ever-increasing number of people. Many of these practices have global ecological effects, and the combination of local and global effects will inevitably affect human health. While advanced technologies can help control some of the environmental impacts, the problem of population growth needs to be vigorously addressed. Fortunately, this need is being recognized, as exemplified by the third United Nations International Conference on Population and Development, which was held in Cairo in September 1994. Other indications of this increasing awareness are the Rio Declaration of 1992, which created the UN Commission on Sustainable Development, and the World Conference on Sustainable Development, which was held in Johannesburg, South Africa, in 2002. Limitations on population growth will of necessity be a strong component of any plans for long-range sustainable development. More details on the nature of population growth and its impacts are provided in Chapter 20.

While advances in modern science and technology have given humans the capability to control much of the natural world, choices will nonetheless have to be made to assure that the controls, as applied, result in an optimal level of health for both the environment and the public. The overall goal should be to achieve the maximum good for the maximum number of people. As part of this exercise, those people who live in the developed countries must decide what changes in their lifestyles they are willing to make to ensure the "greatest good" for the majority of the world's population, a vast number of whom live in the less developed countries.

Once specific environmentally destructive patterns of behavior have been identified and targeted, constructive patterns can be formulated as alternatives. The goal can best be described as application of the principles of environmental stewardship and global bioethics (Morris and Hendee, 1992). The overall objective should be to achieve both sustainable development and a sustainable environment. This will involve many types of trade-offs and, in some instances, could well entail the exchange of one set of environmental problems for another.

The concept of a sustainable environment is based on the premise that renewable resources should only be used at a rate that ensures their continued existence (sustained yield); nonrenewable resources should be used sparingly and recycled wherever possible (conservation); and natural systems should not be polluted to the point where they are no longer able to cope with the resulting damage (pollution prevention). As defined by the World Commission on Environment and Development (1987), sustainable development "meets the need of the present without compromising the ability of future generations to meet their own needs."

The problems of the environment are enormous. Solutions will require the cooperation of government, industry, and commerce, as well as the concern and dedication of individuals throughout the world. Even though the future may at times look bleak, it is noteworthy that of the 10 most significant public health advances made during the twentieth century (Table 1.4), half were due in whole or in part to advances associated with

Table 1.4 The ten most significant public health achievements in the United States during the 20th century

1. Vaccination
2. Motor-vehicle safety
3. Safer workplaces
4. Control of infectious diseases
5. Decline in deaths from coronary heart disease and stroke
6. Safer and healthier foods
7. Healthier mothers and babies
8. Family planning
9. Fluoridation of drinking water
10. Recognition of tobacco as a health hazard

Note: The listing is not intended to represent the relative importance of the individual advances

environmental and occupational health. Improvements in motor-vehicle safety, for example, have resulted from engineering efforts to make both vehicles and highways safer, combined with successful efforts to change personal behavior. Lung diseases, such as coal miners' pneumoconiosis (black lung disease) and silicosis, have been reduced significantly through improved controls within the occupational environment, and severe injuries and deaths in the more hazardous industries, such as mining, construction, and manufacturing, have been reduced through the design of safer machines and other improvements in worker safety (Chapter 11). In a similar manner, clean water and improved sanitation have been major contributors to improvements in the control of infectious diseases, for example, cholera and typhoid, both of which are transmitted by drinking water (Chapter 7). Safer and healthier foods have resulted from developments such as refrigeration and pasteurization and from better design of the equipment and facilities for the preparation and processing of food. Last but not least, fluoridation of drinking water has proved to be an effective measure for the prevention of dental caries (CDC, 1999).

2

TOXICOLOGY

IT HAS been estimated that more than 70,000 chemicals are in common use in the United States, and that the chemical industry markets 200–1,000 new synthetic compounds each year (Ansari, 2004). In Europe where, in contrast to the United States, manufacturers are primarily responsible for compiling and reporting the data required for evaluating the risks of chemicals, an estimated 30,000 are in use (Loewenberg, 2003). Although these materials are manufactured and distributed so that society can take advantage of their benefits, the accompanying processes result in the release of many such materials into the environment. These include a variety of prescription and over-the-counter drugs that are discharged into the environment as components of human and animal wastes. The complexity of the situation is exemplified by the fact that this last group contains antimicrobials, anticonvulsants, antidepressants, and anticancer compounds. Obviously, if these gain access in sufficient quantities to streams and rivers, they can represent a danger to fish and other aquatic life (Service, 2002). As a result of these and other activities, humans and other species are exposed to a wide range of chemicals in the general environment, as well as in the home and in the workplace. In fact, trace quantities of toxic chemicals are present in our food, our air, and our drinking water.

To ensure that human health and the environment are being adequately protected, environmental and public health officials need continuously updated information on the biological effects of these types of compounds. Armed with such information, decision makers can evaluate the appropriateness and recommend limits, where necessary, on various applications of such compounds. The scientific discipline through which such information is developed is referred to as *toxicology,* and the scientists who work in this field are known as toxicologists (Casarett, Klaassen, and Doull, 2001). As defined by the Society of Toxicology, this is "the discipline that integrates all scientific information to help preserve and protect health and the environment from the hazards presented by chemical and physical agents."

As will be demonstrated repeatedly in the discussions that follow, the efforts of toxicologists involve both science and art. The science lies in the observational or data-gathering aspects, and the art is in the projection of these data to situations where there is little or no information (Doull and Bruce, 1986). When the evaluations address the presence of chemicals in the environment, the situation is far more complicated. In these cases, specially trained toxicologists must expand the work of their coworkers, who traditionally deal with the effects of a single chemical in a single animal species, to include assessments of the effects, both direct and indirect, of combinations of chemicals on total ecosystems. This is what is known as *environmental toxicology.* The outcomes of such efforts are increasingly used by regulatory agencies to assess chemical risks, assign priorities to the cleanup of hazardous waste sites, establish government policies, and set levels of allowable exposure (Gochfeld, 1998).

Sources of Information

The information required to assess the health impacts of chemicals is generated through a variety of avenues. One source is epidemiological studies of human populations known to be exposed to certain agents (Chapter 3). But this kind of research is not easy to conduct, the data are difficult to interpret, and the results are available only after the exposures and effects have occurred. Therefore, such studies are not preventive or predictive in nature. A further difficulty is measuring the levels of exposure that occurred, determining whether other toxic agents simultaneously affected the people being studied, and assessing any associated synergistic or antagonistic effects. These voids are partially filled by another source of information, namely, an array of laboratory studies. Traditionally, these types of studies have been performed using small animals. Such studies and the interpretation of the resulting data will be discussed in the sections that follow.

For years, laboratory toxicological studies followed a rather standard format. Today such studies have entered a completely new era. They are no longer confined to evaluations of the effects of toxic agents on animals as complete organisms. Through the application of modern experimental technologies, investigators are exploring the responses and effects of chemicals at the molecular level. This has led to the development of vast amounts of information, including data on the content of the genes in our DNA, on the proteins and molecules made from these genes, and on the small molecules that along with these proteins form the basis of normal biological function. Unraveling the manner in which these parts are assembled into a functioning organism and the ways in which these assemblies become impaired by various stressors not only presents new challenges and opportunities for researchers, but also provides an opportunity for understanding the scientific bases needed to assess the associated biological effects (Greenlee, 2002). Approaching toxicological evaluations at the molecular level will be discussed in a later section of this chapter.

Pathways of Exposure and Excretion

Although protection of other species is important, the discussion that follows will be of the impacts of toxic chemicals on humans. As previously discussed (Chapter 1), the major routes of intake in this case are the lungs (inhalation), the gastrointestinal tract (ingestion), and the skin (absorption). In the case of the respiratory tract, the primary site of uptake is through the alveoli in the lungs—especially for gases such as carbon monoxide, nitrogen oxides, and sulfur dioxide and for vapors of volatile liquids such as benzene and carbon tetrachloride. The capacity of the lungs for absorbing such substances is facilitated by the large surface area of the aveoli, the high blood flow, and the proximity of the blood to the aveolar air. Liquid aerosols and airborne particles may also be absorbed through the lungs. In contrast, the deposition of airborne particles is heavily influenced by their size, the particles of primary interest today being those in the size range 2.5 micrometers or smaller (Chapter 5). Chemicals that are foreign to the human body are known as *xenobiotics.* Such substances can be either naturally occurring or human made.

Once a chemical is absorbed, the nature and intensity of its effects depend on its concentration in the target organs, its chemical and physical form, what happens to it after it is absorbed, and how long it remains in the tissue or organ in question (following the central tenet that "the dose makes the poison"). After being taken up in the blood, a toxic chemical will be rapidly distributed throughout the body. As part of this process, it may be translocated from one organ or tissue to another, and it may be converted into a new compound or metabolite. This process is known as *biological transformation.* Metabolic processes in the cytoplasm, for example, can alter toxic substances through various chemical reactions, including oxidation and reduction. In general, these reactions tend to result in new products that are less absorbable and more polar (charged) chemically and thus are more readily excreted in the urine. The removal of toxic chemicals from the body is thereby enhanced. In certain cases, the new product or metabolite may be more toxic than the parent compound; such reactions are known as *bioactivation* (Lu, 1991). In most cases, however, the newly formed compounds tend to be less toxic (Smith, 1992).

The principal pathway for excretion of chemicals from the human body is the urine, but the liver (via reabsorption from the bile into the blood and excretion through the bowels) and the lungs (via various clearance mechanisms—Chapter 5) can also be important excretory organs. In general, the GI tract is not a major route of excretion of toxicants. Among the less significant routes are the sweat glands (Lu, 1991).

Toxic chemicals may cause injuries at the site of first contact, or they may be absorbed and distributed to other parts of the body where they exhibit their effects. These effects may be considered reversible or irreversible. In general, reversible effects are observed for short-term exposures at low concentrations; irreversible effects are more commonly observed following long-term exposures at higher concentrations. Toxic agents may also produce either immediate or delayed effects. A notable example of the latter is carcinogenesis; many types of cancer do not appear in humans until a decade or more after exposure. The effects of a toxic agent may be influenced by previous sensitization of the exposed person to the same or a similar chemical, for example, beryllium. Such effects are often classified as allergic reactions (Lu, 1991).

Other factors that can modify the response to toxic chemicals include the species and strain of animal being affected, its age and sex, and its nutritional and hormonal status. Because young animals have less effective mechanisms for biotransforming and detoxifying certain chemicals, they may be more susceptible to certain toxic agents. In a similar manner, people with diseases of the liver, which is the major detoxifying and biotransforming organ within the body, are more susceptible to a variety of chemicals (Casarett, Klaassen, and Doull, 2001).

Physical factors can also alter the effects of chemicals. For example, a rise in ambient temperature will increase the toxicity in adult male workers of dinitrophenol, occasionally used as a herbicide. Usually, however, the duration of the response will be shorter when the temperature is higher, apparently because of the temperature-dependent biochemical reactions responsible for biotransformation of the chemical. Social factors also can affect toxicity. Those that have been shown to be important, particularly in laboratory testing, include the types of cages in which the animals are kept, whether they are housed singly or in groups, and the bedding materials provided.

Individual chemicals vary widely in their toxicity. Some, such as botulism toxin, produce death in humans at concentrations of only nanograms (10^{-9} gram) per kilogram of body weight. Others, such as ethyl alcohol, may have relatively little effect even after doses of several grams per kilogram (Table 2.1). Data of these types are often used to rank chemicals in terms of their toxicity (Table 2.2). Under this categorization, botulism toxin

Table 2.1 Approximate concentrations of various chemicals required to produce death in 50 percent of exposed animals

Chemical	LD50[a] (mg/kg of body weight)
Ethyl alcohol	10,000
Sodium chloride	4,000
Ferrous sulfate	1,500
Morphine Sulfate	900
Phenobarbital sodium	150
Picrotoxin	5
Strychnine sulfate	2
Nicotine	1
d-Tubocurarine	0.5
Hemicholinium-3	0.2
Tetrodotoxin	0.10
Dioxin (TCDD)	0.001
Botulinum toxin	0.00001

a. Dose that causes death in 50 percent of the exposed population.

Table 2.2 Toxicity ratings

	Probable lethal dose for humans	
Toxicity rating	Dosage	For average adult
Practically nontoxic	>15 g/kg	More than 1 quart
Slightly toxic	5–15 g/kg	Between 1 pint and 1 quart
Moderately toxic	0.5–5 g/kg	Between 1 ounce and 1 pint
Very toxic	50–500 mg/kg	Between 1 teaspoon and 1 ounce
Extremely toxic	5–50 mg/kg	Between 7 drops and 1 teaspoon
Supertoxic	<5 mg/kg	A taste (less than 7 drops)

would be classified as supertoxic, whereas ethyl alcohol would be classified as slightly toxic. Although such a classification scheme is primarily qualitative, it serves a useful purpose in providing laypeople with answers to the question: How toxic is this chemical? (Klaassen, 1986). Toxic chemicals can also be classified in terms of their target organ (liver, kidney), their use (pesticide, food additive), their source (animal or plant toxin), and their effects (cancer, mutations).

The presence of toxic chemicals in various media within the environment and their uptake by different species can lead to a variety of interesting situations. The concentrations of certain heavy metals, such as mercury, in plankton, for example, will be higher than those in the water in which they live, and the concentrations in fish will be higher still. The concentrations in birds that feed on the fish will be even higher, perhaps by as much as several hundredfold. This phenomenon is known as *biological magnification* or *bioaccumulation* (Moriarty, 1988). Such magnification led, for example, to the harmful effects of DDT on pelicans via a thinning of the shells of their eggs. For these and other reasons, it is unlikely that procedures for the establishment of an acceptable level of intake of a chemical by humans can be directly applied in setting a corresponding limit for the environment. In a similar manner, DDT will concentrate in a human mother's milk to the extent that her baby's intake of this pesticide per unit of body weight may be more than 20 times that in the mother's diet.

Conventional Tests for Toxicity

Depending on the dose, the effects of toxic chemicals on animals may range from death to sublethal effects and to situations in which there are apparently no effects at all. Often the first step in the prediction of effects is to conduct a series of laboratory studies involving a single chemical and a single animal species. Because of legal and ethical limitations, most such studies are conducted on rats or mice rather than humans. To evaluate possible endpoints/effects associated with a range of exposures over various timeperiods, toxicological studies have generally been divided into two categories, *acute* and *chronic.* A typical endpoint for the first category is death, often within only a matter of hours after exposure; a typical endpoint for the second category is the appearance of one or more types of cancer, some months to years after exposure. Chronic studies may be further divided into those of short or long term duration (Lu, 1991). In either case the amounts of the chemicals administrated must be well below the acutely lethal level, since the goal is to simulate environmental exposures of humans and the potential for latent effects.

ACUTE TOXICITY STUDIES

Acute toxicity studies may require only hours to conduct and may involve only a single administration of the chemical being tested. If death is the endpoint being observed, the data are generally analyzed by beginning with a plot showing the relationship between the dose and the percentage of test animals that die. Such a curve often exhibits the pattern shown in Figure 2.1. The portion of the curve between "Minimum" and point "B" represents the range of doses in which the most susceptible animals respond; the portion between "B^1" and "Maximum" represents the range in which the most resistant animals respond. The peak of the curve (directly above point "X") indicates the dose that causes 50 percent of the exposed animals to die. This is designated as the LD_{50} and, in the case of humans, it is often expressed in terms of the LD_{50} at different times, for example, at 30 and 60 days following a single acute exposure. In the case of small animals, the LD_{50} is generally expressed in terms of much shorter time periods after exposure. Since the curve follows a normal or Gaussian distribution, statistical procedures can be used to evaluate the resulting data (Loomis, 1968).

Although the Gaussian distribution is interesting, data resulting from

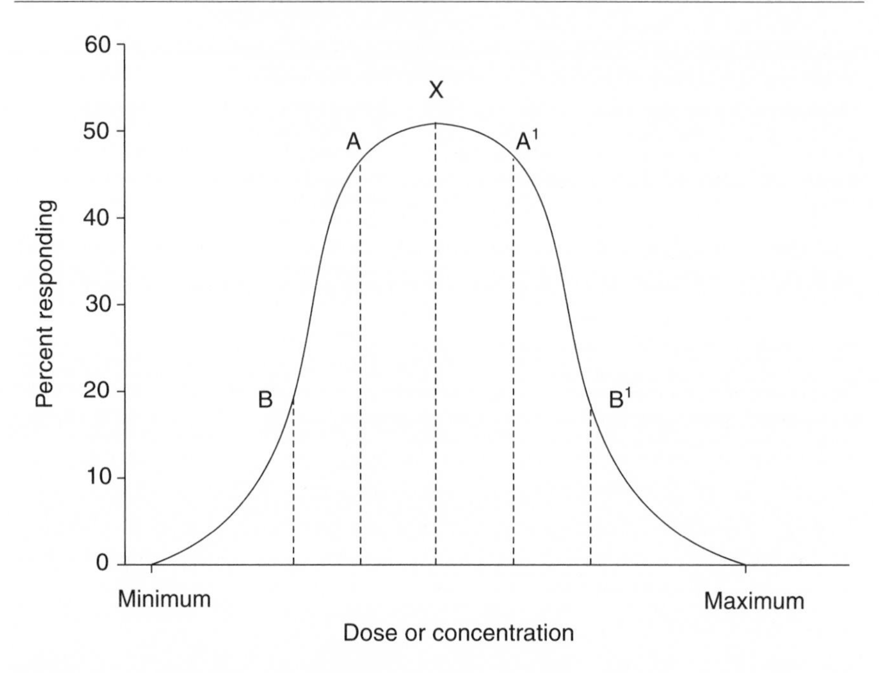

Figure 2.1 Distribution of animal responses to a toxic chemical as a function of dose

toxicological studies are generally plotted in the form of a curve relating the dose or concentration to the *cumulative* percentage of animals exhibiting the given response. The curves in Figure 2.2 show this type of plot for two different chemicals, A and B. The curve to the left represents the more toxic of the two compounds, since the dose (or concentration) required to cause death in 50 percent of the exposed population is lower. Such graphs are commonly referred to as dose-response curves and are plotted using an arithmetic scale on the vertical axis and a logarithmic scale on the horizontal axis. One advantage of this format is that a major portion of the curve is linear; for this portion the response (in this case, death) is directly related graphically to the dose or concentration of the chemical agent (Smith, 1992).

Figure 2.2 also illustrates another approach for determining the LD_{50} for exposed animals. As in the previous case (Figure 2.1), this would be for those deaths that occur within a specified period of time after exposure.

One of the advantages of this approach is that the endpoint is easily measurable; it either occurs or it does not. In fact, in previous years determination of the LD_{50} was one of the primary goals of many acute toxicity studies. This is far less true today, particularly in light of the diminished need for this type of information by regulatory agencies. Another contributing factor is the increased interest in both cancerous and noncancerous diseases, as well as possible behavioral effects, that may be caused by chemical exposures. Nonetheless, studies that use the LD_{50} as an endpoint provide an excellent illustration of the differences in the dose required to reach this endpoint for one animal species versus another, and as a function of age in the same animal (Tables 2.3 and 2.4). As will be discussed later, such differences increase the complexity of extrapolating toxicological information from various animals to humans.

Other benefits of acute toxicity studies are that they can provide information on the probable target organs for the chemical and its specific toxic effect, as well as guidance on the doses to be used in more prolonged (long-term) studies. Acute toxicity studies can also provide information

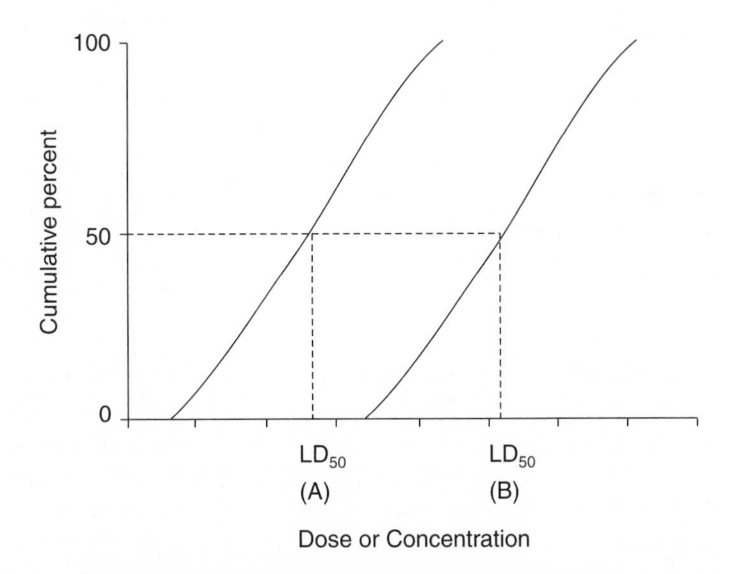

Figure 2.2 Cumulative percentages of animals showing responses to toxic chemicals. The LD_{50} designates the dose that is lethal to 50 percent of the exposed animals. The curve to the left represents the more toxic of the two chemicals.

Table 2.3 Effects of species differences on LD_{50} for TCDD [tetrachlorodibenzo-p-dioxin]

Species	LD_{50} (μg/kg)
Guinea pig	2
Mink	4
Rabbit	50
Monkey	70
Mouse	200
Rat	350
Hamster	2,000

Table 2.4 Influence of age on LD_{50} for DDT [1,1,1-trichloro-2,2-bis (*p*-chlorophenyl) ethane] in rats

Age	LD_{50} (mg/kg)
Newborn	>4,000
10 days	730
2 weeks	440
1 month	360
2 months	250
4 months	190
Adult	220

on the synergistic and antagonistic effects of certain combinations of chemicals. An interaction is described as *synergistic* when exposure to one chemical causes a dramatic increase in the effect of another. The enhanced toxicity of ketone in combination with haloalkane is an example. In this case, the combined response is more than the sum of the responses to the individual chemicals. An interaction is described as *antagonistic* when exposure to one chemical results in a reduction in the effect of another. The protection that selenium provides against mercury is an example of this type of interaction. Such information is very important in the evaluation of environmental exposures.

CHRONIC TOXICITY STUDIES

Chronic toxicity studies are conducted on both a short- and long-term basis. Short-term studies generally involve repeated administrations of a chemical, usually on a daily basis, over a period of about 10 percent of the life span of the animal being tested (for example, about 3 months in rats and 1 to 2 years in dogs); however, shorter durations such as 14-day and 28-day treatments have also been used by some investigators. Long-term studies involve repeated administrations over the entire life span of the test animals (or at least a major fraction thereof). For mice, the time period would be about 18 months; for rats, 24 months; for dogs and monkeys, 7 to 10 years.

For short-term studies, generally two or more species of animals are used, the objective being to have them biotransform the chemical in a manner essentially identical to the process in humans. It cannot be assumed, however, that this will be the case. In fact, differences in the abilities of various species to biotransform chemicals are the basis for the effectiveness of many of the pesticides that have been developed to be selectively toxic to only one insect, plant, or animal (Smith, 1992). Under normal circumstances, the animals selected are the rat and the dog because of their appropriate size, ready availability, and the preponderance of toxicological information on their reactions to a wide range of chemicals (Lu, 1991). Differences in response by gender require that equal numbers of male and female animals be used, and that a control group be maintained for comparison purposes. In addition, the chemical should be administered by the same route of exposure that is anticipated for humans. Other factors that must be taken into consideration include the possibility that exposed population groups may include some people who are unusually susceptible, and that effects may have occurred but were not observed (Moriarty, 1988).

The role of the long-term studies is to ensure that the studies encompass the full range of anticipated outcomes. Such studies generally include administration of the toxic chemical in three dose ranges—one sufficiently high to elicit definite signs of toxicity but not high enough to kill many of the animals; one sufficiently low that it is not expected to induce any toxic effects; and an intermediate dose (Lu, 1991).

Multiple Chemical Sensitivity

Since the early 1990s, some scientists have postulated that certain people have "multiple chemical sensitivity," a condition that they have assigned to individuals who report having developed symptoms in multiple organs as a result of living in an area where the environment contained trace concentrations of a combination of chemicals. The symptoms described are generally nonspecific and frequently involve the central nervous system or the respiratory or gastrointestinal tract. The supposition is that these types of exposures may impair the body's immune system. Confirming any type of a cause-effect association under such circumstances is difficult because (1) the symptoms, namely, headaches, fatigue, memory loss, joint discomfort, and sleep disturbances, are largely subjective; (2) they are reputed to have resulted from exposures several orders of magnitude below those known to cause adverse health effects in individuals without such sensitivity; and (3) the reported symptoms appear to have no relationship to the known effects of the chemicals in question (Schettler and Seeley, 2002).

A major challenge for the toxicologist is to explain why the reported symptoms have occurred. One observation that is fundamental in evaluating such occurrences is that for every noncarcinogenic chemical, no matter how hazardous, there is a level below which there is what is called "no observed adverse effect." A second observation is that the effects of a given chemical are typically specific for one or, at most, only a few organ systems. A third observation is that while there may be differences in the magnitude of the response for a given chemical among individuals, there is usually no difference in the nature of the response. In contrast, in the case of multiple chemical sensitivity, (1) there is no apparent safe level or threshold; (2) the effects involve multiple organ systems; and (3) there are differences in the nature of the responses among individuals. Furthermore, in spite of the numerous reported cases of this type of response, no scientifically defensible mechanisms to explain the associations have been provided. In particular, there are no scientific studies that link pesticide exposure with multiple chemical sensitivity. At this time, low levels of exposure to environmental chemicals, such as pesticides, cannot be clearly implicated as the cause of this condition (Schettler and Seeley, 2002). Nonetheless, it should be recognized that such controversies and the wide span of opinions that have been expressed are typical of the early stages of emerging scientific questions. This is especially true when concerns are being raised but existing information, as well as scientific consensus about the meaning of that information, is insufficient to resolve whether the expressed fears are well founded (Rhomberg, 1996).

In seeking to respond to these types of problems, the Agency for Toxic

Substances and Disease Registry (ATSDR) has developed a *Guidance Manual for the Assessment of Joint Toxic Action of Chemical Mixtures* (ATSDR, 2001). Although the manual is a public document, it was prepared primarily for internal use for the evaluation of the types of chemical mixtures present at hazardous waste sites. The approach outlined, which is consistent with that articulated by the U.S. Environmental Protection Agency (EPA 1986), seeks to integrate information presented in ATSDR's interaction profiles and toxicological profiles, combined with the outcomes of research on the effects of chemical mixtures. The strategies for assessing noncancer and cancer effects are similar.

Endpoints for Toxicological Evaluations

As indicated in the previous discussion, acute and chronic short-term tests served as the principal approaches in earlier toxicological studies. In these cases, only death or tissue damage served as recognized endpoints. As toxicologists sought to obtain information for evaluating a fuller range of effects in humans, the laboratory studies were expanded, and new and different endpoints were adopted. One of the primary stimuli for these changes was the development of what is known as the subfield of molecular toxicology. Today the evaluation of human exposures tends to be directed to studies that involve a full range of endpoints or effects, including noncancer endpoints and effects on behavior. The more prominent of these are discussed here. A discussion of molecular toxicology follows.

1. *Carcinogenesis.* Chemical carcinogenesis is recognized today as a multistage process that involves at least three steps: initiation, promotion, and progression. Although formerly it appeared that various chemical compounds and physical agents were either purely initiators or purely promoters, more recent interpretations suggest that some chemicals and agents are both initiators and promoters. Current theory posits that the development of cancer involves the activation or mutation of oncogenes or the inactivation of suppressor genes, and that this process causes a normal cell to develop into a cancerous cell.

   Because of the time and expense required for related tests using animals, toxicologists have for years experimented with the development of short-term, in vitro tests (experiments conducted outside the body). One of the most widely applied is the Ames test (Ames,

1971), which is a measure of the mutagenicity of chemicals in bacteria. It is based on evidence that deoxyribonucleic acid (DNA) is the critical target for most carcinogens and on the fact that mutagenic chemicals are often also carcinogenic. Although the Ames test provides an indication of the ability of a chemical compound to induce mutations or stimulate other types of biological activities, it does not reflect the complex patterns of uptake, metabolism, detoxification, and excretion that occur in the whole animal or the gene or target-organ specificity—information that can be critical in evaluating cancer responses (Butterworth et al., 1999).

2. *Reproductive toxicity.* Toxic effects on reproduction may occur anywhere within a continuum of events ranging from germ-cell formation and sexual functioning in the parents through sexual maturation in the offspring. For this reason, and because exposure of the mother, father, or both may influence reproductive outcome, the determination of the relationship between exposure and these types of effects is highly complex. In addition, critical exposures may include maternal exposures long before or immediately prior to conception as well as exposure of the mother and fetus during gestation (NRC, 1986).
3. *Developmental toxicity (teratogenesis).* Developmental effects that lead to the formation of congenital defects have been known for decades and are an important cause of morbidity and mortality among newborns. Such effects encompass embryo and fetal death, growth retardation, and malformations, all of which can be highly sensitive to chemical exposures. For some years, no connection was suspected between such effects and chemicals. Toxicologists therefore had a tendency to assume that the natural protective mechanisms of the body, such as detoxication, elimination, and the placental barrier, were sufficient to shield the embryo from maternal exposure to harmful chemicals. These concepts changed dramatically after the clinical use of thalidomide, a sedative first employed in Germany in the late 1950s to relieve morning sickness in pregnant women, led to a host of developmental effects in their fetuses (Smith, 1992).
4. *Neurotoxicity.* Fewer than 10 percent of the chemicals in worldwide use have been tested. Of these, almost 1,000 have been identified as known neurotoxins in humans and/or other animals (Stone, 1993).

The impacts on humans range from cognitive, sensory, and motor impairments to immune system deficits. For this reason, significant efforts are being devoted to the development of techniques for the identification, evaluation, understanding, and classification of chemical neurotoxic actions and, most especially, the application of data from studies in animals to humans. Complicating the situation is that there are often major differences between the degree of neurotoxic responses observed in animals and humans.

5. *Immunotoxicology.* Various toxic substances are known to suppress the immune function, leading to reduced host resistance to bacterial and viral infections and to parasitic infestation, as well as to reduced control of neoplasms. The importance of these effects is well illustrated by the concern about AIDS, in which the infected person often dies due to his/her inability to resist an organism that would not be a problem in a healthy individual. Certain toxic agents can also provoke exaggerated immune responses that lead to local or systemic reactions.

Molecular Toxicology

As noted earlier, the instigation of studies involving molecular toxicology has considerably expanded the nature and range of endpoints being used to assess the potential effects of toxic agents. One of the major stimuli for the development of this subfield was the realization on the part of toxicologists that if they were going to understand the effects of toxic chemicals in the broadest sense, they needed to understand the fundamental mechanisms through which such materials interact with living organisms (Gochfeld, 1998). At the same time, they realized that the endpoints that needed to be developed were, in essence, markers or indicators that signal the types of interactions that are taking place in biological systems or samples. The markers that were developed and are being used today are of three basic types: (1) those that are a measure of exposure or dose; (2) those that signal effects; and (3) those that are indicative of susceptibility. Biomarkers that are indicative of effects record biological responses in individuals who have been exposed to a genotoxic agent. In contrast, biomarkers of exposure (or dose) do not necessarily indicate effects. Superimposed on these two are the biomarkers of susceptibility, which can be used to identify persons who are at increased risk of developing a disease that could be triggered (initiated) by a given exposure. Included in this group might be persons whose ability to repair DNA damage is limited (NRC, 1995).

The use of such biomarkers could revolutionize the way in which toxicologists collect raw data. Rather than depending on animal pathology to identify illnesses, they can use the techniques of molecular toxicology to probe human or animal genetic material, as expressed in DNA arrays. Just as cancer researchers use such arrays to compare gene expression in healthy and diseased cells, molecular toxicologists are using this technology to profile gene expression in cells exposed to toxic agents (Butterworth et al., 1999). The advantages of the DNA tests are that they are fast and efficient and reduce the expense of maintaining live animals. One possible outcome of the application of this technology would be the development of an ability to identify the metabolic precursors of slowly developing diseases without having to wait for latent effects, such as tumors, to develop in animals. Full implementation of these new approaches, however, is dependent on developing clear scientific relationships between the observed results and known toxicological responses (Lovett, 2000).

Extrapolations of Animal Data

The application of animal bioassay data for estimating human responses to environmental exposures involves two types of extrapolations. One is to determine or estimate the relative responsiveness of humans and the animal species used in the bioassays—the so-called extrapolation from small animals to humans. The second is to extrapolate from the biological effects observed at relatively high exposures to the range anticipated in the ambient environment (Lippmann, 1992). In the past, the general approach in making such extrapolations has been to assume that the dose-response relationship in the low-dose range is linear for carcinogenic agents and nonlinear (that is, has a threshold) for non-cancer-producing agents. In the former case, the linear assumption is based on our present understanding of the cancer process as derived from studies involving ionizing radiation and genotoxic chemicals. In the latter case, the approach is based on data generated in numerous studies of the effects of noncarcinogenic chemicals in the low-dose region.

The choice of the appropriate dose-response relationship is further complicated because a chemical that has been tested and found to be carcinogenic may be so simply because the detoxification pathways in the animal being studied were overwhelmed (Schmidt, 2002). Other challenges are the need to account for differences in the pathways of environmental exposures, the rates at which these materials are metabolized, the lengths of time they are retained in the target tissues, and the sensitivities of these tissues. Furthermore, environmental exposures are often tenfold to a thousandfold below the lowest dose administered in the toxicity tests. Since the slope of the dose-response curve becomes increasingly uncertain as one extends it to exposures below the range of data obtained in the laboratory, an assumption of a linear dose-response relationship where it is not justified may yield response estimates that are in error by a large factor. For these reasons, the routine use of a linear dose-response relationship as a default position is being widely challenged. The published literature contains data from an increasing number of studies that show that the relationship between effects and exposure/dose at low levels for a host of toxic agents, including some that are carcinogenic such as ionizing radiation, follows a J-shaped curve, not a straight line (Figure 2.3;

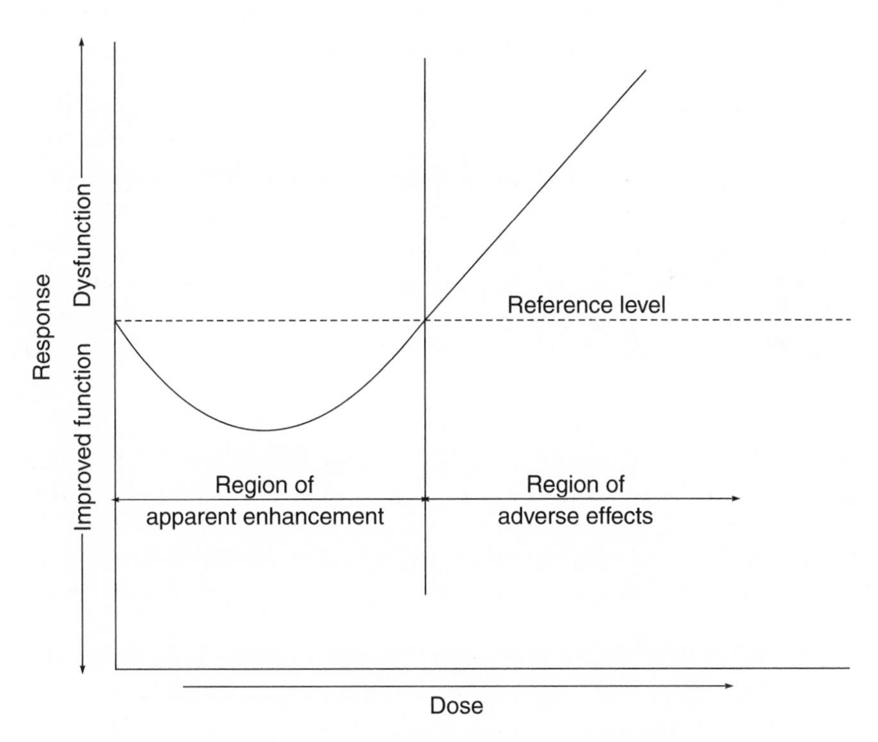

Figure 2.3 General form of the "J-shaped" dose response curve

Calabrese, Baldwin, and Holland, 1999; Kaiser, 2003). This means that the risks posed by some toxic agents in the range of the exposures that occur within the environment are being exaggerated. In fact, the J-shaped response to such toxins implies that at very low doses, the affected organism or animal may be receiving some benefit. It is only after the exposure or dose exceeds a critical level that negative effects appear. In these cases, other models for expressing the dose response in the low-dose region would appear to be more appropriate. This is especially true when the costs of removing such chemicals from contaminated sites or controlling their releases into the environment at low concentrations are taken into consideration.

Establishing Exposure Limits

Two basic principles should be applied in setting health-based exposure limits for human populations. The first is to use human data whenever possible; the second is to use surrogate chemicals or surrogate species only when the scientific evidence indicates that they provide an appropriate basis for such an application (Doull, 1992). When these principles have been taken into account, the next step is to determine whether the information in the database is relevant and appropriate for estimating effects using the existing or anticipated exposure conditions. If these criteria are fulfilled, and there is a threshold or no-effect level for the specific adverse impact on which the estimates will be based, the data can then be used to establish an appropriate exposure limit for humans. In so doing, scientists and regulators generally incorporate a safety factor into the threshold or no-effect level observed in animals. Selection of this factor should not only reflect the confidence of the evaluator in the quality and relevance of the data, but also account for differences in the susceptibility and kinetics between test and target species and between individual members of the exposed population (Doull, 1992). The magnitude of these safety factors is illustrated by the values used by the Safe Drinking Water Committee of the National Research Council in recommending no-response levels for various toxic agents in drinking water (NRC, 1983):

A factor of 10 was used when valid chronic exposure data existed on humans and supportive chronic data were available on other species; the factor was added to assure protection of the more sensitive individuals.

A factor of 100 was used when there were no data on humans but satisfactory chronic toxicity data existed for one or more other animal species; the 100 includes a factor of 10 to protect sensitive individuals, plus a factor of 10 to account for interspecies extrapolations.

A factor of 1,000 was used when the chronic toxicity data were limited or incomplete.

Regardless of how sound these safety factors are thought to be, a basic principle of health protection is to keep all exposures as low as reasonably achievable.

Applying Toxicological Data to the Environment

Whereas the laboratory toxicologist is primarily concerned with the effects of toxic chemicals on individual organisms, evaluation of the effects of these same chemicals in the environment is far more complicated. The complications arise from several sources (Moriarty, 1988):

1. Different species, and different groups and individuals within a single species, may react differently to identical exposures to the same chemical.
2. Some pollutants may occur in more than one form, and the determination of either the details of exposure or the resulting biological effects may be difficult. A further complication is that in many cases, the structure of individual chemicals is being changed by interactions within the environment.
3. Depending on the circumstances, the interactive effects of two or more toxic chemicals may be mutually additive, synergistic, or antagonistic.
4. The indirect effects of the toxic chemical may be as important as or more important than the direct effects. In fact, a chemical that kills no organisms but retards development may have more ecological impact than one that is lethal.

When multiple species are involved, additional complications arise. Even though predictions of the biological effects may be correct for the species under study, other species may be significantly more vulnerable and/or susceptible. In fact, the effects of many pollutants on wildlife may pass completely unnoticed. Ideally, the goal would be to identify the first-affected species. Even when an obvious effect is observed, however, identification of the chemicals that are responsible is often extremely difficult.

The problems of assessing the effects of chemicals within the environment do not end here. Alterations in the physical and chemical characteristics of the environment may have an impact on the ability of a species to survive: witness the releases of sulfur dioxide into the atmosphere that result in acid rain, and airborne discharges of carbon dioxide and other chemicals that affect global temperatures. In a similar manner, lakes and streams may be enriched through the release of sewage and agricultural chemicals, which in turn leads to eutrophication and detrimental impact on the survival of certain types of aquatic life. The analysis of indirect effects of these types must take into account not only the realization that the impacts of certain airborne emissions may be global in nature, but also that their concentrations and resulting impacts can vary significantly from one region of the world to another.

As a result of these complications, it is quite probable that precise predictions of the effects of chemicals within the environment are unlikely to be achieved in the foreseeable future. Nonetheless, continuing guidance is needed to make sound judgments relative to the introduction and use of chemicals, and environmental toxicologists will undoubtedly continue to direct their attention to these problems (Moriarty, 1988). As a general guide, the chemicals that will be most important in terms of the environment are those that have known toxic effects, that are persistent, and that are biologically concentrated by various animals and/or plants.

The General Outlook

Since the early 1990s, a variety of sophisticated analytical technologies have become available for use by toxicologists. As the previous discussion indicates, these advances are enabling them to gather detailed information on the effects of toxic chemicals at the molecular level. One of the primary challenges they face is to dissect and interpret what these data mean in terms of the normal functioning of the human body. Once this is accomplished, they should be able to provide environmental health specialists and regulators a vastly improved scientific basis for establishing permissible limits for toxic chemicals and implementing control measures in the most cost-effective manner (Greenlee, 2002).

Concurrently, events are taking place that demonstrate that the leaders of increasing numbers of industrial organizations are recognizing the benefits of environmental stewardship. What is even more encouraging is that they are responding by providing funds to academic and scientific institutions to support toxicological research on the effects of chemicals. This is exemplified by the U.S. chemical industry, which, in 1999, committed more than $100 million in support of such research. Specific areas being investigated include chemical carcinogenesis; endocrine, reproductive, and developmental toxicology; neurotoxicology; and respiratory toxicology (Henry and Bus, 2000).

Even so, many other types of problems must be addressed. One is the need to develop well-designed systems for summarizing and critically analyzing the existing toxicological information and to assemble a database through which to assess trends in the amounts of toxic agents present in the environment. Another is to implement on a worldwide basis a uniform system for classifying the risks of toxic chemicals. Two of the leading U.S. organizations involved in meeting these challenges are the ATSDR and the EPA. Through its "Toxicological Profiles," the ATSDR provides detailed information on a wide range of hazardous materials. The data and information provided in these documents is exemplified by the report on malathion (ATSDR, 2003). For each such material, the toxicological and adverse health effects of the given substance are succinctly characterized, an assessment is made of the adequacy of the information available for making such judgments, and the key references on which the assessments are based are identified. If the available information is determined not to be adequate, the additional testing needed to fill the voids is identified.

To assemble a database through which to assess environmental trends on toxic agents, the EPA maintains the Toxics Release Inventory. Through this program, more than 23,000 factories, refineries, mines, power plants, and chemical manufacturers self-report to the EPA on an annual basis the identities and quantities of toxic chemicals that they release into the environment. In addition to providing useful data on trends in the quantities of such releases, these reports often make industrial leaders more aware of how their activities are affecting the environment and stimulate them to be more careful. Data show, for example, that such releases were reduced by 10 percent between 1999 and 2000; the reduction from 1988 to 2000 was almost 50 percent. At the same time, however, the EPA has been careful to point out that while such data are extremely useful, they are not sufficient to determine exposure or to estimate the potential adverse effects of such releases on human health and the environment.

In an effort to meet this need, at least on a partial basis, the ATSDR periodically measures the quantities of certain chemicals in the blood and urine of people throughout the nation. Toxic agents being assessed include lead, mercury, and uranium; the breakdown products (or metabolites) of several organophosphate pesticides (representing about half of those used in the United States); phthalate metabolites (additives found in plastics, particularly polyvinyl chloride [PVC]); and cotinine (a metabolite of nicotine). One of the impetuses for this program was a study by the National Academy of Sciences that suggested that as many as one in four developmental and behavioral problems in children in this country may be linked to genetic and environmental factors. Neurotoxic compounds, such as lead, mercury, and organophosphate pesticides, were cited as possibly playing a significant role (CDC, 2001, 2003). Although the latest data confirm continuing reductions in the concentrations of lead in children, almost half a million children between one and five years of age exceed the limit. Nonetheless, progress is being made. The levels of synthetic contaminants, such as DDT and dieldrin, for example, declined by more than 90 percent during the last quarter of the twentieth century. At the same time, it is important to note that while the declines are continuing, the rate of decrease has slowed. In addition, there are some subpopulations that are still exposed to unusually high amounts of these and other contaminants (Kamrin, 2003).

Still to be developed is an international system for classifying the risks of toxic chemicals on a uniform basis. The need for such a system is illustrated in several ways. The International Agency for Research on Cancer (IARC), as well as the European Union, Germany, and Sweden, classify such materials on the basis of their carcinogenic potential in humans. The Netherlands and Norway, in contrast, do not explicitly differentiate between the effects in humans versus those in other animals. The Netherlands classifies carcinogens according to genotoxicity, that is, DNA damage. Norway, on the other hand, classifies carcinogens according to their potency. Other differences are exhibited by the classification approaches used in Germany, where rankings are based on data on malignant tumors only, and Norway, where data on both malignant and benign tumors are considered. Norway uses both published and unpublished data, while the IARC restricts the basis for its classifications to published data. Because of these differences, it is not surprising that a review of the status of a group of eight chemicals, as classified by these countries, revealed a consensus for only two—benzene and vinyl chloride (Seeley, 2001). Nonetheless, the fact that efforts to achieve harmonization are under way is encouraging. The identification of differences such as these represents a sound first step.

3

EPIDEMIOLOGY

FOR WELL over a hundred years, epidemiological studies have played an important role in the investigation of the ways in which infectious diseases spread through the community. With the increasing awareness of environmental pollution and its potential effects on health, the techniques of epidemiology have been expanded to examine the effects of a variety of chemical and physical agents within the environment. The result has been the science of "environmental epidemiology," defined by the National Research Council (NRC, 1991) as "the study of the effect on human health of physical, biological, and chemical factors in the external environment, broadly conceived. By examining specific populations or communities exposed to different ambient environments, it seeks to clarify the relationship between physical, biological, or chemical factors and human health."

In this definition it is important to note that the techniques of environmental epidemiology are generally not designed—nor should they be expected—to prove that a given environmental agent *causes* a given disease or health effect; in most cases, the best outcome that can be anticipated is that the methods of environmental epidemiology will demonstrate a *relationship* or *association* between a given agent and one or more specific health effects. The basic difficulty is that few of the nonbiological agents have unique effects on health; conversely, the effects considered may often be related to a wide range of factors (NCRP, 2001). A classic example is ionizing radiation, which is known to be capable of causing cancer in a multitude of body organs. To "prove" that a given cancer was due to radiation exposure, however, is difficult, as is exemplified by the types of data presented in Table 3.1. In the range of the dose rate from natural background radiation (which, worldwide, yields a dose rate of about 2.5 mSv per year) (Chapter 12), detection of an increase in total cancer mortality would require careful observations throughout their lifetimes of an exposed population group in excess of 30 million people.

For these reasons, when decisions have to be made about the need to control suspected agents within industry or the community at large, many aspects of the situation must be taken into account—the strength and consistency of the association, toxicological and clinical findings, and the economic and social implications of control measures. An ancillary consideration is whether there is a plausible mechanism through which the given

Table 3.1 Size of exposed population group and radiation dose required to detect an increase in total cancer mortality, assuming lifetime follow-up

Mean whole body dose (mSv)	Excess cancers per 10,000 population	Required sample size
2.5	1.9	32,000,000
5.0	3.8	7,900,000
10.0	7.5	2,000,000
20.0	15.0	500,000
30.0	22.5	220,000
40.0	30.0	130,000
50.0	37.5	80,000
60.0	45.0	56,000
70.0	52.5	41,000
80.0	60.0	31,000
90.0	67.5	25,000
100.0	75.0	20,000
120.0	90.0	14,000
150.0	113.0	9,100
200.0	150.0	5,200

chemical or physical agent can cause the suspected effect. Nonetheless, in cases where the evidence of the relationship between a given environmental factor and an effect in humans subjected to that factor is compelling, it may be possible to extend the results of the epidemiological studies and quantify the relationship between the two, as in the cases of the previously cited ionizing radiation, whose ability to cause cancer in humans is undisputed, and cigarette smoking, whose ability to cause lung cancer is similarly undisputed. As will be noted later in this chapter, a similar relationship appears to have been confirmed between the presence of extremely small particulates in the air and deaths from lung cancer and heart- and other lung-related causes.

As contrasted to the field of toxicology, which is experimental in nature and involves laboratory studies ranging from those conducted at the molecular level to those involving animals, the field of environmental epidemiology is nonexperimental and involves studies of existing human population groups who have been exposed to one or more chemical and/or physical agents. In the sections that follow, the general principles of environmental epidemiology will be outlined. The discussion will include a review of some of the precautions that must be taken both in the design of such studies and in the analysis and interpretation of the collected data.

A Classic Example

The founder of epidemiology is often considered to be John Snow, who conducted what is regarded today as a classic study of the transmission of cholera in London in the mid-1800s (Monson, 1990). His work illustrates many of the principles of a valid environmental epidemiological study.

Snow, a practicing physician, observed that people who worked with cholera patients did not always contract the disease, and that people who did not have contact with infected patients often did contract the disease. He postulated the existence of some vehicle that transmits the disease and, with support from other physicians and local laypeople, hypothesized that one possibility was the presence of sewage (fecal) contamination in drinking water. Snow conducted a study of population groups in different parts of the city who obtained their drinking water from different suppliers. Recognizing that other factors could influence the spread of the disease, he analyzed the mortality rates in a single subdistrict, where the only observable difference was that one portion of the population obtained its drinking water from one supplier and the other obtained its water from a second supplier. Using a chemical test that took advantage of a difference in the chloride content of the two water supplies, he was able to identify the supplier of each individual household. From these data, he confirmed that the disease was transmitted by sewage in the drinking water supplied by one of the companies (Goldsmith, 1986; Monson, 1990).

As pointed out by Monson, several factors make Snow's study a model of environmental epidemiology:

1. Snow recognized an association between exposure and disease—that is, between the source of the drinking-water supply and the incidence of cholera.
2. He formulated a hypothesis—that fecal contamination of drinking water was the specific agent of transmission of the disease.
3. He collected information to substantiate his hypothesis—in subdistricts where the drinking water was supplied by only one company, the association was stronger.
4. He recognized that there could be an alternative explanation for the association—that social class or place of residence might influence transmission of the disease.
5. He applied a method to minimize the effects of the alternative explanation—he compared cholera rates within a single district or neighborhood, rather than between neighborhoods, on the basis of their water supply.
6. He effectively minimized the collection of biased or false information—since most residents were not aware of the name of the company that supplied their water, he applied a chemical test to make this determination in a positive manner (Goldsmith, 1986).

These criteria have withstood the test of time and are regarded today as fundamental to the design of all types of epidemiological studies.

Modern Environmental Epidemiology

As noted in the Introduction and exemplified by the work of Snow, early epidemiological studies were "disease centered," and the diseases primarily involved were infectious in nature. As a result, investigators at that time relied primarily on laboratory investigations and paid little attention to study design. Their basic principles were that a microorganism should be considered as causally related to a disease when it was present in all subjects affected and when it was absent, or found as a fortuitous parasite, in other diseases. The implication that a given agent was the source of the disease was then confirmed by isolating it in the laboratory, inoculating it into animals, and demonstrating that the animals developed the disease (Terracini, 1992).

Today the trend is to employ epidemiological studies that are "exposure centered." This approach is an outgrowth of the realization of a multitude of factors. One is that in the developed countries of the world, degenerative diseases such as cancer, whose etiology is multifactorial, have become the prevailing pathology. The result has been an increasing awareness of the need for a rational, systematic, explicit, and reproducible approach to evaluating the associations between various diseases and environmental agents. Meeting this need requires the consideration of certain basic criteria, enumerated by Hill (1965):

1. The strength and specificity of the association;
2. The consistency of the findings in different studies;
3. The existence of a dose-response gradient between the exposure and the occurrence of the disease;
4. The biological plausibility of the proposed association;
5. The coherence of the evidence with the natural history of the disease;
6. The supporting experimental, or quasi-experimental, evidence.

Although subsequent investigators have expanded on these criteria, they have served as one of the foundations of modern epidemiology, much as Snow's principles did during the early years. The primary changes have been to emphasize the control of confounding variables and to improve study design (Terracini, 1992).

DESIGN OF AN EPIDEMIOLOGICAL STUDY

One of the first considerations in the design of an environmental epidemiological study is the definition of its objectives and scope. As an extreme, one might consider monitoring the health records of the whole population and linking that information with as many data on environmental factors as possible. Basic to such a study would be national death statistics and records on morbidity. To extend this type of study to include inquiries into the "health and habits" of individual members of the population on a national scale, however, might be considered an intrusion on privacy, and the financial costs would be prohibitive. Nonetheless, if success is to be achieved, some form of additional data gathering may be required (WHO, 1983).

An alternative approach is to focus on small groups of people considered to be at risk. The objective in this case would be to consider a specific disease or effect and to compare the available information on exposures in this group to those in a control group. Depending on the type of study, the control group is generally one that either has not been exposed to the agent in question or does not have the disease being investigated. Because it is unethical to expose people to potentially hazardous environmental agents solely for purposes of epidemiological study, essentially all such studies are nonexperimental. As a result, it may be difficult to define or quantify the exposures received by the population group that is being evaluated.

Two of the multitude of ways in which environmental epidemiological studies can be classified are as follows:

COHORT STUDY

A group of persons who has received unusual exposures is followed over time to determine what diseases they develop and whether there is an increase in the incidence of those diseases that might be presumed to have been caused by the exposures. The epidemiological studies of the survivors of the World War II atomic bombings in Japan exemplify a cohort study (Shigematsu, 2000).

CASE-CONTROL STUDY

People who are known to have a specific disease are examined to determine what, if any, exposures that they are receiving now or have received with unusual frequency in the past might have been the source of the disease. Early epidemiological studies of the relationship between cigarette smoking and lung cancer (Doll and Hill, 1950) are examples of case-control studies. This approach has also been used to evaluate various diseases in occupational settings. One example is the associations between certain illnesses and pesticide exposures (Cantor et al., 1992).

A cohort study may be either prospective, in which case the disease has not yet occurred at the time the exposed and nonexposed groups are defined, or retrospective, in which case the disease has already occurred. The study by John Snow was an example of a retrospective cohort study. Case-control studies, in contrast, are frequently termed retrospective because the investigator is looking backward from disease to exposure. That is, individuals are included on the basis of whether they have or do not have the disease being evaluated. Simply referring to a study as retrospective or prospective, however, leads to confusion, especially in the case of retrospective cohort studies, where the investigator is looking forward from exposure to disease but is basing the analysis on data collected in the past. As might be anticipated, various combinations of approaches are often included within a single study.

The basic differences in the various types of epidemiological studies can be summarized as follows (Monson, 1990):

In a cohort study, individuals are included on the basis of whether they have been exposed; in a case-control study, individuals are included on the basis of whether they have the disease being evaluated.

In a prospective cohort study, the disease has not occurred at the time the exposed and nonexposed groups are defined; in a retrospective cohort study, the disease has already occurred.

In a prospective cohort study, the investigator usually compares the disease rates of two or more groups (for example, smokers and nonsmokers); in a retrospective cohort study, mortality rates among the exposed group are compared to mortality rates of some general population (no formal comparison group).

Since in a case-control study the past history of exposure is the primary information that is collected, such studies can be completed relatively quickly; because time must pass in order for the disease to develop, completion of a prospective cohort study often requires a relatively long period of time.

In a case-control study, the general approach is to evaluate a number of exposures in relation to one disease; in a cohort study, one exposure is evaluated in relation to a number of diseases.

In the conduct of current environmental epidemiological studies, the general approach is not to compare an exposed and a presumably nonexposed group; rather, it is to compare the incidence of a given disease as a function of the degree, extent, or amount of exposure. This approach is taken because it is often difficult to identify persons who have not been exposed at all to a given physical or chemical agent.

Major Challenges

A variety of challenges face environmental epidemiologists. Some of the more important are enumerated here.

EXPOSURE ASSESSMENT

As mentioned earlier, assessment of the exposures to which a population study group has been subjected is a crucial, often inadequately addressed component of epidemiology. A difficulty is that assessments of exposures in the workplace, versus the home or ambient environment, require entirely different approaches. The same is true in assessing exposures to different types of agents. Regardless of these challenges, valid monitoring and accurate estimates of exposures, particularly those in the ambient environment, are essential if confidence is to be placed in the associations that are developed between exposures and observed adverse consequences to human health (NRC, 1991).

A further complication is that exposures to physical agents, such as noise or vibration, may be transitory, and the resulting assessments must therefore be made on a real-time basis. Unfortunately, in some cases (as with electric and magnetic fields), assessment personnel do not yet fully understand which parameters are indicative of exposure (Brain et al., 2003). Nor do they know whether it is average or peak exposures that are important. In the case of chemical and radioactive contaminants, the field of environmental monitoring and exposure assessment requires consideration of the source of the contaminant, its associated media or pathways of exposure, its avenues of transport through each medium, its routes of entry into the body, the intensity and frequency of contact with the contaminant of the persons exposed, and its spatial and temporal concentration patterns. The importance of such movement and interactions is exemplified by the fact that the composition as well as the physical form of chemical contaminants can be readily altered thereby. The progression from the release of a contaminant, its movement through the ambient environment, and its uptake by humans to the production of associated health effects, is depicted in Figure 3.1. Additional information on monitoring within the workplace is provided in Chapter 4; similar information on monitoring the ambient environment is provided in Chapter 16.

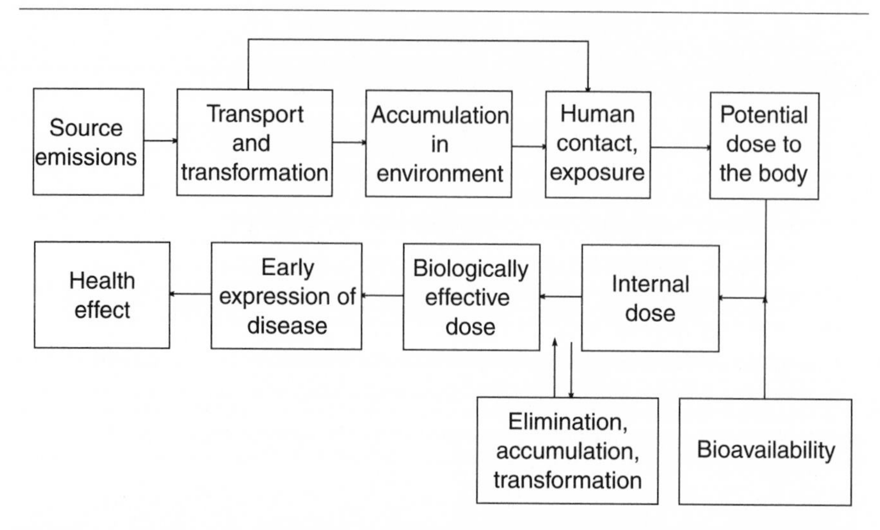

Figure 3.1 Progression of factors that influence the behavior of a contaminant within the environment, its uptake by humans, and the resulting health effects

Accurate assessment of exposures from airborne particulates requires, for example, identifying not only what the contaminant is, its physical form (amorphous, crystalline, discrete particulate, or fibrous), and its particle-size distribution, but also, in many cases, its physicochemical surface properties. Especially complex is the assessment of exposures necessary for cross-sectional studies of the effects of environmental air pollution. In earlier times, such studies typically involved a comparison of the health of populations in communities that had different ranges of specific contaminants in the outdoor air. It is now recognized that people spend 90 percent or more of their time indoors. Thus the assessment of their exposures must include determination of the concentrations of airborne contaminants inside their homes and places of work.

As will be noted later (Chapter 5), factors that can contribute to airborne exposures within a home include personal or family eating habits, the type of cooking facilities (natural gas or electricity), personal hobbies and recreational activities, pesticide applications within the home and garden, and the nature of the domestic water supply. Contaminants released into the air during showering, bathing, and cooking may become sources of exposure through inhalation (NRC, 1991). The hierarchy of exposure data or surrogates, ranging from quantified measurements of individual exposures to simply knowing the person's residence or place of employment, is depicted in Table 3.2.

In the case of environmental epidemiological studies that are retrospective in nature, it is often difficult to obtain records of the required exposure data. In such cases, pollutant concentrations at remote locations are frequently estimated using environmental transport models based on measurements of the concentrations of the contaminants as they were discharged from nearby stacks (NRC, 1991). Because airborne pollutant concentrations do not always have a direct correlation with the total dose an exposed person receives, it is standard practice in many cases not to attempt to convert the pollutant concentrations into doses. Instead, the goal of the study is simply to determine if there is an association between the pollutant concentration (or "exposure") and a particular health effect.

HEALTH ENDPOINTS

A second major challenge in the design and implementation of an environmental epidemiological study is the selection of the health endpoints to be evaluated. Formerly it might have been adequate simply to determine whether the chemical or physical agent in question was causing an increase in the number of deaths (mortality) or hospital admissions (mor-

Table 3.2 Hierarchy of exposure data or surrogates

Types of data	Approximation to actual exposure
Quantified personal measurements	Best ↑
Quantified area or ambient measurements in vicinity of residence or other sites of activity	
Quantified surrogates of exposure (e.g., estimates of drinking water and food consumption)	
Residence or employment in proximity to site of source of exposure	
Residence or employment in general geographic area (e.g., county) of site of source of exposure	↓ Poorest

bidity) among the exposed population. Subsequently, the potential increase in the incidence of cancer became the health endpoint (or indicator) of primary importance. Today environmental and public health officials, as well as the public at large, are concerned with the possible impacts of environmental agents on the quality of life. They are demanding that a variety of possible pathological conditions—biochemical, physiological, and neurological dysfunctions—also be considered. These include effects on the respiratory and cardiovascular, central nervous, and musculoskeletal systems, as well as behavioral, hemopoietic, growth, and reproductive effects.

Such considerations add enormous complexity to the studies. If, for example, the only effect of an agent at a given intensity is a small change in bodily function, well within an individual's normal physiological range of variation, then its importance in comparison with other factors affecting health must be carefully weighed. Competing factors that must be considered include the duration of the effects and the number of persons likely to be affected. The relative importance of a minor immediate effect versus a potentially more serious but delayed effect must also be evaluated. (WHO, 1983). A key criterion is whether the chemical or physical agent being evaluated has been demonstrated to be capable of causing the suspected effect. Unless it has, successful conduct of the proposed study may be seriously impaired.

Assessment of any of these endpoints requires some standardized measure of effects. The indicators that have been developed for measuring behavioral effects of noxious environmental agents, for example, fall into two broad groups: (1) measures of psychological and psychophysiological functioning and (2) measures of mental state and behavior. Psychological tests have proved effective in the detection and measurement of organic brain damage. In a similar manner, relatively simple techniques, such as Raven's progressive matrices and vocabulary and memory tests, have proved both reliable and practical in field studies involving the screening of large numbers of individuals (WHO, 1983).

An emerging development is the use of biological measurements made at the molecular level as indicators (or biomarkers) of the effects of a particular environmental agent or stress. This is an outgrowth of the field of molecular toxicology (Chapter 2) and has led to what is called *molecular epidemiology*. Because of its significance, this topic will be discussed separately in the section that follows.

POTENTIAL BIASES

Another challenge in the conduct of environmental epidemiological studies is the variety of potential biases that can cause the outcome to be in error. As noted earlier, measurement errors occur because of the assumption that all subjects within a given population group receive the same exposure. There can be undetected differences among communities in risk factors such as illness, tobacco use, or occupational exposure. The lack of standardization in the equipment used to measure exposures at different locations also contributes to biases (NRC, 1991). As indicated later, "recall bias" can also be present. Monson (1990) has separated the biases that can influence epidemiological studies into three categories:

Selection bias occurs as a result of deficiencies in the study design. If two groups of persons, one exposed and one not exposed, are identified today and followed forward in time until disease occurs (as in a prospective cohort study), no selection bias is possible. If, however, a group of people with a disease is identified and a group of controls (for example, in a case-control study) is selected, selection bias is possible because of the fact that the disease had occurred prior to the initiation of the study. Once selection bias has occurred, no amount of data manipulation can correct its effects—the two groups are forever noncomparable. In essence, selection bias cannot be controlled; it must be prevented.

Observation bias is also a result of deficiencies in the study design. In a cohort study, for example, observation bias occurs when information on disease outcome is obtained in a noncomparable manner from exposed and nonexposed groups. In case-control studies, observation bias occurs when information on exposure is obtained in a noncomparable manner from cases and controls. An obvious way to prevent observation bias in a cohort study is for the data collectors or interviewers not to know the exposure status of study individuals when they gather information. Likewise, in a case-control study, no observation bias is possible if neither the patient nor the data collector knows the diagnosis. Again, this type of bias must be considered in the design of the study, and efforts must be made to minimize its effects. Underlying the prevention of observation bias is the need for all concerned with a study to be impartial. A further influence on the types of data obtained through interviews is what is called "recall bias." Often the input data are based on past experiences of people, which in turn depend on the accuracy of their memories and the information they are willing to share.

Confounding bias is inevitable in all types of studies. For example, an evaluation of two variables (exposure and disease) is influenced by a third variable that is a cause of the disease and is also associated with the exposure. Specifically, cigarette smoking is a cause of lung cancer. Cigarette smoking is also associated with high alcohol consumption. If one examines the relationship between high alcohol consumption and lung cancer, it will be observed that the rate of lung cancer in high alcohol consumers is higher than in nonconsumers. All that can be done is to collect information on known or suspected confounding factors, so as to be able to measure any bias that is introduced. Confounding bias does not result from any error of the investigator; it is a basic characteristic of existence.

Studies of geographic variations in disease rates, often referred to as ecological studies, illustrate the potential biases in environmental epidemiological studies. One of the principal goals of such studies is to formulate hypotheses about the etiology of disease by taking into account spatial variations in environmental factors. Testing such hypotheses on the basis of geographic variations, however, is generally not possible. The hypotheses need to be tested by more rigorous methods, using cohort and case-control studies. The primary reason is that in geographic studies the exposure to a particular environmental agent (for example, water containing a specific contaminant) and the suspected effect (an increase in cancer) are not measured with respect to the same individuals. Nonetheless, because they take advantage of large differences in both the frequency of disease and the prevalence of exposure, geographic studies at an international level have been successful in identifying a number of possible risk factors for disease (English, 1992).

One of the basic epidemiological applications of geographic studies is in conjunction with simple descriptive studies of geographic variation, where the goal is to determine if variations in disease rates are associated with variations in the accompanying levels of exposure. The place and time of residence of the affected populations are often used as surrogates for the exposures of interest. Although the relationship between surrogate and exposure may be direct (say, between cosmic radiation and altitude or between ultraviolet radiation and latitude), in most cases it is indirect.

The validity of the geographic approach depends on how well the surrogate serves as a measure of the actual exposure of an individual who develops the disease.

The Developing Field of Molecular Epidemiology

As noted earlier, the field of molecular epidemiology is based on the use of various types of biomarkers, each of which is typically designed to detect damaged or naturally variant molecular structures. Such biomarkers may also be designed, through the use of immunologically based techniques, to detect particular gene-product molecules. One of the advantages of the incorporation of such techniques into the field of epidemiology is that they can be used to measure exposure, early biological response, or host characteristics that influence susceptibility. They may also elucidate mediating biological events and enable adverse health outcomes to be differentiated. Molecular biomarkers, for example, have been widely applied to the field of cancer epidemiology. In this case, the principal objectives have been to measure DNA damage, heritable genetic polymorphisms that influence susceptibility, and "cancer family" genes. The primary goal is to enhance the measurement of exposure, effect, or susceptibility; it is not to formulate new etiologic hypotheses (McMichael, 1994). At the same time, it must be recognized that knowledge of the relationships between specific markers and specific disease endpoints is required if the use of markers is to have value. There must be more than a simple correlation between a marker and an effect; that is, studies must have demonstrated that the marker is correlated with the effect (such as cancer) in specific persons.

Biochemical measures are not new to epidemiologists. The concentrations of lipids in blood, hormones in urine, and steroids in fecal samples have been used for several decades as indicators of potential effects. What is new is the emergence of a category of biomarkers based on the detection of specific aberrant or variant molecular structures. This permits biologically based measurements to be made at higher resolution or in relation to different criteria. Benefits of the application of this approach to the field of epidemiology are illustrated by the following examples: (1) monoclonal antibody subtyping of the environmental and clinical isolates of the Legionnaires' bacterium enabled investigators to trace an outbreak of this disease to a specific decorative fountain in a hotel lobby (McMichael, 1994); and (2) measurements of cotinine in the blood of participants before, during, and after a smoking-cessation program provided conclusive evidence of the success of the individual participants in having ceased to smoke (Perera, 2000). Especially exciting is the discovery that it may be possible to identify environmental carcinogens, including radiation, based on the genetic fingerprint they leave on cells that become cancerous (Hande et al., 2003).

During the initial phases, a major goal in the development of molecular epidemiology was to provide methodologies that would make it possible to predict human risks far more precisely. Experience has demonstrated, however, that the use of molecular endpoints also increases the tools available for providing early identification of environmental agents capable of causing cancer. As a result, epidemiologists now have the capability of identifying risks before malignancies develop. Another benefit is that they are now able to address both the range of risks across a population and the risks to subgroups, such as children, who may be more susceptible (Perera, 2000). Still another is that this methodology has provided novel approaches for defining the role of genetic susceptibility in epidemiological studies of cancer etiology (Ishibe and Kelsey, 1997). Last, but not least, this approach has proved useful, as noted earlier, in the development and evaluation of new intervention strategies for controlling specific cancer risks (Perera, 2000). A distinguishing feature of these new markers is their increased analytical sensitivity and their ability to describe events all along the continuum between exposure and clinical disease (Figure 3.2). Health

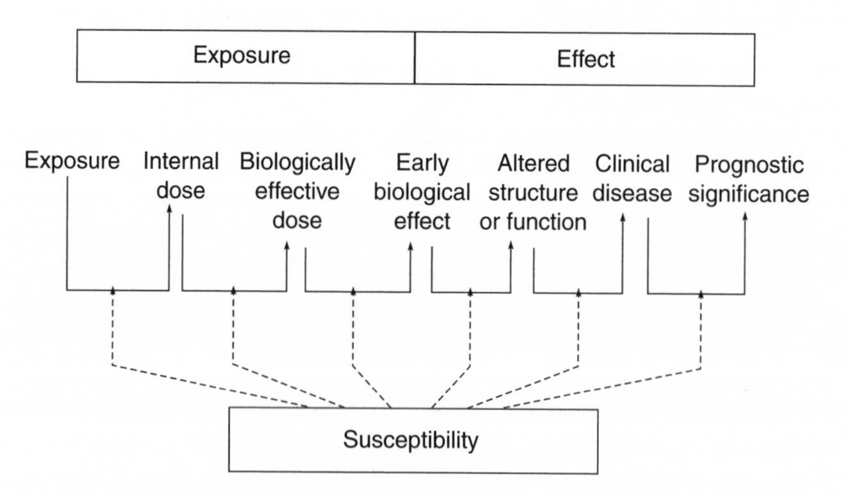

Figure 3.2 Relationship of susceptibility, exposure, and effect

events are now far less likely to be viewed as binary phenomena (presence or absence of disease) than as a series of changes through homeostatic adaptations and dysfunction to disease and death (NRC, 1991).

Conduct of an Environmental Epidemiological Study

The many practical problems in the organization of an environmental epidemiological study include both the level of study to be conducted (simple to complex) and the resources required. One of the first objectives is to identify the population group to be examined. It is often helpful to consider the conduct of an initial study among workers who may be exposed to the same agent. One advantage of this approach is that exposures in occupational settings are often higher than in the general environment. At the same time, it should be borne in mind that a working population is preselected: it excludes children, the elderly, and those whose health is already impaired, as well as individuals who may be hypersensitive to certain agents. For certain occupations, the working population also frequently includes a disproportionately low number of women. Furthermore, exposures of workers are limited in most cases to eight hours a day. As a result, caution must be exercised in extrapolation of the resulting observations to the general population (WHO, 1983).

Once the study group has been identified, contacts need to be established with individuals within the group to guarantee their interest and cooperation. Where individuals decline to participate, care must be taken that their response does not bias the results of the study. If a number of people are engaged in collecting information, joint training sessions are required to ensure uniformity of approach, and it may be necessary and beneficial to interchange the teams periodically during the data-collection period. Experience has shown that the most effective approach when the effects of common environmental agents on individuals within the general population are being studied is to have the data collectors visit the subjects in their homes. Although this technique is labor intensive, it is often justified by the improved quality of the results. Any instruments used to collect data need to be calibrated on a regular basis, and all related methods should be standardized. If biological indices of effects are used, it may be necessary to have all measurements performed in a single laboratory.

Ethical problems may also arise. If some tests are intrusive (for example, the collection of blood samples), prior permission will be required. Conidentiality is another issue. Thus it is common practice to include the iames and addresses of those interviewed only on the original survey form.

As indicated by the World Health Organization (1983) and Monson (1990), the computer has had a revolutionary impact on the conduct of environmental epidemiological studies. In fact, the dramatic increase in the number of such studies since the 1950s is directly related to the development and wide availability of these devices. As Monson has stated, the ability to collect large amounts of data, to store them, and to conduct extensive analyses is "the hallmark of epidemiology today." This is especially true of data that show weak associations between exposure and effects. Still, the computer has separated many epidemiologists from the data they are analyzing. They may not be familiar with weaknesses inherent in the collection of the data or with limitations in the computer programs that are used.

Case Studies

Many environmental epidemiological studies have served as examples of the beneficial uses and applications of this methodology. One of the earliest documented the fact that the intake of fluoride in drinking water led to a reduction in dental caries (Terracini, 1992); later studies (Chapter 15) led to a quantification of the relationship between the dose from ionizing radiation and the induction of a fatal cancer (Shigematsu, 2000). Two others, the relationship between cigarettes and lung cancer and between airborne concentrations of extremely small airborne particles and population death rates, are summarized here.

CIGARETTES AND LUNG CANCER

The determination of a definitive association between cigarette smoking and lung cancer is a classic example of the useful application of environmental epidemiology. It is also an example of how the personal choices of individuals can have an extremely detrimental effect on their health and of how difficult it is, even when a relationship has been thoroughly demonstrated, to implement effective control measures.

In the middle to late 1940s, physicians in several of the industrialized countries of the world, including the United States and the United Kingdom, noted an increasing number of diagnoses of men with lung cancer. A decade earlier, such cancers had been a medical curiosity. Although cigarette smoking was immediately suspected as a cause, the presumption had to be confirmed. Two types of studies were undertaken—case-control studies in which persons with and without lung cancer were asked about past habits, including smoking; and cohort studies in which smokers and nonsmokers were followed and the rates of development of a variety of diseases, including lung cancer, were measured (Monson, 1990).

One of the leading epidemiologists who conducted such studies was Richard Doll, working first with A. Bradford Hill and later with Richard Peto. On the basis of an initial case-control study, Doll and Hill (1950) concluded that "smoking is a factor, and an important factor, in the production of carcinoma of the lung." They admitted, however, that they had no evidence about the nature of the carcinogen. Nonetheless, their research and that of other scientists led to the issuance in 1964 of the surgeon general's report on *Smoking and Health* (USPHS, 1964), a major breakthrough in the campaign against smoking. On the basis of a subsequent series of longer-term cohort studies, Doll and Peto (1976) concluded that the death rate from lung cancer in smokers was ten times that in nonsmokers. These studies and related stimulated the subsequent development of a broad range of antismoking campaigns, including the banning in the United States of cigarette commercials on television stations (Surgeon General, 1989). Interestingly, one of the principal actions that finally brought about a noticeable reduction in cigarette smoking in the United States was publication of the results of epidemiological studies that showed that nonsmokers were harmed by "secondhand" (sidestream) smoke (Trichopoulos, 1994).

EFFECTS OF AIRBORNE PARTICULATES

In 1974, researchers at the Harvard School of Public Health initiated a study of the relation between human respiratory health and survival and the concentrations of particulate matter (PM) and sulfates (a component of smaller particles) in ambient air within the United States. The study involved a random sample of more than 8,000 people living in six eastern cities. One of the major findings, based on the results of 15 years of observations, was that death rates among the study populations correlated with the concentrations of fine particulate air pollution in the communities in which they lived (Dockery et al., 1993; Pope et al., 1995). The particles that proved to be most significant were those 2.5 micrometers ($PM_{2.5}$) or less in size. Primary sources include motor vehicles and power plants;

they are also formed by photochemical transformations in the air (Chapter 5).

A similar study was undertaken by the American Cancer Society in 1980. In this case, the study population included more than 500,000 people residing in 154 cities. Increased deaths were found as a result of exposures to particles in the same size range even though the concentrations of particles smaller than 10 micrometers (PM_{10}) in the air in the cities in which the people lived complied with the 1987 air quality standards established by the Environmental Protection Agency (EPA). Even though the increase in mortality was small, nationwide it was estimated to be producing as many as 60,000 deaths per year. Reacting to these findings, the American Lung Association sued the EPA seeking a review of the air quality standards for particulate matter which are mandated by Congress to be set at a level so as to protect the public health. In 1997, the EPA announced new regulations to limit the concentrations of $PM_{2.5}$. Although these regulations were subsequently challenged, in 2001 the U.S. Supreme Court unanimously ruled in favor of the EPA's action.

Subsequent studies at the Johns Hopkins University not only confirmed these observations but also provided compelling evidence of the relationship between air pollution and lung cancer and heart disease. On the basis of analyses of the data, it was estimated that there is a 6 percent increase in deaths from heart- and lung-related causes and an 8 percent increase in deaths from lung cancer for each 10 micrograms per cubic meter (10 $\mu g/m^3$) increase in the concentrations of fine particulates ($PM_{2.5}$) in the air (Pope et al., 2002).

The General Outlook

The importance of the role of epidemiological studies in environmental and public health continues to increase. Initially, the primary outcome of such studies was to provide information on the relationships between a given environmental stress and one or more diseases. Since this was the case, one possible response would have been to view this field as a "negative" science; that is, the outcomes or information it provided were predominantly "bad." Today, however, the use and application of epidemiological studies have broadened considerably. This is illustrated in several ways. Through applications of molecular epidemiology, as described earlier, it may ultimately be possible to correlate early changes in the human body, as denoted by biomarkers, to longer-range health effects, with the result that the information developed can be used not only in preventing disease, but also possibly in curing it. Equally important are the studies of the correlations between factors such as diet and exercise and health through which epidemiologists are illuminating steps that can be applied for enhancing personal health and well-being. Another outcome of epidemiological studies is the determination of voids in our knowledge. The information can then be used to identify areas in which additional research would be beneficial (Muirhead, 2001).

If the benefits of epidemiological studies are to be realized, however, it is important that their findings and implications be adequately communicated to the media and the public. This is especially true with regard to life-style choices, such as diet and exercise (that were discussed in Chapter 1). Otherwise, these benefits may not be realized by major segments of the population. Such communication is also important in those cases where there are doubts and uncertainties in the outcomes of certain studies, particularly as epidemiologists search for subtler links between diseases and environmental agents. A key example where communication is important involves those cases in which members of the public, upon learning about a cluster of cancers within a given locality, immediately attribute them to one or more suspected environmental contaminants. What they fail to realize and need to understand is that even when such cases occur on a random basis, some will tend to be localized. While, their judgment may be correct in some instances, such claims are seldom supported by the scientific facts. In fact, in 85 percent or more of such cases, detailed statistical analyses subsequently reveal that in actuality, the cancer rates are not elevated (Robinson, 2002).

Concurrent with these needs, the range and nature of the environmental stresses that potentially could be important and therefore need to be evaluated continue to increase. One example is an elucidation of the impacts of climate change on human health. With so much evidence of the nonhuman impacts of global warming (Chapter 20), one can question why this facet of the environment initially received so little attention on the part of epidemiologists (McMichael, 2001). Fortunately, this situation is changing. Studies have demonstrated, for example, that periods of extreme rainfall and accompanying runoff are associated with outbreaks of both waterborne and foodborne illnesses (Curriero et at., 2001). What is needed is to expand these efforts to provide the analogs that will make it possible to estimate with confidence the impacts on human health of potential longer-term changes in climate. A primary goal of such efforts should be to ensure that the necessary data will be available for input into the mathematical models that are being used to assess such impacts (McMichael, 2001).

4

THE WORKPLACE

As early as the fourth century B.C., Hippocrates apparently observed adverse effects on miners and metallurgists caused by exposure to lead. In 1473, Ulrich Ellenbog recognized that the fumes of some metals were dangerous and suggested preventive measures. In the early 1500s, Georg Bauer (known as Georgius Agricola), a physician and mineralogist, attributed lung disease among miners in the Carpathian Mountains to the inhalation of certain kinds of mineral dusts, observing that so many miners succumbed to the disease that some women were widowed as many as seven times. In 1700, Bernardino Ramazzini published the first complete treatise on occupational diseases, *De morbis artificum diatriba* (Diseases of workers). As a result of this pioneering effort, Ramazzini is known as the "father of occupational medicine" (Franco, 2001). In the mid-1880s, Karl Bernhard Lehmann, whose work continues to serve as a guide on the effects of exposure to airborne contaminants, conducted experiments on the toxic effects of gases and vapors on animals (Patty, 1978). During the same period, the first occupational cancer, scrotal cancer in chimney sweeps, was observed in England.

In the United States, occupational health problems received little attention until the twentieth century. The U.S. Bureau of Labor was created in 1885, but even when it became the Department of Labor in 1913, its stated goals included no mention of workers' health beyond "promoting their material, social, intellectual, and moral prosperity" (U.S. Congress, 1913). Alice Hamilton's classic work *Exploring the Dangerous Trades,* now perhaps the most widely quoted book on the field in the world, was not published until 1943 (Hamilton, 1943).

Today the profession that has primary responsibility for recognizing, evaluating, and controlling hazards in the workplace is known in most countries of the world as *occupational hygiene,* the exception being the United States, where it is identified as *industrial hygiene.* In either case, the primary responsibility of those working in this field is to address the full range of chemical, biological, and physical hazards, including the musculoskeletal problems that are becoming increasingly common in the modern technological world. As the scope of the challenges implies, if industrial hygienists are to address these problems effectively, they must have the abilities to combine the skills and knowledge of people working in the physical sciences and engineering, as well as in the health sciences and medicine. They must also be able to apply relevant information being generated in the fields of toxicology and epidemiology (Herrick, 1998). If successful in these efforts, industrial hygienists have an opportunity of demonstrating the effectiveness of the public health intervention model (Chapter 1) which emphasizes the prevention of the development of disease as contrasted to waiting until people become ill and then seeking to cure them.

Protective Legislation

Protective legislation came piecemeal and slowly (Table 4.1). In 1908, the federal government provided limited compensation to civil service employees injured on the job. In 1911, New Jersey became the first state to enact a workers' compensation law; although many other states rapidly followed suit, it was not until 1948 that all the states in the United States required such compensation (Patty, 1978).

Workers' compensation laws passed in France, Germany, and the United Kingdom in the nineteenth century were one of the earliest forms of social insurance provided on a prepaid basis. One unique feature was that these laws required no direct monetary contribution from the workers. As later enacted at the federal level in the United States, these laws removed from the courts the determination of compensation for occupational injuries. This revolutionized the approach for controlling workplace hazards and did more than any other measure to reduce occupational injuries in this country. Also playing a major role was the passage of similar laws at the state level. The associated contributions included expanding the coverage to include occupational disease, and requiring that compensation for occupational injuries be paid on a no-fault basis, that settlements be reached

Table 4.1 Significant federal legislation pertaining to occupational health and safety

Year	Act	Content
1908	Federal Workers' Compensation Act	Granted limited compensation benefits to certain U.S. civil service workers for injuries sustained during employment
1936	Walsh-Healey Public Contracts Act	Established occupational health and safety standards for employees of federal contractors
1969	Federal Coal Mine Health and Safety Act	Created forerunner of Mine Safety and Health Administration; required development and enforcement of regulations for protection of mine workers
1970	Occupational Safety and Health Act	Authorized federal government to develop and set mandatory occupational safety and health standards; established National Institute for Occupational Safety and Health to conduct research for setting standards
1976	Toxic Substances Control Act	Required data from industry on production, use, and health and environmental effects of chemicals; led to development of "right-to-know" laws, which provide employees with information on nature of potential occupational exposures
1990	Pollution Prevention Act	Established policy to ensure that pollution is prevented or reduced at source, recycled or treated, and disposed of or released only as last resort; led to substitution of less toxic substances in wide range of industrial processes, with significant reductions in worker exposure

promptly through administrative tribunals; and that payments be made in accordance with a system of scheduled benefits.

The Walsh-Healey Public Contracts Act of 1936 established safety and health standards in industries that were conducting work under contract to the federal government. This forerunner of modern occupational health regulations stimulated research on occupational diseases and the development of occupational health programs by state and local agencies, insurance companies, foundations, management, and unions. The first significant federal legislation for workers not involved in government projects, however, was not enacted until 1969, when the Federal Coal Mine Health and Safety Act was passed. This legislation was followed by the landmark Occupational Safety and Health Act of 1970, whose principal purpose, as described, was "to assure so far as possible every working man and woman in the Nation safe and healthful working conditions and to preserve our human resources." Among its provisions were the establishment of the Occupational Safety and Health Administration (OSHA) and the creation of the National Institute for Occupational Safety and Health (NIOSH). The stipulated purposes of OSHA were to encourage the reduction of workplace hazards, to provide for occupational health research, to establish separate but dependent responsibilities and rights for employers and employees, to maintain a reporting and record-keeping system to monitor job injuries, to establish training programs, to develop mandatory safety and health standards, and to provide for development and approval of state occupational safety and health programs. The responsibilities of NIOSH were to conduct the research necessary to establish a scientific foundation on which to base such standards and to implement education and training programs to ensure the availability of adequate numbers of qualified people to implement and enforce the Occupational Safety and Health Act.

In a later development, Congress incorporated "right-to-know" provisions in amendments to the 1976 Toxic Substances Control Act that required employers to provide workers with information about the health hazards of their occupational environments. This stipulation has made it much easier for workers to be aware of the hazards they face and to raise questions about the protection being provided. Another law that has significantly reduced occupational exposures is the Pollution Prevention Act of 1990 (U.S. Congress, 1990). This law, which established a national policy to encourage the prevention of pollution at the source, with disposal to the environment acceptable only as a last resort, has led to the substitution of less toxic substances for once previously used in a wide range of industrial processes. These actions, in turn, have significantly reduced workplace exposures.

Identification of Occupational Health Problems

Today well over 135 million men and women are gainfully employed in the United States (NSC, 2001). To some degree, all of these people are exposed to occupational hazards, and all are at risk of job-related adverse health effects. These problems are compounded by the fact that more than 25 percent of U.S. workers are employed in businesses that have fewer than 20 employees, and more than 50 percent in companies with fewer than 100 employees. Unfortunately, smaller companies often lack the knowledge to identify occupational health hazards and the funds to finance associated control programs; moreover, many are exempt from state and federal occupational health and safety regulations.

The effects of occupational exposures range from lung diseases, cancer, hearing loss, and dermatitis to more subtle psychological effects, many of which are only now being recognized (Table 4.2). Workplace exposures include those to airborne contaminants, ionizing radiation, ultraviolet and visible light, electric and magnetic fields, infrared radiation, microwaves, heat, cold, noise, extremes of barometric pressure, and stress. Each of these may also interact with other chemical, physical, or biological agents. For example, cardiovascular diseases may be related to a combination of physical, chemical, and psychological job stresses. The workplace can also be the source of a wide range of infectious diseases. As noted in Chapter 1, hospital workers in particular must be concerned with protection against hepatitis B, tuberculosis, influenza, and other viral infections, including acquired immune deficiency syndrome (AIDS) and severe acute respiratory syndrome (SARS).

Crude extrapolations based on reports by hospital emergency departments indicate that about 10 million occupational injuries and illnesses occurred in the United States in 1998 (CDC, 2001a). Some 35 to 40 percent of these injuries were disabling, resulting in the loss of about 80 million days of lost time (NSC, 2001). The number of U.S. workers who die each year as a result of occupational injuries and illnesses is about 65,000; some 5,000 of these are due to injuries. Worldwide, work-related illnesses and injuries are estimated to kill more than 1 million people each year. These deaths fall into two broad categories: (1) those due to chronic diseases,

Table 4.2 The ten leading work-related diseases and injuries, United States, 2000

Type of disorder or injury	Examples
Occupational lung diseases	Asbestosis, byssinosis, silicosis, coal workers' pneumoconiosis, lung cancer, occupational asthma
Musculoskeletal injuries	Disorders of the back, trunk, upper extremity, neck, lower extremity; traumatically induced Raynaud's phenomenon
Occupational cancers (other than lung)	Leukemia, mesothelioma, cancers of the bladder, nose, and liver
Severe occupational injuries	Amputations, fractures, eye loss, lacerations, traumatic deaths
Cardiovascular diseases	Hypertension, coronary artery disease, acute myocardial infarction
Reproductive disorders	Infertility, spontaneous abortion, teratogenesis
Neurotoxic disorders	Peripheral neuropathy, toxic encephalitis, psychoses, extreme personality changes
Noise-induced loss of hearing	
Dermatologic conditions	Dermatoses, burns, chemical burns, contusions
Psychological disorders	Neuroses, personality disorders, alcoholism, drug dependency

such as cancer, asbestosis, and silicosis; and (2) those due to workplace injuries, such as motor-vehicle accidents, machinery-related accidents, homicides, falls, and electrocutions. These events, however, are not evenly distributed among various industries (Herbert and Landrigan, 2000). In the United States, the highest rate of death per 100,000 workers (22.5) occurs in forestry, fishing, and agricultural services. The second-highest rate (21.2) occurs in mining and quarrying and oil and gas extraction (Chapter 11) (NSC, 2001). The estimated total cost of fatal and nonfatal unintentional work-related injuries in 2000 was more than $130 billion (CDC, 2002). Trends in worker deaths and death rates in the United States from 1980 through 2000 are shown in Figure 4.1. For workers in the less developed countries of the world, the rates in essentially all comparable industries are higher, frequently dramatically higher, than those for the United States.

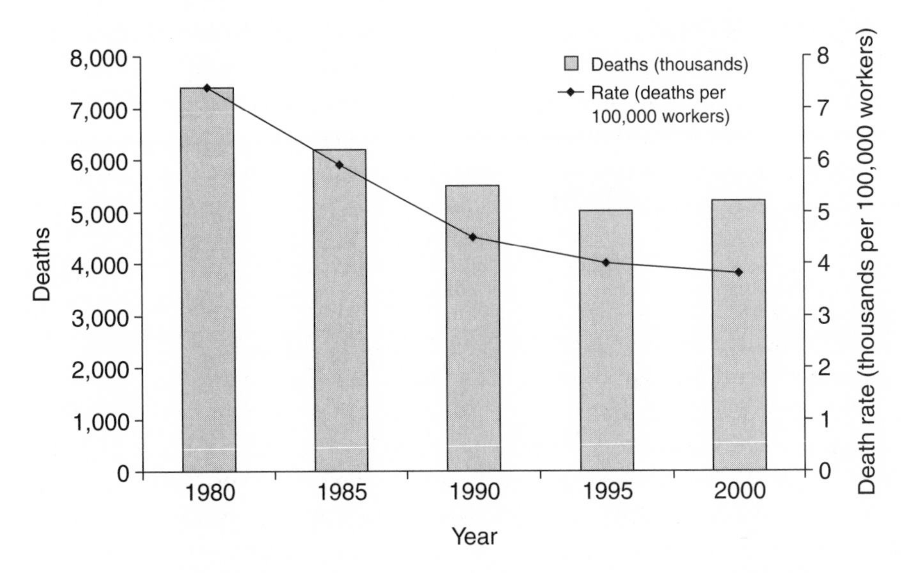

Figure 4.1 Trends in occupational injuries and deaths in the United States

Large though these numbers are, the true magnitude of the health and economic impacts of occupational disease and injury in the United States remains unknown. First of all, the recording of data on workers' illnesses and deaths is often incomplete or erroneous. Physicians frequently fail to relate observed diseases to occupational exposures. This is particularly true for neurologically based illnesses and for chronic degenerative diseases, such as atherosclerosis and chronic obstructive respiratory ailments. In other cases, the diagnosed cause of death may not be coded onto the death certificate. Even when the required information is available, it may not be used to promote worker protection. Second, because the appearance of the health effects caused by chronic exposures in the workplace is delayed, and because many workers change jobs frequently, by the time a disease manifests itself, it may be difficult to relate it to a specific exposure or combination of exposures. Third, even if an association between a specific disease and a given toxic agent is known to exist, it is often difficult to quantify the concentration of the toxic agent to which the worker was exposed and to estimate the intake and the accompanying dose.

Economic considerations also tend to delay or reduce attempts to address occupational health problems. In some corporations, for example, the directors and officers are frequently pressured by stockholders to show a profit even during times of economic recession. To satisfy these demands, senior company officials may insist that a plant be kept in operation with minimum downtime for maintenance or overhaul. This can, in turn, seriously reduce attention to worker health and safety. Sensing the urgency to maintain production and fearing the loss of their jobs, workers may in turn disregard controls designed to enhance health and safety, especially in cases when such measures slow production or interfere with comfort. If there is an accompanying reduction in federal funding for regulatory agencies, such as OSHA, the number and frequency of inspections may concurrently be curtailed. Even today, records show that OSHA inspectors annually visit only about 30,000 of the more than 1 million workplaces in the United States. In fact, three-quarters of the facilities in which workers were injured by serious accidents in 1994 had not been inspected during the 1990s. The situation, however, is not as bad as it may seem since the companies that insure workplace establishments make hundreds of thousands of visits each year. If they discover that good practices are not being observed, they immediately increase the premiums on the insurance policies carried by operators of that facility. This provides a strong economic incentive to industrial organizations to pay attention to worker safety. Additional inspections are conducted by state and local agencies.

Another problem is that the patterns of occupational disease are constantly changing and require ever more refined methods to reveal the subtle injuries and disabilities that result from low-level exposures to chemicals and physical, nonphysical, and on-the-job psychological stresses. One possible help in alleviating this problem is the rapidly developing field of molecular toxicology (Chapter 2). One of the major contributions of this field is the expanded nature and range of endpoints it makes available for assessing the potential effects of all types of toxic agents. Of particular interest are the endpoints or biomarkers that are becoming available for measuring exposure or dose. Some predict that it may be only a matter of time before biomarkers will be routinely applied by industrial hygienists in assessing a wide range of workplace risks.

Without such advances, it will continue to be difficult to identify occupational risks, determine their magnitude, and judge the adequacy of control measures. The success of these efforts has implications far beyond the occupational environment. Because higher-level exposures to hazardous agents frequently occur first in the workplace, and the associated health effects are initially identified and observed among workers, the monitoring of occupational exposures can, and often does, provide the first warning of the presence of potential hazards in the general environment. Clearly, our ability to keep abreast of occupational diseases and injuries has consequences not only for workers but also for their families and their communities.

Types and Sources of Occupational Exposures

Years ago, most of the people who were classified as workers were employed in manufacturing. During the past several decades, this situation has undergone major changes. Today only about 20 million of the workers in the United States are employed in this category; the remainder are in service industries. Even so, both types of employment have associated occupational health problems, and, as would be expected, many problems are common to both. One of the most common in manufacturing is the presence of contaminants in the air that result from various industrial processes. Other problems include noise, vibration, and ionizing radiation. Common problems in the service industries include inadequate indoor air quality, low-back pain, and cumulative trauma disorders. In certain situations, problems not heretofore recognized are assuming importance. These include the need to protect workers from potential exposures to biological agents and to provide them with safe (nonslip) floors and stairs and comfortable, employee-friendly workstation environments. Three of the primary agents or factors to which workers are exposed today are discussed in the sections that follow.

TOXIC CHEMICALS

As would be anticipated, toxic chemicals play a major role in occupationally related diseases. Their two primary portals of entry are the skin and the respiratory tract. Once inside the body, such agents can affect other organs, such as the liver and kidneys. The ideal way to assure that chemical exposures are properly controlled is to ensure that the techniques necessary for assessing their toxicological risk are available and applied before they are introduced into the workplace (Burgess, 1995). Despite the advances described earlier, much work needs to be done to develop practical and reliable screening systems to identify chemicals that have a potential for harming human health.

Typical of the chemicals that can gain access to the body through the respiratory tract are those that are released into the air as a result of activities associated with metal fabrication, machining, welding, brazing, and follow-up operations involving the cleaning, electroplating, or painting of the finished product. Included in such releases are mineral

Table 4.3 Types, sources, effects, and control of typical airborne contaminants in the construction industry

Contaminant	Examples	Sources of exposure	Effects	Methods for Control
Volatile organic compounds	Aeromatic hydrocarbons, chlorinated solvents, formaldehyde, toluene diisocyanate	Use as solvents and additives in paints to enhance color and spreadability	Headaches, respiratory problems, allergic reactions	Ventilation, reduce or eliminate use in paints
	Urea formaldehyde	Use as binder in particle board and hardwood plywood paneling	Brain impairment, lung cancer, nasopharyngeal cancers	Add scavengers to prevent formaldehyde from volatilizing
Toxic metals	Lead	Renovation and demolition of old buildings and metal structures, use of lead-based paints	Nausea, fatigue, aches and pains, damage to central nervous system	Ban use of lead paint indoors, respiratory protection, periodic tests for lead levels in blood
	Cadmium, chromium, copper, nickel, and zinc	High-temperature welding of metals such as stainless steel	Metal fume fever, chemical pneumonia	Ventilation, use of air-supplied respirators
Silica	Sand, flint, agate, and quartz	Cleaning buildings and bridges using sandblasting equipment	Silicosis	Use of non-silica-containing abrasives, respiratory protection
Asbestos	Floor tile, pipe insulation, fireproofing materials	Refinishing tile floors, maintenance of heating systems	Asbestosis, mesothelioma, lung cancer	Isolation, personal protective equipment

dusts, metal fumes, and resin systems used in sand bonding agents, as well as carbon monoxide. Specific examples of airborne contaminants produced in the construction industry, their effects, and methods for their control are shown in Table 4.3. The last aspect, control, will be discussed in more detail in a section later in this chapter. The operations that produce some of these contaminants also generate a host of physical hazards, such as noise, vibration, and heat stress, as well as dermal exposures to cutting fluids and coolants. The last two items produce upwards of half a million cases of dermatitis in this country each year.

BIOLOGICAL AGENTS

The presence of biological agents (bioaerosols) in the air of the workplace is increasingly recognized as a common problem. This is especially true in the health-care industry, where respirable aerosols that contain blood are routinely produced in the operating room during surgical procedures. Similar exposures have been observed in dental offices. In like manner, flax dust in the linen industry has been shown to contain microbial contaminants. A further problem is that fungi may grow in certain types of respirators, especially those in which the filters contain cellulose and/or fiberglass. A closely related problem is the presence in some workplaces, such as those associated with the agriculture and food industries, of airborne dusts that can cause respiratory allergies, such as asthma and allergic rhinitis.

Related instances of exposures to infectious disease agents include health-care workers who are exposed to bloodborne pathogens, such as hepatitis B virus and the human immunodeficiency virus, which causes AIDS. In contrast to the examples cited earlier, the primary sources of these exposures are accidental punctures of the skin with contaminated needles. Overall, almost 6 million U.S. workers are subject to these types of workplace hazards. In addition to those employed in the health-care industry, these include people employed in funeral services, linen services, medical equipment repair, correctional facilities, and law enforcement and at hazardous waste sites. The accompanying exposures are estimated to lead to more than 9,000 infections and more than 200 deaths in the United States each year.

PHYSICAL FACTORS

Many health and safety problems in the workplace are caused by inadequate attention to the complex relationships among people, machines, job demands, and work methods, a specialty area that is designated by the term *ergonomics.* Such relationships include repetitive motions, forceful motions, static or awkward postures, mechanical stresses, and local vibration (Figure 4.2). If such relationships are properly addressed, the performance and health of the involved workers will not be jeopardized. Otherwise, there can be a range of undesirable outcomes. These include an increase in the rates of errors, accidents, and injuries as well as serious impacts on health, the most common of which are musculoskeletal disorders (Keyserling and Armstrong, 1998). For the United States, the lack of attention to ergonomics causes an estimated 250,000 new cases of repeated trauma associated disorders each year (NSC, 2001). Although discussions during the past decade would imply that the recognition of such problems is new, in reality it is not. In the previously cited *De morbis artificum diatriba,* Ramazzini noted that a variety of common occupational diseases were caused by prolonged, violent, and irregular motions and prolonged postures (Franco, 2001).

Data show that nearly two-thirds of the illness cases reported among U.S. workers are associated with factors that involve problems of the human-machine interface. Overall, it is estimated that approximately 1

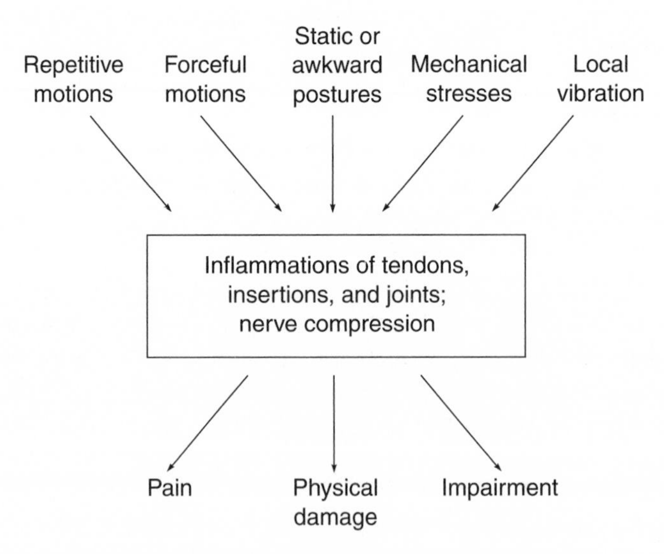

Figure 4.2 Sources and effects of physical factors in the occupational environment

million people in this country miss work each year due to disorders resulting from such relationships, the most common of which is carpal tunnel syndrome. According to the Bureau of Labor Statistics, the median time (24 days) away from work in 1998 was higher for this particular disorder than for any other major disabling injury or illness. If compensation, lost wages, and decreased productivity are taken into account, such disorders cost the United States an estimated more than $45 billion each year (NRC, 2001).

About 25 percent of all injuries in the workplace occur in the process of lifting and moving objects. Another 15–20 percent are caused by slips and falls. Inadequate lighting is often a contributing factor in the case of the latter events. Overall, these two categories represent almost half of all such injuries. Furthermore, data indicate that if a worker who has suffered a low-back injury has not returned to work within six months, he or she will probably never return (Snook, 1989).

Another major source of physical stress is noise, one of the most common of all occupational problems. Because noise-induced hearing loss occurs gradually, invisibly, and often painlessly, many employers and employees do not recognize the problem early enough to provide protection; indeed, for years hearing loss was considered a "normal" hazard of employment. Today most people recognize that noise can interfere with communication, can disturb concentration, and can cause stress. In fact, people subjected to excessive noise have elevated blood pressure, an increased pulse rate, and a higher respiratory rate. By-products of these effects are increased levels of fatigue and higher rates of injuries.

Another pervasive problem is heat stress, especially among workers who wear protective clothing. As body temperature increases, the circulatory system seeks to cool the body by increasing the heart's pumping rate, dilating the blood vessels, and increasing blood flow to the skin. If these mechanisms do not provide sufficient cooling, the body perspires; the evaporation of sweat will cool the skin and the blood and reduce body temperature. Because sweating causes a loss of both water and electrolytes, some form of heat stress, including heatstroke, may develop if the body temperature is not reduced. The degree to which a given worker is affected depends on his/her level of physical activity, the velocity of air movement, the dry-bulb air temperature, and the relative humidity (which influences the effectiveness of perspiration as a mechanism for cooling the body). Overall, an estimated 5 million or more U.S. workers are affected by heat stress each year.

Occupational Exposure Standards

The American Conference of Governmental Industrial Hygienists (ACGIH), established in 1938, has played a major role in reviewing and assessing the literature and recommending limits for the control of workplace exposures in the United States. One of its early contributions was the development of what are called threshold limit values (TLVs), which provide guidance on permissible concentrations of airborne contaminants (Table 4.4; ACGIH, 2004). Through its efforts, TLVs now exist for more than 600 chemical substances. The American Industrial Hygiene Association (AIHA) (established in 1939), OSHA, and NIOSH have subsequently also been involved in developing related standards. The primary difference is that the latter two agencies have regulatory authority, whereas the ACGIH and the AIHA do not. In more recent years, NIOSH has issued health standards for a variety of chemicals used in industry. This agency has also periodically issued a series of "Alert" bulletins requesting the assistance of industrial personnel in the control of specific problems that range from lead poisoning, organic dust, and silicosis to the development of a strategy for prevention of and research on homicide in the workplace.

In evaluating the validity of the TLVs, it is important to note that they are derived on the basis of data from a variety of sources. These include industrial experience, experiments involving humans, and experiments involving other animals. Whenever possible, they are based on a combination of information from all three sources (ACGIH, 2004). Nonetheless, the basis on which the values are established may differ from substance to substance. Protection against impairment of health may be a guiding factor for some TLVs, whereas reasonable freedom from irritation, narcosis, nuisance, or other forms of stress may be the basis for others. Threshold limit values, however, do not always represent valid thresholds for adverse effects on health. A small percentage of workers, because of age, genetic factors, personal habits (such as cigarette smoking or the use of alcohol or other drugs), medication, or previous exposures, may be affected by some substances at concentrations at or below the threshold limit. For most workers, however, maintaining exposures below the threshold should provide protection. If the relationship between exposure (or dose) and the associated health effects is linear, there will be some effect, however small the exposure (Figure 4.3). For this reason, TLVs do not designate fine lines between safe and dangerous concentrations, nor should they be interpreted as a definitive index of the relative toxicity of various substances (ACGIH, 2004).

Table 4.4 Threshold limit values (TLVs) for selected chemical substances in the air

Substance	Typical industrial uses or sources	Time-weighted average[a]		Short-term exposure limit[b]	
		ppm[c]	mg/m^2	ppm[c]	mg/m^3
Benzene	Gasoline refining	0.5		2.5	
Cadmium	Welding		0.01 0.002[d]		
Carbon monoxide	Blast furnaces	25		—	
Chlorine	Water disinfection	0.5		1	
Formaldehyde	Embalming	—		0.3[d]	
Lead	Battery manufacturing		0.05[e]		—
Manganese	Steel making		0.2		—
Mercury, inorganic	Fungicide applications		0.025		—
Silica, crystalline	Sand blasting, granite quarying		0.05		—
Toluene (skin absorption)	Petroleum refining	50		—	
Trichloroethylene	Metal degreasing	50		100	

a. For normal 8-hour day, 40-hour workweek.
b. Not to exceed 15 minutes more than four times per day.
c. Parts per million.
d. Respirable fraction.
e. Ceiling limit.

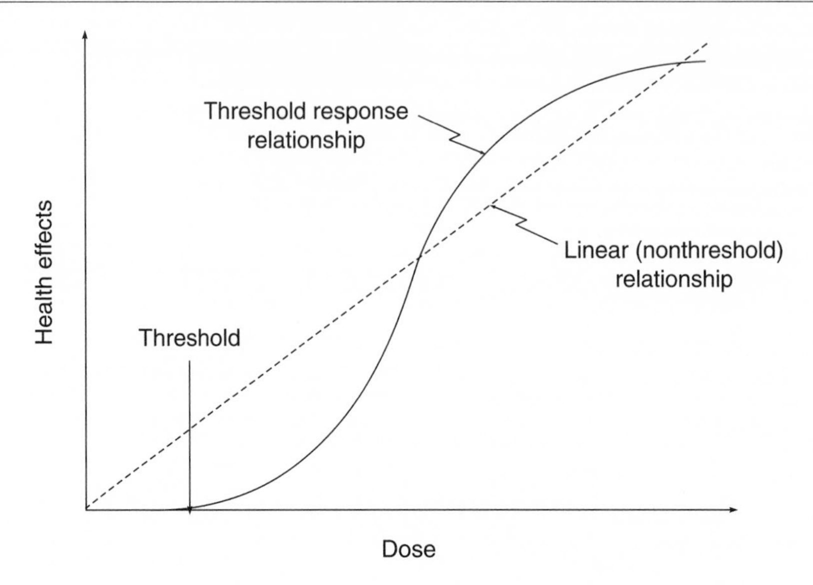

Figure 4.3 Possible relationships between dose and health effects

In recent years, the ACGIH has supplemented its TLVs by providing biological exposure indices (BEIs) for more than three dozen chemicals. Recommended limits and indices for selected chemicals are shown in Table 4.5. By establishing both TLVs and BEIs, the ACGIH offers a two-step approach for assessing the importance of chemicals in the workplace: first, monitoring the air being breathed; second, monitoring the chemicals themselves or their metabolites in biological specimens (such as urine, blood, and exhaled air) collected from the exposed workers at specified intervals. The first step provides data on exposures of workers. The second provides data on the amount of a given contaminant that has been taken into the body, which, in turn, can be used to estimate the accompanying dose. Although the correlation between measurements of the exposure and the amount of a contaminant in the body is generally close, as is explained in Chapter 5, the resulting data are not synonymous. The ACGIH has also developed TLVs for physical agents, including heat and cold, noise (Table 4.6), vibration, lasers, radio-frequency/microwave radiation, magnetic fields, and ultraviolet and ionizing radiation (ACGIH, 2004).

Table 4.5 Biological exposure indices (BEIs) for selected chemicals

Chemical	Sampling time	Biological exposure index
Acetone		
Acetone in urine	End of shift	50 mg/L[a]
Arsenic, elemental and soluble inorganic compounds		
Inorganic arsenic plus methylated metabolites in urine	End of workweek	35 μg As/L
Cadmium and inorganic compounds		
Cadmium in urine	Not critical	5 μg/g creatinine[b]
Cadmium in blood	Not critical	5 μg/L
Carbon monoxide		
Carboxyhemoglobin in blood	End of shift	3.5% hemoglobin
CO in end-exhaled air[c]	End of shift	20 ppm
Lead		
Lead in blood	Not critical	30 μg/100 ml
Mercury		
Total inorganic mercury in urine	Prior to shift	35 μg/g creatinine
Total inorganic mercury in blood	End of shift at end of workweek	15 μg/L
Trichloroethylene		
Trichloroacetic acid in urine	End of workweek	100 mg/g creatinine
Trichloroacetic acid and trichloroethanol in urine	End of shift at end of workweek	300 mg/g creatinine
Free trichloroethanol in blood	End of shift at end of workweek	4 mg/L

a. Milligrams per liter.
b. Micrograms per gram.
c. Usually represents alveolar air from lower respiratory system.

Table 4.6 Threshold limit values for noise in the workplace

Typical industrial source	Exposure time (hours per day)	Sound level (decibels)
Textile plants, forge shops, machine shops, jackhammer operators	1/8	103[a]
	1/4	100
	1/2	97
	1	94
	2	91
	4	88
	8	85
	16	82
	24	80

a. No exposure should be permitted to continuous, intermittent, or impact noise in excess of 140 decibels.

Monitoring the Workplace

Workplace monitoring can be done to assess exposures of workers under routine conditions, to alert workers to abnormal (accident) situations, or to design a control strategy. The type of monitoring program depends to a large extent on the nature of the stress being evaluated.

AIRBORNE CONTAMINANTS

If the source of exposure is an airborne contaminant, air sampling may be the only approach necessary for assessing exposure. This is particularly true if a technique that has the necessary sensitivity is readily available. If, however, both the measurement and its interpretation are difficult, a combination of monitoring techniques may be required. As a general rule, analyses of excreta, primarily of urine and sometimes of feces, provide more accurate information on workplace exposures. The information they provide, however, is "after the fact." Since air monitoring provides a warning of potentially unacceptable conditions, it is almost always mandatory in cases of the presence of airborne contaminants in the workplace. If the contaminants are particulate in nature, the information obtained should include their concentrations, size, chemical form, and solubility, since these factors affect where they will be deposited within the lungs, how effectively they will be retained, and their rate of uptake and metabolism by the body (Chapter 5).

Essentially all air samplers consist of a filter or sorbent collector, an air mover or fan to pull the air and associated contaminants through the collector, and a means of controlling the rate of flow. The system selected depends on the purpose of the monitoring program and the type (particulate or gas) and concentration of the contaminant. The collection medium depends on the physical and chemical properties of the materials to be collected and analyzed. Particles are generally collected by means of various types of filters. Gases and vapors are generally collected via solid sorbents and liquid reagents. The air mover may be small and serve only one sampler, or it may be a central vacuum system that serves a number of air-sampling stations. Once a sample has been collected, its identification and quantification commonly require laboratory chemical or physical analysis.

A variety of sampling schemes are in use. The most common approach is the use of personal air samplers. These include small, lightweight units that are battery powered and can be worn by individual workers. The small size of such samplers permits them to be positioned on the lapel or collar, so they collect samples representative of the air actually being breathed. At the same time, however, the sensitivity of such samplers is often limited by the relatively low rate at which air can be pumped through them. For this and other reasons, passive personal samplers that do not require an air mover have also been developed. In the main, the active component of these units is a material that collects the contaminant through diffusion or direct absorption. Another approach, but one that is used less often, is to place samplers at fixed locations in the workplace that have been selected so as to be as representative as possible of the breathing zones of the workers. Because they need not be portable, these units can be provided sufficient power to sample at a much higher rate.

As is implied by the discussion of BEIs, a monitoring program for airborne contaminants may be supplemented by a variety of measurements of biological indicators of contaminants within the bodies of the exposed workers. Generally, this method of monitoring requires the collection of prescribed samples of urine, blood, sputum, hair, and body fluids and/or tissues that are analyzed for specific contaminants or their metabolites. To foster the use of the latest techniques, NIOSH has developed and made available a manual of acceptable analytical methods (Schlecht and O'Connor, 2003). Where the assessment of exposures and/or intake requires the analysis of urine, a 24-hour collection is preferred. If this is not possible, the uncertainties associated with the analysis and representativeness of smaller samples must be recognized. This is almost always a problem if the sensitivity of the analysis requires a large sample. In cases involving analyses for the intake of radionuclides, fecal samples may also be collected and analyzed. This is primarily limited, however, to cases in which the contaminant being evaluated is preferentially excreted via this avenue.

One of the advantages of such measurements is that the resulting data can be used to complement and/or confirm the adequacy of other types of workplace monitoring programs. For a variety of reasons, however, it is important to recognize that there may be variations in the uptake of individuals within a workgroup who are exposed within a common environment. This will, in turn, lead to what may appear to be inconsistencies between the information obtained through air monitoring and biological monitoring. Sources of such inconsistencies include the location of the air-monitoring device and differences in the effectiveness of personal protective devices, such as respirators, worn by individual workers. Other contributing factors include the physiological makeup and health status of the worker, such as body build, diet (water and fat intake), metabolism, body fluid composition, age, gender, and pregnancy, medication, and disease state. Also of possible importance are nonoccupational exposure factors, such as community and home airborne contaminants, alcohol and drug intake, and exposure to chemicals used, for example, in woodworking and related hobbies (ACGIH, 2004).

BIOLOGICAL AGENTS

For those cases in which the transmission of biological agents is through the air, associated monitoring techniques closely parallel those developed for airborne gases and particulates. Because of the many different types of bioaerosols that must be evaluated, no single sampling method or analytical procedure is optimal. Once a sample is collected, the contaminants must be identified. Usually, culturing is required in the case of microorganisms, and microscopic examination in the case of contaminants such as pollen grains, fungal spores, and house dust mites. Such techniques are now being expanded to include newer technologies such as gene probes and DNA amplification.

PHYSICAL AND PSYCHOLOGICAL FACTORS

For certain physical factors such as heat and noise, a variety of measuring instruments are available for collecting real-time data in the workplace. Ergonomic hazards obviously present a different type of problem. The assessment of ergonomic factors is complicated by the multitude of settings in which workers are employed, the large number of interfaces between them and the equipment they use, and the increasing recognition that organizational and psychological factors may be as important as physical factors in terms of the resulting impact on health. A further obstacle is the scarcity of data that can be used to quantify dose-response relationships for the specific physical factors involved and the total lack of data to quantify dose-response relationships for the associated psychosocial or organizational factors. The seriousness of these deficiencies is illustrated by observations that jobs that place high psychological demands on workers and give them little control over the work process are causally related to atherosclerosis of the coronary arteries (Fine, 1996).

CONTROL OF OCCUPATIONAL EXPOSURES

A complete and effective control program requires process and workplace monitoring systems, education, and commitment of both workers and management to appropriate occupational health practices. Obviously, steps must be taken to ensure that protection is provided not only under normal operating conditions but also under conditions of process upset or failure, particularly in systems for controlling airborne contaminants. Although a majority of the problems associated with toxic chemicals can be controlled by ventilation, those associated with biological agents, particularly in the case of health-care workers, often require personal protective equipment. The situation is similar when workers must be protected in the presence of physical stresses, such as noise and heat. To assure that the best available technologies are applied, supervisory personnel must be knowledgeable about the full range of control measures available.

TOXIC CHEMICALS

In the control of exposures from toxic chemicals, emphasis since the early 1990's has been on designing each element in the manufacturing or production process to eliminate the generation of the contaminant. If this proves impossible, the second or supplementary approach is to prevent dispersal of the contaminant. If this cannot be achieved or the degree of control is inadequate, the backup is to collect and remove the contaminant by exhausting the air into which it is released. As will be noted in the discussion that follows, there are six basic approaches that can be used to implement one or more of these goals.

Elimination or substitution. This approach involves control at the source by completely eliminating the use of a toxic substance or substituting a less toxic one. Examples include discontinuing the use of mercury in Leclanche-type batteries and using toluene or xylene instead of the more toxic benzene in paint strippers.

Process or equipment modification. The goal in this case is to design processes so that, as far as practical, the hazardous materials involved are contained within sealed or enclosed equipment and maintenance requirements and associated exposures are minimized. This is frequently applied to older processes that do not meet existing or proposed occupational health standards and can effectively be modified and upgraded.

Isolation or enclosure. Operations involving highly toxic materials can be isolated from other parts of the facility by constructing a barrier between the source of the hazard and the workers who might be affected. The barrier can be a physical structure or a pressure differential. A common approach is to place toxic or radioactive materials in an enclosure with a negative pressure or to cause the space occupied by workers to be at a positive pressure. Often the isolation of a process from a worker is made possible by the use of robots.

Local exhaust ventilation and air cleaning. Airborne gases or particulates produced by essentially all industrial operations can be captured at the point of generation by an exhaust ventilation system. Two possible types of equipment are a glove box (Figure 4.4) and a laboratory hood. Before the exhaust air is released to the environment, however, it should be passed through an air-cleaning device (such as a filter, adsorber, or electrostatic precipitator) to remove any contaminants present. In some cases, the contaminant that is removed can be recycled within the process itself.

Personal protective equipment. Controls can also be applied to individual workers. The concept is to isolate the worker rather than the source of exposure. People working with heavy equipment, for example, should be provided with protective helmets, goggles, and safety shoes. Those working with corrosive and toxic chemi-

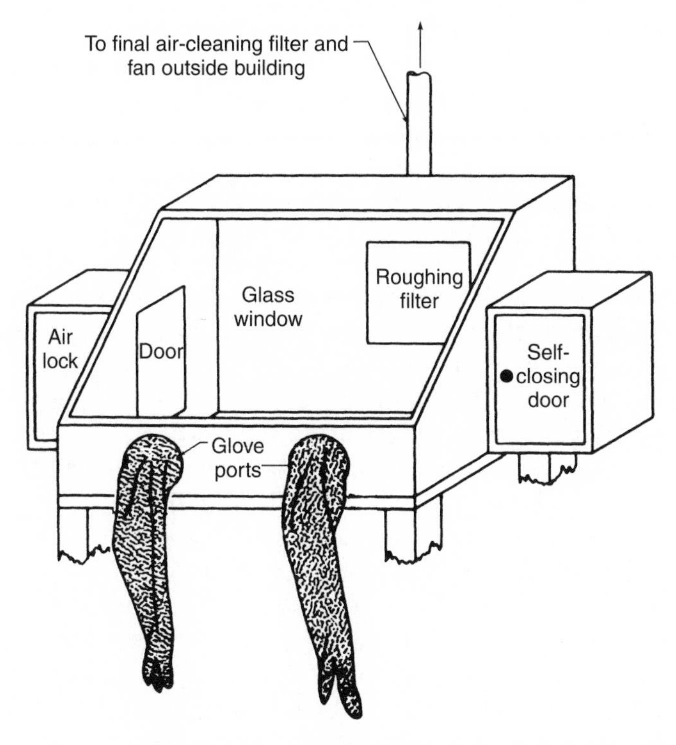

Figure 4.4 Glove box for handling highly toxic or radioactive materials

cals should be provided with face shields and protective clothing. The choice of clothing is based on the exposure hazard, the amount of body coverage required, and the permeability of the hazardous agent. As noted earlier, however, the use of protective clothing can readily be a source of heat stress. For this reason, the use of such equipment should generally be considered a last resort. A similar comment applies to respirators. Another factor to consider is that unless the use of such equipment is limited to situations in which control at the source or in the workplace is impractical or has failed, procedures for maintaining the workplace free of contamination may either not receive adequate attention or be ignored.

Proper work practices and housekeeping. Both of these activities are important components of an effective control strategy. The first involves proper equipment design coupled with operating and maintenance procedures that minimize exposures and emissions. Examples include the use of handheld quick-response instruments to conduct periodic leak-detection surveys, the requirement that safe-work permits be obtained before a task is begun, and the use of "lockout" systems, which prevent operation of a facility except when conditions are safe. Appropriate housekeeping practices include chemical decontamination, wet sweeping, and vacuuming.

Before discussion of the control of biological agents and physical factors, the use of respirators, one of the most common protective devices for reducing the inhalation of airborne contaminants, deserves special comment. Even though, as noted earlier, dependence on respirators should not be a primary approach in the control of airborne contaminants, an estimated 5 million workers at approximately 1.3 million U.S. establishments currently wear such devices. Recognizing this fact and seeking to improve the effectiveness of the protection being afforded, OSHA promulgated a new Respiratory Protection Standard in 1998. One of the primary goals of this effort was to incorporate into one document updates of all the relevant information and requirements pertaining to these devices. Equally important was the promulgation of a requirement that employers affected by the new standard establish and maintain a respiratory protection program that ensures proper workplace practices. These include ensuring that employees are provided adequate training and guidance on respirator selection, fit testing, cleaning, maintenance, and repair. Another requirement applies to the use of a self-contained breathing apparatus in atmospheres deemed to be "immediately dangerous to life or health." In cases where two firefighters must enter a burning building, the standard requires that two firefighters must be on standby to provide assistance or perform rescue if something unanticipated happens. The goal is to accomplish a significant reduction in the 900 deaths and estimated 4,000 injuries and illnesses that occur each year in the United States among workers whose job requires that they wear respirators (OSHA, 1998).

BIOLOGICAL AGENTS

One of the best approaches to controlling airborne biological agents in the workplace is to limit the types of environments, namely, wet spots and pools of water, that promote the growth of organisms. Another key step is proper maintenance of the air-handling system, especially the humidifier.

When exposures to biological agents arise primarily through puncture wounds from contaminated needles, as in the health-care setting, the principal controls are to ensure that used needles are placed in puncture-resistant containers, hands are washed to reduce contamination, and appropriate personal protection, such as gowns, gloves, and goggles, is worn. Control in this case is also dependent on careful housekeeping, with specific requirements for discarding contaminated needles and other sharp instruments and for proper handling of the accompanying wastes.

PHYSICAL FACTORS

As noted earlier, ergonomic-related problems have become of increasing concern in recent years. Contributing factors often include the lack of visual indications of a problem. Nonetheless, steps are available to control such problems. As is the case with airborne contaminants, such procedures range from the simple to the complex, depending on the nature of the problem.

One of the most common sources of such problems is the desktop computer, the use of which has resulted in an increasing number of reported cases of carpal tunnel syndrome. Steps that can be taken to reduce this and other disorders associated with such equipment are illustrated in Figure 4.5. Strategies for reducing the occurrence of specific types of musculoskeletal disorders, such as low-back pain, can be far more complex. These include the use of mechanical aids to lift heavy weights, rearranging the workplace layout to help workers avoid unnecessary twisting and reaching, modifying seat design to permit adjustments in the height and lumbar support, and establishing new guidelines for the packaging of products so their weights are compatible with human capabilities. At the same time, the ability to address these types of problems in an adequate manner will depend on understanding the complex process of how exposures and reactions to physical risk factors may lead to injury or disease (Keyserling, 2000). That progress is being made is illustrated by studies that show a correlation between the accuracy of vision prescriptions and neck, back, and shoulder muscular skeleton problems among computer operators. Because even slightly inaccurate vision prescriptions can be important, OSHA recommends that all such operators have regular vision examinations (Daum, 2004).

Although back supports or belts are popular, their use to avoid workplace injuries is of questionable value. The most recent epidemiological studies, for example, reveal that there is no statistically significant differ-

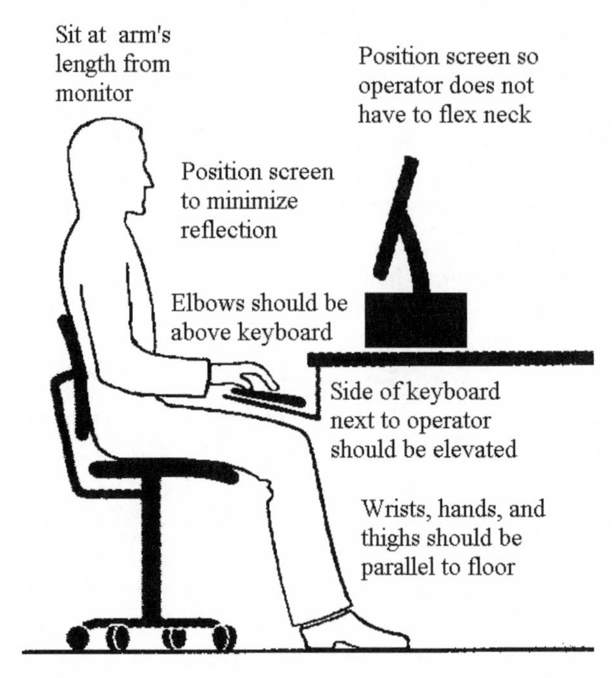

Figure 4.5 Key factors in a properly designed computer-operator interface

ence between the incidence of workers' compensation claims among employees who reported using back belts every day and those who reported never using such belts or using them no more than twice a month (Wassell et al., 2000). The better approach, as described earlier, is to focus on job design. Another example of physical challenges is the previously discussed use of protective clothing and the possibility of heat stress. Control measures in this case include reducing humidity to improve evaporative cooling, increasing air movement via natural or mechanical ventilation, providing radiant-reflecting shields between workers and the heat source, reducing demands in terms of workload and duration, or some combination of these elements.

In a similar manner, noise can be controlled at the source by damping, reducing, or enclosing the vibrating surface that produces it. For instance, low-speed, high-pitch fan blades can be substituted for high-speed, low-pitch ones; sound absorbers can be placed between the source and the employees; and hearing protection can be provided to individual workers.

One innovation is the development of headsets that contain a small computer capable of analyzing incoming noise. The information is fed to an electronic controller that generates an opposing sound wave that in essence cancels out a portion of the incoming noise. Although such systems are primarily effective in controlling low-frequency sounds, they have proved capable of reducing incoming noise by up to 95 percent while permitting the wearer of the headset to hear desired sounds, such as speech and music. Electronic controllers can also be incorporated into the source of the noise itself, so that it is canceled before it can be emitted.

Special Groups

Although progress is being made in the United States in the overall reduction of workplace injuries and illnesses, as is often the case, certain age and occupational groups appear not to have received the attention they deserve. Three such groups are discussed here.

TEENAGERS

Some 70 to 80 percent of all teenagers in the United States work at some time during their high school years (NIOSH, 2003). Typical places of employment include the services and retail trades. During an average summer, upwards of 4 million people in this age group will be involved in such activities. Because they frequently lack work experience, safety training, and appreciation of the need to observe safe practices, this group is at a particularly high risk for injuries. For those employed in eating establishments, the most common sources of injuries are burns due to exposures to hot oil and grease, hot water and steam, and hot cooking surfaces. These and related events lead to more than 200,000 injuries each year. Even more troublesome is that 70 to 80 members of this age group die due to their injuries. Of these, the largest sources of deaths are in jobs that involve deliveries and driving; another important source is the construction industry (CDC, 2001a; Pratt, 2003).

OUTDOOR WORKERS

Many people work outdoors. These include those involved in abrasive blasting to remove surface coatings, scale, and rust in preparing large metal structures for finishing operations, such as in the repair of bridges, buildings, and ships; those employed at hazardous waste sites, who may be exposed to toxic chemicals; airport workers, who may exposed to air pollution and noise; building and highway construction workers subjected to the hazards of lifting cranes and earth-moving machinery; and farmworkers, who are exposed not only to higher rates of injury from accidents involving moving machinery, but also to a wide range of toxic chemicals and pesticides (CDC, 2001b). Ironically, one of the interesting aspects of the Occupational Safety and Health Act of 1970 is that it exempted farmers from the enforcement efforts of the Occupational Safety and Health Administration. Experience to date certainly supports a reevaluation of this exemption.

HEALTH-CARE WORKERS

One of the most hazardous occupations in the United States is the health-care industry. From the late 1980s to the late 1990s, the rates of injuries and illnesses in this industry doubled. The same was true for the number of lost work days for nursing-home workers. Some of the specifics are revealing. During one or more years during the 1990s, the health-care industry accounted for one-third of all instances of workplace violence and musculoskeletal injuries. Nurses and nurses' aides, for example, had the highest claims rate of any occupation because of back injuries. In fact, the overall injury and illness rate for health-care workers was higher than that for mining. Exacerbating the problem was the revelation that the number of occupational health and safety professionals employed in this profession was lower than would have been anticipated (Levine, 2001). If ever the adage "Physician, heal thyself" was applicable, this appears to be the case.

Integrated Safety Management (ISM)

To ensure the adequacy and effectiveness of their efforts to protect workers, many industrial leaders are taking steps to ensure that safety management is an important component of their occupational health programs. One approach is the adoption of what is called an "integrated safety management" (ISM) system. The foundation of this approach is the development of procedures for the identification and control of worker risks and hazards. These procedures, in turn, are strongly supported by the adoption of a corporate policy designed to protect workers, the public, and the environment, not as an afterthought but as a part of the processes for planning and conducting the work (Kenoyer et al., 2000).

The guiding principles of the ISM system include the stipulations that

(1) the responsibility for worker safety and health (S & H) rests with management; (2) the roles and responsibilities for worker S & H must be clearly defined; (3) priorities for coping with the various components of worker S & H and the environment must be balanced; (4) controls to ensure worker protection must be tailored to the task being performed; and (5) authorization must be in place prior to initiation of a new process or procedure. Experience demonstrates that the implementation of these principles can lead to continuous improvement when employee involvement is combined with the proper applications of behavioral and statistical science. If success is to be achieved, however, two factors are essential: (1) all employee levels of workplace behavior must be scientifically measured and managed; and (2) all employees must be involved in the ongoing feedback and problem-solving process.

The General Outlook

As discussed earlier, the problems of musculoskeletal disorders have reached almost epidemic proportions in the United States. After years of turmoil and controversy, the Occupational Safety and Health Administration in 2002 promulgated a rule for addressing these problems. Yielding to pressures from industrial and other groups, however, the U.S. Congress shortly thereafter passed legislation revoking the rule. This action was subsequently signed into law by the President. Since the Congressional Review Act prohibited OSHA from proposing a new rule substantially similar to the one that was repealed, that agency subsequently began the development of a series of voluntary guidelines that will be targeted to specific industries and tasks. Among the first groups to be addressed were the construction, agriculture, and maritime industries. Guidelines targeting other industries are to be issued on a continuing basis. They will be designed to assist employers in a cooperative effort to identify ergonomic hazards and to establish feasible means for controlling them. Although failure to implement the guidelines will not be interpreted by OSHA as a violation, where inspections reveal serious problems, OSHA will be able to prosecute under the General Duty clause of the 1970 Occupational Safety and Health Act, which requires employers to provide workplaces safe from recognized serious hazards (OSHA, 2002).

In the meantime, those responsible for addressing these issues are not without sound guidance for the control of such problems. Not hampered by the congressional action, the ACGIH has moved forward with the development of TLVs for a number of work activities. These include activities involving repetitive lifting, movements of the hand, wrist, and forearm, and vibration of the hand and arm (ACGIH, 2004). In addition, NIOSH has adopted a four-pronged approach that includes the development of industry-specific and task-specific guidelines on ergonomic problems, including procedures for enforcement, outreach, assistance, and research. The first guidelines will cover the nursing-home industry; those for poultry processing and retail groceries will follow. Enforcement will be based on the previously mentioned OSHA General Duty clause and will focus on those industries that choose to ignore ergonomic hazards. OSHA will also provide seminars, training sessions, and workshops to assist employers in preventing these types of injuries.

Concurrently, there has been the previously mentioned shift in employment in the United States from manufacturing to the service industries. Other significant changes are taking place as well, for example, the increasing tendency of people to work at home, supported through such technological innovations as computers and e-mail. Another factor that supports this trend is the increasing number of women workers, two-thirds of whom have children. Additional developments in electronic communications and robotics will further decentralize our workforce. Also relevant are the facts that the U.S. workforce includes an increasing number of minorities, is growing older as a whole, and faces ever-stronger competition from overseas production. Witness, for example, the impacts of the North American Free Trade Agreement (NAFTA). All of these factors may have a tendency to weaken the resolve to maintain high workplace standards in the United States, as demonstrated by the response of Congress to the proposed ergonomics rule.

At the same time, increasingly complex arrays of materials, processes, equipment, and technologies are being introduced into industrial operations. Some of these involve the use of new and less toxic chemicals and/or the introduction of less hazardous processes and equipment. Recognizing the potential benefits, NIOSH is actively seeking to identify and promote these types of activities (Burgess, 1997). Other proposed changes, however, have the potential for introducing new hazards. To ensure that these are being adequately evaluated and that effective measures are available and will be applied to reduce their workplace and environmental impacts to acceptable levels, the Environmental Protection Agency has mandated that manufacturers document the types of controls that will be applied in what are called "Pre-Manufacturing Notifications." This activity, in concert with the earlier cited Pollution Prevention Act, holds promise of significantly reducing toxic chemical exposures to both workers and the public.

Another change to be recognized is that many current workplace hazards are less obvious and less clearly related to the job. These include effects on the reproductive system and a host of subtle injuries, diseases, and disabilities resulting from low-level, on-the-job psychological stresses. The development of methods for assessing the impacts of effects of this nature will require input from a wide range of specialists, including social and behavioral scientists, public health research workers, medical-care specialists, and many others who may currently view the problems of occupational health as being outside their profession. Also important are the changes that are occurring in workers' attitudes and needs, including their demands that the provision of a safe workplace is not only important but to be expected. Corporate leaders are increasingly recognizing that an effective occupational health program can have long-term economic benefits. In fact, improvements in protecting the health of the workforce often depend more on the manner in which corporate management views this subject than on technical factors related to the control measures that must be applied. Still another change is the increasing awareness of many corporate leaders of the important role of physical activity, nutrition programs, stress management, and other positive lifestyle behaviors on the health and productivity of their workforce.

5

AIR IN THE HOME AND COMMUNITY

PROBLEMS stemming from air pollution were noted during the Roman Empire, a time during which some 80,000–100,000 metric tons of lead, 15,000 tons of copper, 10,000 tons of zinc, and more than 2 tons of mercury were used annually in industrial operations. Uncontrolled smelting of large quantities of related ores resulted in substantial emission of these materials into the atmosphere. In the 1300s, authorities in England banned silver and armor smithing because they realized that they contributed to air pollution. Nonetheless, it was not until about 500 years later that regulatory control of air pollution was initiated in the United States. That occurred in 1881 when the cities of Chicago and Cincinnati, recognizing the need to control emissions of smoke and soot from furnaces and locomotives, passed the first air-pollution statutes in this country. Pittsburgh followed in 1895 by passing similar ordinances to reduce emissions from local steel mills. An ordinance passed in Boston in 1911 was the first to acknowledge that air pollution has regional and national as well as local effects. During the early 1900s, a number of county governments in the United States passed similar laws (EPA, 2000).

As is frequently the case, several major, acute episodes were required to demonstrate conclusively to policy makers and the public that air pollution could have significant effects on health. In 1930, for example, in Belgium's Meuse River valley, high concentrations of air pollutants held close to the ground by a thermal atmospheric inversion during a period of cold, damp weather led to the deaths of 60 people. The principal sources of pollution were industrial operations, including a zinc smelter, a sulfuric acid plant, and glass factories. Most of the deaths occurred among older people with a history of heart and lung disease. In 1948, in another river valley in Donora, Pennsylvania, about 20 people died due to exposures to air pollution from iron and steel mills, zinc smelters, and an acid plant (Helfand, 2001). Again, cold, damp weather was accompanied by a thermal atmospheric inversion. In London in 1952, 4,000 people died as a result of domestic coal burning during similar meteorological conditions (Figure 5.1). Most of those admitted to hospitals were elderly or already seriously ill and were affected by shortness of breath and coughing. Half a century later, controversy remains on whether the estimate of the number of people killed was accurate, since during the next several months additional thousands may have succumbed due to delayed effects (Stone, 2002). Similar episodes occurred in London in 1959 and

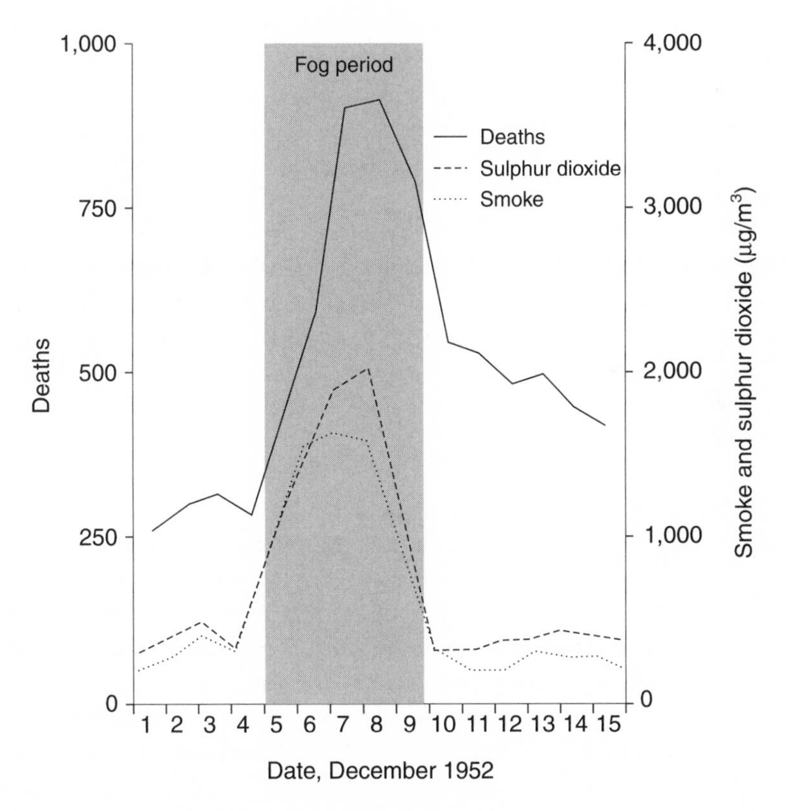

Figure 5.1 Daily mean pollution concentrations and number of deaths during the London fog episode of 1952

1962, and analyses of death records have shown that additional episodes took place in 1873, 1880, 1882, 1891, and 1892.

Concern is mounting over the effects of decades of environmentally blind industrial development in eastern Europe and the former Soviet Union, which appears to have produced widespread threats to health and life from air pollution. Although specific data are lacking, reports indicate that high concentrations of airborne contaminants may have caused tens of thousands of people to develop respiratory and cardiovascular ailments. In some cases, air pollution was so severe that drivers had to use their headlights in the middle of the day; and in many industrial areas 75 percent of children now have respiratory disease. Outrage over environmental pollution is even said to have been a catalyst in the 1990 revolution against Communist rule in Poland (French, 1991).

Today the effects of air pollution on human health and on the global environment are widely recognized. Most industrialized nations have taken steps to prevent the occurrence of acute episodes and to limit the long-term, or chronic, health effects of airborne releases. All the same, estimates suggest that up to 8 percent of Americans suffer from chronic bronchitis, emphysema, or asthma either caused or aggravated by air pollution. Newer epidemiological data suggest that tens of thousands of people in the United States may be dying annually from effects of air pollution, even though the concentrations of most airborne contaminants are within federal limits (Pope et al., 2002). The costs to society are enormous: a lower quality of life for the affected individuals, shorter life spans, and less productivity and time at work.

The Body's Responses to Air Pollution

The intake of pollutants into the lungs and their retention at potential sites of injury depend on the physical and chemical properties of the pollutant as well as the extent of activity of the subject exposed. Gases, such as sulfur dioxide and formaldehyde, that are highly water soluble are almost completely removed in the upper airways. Less soluble gases, such as nitrogen dioxide and ozone, penetrate to the small airways and alveoli.

The ease of entry and the sites for deposition of particulates are heavily influenced by their aerodynamic size and the anatomy of the space through which they are moving. Relatively large particles are susceptible to inertial impaction in the airways, where the flow rate is high and the passageways change direction frequently. Particles that penetrate to the small bronchiolar and alveolar region can rapidly deposit in the lungs through settling and diffusion. Fractional depositions in various regions of the respiratory tract of inhaled particles within a range of sizes are shown in Figure 5.2. As may be noted, the total collection efficiency is lowest for particle sizes of about 0.5 micrometer. The reason for this is that such particles do not settle rapidly and are too large to diffuse effectively. Another factor that influences particle delivery and deposition is the aerodynamics of respiration. Total deposition is higher and is more uniformly distributed with slow, deep breathing, as contrasted to rapid, shallow breathing.

As with all kinds of environmental stresses, the human respiratory system has a variety of protective mechanisms against airborne pollutants. Particles ranging in size from 5 to 10 micrometers or larger are effectively removed by the nose, which acts as a prefilter. Particles that are inhaled and deposited in the upper respiratory tract can be removed by mucociliary action. Those that are deposited in the lower parts of the lungs can be engulfed and destroyed by cells called macrophages. Usually the cilia

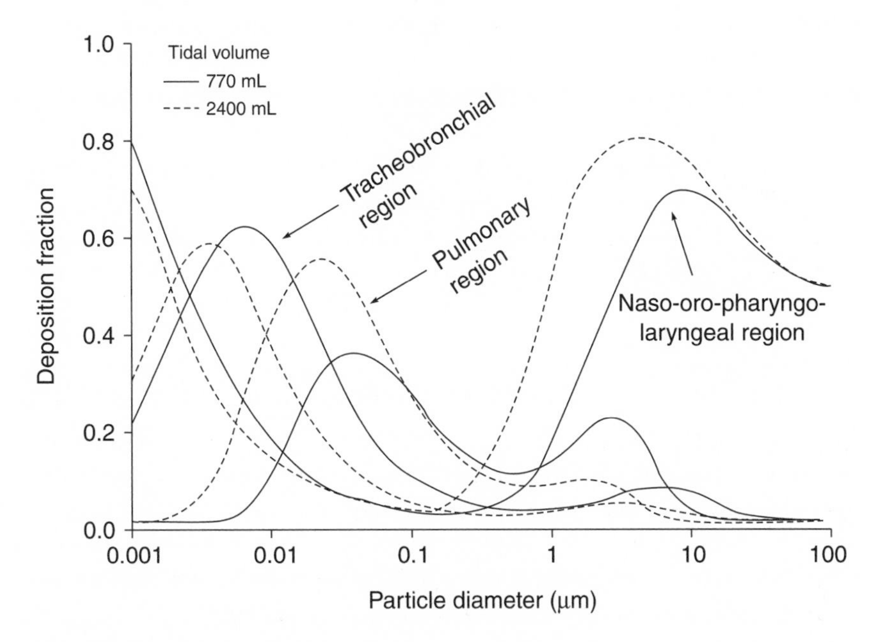

Figure 5.2 Relation of particle diameter to calculated regional deposition in the lungs for spherical particles of density 1 gram per cubic meter

sweep the macrophages, along with dirt and bacteria-laden mucus, upward to the posterior pharynx, where they are expectorated or swallowed. Exposure to airborne gaseous irritants may cause sneezing or coughing and thus prevent their entry into the deeper parts of the lungs. Even if gases are taken into the lungs and absorbed, the body has mechanisms that detoxify most of them. A notable exception is carbon monoxide. Where the detoxification takes place depends on how soluble the gas is in various tissues and organs, and how and with what it reacts chemically.

Despite these mechanisms, some pollutants will still be deposited in the body. If they remain in the lungs, they may cause constant or recurrent irritation and lead to long-term illnesses. If they are transported by the bloodstream to other parts of the body, they can cause chronic damage to organs such as the spleen, kidneys, or liver.

The likelihood of an adverse response to an inhaled pollutant depends on the degree of exposure, the site of deposition, the rate of removal or clearance, and the susceptibility of the exposed person. Recent epidemiological data suggest that particulate matter 2.5 micrometers or less in size has far more impacts on health than heretofore recognized (Pope et al., 1995). Although the reasons for this are yet to be confirmed, one major factor could be, as noted earlier, the relatively high deposition in all regions of the lungs of particles in the smaller size ranges. Another factor may be the larger amounts of toxic and carcinogenic compounds that smaller particles can adsorb per unit mass. Although the mass of one 10-micrometer particle is equivalent to that of 1,000 1-micrometer particles, the total surface area of the 1,000 smaller particles will be at least 100 times that of the larger particle. This is important because the chemical contaminants that attach to these smaller particles may be more important than the particles themselves (Weinhold, 2002b).

Standards for Air Pollution

Air pollution has been defined as the presence in the air of substances in concentrations sufficient to interfere with health, comfort, safety, or the full use and enjoyment of property. Substances released into the air therefore are considered potential pollutants in terms of their effects not only on human health but also on agricultural products and on buildings, statues, and other public landmarks. In fact, concentrations of some air pollutants considered acceptable for avoiding damage to agricultural products (so-called secondary standards) are lower than those considered acceptable for humans (so-called primary standards). That standards are needed to protect property is confirmed by many instances of damage, one example being the extensive discoloration of the marble of the Taj Mahal in India.

Under the requirements of the Clean Air Act Amendments of 1970, 1977, and 1990 (Chapter 14), the administrator of the Environmental Protection Agency (EPA) is required to identify a set of criteria pollutants that play key roles in terms of their effects on human health and the environment. To date, six such pollutants have been identified. Their identities and physical and biological characteristics are summarized here (CEQ, 1997; Findley and Farber, 2000).

Carbon monoxide (CO) is a colorless, odorless, poisonous gas that is slightly lighter than air. It is produced through the incomplete combustion of carbon, primarily by the operation of internal combustion engines, such as those in automobiles. CO enters the bloodstream and reduces the delivery of oxygen to the body's organs and tissues. The health threat is most serious for people who suffer from cardiovascular disease, particularly those with angina or peripheral vascular disease. Exposures to elevated carbon monoxide concentrations are associated with impairment of visual perception, work capacity, manual dexterity, learning ability, and performance of complex tasks.

Lead (Pb) is a heavy, comparatively soft metal that for years was used as an additive to gasoline and household paint and in shotgun pellets and stained-glass windows. When it is taken into the body, it accumulates in the blood, bones, and soft tissues. Because it is not readily excreted, it also affects the kidneys, liver, nervous system, and blood-forming organs. Excess exposure may cause neurological impairments such as seizures, mental retardation, and/or behavioral disorders.

Nitrogen dioxide (NO_2) is produced when fuels are burned at high temperatures. The main sources are transportation vehicles and power plants. When NO_2 and other oxides of nitrogen are inhaled, they can irritate the lungs and lower resistance to respiratory infections such as influenza. Although the effects of short-term exposure are not yet clear, continued or frequent exposure to high concentrations causes increased incidence of acute respiratory disease in children. Nitrogen oxides are also an important precursor of both ozone and acidic precipitation and may affect both terrestrial and aquatic ecosystems. Therefore, the limit for nitrogen dioxide is also designed to support the limit for ozone.

Ozone (O_3) is formed in the atmosphere as a result of chemical reactions between oxides of nitrogen and volatile organic compounds, such as hydrocarbons (HCs). If it is inhaled, it damages lung tissue, reduces lung function, and sensitizes the lungs to other irritants. Scientific evidence indicates that ambient levels of ozone not only affect people with impaired respiratory systems, such as asthmatics, but healthy adults and children as well. Specific effects, particularly at elevated concentrations, include eye and lung irritation. Ozone is also responsible for several billion dollars of agricultural crop loss in the United States each year.

Particulates are solids or liquids that are produced by the combustion of fuel in stationary power plants, diesel-powered vehicles, and various industrial processes. They are also produced by plowing and burning of agricultural fields. If particulates are inhaled, they can lead to respiratory symptoms, aggravate existing respiratory and cardiovascular disease, alter the defenses of the body against foreign materials, damage lung tissue, and produce latent cancers and premature mortality.

Sulfur dioxide (SO_2) is a corrosive, poisonous gas that is produced in power plants, particularly those that use high-sulfur coal as a fuel. SO_2 and oxides of nitrogen after being released into the atmosphere, can be chemically converted into sulfates and nitrates. These, in turn, may later be deposited on the ground in the form of so-called acid rain or snow. At high concentrations, SO_2 affects breathing and produces respiratory illness, alterations in the defenses of the lungs, and aggravation of existing respiratory and cardiovascular disease. Sulfur dioxide can also produce foliar damage on trees and agricultural crops.

The EPA is also required to establish and revise, when deemed necessary, national ambient air-quality standards (NAAQS) for each of the criteria pollutants. The current standards are summarized in Table 5.1. Although those for ozone and particulates have been changed over the years, those for carbon monoxide, nitrogen dioxide, and sulfur dioxide are the same as they were when they were originally established in 1971. In a

Table 5.1 National primary U.S. air-quality standards

Pollutant	Primary standard	
	Applicable period	Limit[a]
Carbon monoxide (CO)[b]	8-hour average	9 ppm (10 mg/m^3)
	1-hour average	35 ppm (40 mg/m^3)
Lead[c]	Quarterly average	1.5 μg/m^3
Nitrogen dioxide (NO_2)[c]	Annual arithmetic mean	0.053 ppm (100 μg/m^3)
Ozone[c]	8-hour average	0.08 ppm (157 μg/m^3)
	1-hour average	0.12 ppm (235 μg/m^3)
$PM_{2.5}$[c]	Annual arithmetic mean[d]	15 μg/m^3
	24-hour average[e]	65 μg/m^3
PM_{10}[c]	Annual arithmetic mean[d]	50 μg/m^3
	24-hour average	150 μg/m^3
Sulfur dioxide (SO_2)[f]	Annual arithmetic mean	0.03 ppm (80 μg/m^3)
	24-hour average	0.14 ppm (365 μg/m^3)
	3-hour average	0.50 ppm (1,300 μg/m^3)

a. Values in parentheses are approximately equivalent concentrations.

b. There are no secondary standards for carbon monoxide.

c. The secondary standards for these contaminants are the same as the primary standards.

d. Compliance requires that the 3-year average of the annual arithmetic mean from single or multiple community-oriented monitors must not exceed this value.

e. Compliance requires that the 3-year average at the 98th percentile of 24-hour concentrations at each monitor within an area must not exceed this value.

f. SO_2 is the only contaminant for which the secondary standard differs from the primary standard. The annual arithmetic mean and the 24-hour average are primary standards; the 3-hour average is a secondary standard. The 24-hour standard is not to be exceeded more than once per year.

similar manner, the standard for lead has remained the same since it was selected as a criteria pollutant in 1978 (Larsen, 2002). A case study that provides background on the establishment of standards for airborne particles is presented in Chapter 3.

While the benefits of the NAAQS are obvious, it is important to recognize that the regulated pollutants serve primarily as surrogates for many other more toxic materials known to be present in the air. These include carcinogens, mutagens, and reproductive toxins. Specific examples not covered by the NAAQS are acid aerosols, polynuclear aromatic hydrocarbons, many toxic metals, and volatile organic compounds. Some of these, such as mercury, asbestos, beryllium, vinyl chloride, benzene, arsenic, and radioactive materials, are controlled by emission standards, that is, through the establishment of limits on the amounts of these substances that can be released into the atmosphere through industrial operations. In a similar manner, the amount of ozone in the ambient air is regulated primarily through the establishment of limits on releases of oxides of nitrogen and volatile organic compounds (VOCs). The New Source Performance Standards and limits on the amounts of VOCs permitted in, for example, varnishes and paints play a major role in such controls (Larsen, 2002).

It is anticipated that the list of criteria pollutants will be expanded as scientists identify additional pollutants that have significant impacts on human health and the environment or are deemed to be capable of serving as surrogates for other pollutants that do. In fact, the Clean Air Act Amendments of 1990 require that the EPA review and evaluate some 200 additional air pollutants. For each such pollutant, the EPA must identify and quantify its major sources and specify the control technologies that are acceptable.

Also under way is a range of activities that are being conducted in response to the Regional Haze rule, which was promulgated by the EPA in 1999. As the name implies, this rule was designed to remedy the effects on visibility of human-made air pollution in so-called Class I areas. These include all national parks, wilderness areas, and memorial parks larger than 5,000 acres. Specific goals of the rule, which is being implemented by the states, require that (1) visibility be improved during the 20 percent most impaired days, (2) there be no degradation in visibility during the 20 percent clearest days, and (3) the annual rate of visibility improvement that would lead to "natural visibility" conditions within 60 years (i.e., by 2064) be determined. The rule further requires that all major stationary sources subject to best available retrofit technology (BART) requirements be identified, using as a basis the collective benefits to visibility that would accrue through the control of such sources in Class I areas. In 2002, however, a three-judge panel of the U.S. Court of Appeals for the District of Columbia Circuit issued an opinion that the portion of the rule dealing with BART contravened the language of the Clean Air Act. It was also judged that the rule did not provide enough discretion to the states in applying the BART requirements. The latter judgment was based on the fact that the rule applied to the effects of a combination of sources. Although the court's decision left the precepts of the rule generally intact, the decision to vacate the BART provisions leaves several aspects of the rule yet to be confirmed (Jezouit and Frank, 2002).

Outdoor Air Pollution

As noted in Chapter 1, there are many components of the environment, two of these being the outdoor and indoor environment. In a similar manner, air pollution can be divided into that which is outdoors and that which is indoors. The first of these will be discussed in the sections that immediately follow.

SOURCES

Sources of the major pollutants vary significantly in terms of their relative contributions. While the burning of fossil fuels in electricity-generating plants is a primary source (85 percent) of sulfur dioxide, other sources, such as paper and pulp mills, smelters, and food-processing plants, can be significant contributors. The combustion of fuel also represents a major source of oxides of nitrogen (41 percent) and volatile organic compounds (37 percent). Petroleum refineries, solvent manufacturers, and distributors and users of their products, such as gasoline stations and dry cleaners, are also major contributors of these compounds. As will be noted in the discussion that follows, motor vehicles, especially diesel-powered vehicles, are another major source of volatile organic compounds, as well as carbon monoxide and oxides of nitrogen. As in addressing any such problem, there are many factors to be considered.

Diesel engines have long been recognized for their power and durability. Since the fuel they burn has a higher energy content, this type of engine is more efficient than those fueled by gasoline. On the negative side, airborne emissions associated with the operation of diesel engines have a host of adverse effects on human health. These include the fact that the particulate matter they release contains a number of contaminants, such as oxides of sulfur, volatile organic compounds, and aromatic hydrocarbons. The extent of the impacts of these contaminants on human health was confirmed by the EPA in 2002 when it concluded that although uncertainties remain, diesel exhausts from large trucks and other sources probably cause lung cancer. This conclusion agrees with that of various world health agencies. When such factors are taken into account, diesel engines are estimated to be the source of at least 70 percent of the total toxic risk posed by air pollutants in the United States. Exacerbating this situation, the rate of increase in the number of diesel-powered vehicles in use in this country during the last few years has been dramatic. While such vehicles consumed about 29 billion gallons of fuel in 1996, this had increased to almost 36 billion gallons by 2000. A continued increase of 2 percent per year is projected for the next several decades (Weinhold, 2002b).

GENERIC METHODS FOR CONTROL

Experience has repeatedly demonstrated that it is better both economically and scientifically to prevent or reduce the production of a pollutant than to concentrate on controlling the amount that is being released. Regardless of the care that is taken, some amounts of any toxic substance that is produced will escape through one route or another. Even if a contaminant is captured prior to its release, it still must be isolated and destroyed.

Some controls can be incorporated during the manufacture of a product. This is exemplified by the installation of emission-control devices in automobiles. In a similar manner, emission controls can be specified nationwide for major industrial operations (such as power stations, solid-waste incinerators, and metallurgical plants) that have uniform characteristics. In other cases, however, controls must be tailored to the specific characteristics of a particular industrial operation. These include the size of the operation, the processes used, and the age and condition of the facility. On a generic basis, approaches for controlling releases of air pollutants from industrial and commercial operations can be categorized as shown here. As will be noted, many of these approaches are similar to those used to control airborne contaminants in the workplace.

Atmospheric dilution. This minimal form of control is designed to take advantage of the diluting capacity of the local atmosphere to reduce the concentrations of the pollutants to acceptable levels. A common approach is to discharge the releases through a very high stack. This, in reality, simply spreads the risk over a larger area. Due to the rapid growth in the population and the ever-increasing number of air-pollution sources, application of this approach is being increasingly restricted.

Substitution or limitation. This approach involves either eliminating a pollutant by substituting materials or methods that do not produce it or restricting the amounts of key chemical elements available for pollutant production. Examples are using substitutes for lead to improve the octane rating of gasoline and limiting the permissible sulfur content in coal and oil burned in electric power plants. One application of the latter approach is the use of biodiesel fuel in diesel-powered vehicles. Because it contains less sulfur, its use automatically leads to reduced emissions of this contaminant. Adding to the benefits of this fuel is that its characteristics are such that it also reduces emissions of heavy hydrocarbons, particulate matter, carbon monoxide, and carbon dioxide (Dooley, 2002).

Reduction in quantity produced. This approach includes improving the combustion efficiency of furnaces and adding exhaust and emission controls to motor vehicles. To ensure that such reductions take place on a systematic basis, the EPA, under the New Source Review (NSR) requirements of the Clean Air Act, applied more stringent requirements for the control of emissions from new power plants and certain other sources than for existing facilities. The concept was that as industries replaced their existing plants with more modern facilities, there would be reductions in the accompanying airborne emissions. As will be noted later, this goal has not been achieved.

Process or equipment change. Typical approaches include the use of fully enclosed systems for processes that generate vapors, floating covers on tanks that store volatile fluids, and electric motors instead of gasoline engines.

Air-cleaning technologies. Common examples are the applications of filters, electrostatic precipitators, scrubbers, adsorbers, or some combination of these to remove pollutants from airborne exhaust systems.

Administrative, economic, and regulatory approaches. These include programs that incorporate the provision of economic incentives to promote mass transportation and carpooling and encourage land-use management to ensure that designated areas are restricted to residential, commercial, or industrial use. Another economic incentive is the concept of tradable emission credits, through which the operator of an industrial facility who reduces emissions below the standard or ahead of the timetable set by the law can earn credits that can be applied to future emissions or sold to an operator of another facility (Chapter 13).

Applications of essentially all such strategies or approaches involve trade-offs. Reductions in one type of pollutant, for example, frequently lead to increases in other types of environmental and public health problems. While the use of electric-powered automobiles reduces airborne emissions in metropolitan areas, certain aspects of their operation—for example, the use of lead-acid batteries—can increase an existing or create a new source of contamination. In addition, the provision of electricity to recharge such vehicles can increase the quantity of contaminants discharged from nearby electricity-generating stations. In a similar manner, scrubbers that remove sulfur from power-plant emissions produce large quantities of solid waste.

Case Study: Motor Vehicles

A review of some of the steps that have been undertaken to control emissions from motor vehicles illustrates the wide range of problems entailed in such efforts. For years, lead was added to gasoline to improve its octane rating. After multitudes of studies showed that it was extremely toxic, and that its use in gasoline was leading to an alarming increase in its concentrations in the ambient air and other components of the environment, legislation was passed requiring that its use be phased out. The results of this action, supported by the development of suitable substitute additives, were dramatic. In fact, concentrations of lead in the ambient air in the United States were reduced by 98 percent between 1970 and 1994 (CEQ, 1997).

Applying similar approaches to reduce other motor-vehicle emissions, however, has been far less successful. Faced with the need to reduce the emission of compounds that serve as precursors to the production of smog, Congress incorporated into the 1990 amendments to the Clean Air Act a requirement that distributors in nonattainment areas add oxygen-rich compounds to gasoline to help it burn more cleanly. Initially, one of the more widely used additives was methyl tertiary-butyl ether (MTBE). Leaks and spills soon led to the contamination of groundwater supplies in an estimated 5 to 10 percent of the areas in which MTBE-treated gasoline was being used. Also, because MTBE is a relatively stable compound, any natural biodegradation that might take place proved to be very slow.

The selection of an acceptable alternate additive, however, proved to be equally challenging. One that was considered was dimethyl carbonate (DMC). Although DMC burns even more cleanly than MTBE, its manufacture involves the use of phosgene, a poisonous gas whose handling encompasses a host of environmental problems (Service, 2002). Another possible replacement was ethanol, which can be made from corn. Exploration of this matter in more detail revealed that plants in which this compound was to be manufactured would emit a range of air pollutants, including formaldehyde, acetic acid, carbon monoxide, and methanol, in quantities higher than anticipated.

Another approach for reducing the discharge of toxic compounds from motor vehicles was the installation of emission controls. A prime early example was the catalytic converter. If this and other similar devices are to be effective, however, it is important that they continue to perform as designed. To ensure that this is the case, inspection programs have been established in all regions of the United States in which certain of the federal clean air standards are being violated. Pollutants that are analyzed in such inspections typically include carbon monoxide, hydrocarbons, and, in some cases, nitrogen oxides. Owners of vehicles that fail to meet the standards are required to have their pollution-control systems repaired. Even this approach, however, is not without its challenges. Evaluations soon revealed that the inspection programs were targeting the wrong vehicles. Most state regulatory agencies devote too much of their resources to the inspection of newer cars, which have the latest control technologies and are therefore far less polluting. Instead, they should be concentrating on the older models, which are commonly much more important sources of pollution. In fact, data show that older, often malfunctioning vehicles, which represent only about 10 percent of those on the road, typically emit about 50 percent of the most harmful pollutants. Obviously, if inspection programs are to be cost effective, these are the vehicles of importance. In fact, it would be better to exempt newer cars from being tested and use the funds thus saved to provide financial relief or other incentives to owners of older cars to enable them to have their control devices repaired. The wisdom of this approach is confirmed by the fact that the owners of these vehicles are generally people of limited economic means (NRC, 2001).

Once again, the lesson to be learned is that prior to the implementation of any type of air-pollution control strategy, care must be taken to evaluate all possible ramifications of the proposed actions. Even when programs appear to be successful, it is important that they be subjected to periodic reassessments, followed by midcourse corrections if necessary.

Progress in Outdoor Air-Pollution Control

There are two primary indicators of progress in the control of air pollution. One is the reduction in emissions; the second is the reduction in the concentrations of the key pollutants in the ambient air. The latter is commonly referred to as the "quality" of the air. Data on both of these factors are collected at thousands of air monitoring stations across the country. An analyses of these data from 1970 through 2002 show that, during that time period, *emissions* of the six principal pollutants were reduced 48 percent. A more detailed summary of changes in the emissions and ambient air concentrations for each of the six principal air pollutants during the time periods from 1983 and 1993 through 2002 is presented in Table 5.2 (EPA, 2003). In terms of emissions, the reductions in releases of NO_x and VOCs are of particular importance since these two compounds, as indicated earlier, when present in the atmosphere and exposed to sunlight and heat, undergo chemical reactions and form ozone. As may be noted, there was little, if any, change in the ambient average 1-hour and 8-hour ground level air concentrations of this compound during the 10-year period from 1993 through 2002. Also of interest are the concentrations of SO_2 since sulfates formed from this compound are a major component of extremely fine particles, known as $PM_{2.5}$. Particles in this size range, as indicated earlier, are thought to be of major health significance. In this regard, it is significant that ambient concentrations of SO_2 were reduced by 54 percent from 1983 through 2002, and by 39 percent between 1993 and 2002.

While most of these trends are favorable, about 160 million tons of pollutants continue to be released into the air each year in the United States and about 146 million people live in counties where the air during 2002 was in excess of the NAAQS at times because of higher concentrations of at least one of the six principal pollutants. Most of these occurrences were due to excessive levels of ozone and particulate matter. Nonetheless, progress is being made, especially when one recognizes that, from 1970 through 2002, the gross domestic product in this country increased 164

Table 5.2 Percent changes in emissions and air quality, United States

Pollutant emissions			Ambient air concentrations		
Pollutant	1983–2002	1993–2002	Pollutant	1983–2002	1993–2002
NO_x	−15	−12	NO_2	−21	−11
VOC	−40	−25	O_3[a] 1–h[b]	−22	−2[d]
			8–h[c]	−14	+4[d]
SO_2	−33	−31	SO_2	−54	−39
PM_{10}[e]	−34[f]	−22	PM_{10}	na[g]	−13
$PM_{2.5}$[e]	na[g]	−17	$PM_{2.5}$	na[g]	−8[h]
CO	−41	−21	CO	−65	−42
Pb	−93[i]	−5	Pb	−94	−57

a. Ozone is produced by interactions of NO_x and VOCs.
b. One hour maximum concentration.
c. Eight hour maximum average concentration.
d. Not statistically significant.
e. Includes only directly emitted particles.
f. Based on percentage change since 1985.
g. Data not available.
h. Based on percentage change since 1999.
i. Based on change from 1982–2001.

percent, energy consumption increased 42 percent, and vehicle miles traveled increased 155 percent.

Although the ambient air concentrations of individual contaminants are of interest, most people are primarily interested in the potential effects of these materials on their health. In response, the EPA is developing an air-quality index (AQI) that is to be designed to provide this type of information. The goal is to develop a classification system that will make it possible to express the quality of the air in terms ranging from "good" to "hazardous," and to do so on a nationwide basis. Since the system can be tied into data generated by continuously operating air monitors, the AQI is to be reported on a real-time basis (Bortnik, Coutant, and Hanley, 2002).

Indoor Air Pollution

Until several decades ago, air pollution was addressed almost exclusively as an outdoor problem. This is no longer the case. Today similar attention is being focused on the assessment and control of airborne contaminants in residential, office, and commercial buildings. There are two major reasons for this change. First, average members of the U.S. public spend from 87 to 90 percent of their time in the home or some other type of building. In fact, urban populations and some of the most vulnerable people (the young, the infirm, and the elderly) typically spend more than 95 percent of their time indoors. Second, not only can indoor air pollutants encompass a range of toxic materials, but in many cities, the indoor concentrations of compounds such as nitrogen oxides, carbon monoxide, airborne particulates, and other volatile organics exceed those outdoors. Even if indoor concentrations proved to be low, the longer duration of indoor exposures could render them significant when they are evaluated on an integrated time-exposure basis.

Although there is clearly a range of reasons for this heightened interest in indoor pollution, one of the primary early factors was the increasing number of complaints by office workers and the response of the media to what was called the "sick building syndrome." Other stimuli were the discoveries that the naturally occurring radioactive gas radon was present in relatively high concentrations in many homes in this country, as were toxic materials such as asbestos and formaldehyde. These conditions were due, in part, to the responses to the so-called U.S. energy crisis of the 1970s, wherein new homes and buildings were tightened up as a means for conserving energy. In many cases, these responses included reducing the amount of air being circulated and the amount of fresh air being brought in from the outside. These steps, in turn, increased the moisture content of the air and promoted the growth of molds. This led to the release of their spores, mycotoxins, and glucans, as well as various volatile organic compounds, into the indoor air. An increase in the use of synthetic building materials and furnishings inside homes and offices also contributed to the problem. Concurrent with these developments, advances in measurement techniques have increased the number of indoor contaminants now being identified and evaluated (Long, 2002).

SOURCES

As noted earlier, airborne contaminants are generated by a variety of activities inside buildings. In a broad sense, there are six major types and/or sources of such pollutants (Spengler and Sexton, 1983).

Combustion by-products. These are generated through the burning of wood, natural gas, kerosene, wax candles, or any similar materials.

Microorganisms and allergens. Sources include detergents, humidifiers, air-cooling towers, household pets, and insects that live in dust and ventilation ducts. Specific contaminants include pollens, molds, mites, chemical additives, animal dander, fungi, algae, and insect parts.

Formaldehyde and other organic compounds. Sources of formaldehyde include building materials (such as plywood and particleboard), furnishings (draperies and carpets), and some types of foam insulation. Other sources include unvented gas combustion units and tobacco smoke. Personal-care products, cleaning materials, paints, lacquers, and varnishes may also generate chlorinated compounds, acetone, ammonia, toluene, and benzene.

Asbestos fibers. Until 1980 asbestos was used in many building materials, including ceiling and floor tiles, pipe insulation, spackling compounds, concrete, and acoustical and thermal insulation. Asbestos is a source of fibers only if it is friable (shedding). In most cases today, exposures are minimal.

Tobacco smoke. As the name implies, the primary source is cigarette smoking, which serves as a source of fine airborne particles and ^{210}Po, a naturally occurring radionuclide, plus more than 2,000 compounds that are known to be carcinogens and/or irritants.

Radon. Although drinking water can be a source of radon in homes where the supply is obtained from the ground (such as from a well), in most circumstances the primary origin is diffusion of the gas from the ground beneath a building. Radon is produced by the decay of radium, which, in turn, is produced by the decay of naturally occurring long-lived uranium.

Consideration of indoor environments, however, should not be restricted to homes and office buildings. Relatively high concentrations of nitrogen dioxide have been observed in the air at hockey rinks because of the use of gasoline- or propane-powered vehicles to resurface the ice. The combustion of gasoline can also lead to concentrations of volatile components inside a passenger car during rush-hour traffic that can be 6 to 10 times higher than at standard urban outdoor monitoring sites. Related studies show that subways frequently contain relatively high concentrations of airborne particles. Concentrations of carbon dioxide (CO_2) in commercial transportation vehicles, such as trains and subways, can also be relatively high when passenger loads are high. Similar studies in commercial airliners showed that while CO_2 concentrations remained stable during the cruise portion of the flights, they were significantly higher during pre- and postflight periods. This reflects what appears to be lower ventilation rates coupled with increased passenger activity at those times (Dumyahn et al., 2000). These and other conditions, combined with the low relative humidity of the air, often lead to complaints of eye irritation and respiratory problems.

CONTROL

Effective control of indoor air pollution depends on an understanding of several factors. The first relates to the characteristics of the contaminant (concentration, reactivity, physical state, and particle size). All such characteristics affect its removal. A second factor is the nature of the emissions. Are they continuous or intermittent, from single or multiple sources, primarily inside or outside? A third is the quantitative relationship between the exposure and the resulting health effects. Are individuals to be protected primarily from long-term chronic exposures to low concentrations, or from periodic short-term exposures at peak concentrations? A fourth is the nature of the facility. Some controls are more readily applied in residential buildings, others in commercial or office buildings. Also influential are the age and condition of the building.

Control measures for some of the more important indoor air contaminants closely parallel those previously described for the control of ambient (outdoor) air pollution. As is frequently the case, however, different pollutants require different control measures. A further complexity is the fact that indoor air pollution often arises through the interaction of a host of factors that are constantly changing: the temperature and humidity of the air, as well as any contaminants it may contain. Various environmental factors may also impact on the building occupants: improper lighting, noise, vibration, and overcrowding. In addition, ergonomic factors and job-related psychosocial problems (such as job stress) may be important. Each of these alone or in combination can readily produce symptoms that are similar to those associated with poor air quality.

Since the exposures may involve mixtures of pollutants, it is often difficult to relate complaints of specific health effects to a given indoor contaminant. The problem is further complicated by the fact that even small problems can have disruptive and potentially costly consequences if the building occupants become frustrated and mistrustful. Effective communication among facility managers, staff, contractors, and building occupants is the key to cooperative problem solving. Another key is recognition that the expense and effort required to prevent most indoor air-quality problems are much less than those required to resolve them after they develop. This is especially the case in seeking to control such problems on a longer-range basis. One important step in achieving such a goal is to provide adequate guidance to the people responsible for the construction of new buildings, as well as those who manufacture the machinery and appliances that are used within them. Many existing indoor air problems can be controlled by following commonsense recommendations: maintaining proper sanitation, providing adequate ventilation, and isolating pollutant sources.

Case Studies: Radon and Mold

As noted earlier, there are many indoor contaminants whose production is directly influenced by a building's occupants and their activities. There are others, however, whose presence may have little to do with the activities of the occupants. Two of these, radon and mold, which are discussed here, illustrate different aspects of the problems that face building occupants and/or owners.

RADON

Radon is not detectable by the human senses. Although it is known to have been one of the causes of increased lung cancer among underground uranium miners, the public has generally chosen to ignore its presence. There are several reasons for this. One is that epidemiological studies have failed to provide convincing evidence that the presence of radon has led to an increase in lung cancer within the general population, even for those who live in homes with relatively high concentrations. The problem is compounded by the fact that the onset of the primary effect (lung cancer), if it occurs, will not take place until years into the future. Perhaps the key factor, however, is that it will be necessary for individual homeowners to bear the costs of remediation, which can be as much as several thousand dollars. Another factor is that people view their homes as their "castle." Therefore, many members of the public object to any group, particularly a governmental agency, mandating that they must spend money to correct a problem that in many cases (1) they did not know existed, and (2) so far as they can tell, has had little or no impact on their health.

Nonetheless, slow but steady progress is being made in addressing the radon problem. This is being accomplished primarily through (1) the adoption of codes requiring that all new buildings in radon-prone areas be designed and constructed to include radon exhaust systems and (2) the requirement that existing homes be certified as radon free prior to being sold to another party.

MOLD

Mold is also a naturally occurring airborne toxin, but its characteristics and control are far different from those of radon. So long as moisture is present, mold can grow on any substance, including wood, paper, carpet, and food. Since the toxins produced by mold can readily become airborne, their presence can often be detected by their characteristic odor. If mold is pervasive, it may be detected by the discoloration produced by its growth on ceilings, walls, and floors. However, if mold is growing in areas that are not readily accessible, such as behind a wall or within the heating, ventilating, and air conditioning (HVAC) system, it may be very difficult to detect by either sight or smell. For these reasons, confirmation of its presence (or absence) generally requires the collection of air samples or taking wipes of room surfaces and submitting them to a microbiological laboratory for analysis.

In contrast to radon, the presence of mold can rapidly lead to unpleasant reactions, particularly among those people who have allergies and/or asthma. Since mold (as noted above) requires moisture to grow, it can often be controlled by repairing leaks in the plumbing system, installing drainage systems to transport water away from areas near a building, replacing any portions that are water damaged, and/or using air conditioners or dehumidifiers to reduce indoor humidity.

A listing of other indoor air contaminants, their sources, and their acute effects is presented in Table 5.3.

Assessments of Control Programs

While, as noted in the case study on motor vehicles, MTBE was found not to be acceptable in the reformulation of gasoline to reduce emissions from automobiles, this type of approach has several distinct advantages. One is that the reformulated gasoline is automatically used by all vehicles and, as such, generally proves to be cost effective. In contrast, compliance with the standards for emissions from the tailpipes of cars has increased their cost and appears to have served as a stimulus for lengthening the time that owners continue to use older, higher-emitting vehicles. In a similar manner, vehicle maintenance and inspection programs have often (as the earlier case study demonstrated) led to far less emissions reductions than projected (Krupnick, 2002).

One of the most successful control programs is the concept of allowance trading (Chapter 13). A prime example is the sulfur dioxide (SO_2) trading program being used by the electric utilities. It has yielded benefits well in excess of the expected costs. Although some analysts feared that such trades would lead to hot spots or unfavorable rearrangements in emissions, this has not proved to be the case. Buoyed by this experience, the EPA now considers market-based instruments to be equal in effectiveness to command-and-control methods as regulatory procedures for reducing air pollution from these sources (Krupnick, 2002).

In contrast to the success of allowance trading, the NSR requirements of the Clean Air Act have had almost the opposite effect. Under the regulations, older plants were, in effect, "grandfathered" and did not have to comply with the newer emission restrictions so long as they were not modified in a significant manner. That is, routine upkeep was permitted, but improvements were not. Unfortunately, guidance on how to distinguish between normal upkeep and improvements was not clear. As a result, the NSR requirements proved to be both excessively costly and environmentally counterproductive, and investments in new, cleaner power-generating technologies did not occur as anticipated. Another negative impact was that the NSR requirements impeded the adoption of cleaner and more efficient energy technologies, such as cogeneration, wherein the waste heat from one industrial operation is used in other processes at the same site versus being vented to the atmosphere or a nearby body of water. The reasons for the delays were essentially the same as those in the case of applying for approval for improvements in existing plants.

In response to these problems, the EPA subsequently modified the NSR requirements to remove these impediments. Even so, some analysts have concluded that in reality, the development of national and regional allowance-trading programs has made NSR redundant. They suggest that the ultimate solution is to place a limit or cap on total pollution emissions and use an allowance-trading system to ensure that emission increases at one plant are balanced by offsetting reductions at another. The SO_2 program, which, as noted earlier, has successfully achieved targeted emissions

Table 5.3 Indoor air pollutants, sources, and acute effects

Pollutant	Source	Acute effects/symptoms
Gases		
NO_2	Gas stoves, malfunctioning gas or oil furnaces/hot-water heaters, fireplaces, wood stoves, unvented kerosene heaters, tobacco products, vehicle exhausts (garages)	Respiratory tract irritation and inflammation; increased air-flow resistance in respiratory tract; increased risk of respiratory infection
CO	Garages, transfer of outdoor air indoors, malfunctioning gas stoves and heaters, tobacco smoke	Impairment of psychomotor faculties; headache, weakness, nausea, dizziness, and dimness of vision; coronary effects at high concentrations
SO_2	Kerosene heaters	Bronchoconstriction, often associated with wheezing and respiratory distress; impairment of lung function; increased asthmatic attacks
Volatile organic compounds		
Formaldehyde	Tobacco smoke, glues, resins	Irritation of eyes and respiratory tract; headaches, nausea, dizziness; bronchial asthma at high doses; allergic contact dermatitis and skin irritation (occupational)
Reactive chemicals		
Isocyanates	Paints, foams, structural supports	Upper and lower respiratory tract irritation; bronchoconstriction; contact dermatitis; pulmonary sensitization

Table 5.3 (continued)

Pollutant	Source	Acute effects/symptoms
Trimellitic anhydride	Plastics, epoxy resins, paints	Bronchial asthma, asthmatic bronchitis, rhinitis; contact dermatitis
Environmental particulates		
Biologic allergens: dust mites, cockroaches, animal dander, protozoa, insects (dusts, fragments), algae, pollen	Pets, insects, plants	Hypersensitivity pneumonitis, causing cough, dyspnea, and fatigue; allergic rhinitis; asthma
Toxins: fungi (including molds) and bacteria (endotoxins)	Fungi and bacteria (especially in high-humidity environments)	Hypersensitivity pneumonitis, causing cough, dyspnea, and fatigue, allergic rhinitis; humidifier fever, causing flulike illness with fever, chills, myalgia, and malaise
Airborne infectious agents		
Legionella pneumophila	Bacteria (in contaminated water sources such as humidifiers and cooling systems)	Pneumonia, Pontiac fever (flulike symptoms including fever, chills, myalgia, and headache)
Complex mixtures		
Tobacco smoke	Indoor smoking	Eye, nose, and throat irritation; nasal congestion, rhinorrhea; inflammation of lower respiratory tract

reductions, can serve as an excellent model for implementing such a program (Gruenspecht and Stavins, 2002).

As implied by the earlier discussion, programs established to control indoor air pollution have, in general, suffered the same range of fates as those for outdoor pollution. That is to say, some have been highly successful, others have not. As noted in the case study on radon, progress has been slow, and what remediation is being achieved appears to be due primarily to actions stimulated by factors other than the concerns of homeowners. Although it is difficult to judge, progress in the control of mold may be better, primarily due to the fact that some of its effects are rather immediate. Another influencing factor is that the presence of mold is indicative of basic problems with construction and/or maintenance of a home. As a result, the homeowner has more than one incentive to initiate remedial actions.

The General Outlook

The preceding discussions of outdoor air pollution show that significant progress is being made in reducing the concentrations of many airborne contaminants within the United States. Nonetheless, much work remains. As noted earlier, the latest EPA data show that almost 150 million people in the United States live in areas where air contaminants pose significant health concerns (Weinhold, 2002a). On a worldwide basis, the problems are much larger in magnitude. This is exemplified by the "Asian brown cloud," produced by the combustion of wood and fossil fuels, that covered southern Asia in 2002. The cloud was estimated to be more than three kilometers (two miles) thick, and the accompanying pollutants may be producing hundreds of thousands of excess deaths annually due to respiratory illnesses. Although one would expect that the cloud would slowly dissipate, the production of new pollutants appears to be replacing the losses as rapidly as they take place.

It must also be remembered that the harmful effects of air pollution are not restricted to humans. It poses an equal, or perhaps higher, risk to forests, natural vegetation, and agricultural crops, as well as its harmful effects on buildings, statues, and other types of physical structures. Also, air pollution readily crosses national boundaries to affect areas far distant from the emission sources (Wilkening, Barrie, and Engle, 2000). This is a critically important consideration in addressing long-term problems, such as acidic deposition and global warming (Chapter 20).

While increasing attention is being directed to the problems of indoor air pollution, these activities continue to be hampered by several factors. A major one is the lack of resolution of certain public policy and public health questions relative to the proper role of the government in safeguarding air quality inside public and private spaces. As long as these issues remain unresolved, it could well be that the problems of indoor air pollution will not be effectively addressed. Even so, one could reasonably ask why members of the U.S. Congress have not addressed these problems in the same forceful and constructive manner that has been applied to the ambient environment. Why not, for example, simply pass a law that the air inside buildings must be equal to, or of higher quality than, that outdoors?

The difficulties in answering this question involve at least one major consideration and one major challenge. From the standpoint of private residences, a major consideration is that many people, as noted earlier, view their homes as their "castle." Another almost overwhelming challenge is the sheer magnitude of the problem. There are literally tens of millions of houses in the United States that have indoor pollution problems. In some respects, any type of legislative approach to control these problems is not likely to be enforceable. Nonetheless, as a minimum, it would appear beneficial to assign primary responsibility for addressing these problems to a single federal agency. Other possible steps are discussed later.

The situation is not significantly different in commercial buildings. One of the exceptions is the increasing prohibition of cigarette smoking inside essentially all types of facilities. Interestingly, this type of action had its genesis in the restrictions placed on passengers in commercial aircraft. Another exception is the quality of the air in the workplace. In this case, the Occupational Safety and Health Administration has clear responsibility at the federal level for ensuring that workers are not unnecessarily exposed. Multitudes of similar groups at the state and local levels strongly support OSHA. This is not to imply, however, that no one is addressing similar problems with respect to buildings that are used for other purposes. The American Society of Heating, Refrigeration, and Air-Conditioning Engineers (ASHRAE), for example, has developed uniform practices for designing and installing the equipment necessary to ensure acceptable indoor air quality. In a related manner, the American Institute of Architects has issued guidelines for the design and construction of hospitals and health-care facilities. These include recommendations for acceptable air-exchange rates. Related reports and recommendations have been issued by the Underwriters Laboratories, the American Industrial Hygiene Association, the American Conference of Governmental Industrial Hygienists, the International Society on Indoor Air Quality and Climate, and the Association of Energy Engineers (Latko, 2000).

With advances in computers and systems of electronic communications, increasing numbers of people in the more industrialized countries are using their homes as a secondary, or even a primary, place of work. For these and other reasons, the importance of indoor air pollution in homes is likely to increase. One approach would be to implement requirements similar to those that have proved successful in controlling radon in homes. These would establish regulations that all new homes and commercial buildings must meet requirements, such as those recommended by ASHRAE and related groups, to ensure acceptable indoor air quality. The same requirements might also be applied each time an existing building is being sold. Although care would need to be exercised, and a wide range of factors would need to be considered in developing such regulations (especially in terms of private residences), this could represent a place to begin. Such actions could be initiated at the local level and gradually be expanded to the state and federal levels. As a forerunner to these actions, the previously mentioned designation of a single federal agency to oversee indoor air pollution would appear to be mandatory.

One item worthy of comment in closing is the leadership demonstrated by California in the control of airborne emissions from automobiles. Because of the unique nature of air-pollution problems in the Los Angeles basin, regulators in that state continue to impose requirements more stringent than those proposed by the EPA. Although automobile manufacturers complain vociferously, the net result inevitably is that the proposed standards are met, and ultimately the whole nation benefits. The state of California has similarly been a leader in promoting the development of energy-efficient transportation vehicles (Chapter 18).

6

FOOD

GIVEN the central importance of food in our personal environment, one would expect it to be an aspect of our lives that we control. This is far from the case. The production, preparation, and handling of food continue to present new and novel challenges. These include the introduction of new agricultural and food technologies, such as genetically modified food crops; an increasing globalization of the food supply; changes in human demographics and food preferences; and intense public and media scrutiny of issues such as mad cow disease and biotech foods (Taylor and Hoffmann, 2001). Concurrently, estimates are that the consumption of contaminated food in the United States leads each year to the illness of about 76 million people, of whom more than 325,000 are hospitalized and about 5,000 die (Tick, 2004). Even so, this may still be a gross underestimate of the true magnitude of the problem. One of the contributing factors is that most foodborne diseases occur as isolated or sporadic events. In fact, many victims do not seek medical care.

The focus in this chapter is on contaminants that are commonly found in food, their effects on health, and the steps that must be taken in the preservation and handling of food to assure its safety. Aside from objectionable materials, such as rust, dirt, hair, machine parts, nails, and bolts, such contaminants fall into two broad categories: (1) biological agents, such as bacteria, viruses, molds, antibiotics, parasites, and their toxins, which can cause a wide range of illnesses; and (2) chemicals, such as lead, cadmium, mercury, nitrites, nitrates, and organic compounds, which can have both acute and chronic health effects (Figure 6.1). Such contaminants can gain access to the food chain at any of a multitude of stages during

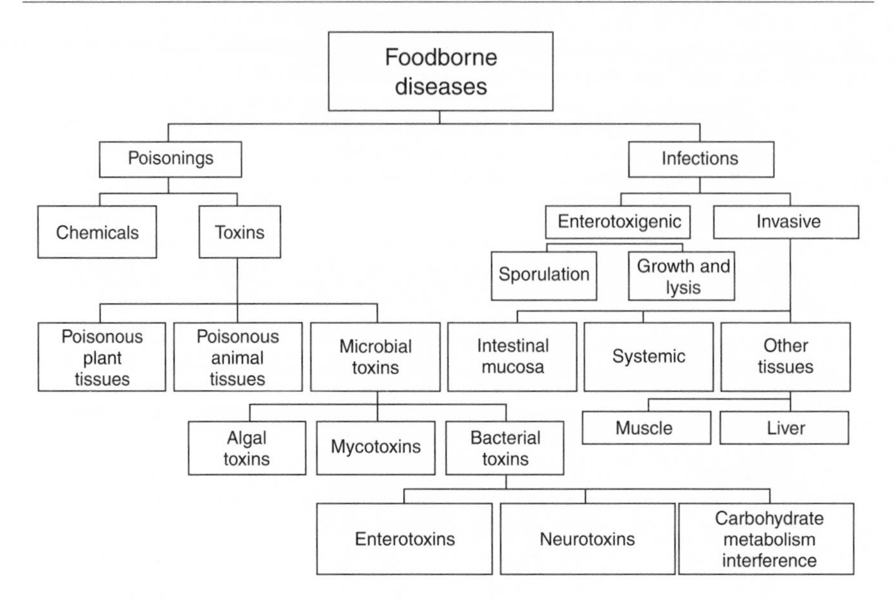

Figure 6.1 Classification of foodborne diseases

growing, processing, preparation, or storage. Of the two, microbial sources account for upwards of 95 percent of all reported outbreaks (97 percent of all cases) (Marshall and Dickson, 1998). Accordingly, most of the attention in this chapter is devoted to illnesses of this type.

Food and Health

For well over a decade, the primary source of dietary guidance for the U.S. public has been the "food pyramid," developed in 1992 by the U.S. Department of Agriculture (USDA). Basic to this guidance is the recommendation that certain food groups, such as grains, vegetables, and fruit, be consumed in larger quantities than others, such as meat and dairy products, fats, oils, and sweets. In recent years, this guidance has been subjected to increasing criticism. A common complaint is that in discouraging the consumption of fats, the USDA appeared to convey a sense that all carbohydrates were harmless. As nutritionists and health experts gained new information and insights, it became increasingly clear that the USDA pyramid had become obsolete. One deficiency is that it fails to point out that there are good sources of dietary intake in all food groups.

Examples are grain foods that are good sources of carbohydrates, plant oils that are good sources of fats, and nuts and legumes, followed by fish, poultry, and eggs, that are good sources of protein (Lambert, 2004). For this reason, a revised pyramid (Figure 6.2) is now being proposed. One of the primary goals of its developers (Willett, 2001) is to provide dietary guidance that is designed not for short-term weight loss but for lifelong health. Therefore, they emphasize that any dietary program should be accompanied by daily exercise and weight control.

Numerous epidemiological studies support the changes that are being suggested. Such studies have confirmed, for example, that the consumption of fruits, vegetables, and fiber protects the heart, and that the consumption of whole grains reduces the risk of stroke and diabetes. Although it has been known for some years that saturated fats in red meat, butter, and cheese are contributors to coronary heart disease, epidemio-

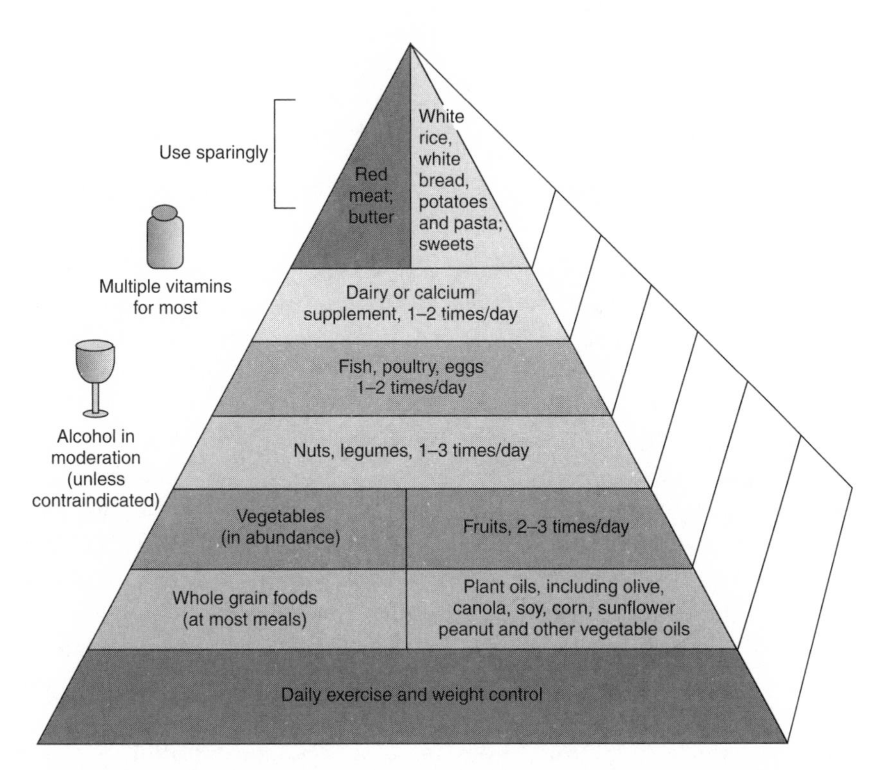

Figure 6.2 Proposed revision of food pyramid—guides for a healthy diet

logical studies now show that the consumption of foods containing trans fatty acids plays a key role, particularly in relation to the types and quantities of cholesterol in the blood. In fact, some nutrition experts have postulated that these acids may be responsible for the epidemic of heart disease that began in the United States during the 1930s and 1940s. Such acids are produced, for example, when partially hydrogenated oils are solidified in the production of margarine and shortening and are commonly present in baked goods, chips, and so-called fast foods. Responding to these findings, many U.S. fast-food restaurants have modified the ways in which they process the foods they serve. Responding to this and related information, the Food and Drug Administration (FDA) announced that beginning in 2006, U.S. food producers will be required to add to the labels on their products the amounts of trans fatty acids they contain (Sheehan, 2003).

Foodborne Illnesses and Their Causes

Table 6.1 summarizes the major foodborne illnesses, the causative agents, the food usually involved, and the incubation period. The illnesses described may be caused by parasites, bacterial infections, viral infections, or toxins.

PARASITES

Two of the more common parasitic diseases in the United States are amebic dysentery (caused by *Entamoeba histolytica*) and giardiasis (caused by *Giardia lamblia*). *Entamoeba histolytica,* a parasitic organism, can exist as a hardy, infective cyst or a more fragile, potentially pathogenic trophozoite. The parasite can coexist with its host without injury to either, or it may invade the tissues of the host, giving rise to intestinal or extraintestinal disease. In these cases, the effects may range from acute dysentery, accompanied by fever, chills, and bloody diarrhea (amebic dysentery), to mild abdominal discomfort with diarrhea, alternating with periods of constipation or remission. Transmission occurs primarily through the ingestion of fecally contaminated food. Because amebic cysts are relatively resistant to chlorine, water can also be a source of intake. If the water is filtered through sand, most of the cysts will be removed. Diatomaceous earth filters will remove them completely. Individuals infected with *E. histolytica* should be excluded from handling food and from direct care of hospitalized and institutionalized patients (Chin, 2000).

Giardiasis is a gastrointestinal illness caused by the flagellated protozoan *Giardia intestinalis,* also known as *G. lamblia* or *G. duodenalis.* It is the most commonly diagnosed intestinal parasite in the United States, with upwards of 5,000 infected people being hospitalized due to this disease each year. The total number of cases, however, is estimated to be as high as 2.5 million. Symptoms include diarrhea, fever, or both—and flatulence, nausea, malaise, or abdominal cramps. *Giardia* organisms are present worldwide and infect both domestic and wild animals, including cats, dogs, cattle, deer, and beavers. Like *E. histolytica,* these organisms are spread from person to person and from animals to humans through fecal-oral transmission, with either food or water serving as a typical route of intake. Children are infected more frequently than adults. Although the infectious dose is low, the causative organism is only moderately resistant to chlorine. Because the symptoms may be mild, giardiasis is often regarded as a benign gastrointestinal illness. Nonetheless, chronic or debilitating giardiasis has been reported (CDC, 2000; Chin, 2000).

BACTERIAL INFECTIONS

Certain bacteria can gain access to foods and be ingested and transported to the digestive tract, where they can multiply and cause illnesses. Two common examples are discussed here.

Salmonella infections occur in an estimated 1.4 million people in the United States each year. The majority suffer diarrhea, fever, and abdominal cramps one to three days after exposure. The illness usually lasts four to seven days, and the majority of those infected recover without treatment. The causative agent, the *Salmonella* organism, exists in the intestines of chickens, dogs, and rodents. It can also live in the ambient environment and can survive conditions that many other organisms cannot. This accounts for its transmission through food as well as drinking water. Common foods involved include chicken, pork, and beef, with eggs and poultry being primary sources of infection. In fact, about 12 percent of chickens marketed in U.S. supermarkets are estimated to contain this organism (Consumers Union, 2003). In the case of eggs, *Salmonella* transmission was originally primarily due to contamination on the outside of the shells. Today these organisms are often present inside the eggs because of infections in the ovaries of chickens. All protein foods requiring a large amount of handling are subject to contamination. Low-acid foods, such as meat pies, custard-filled bakery products, and improperly cooked sausages, are also common sources of outbreaks. *Salmonella* can also be trans-

Table 6.1 Examples of important foodborne illnesses, United States

Illness	Causative agent	Food usually involved	Incubation period
Foodborne parasites			
Amebiasis (amebic dysentery)	*Entamoeba histolytica*	Food contaminated with fecal matter	2–3 days to 1–4 weeks
Cryptosporidiosis	*Cryptosporidium parvum*	Vegetables, unpasteurized milk, fruits	2–28 days, 7 days average
Cyclosporiasis	*Cyclospora cayetanensis*	Imported berries, lettuce	1–11 days
Giardiasis	*Giardia lamblia*	Raw salads and vegetables	1–4 weeks
Trichinosis	*Trichinella spiralis*	Raw or undercooked meat, usually pork or wild game	1–2 days to 2–8 weeks
Foodborne Bacteria			
Dysentery	*Shigella dysenteriae*	Food contaminated with fecal material, raw vegetables, egg salads	Up to 1 week
Gastroenteritis	*Shigella sonnei*	Food contaminated with fecal material, person-to-person contact, raw produce, parsley	2–4 days
Salmonellosis	*Salmonella* spp.	Eggs, poultry, unpasteurized milk or juice, cheese, raw fruits and vegetables	1–3 days
Foodborne Viruses			
Gastroenteritis	Norwalk-like viruses	Fecally contaminated food, salads, sandwiches, ice, cookies, fruit, poorly cooked shellfish	24–48 hours

Table 6.1 (continued)

Illness	Causative agent	Food usually involved	Incubation period
Viral hepatitis	Hepatitis A virus	Shellfish from contaminated waters, raw produce, uncooked foods	15–50 days, 30 days average
Foodborne toxins			
Botulism (in infants)	*Clostridium botulinum*	Honey, home-canned fruits and vegetables	3–30 days
Brucellosis	*Brucella abortus, B. melitensis, and B. suis*	Raw milk, goat cheese from unpasteurized milk, meats	7–21 days
Diarrhea	*Escherichia coli* O157:H7, and other shiga-toxin producing *E. coli*	Undercooked beef, unpasteurized milk and juice, raw fruits and vegetables	1–8 days
Paralytic shellfish poisoning	Dinoflagellates (neurotoxins)	Scallops, mussels, clams, cockles	30 minutes–3 hours
Staphlococcal food poisoning	*Staphylococcus aureus*	Improperly refrigerated meats, potato and egg salads, pastries	1–6 hours

mitted to humans, especially children, through direct or indirect contact with reptiles, such as lizards, snakes, and turtles.

Shigella dysenteriae, which causes bacillic dysentery, is another common source of foodborne illness. Two-thirds of all cases, and most deaths, occur in children under 10 years of age. Illness in infants less than 6 months old is unusual. Secondary attack rates in households can be as high as 40 percent. *Shigella sonnei,* another species, is a common cause of gastroenteritis. In fact, it accounts for almost three-quarters of the approximately 15,000 laboratory-confirmed *Shigella* infections in the United States each year. In both cases, the organisms are present in human feces, and transmission is favored by crowded conditions where personal contact is unavoidable. Food handlers can readily spread the infection. Flies can also transfer the organisms to nonrefrigerated food, where they can multiply. In the case of *S. dysenteriae,* ingestion of a large number of organisms is required in order for a person to become infected; in the case of *S. sonnei,* as few as 10–100 organisms can cause infection. As a result, person to person is a viable method of transmission for this organism. The incubation period in the case of *S. dysenteriae* is up to a week, while the bacteria in the body multiply; in the case of *S. sonnei,* it is only two to four days (Chin, 2000).

VIRAL INFECTIONS

Prominent sources of viral infections are the Norwalk-like viruses (NLVs) and those that cause infectious hepatitis and bovine spongiform encephalopathy, so-called mad cow disease. NLVs annually cause an estimated 23 million episodes of gastroenteritis, 50,000 hospitalizations, and 300 deaths in the United States. These viruses can be transmitted by fecally contaminated food, such as salads, sandwiches, fruit, and improperly cooked shellfish, and by direct person-to-person contact. They are extremely contagious for two reasons: (1) the dose that will cause infection is low, and (2) patients continue to be infectious for up to two weeks after recovery. Exacerbating the situation is that the viruses are resistant to chlorination as well as to temperature variations ranging from 0°C (32°F) to 60°C (140°F). Outbreaks of gastroenteritis due to Norwalk-like viruses are common in settings in which people are crowded and sanitation facilities are inadequate, such as summer camps. During the summer of 2002, there were also multiple outbreaks among passengers and crews on ships operated by several of the major pleasure cruise lines. One of the problems in the control of this illness is the lack of a simple and sensitive technique for detecting its presence (CDC, 2002c).

Infectious hepatitis (hepatitis A) is a highly contagious disease caused by a virus whose symptoms are fever and general discomfort. The disorder occurs most frequently among school-age children and young adults, and the infectious agent commonly is present in feces. Adults are usually immune. Common sources include foods, such as sandwiches and salads, that are not cooked or foods that are handled after cooking by infected handlers. Raw or undercooked mollusks harvested from contaminated waters, as well as contaminated produce, such as lettuce and strawberries, and contaminated drinking water, may also be sources of infection (Chin, 2000).

Still another virus-related disease that has gained prominence in recent years is the previously cited bovine spongiform encephalopathy, which leads to progressive neurological degeneration in cattle. There was a major outbreak of this disease, in the United Kingdom during the 1990s. The first cases were observed in 1986. As a consequence, about 100 people developed the variant Creutzfeldt-Jakob disease and died (Gray, Cohen, and Kreindel, 2002). It has been hypothesized that the source of their infections was the ingestion of beef from infected cattle (Chin, 2000). Fortunately, due to a well-organized program that included quarantining areas where infected cows were present and destroying diseased animals, the epidemic was brought under control, and possible transmission to other countries of the world was minimized. Nonetheless, a cow imported from Canada into the United States was discovered in 2003 to be infected with mad cow disease. Obviously, continuing vigilance will be required. Additional information on this episode is presented in Chapter 17 (Gray, Cohen, and Kreindel, 2002). A similar disease in sheep is known as scrapie.

TOXINS

In contrast to bacterial infections, which are caused directly by the organisms, some foodborne illnesses are caused by toxins produced by bacteria that are not in themselves harmful. Toxins can similarly be produced in food by viruses and fungi. Toxins can be introduced into food through improper handling, or they may be naturally present. In either case, the ingestion of the accompanying toxins can readily lead to illnesses and, in some cases, death.

Toxins resulting from improper handling. The most common toxins introduced into food through improper handling are produced by bacteria. Three of the more common such organisms are discussed here.

1. Under favorable conditions, *Staphylococcus aureus* can produce one or more enterotoxins that, if ingested, can abruptly (within one to six hours) lead to severe nausea, cramps, vomiting, and prostration, often accompanied by diarrhea and sometimes by subnormal temperature and reduced blood pressure. This organism can readily be transmitted to food from infected cuts, boils, sores, postnasal drip, or sprays expelled during coughing or sneezing. It is also present in air, water, milk, and sewage, and an estimated one-quarter of the population are carriers. Meat (especially ham), meat products, poultry, poultry products, and poultry

dressing, as well as custards used for pastry fillings, are common sources. *Staphylococcus aureus* grows rapidly, especially in food held at room temperature for several hours before being eaten (Chin, 2000).

2. Botulism is a paralytic illness caused by the neurotoxin produced by the bacterium *Clostridium botulinum.* Its spores are present in the soil throughout the world. Conditions that promote their germination and growth include the absence of oxygen (that is, anaerobic conditions), low acidity (pH approximately 4.6), temperatures higher than 4°C (39°F), and high moisture content. Although rare and sporadic, foodborne botulism is a persistent cause of morbidity and mortality in the United States. It has been a common problem for decades in Alaska, where several hundred Natives have become ill. The source in this case was beaver meat that was being fermented in plastic or glass containers (CDC, 2001b). Most poisonings in the continental United States result from the consumption of vegetables and fruits that have been improperly canned at home. The toxins are extremely potent; a few nanograms (10^{-9} gram) can cause illness. While the toxins can exist for long periods, they can be destroyed by boiling. Inactivation of the spores, however, requires higher temperatures (Chin, 2000).

3. *Escherichia coli,* another toxin producer, is present in the lower intestinal tract of most warm-blooded animals and is the most prevalent oxygen-tolerant bacterium in the large intestine of humans. Foods of bovine origin, particularly ground beef, are common sources of sporadic infections and outbreaks. Unless care is exercised, it is relatively easy for *E. coli* to be transferred from feces and intestinal contents to carcasses and meat during processing. The prevalence of *E. coli* O157:H7 in animal feedlots in the United States can be as high as 60 to 100 percent. Symptoms of infection include bloody and nonbloody diarrhea, vomiting, and abdominal cramps, with onset ranging from one to eight days after ingestion. In 2002, simultaneous outbreaks occurred in Colorado and six other states. This led to a nationwide recall of almost 19 million pounds of fresh and frozen ground-beef-related products (CDC, 2002b). As part of its control program, the USDA now requires that all raw meat and poultry be labeled with instructions regarding proper cooking and handling. *Escherichia coli* can also be transmitted to humans through the ingestion of shellfish and watercress that become contaminated while being grown in sewage-contaminated waters.

Naturally occurring toxins. Laboratory studies of foodstuffs and cooked food show that they contain a surprising array of naturally occurring toxins that would normally not be permitted as regulated additives. Carrots, for example, contain *carotatoxin,* a fairly potent nerve poison; *myristicin,* a hallucinogen; and *isoflavones,* which have an estrogenic effect similar to that of female hormones. Peanut butter contains *aflatoxins,* some of which (aflatoxin B, for instance) have been shown to be acutely poisonous and carcinogenic in animals. The common assumption that "natural" is safe and "human-made" is suspect is contrary to current scientific knowledge. In fact, a typical diet contains far more natural carcinogens than synthetic ones.

Of the many toxins that occur naturally in plants and animals, only a few have been specifically associated with human illness. The most dramatic example is paralytic shellfish poisoning, which is caused by a highly potent neurotoxin that is a metabolite of certain marine dinoflagellates. One of the most common is *Karenia brevis,* which produces what are known as brevetoxins. Because the blooms of the dinoflagellate impart a characteristic red color to the water, their appearance is referred to as red tide. It is a common problem in the Gulf of Mexico and the western North Atlantic Ocean. It is also a problem in the waters along the west coast of Florida, where episodes occur essentially every year. Poisonous concentrations of the brevetoxins can accumulate in shellfish (mussels, clams, and occasionally scallops and oysters) that feed in areas where the dinoflagellates are blooming. The toxin appears to produce no effect in the shellfish (Red Tide Research Group, 2002). When the contaminated seafood is consumed by humans, however, symptoms of toxicity usually develop within one to three hours—numbness of the lips and fingertips, ascending paralysis, and finally, in cases of severe poisoning, death from respiratory paralysis. Should the victim survive the first 24 hours, recovery is generally uneventful (CDC, 2001a). Control can be accomplished by monitoring the potentially affected waters and discontinuing seafood harvesting when dinoflagellates are present (Marshall and Dickson, 1998).

Data show that humans may also be exposed to brevetoxins through inhalation. People who have frequented beaches in Florida during red tide events have reported a number of symptoms, including respiratory complaints. Although the hypothesis is yet to be confirmed, scientists believe that these symptoms are caused by exposure to aerosolized brevetoxins and perhaps airborne *K. brevis* cellular debris (Red Tide Research Group, 2002).

The problems of shellfish toxins, however, are not limited to *K. brevis.* Another example is the ciguatera toxin that is produced by dinoflagellates that live in the Caribbean and tropical Pacific regions. In this case, the dinoflagellates are consumed by herbivorous fish that, in turn, are eaten by large predatory reef fish, such as barracuda, grouper, and amberjacks. Although the fish do not appear to be harmed by the toxin, humans do not share this immunity. More than 100 poisoning events occurred in Florida and Hawaii between 1983 and 1992, 90 percent of them in the latter state. No deaths were reported (CDC, 1998). Other organisms, most particularly *Chattonella cf. verruculosa,* have similar capabilities. In this case, fish are also susceptible. This organism is believed, for example, to have been the cause of the deaths of several million fish in estuarine waters in coastal Delaware in recent years (Bourdelais et al., 2002).

Inorganic and Organic Chemical Contaminants and Additives

Foods can contain a variety of inorganic and organic chemicals. Some of these are purposefully added, and others result from human actions. Heavy metals, such as lead, copper, tin, zinc, or cadmium, can leach from containers or utensils, particularly in cases in which acidic foods are being prepared or stored. Other chemicals can be introduced through accidental or inadvertent contamination with detergents or sanitizers. Pesticides, herbicides, fungicides, fertilizers, and veterinary drugs and antibiotics can be introduced into vegetables, poultry, and livestock as a result of the conditions under which they are grown (Marshall and Dickson, 1998). Relatively speaking, however, such contaminants are not a problem today, particularly in the developed countries of the world.

INORGANIC CHEMICALS

Mercury discharged into rivers, lakes, and oceans in the form of inorganic salt or as the metallic element (which is not harmful to humans) can be converted by microbes to methyl mercury. In this form, it can pose a significant health risk. Large-scale poisonings by these compounds have caused deaths and cases of permanent damage to the central nervous system. In a classic episode in Japan in the early 1950s, industrial wastes containing mercury were discharged into Minamata Bay. More than 100 people who ate contaminated fish were poisoned, and 46 died. Other sources of mercury include volcanic eruptions, which account for about one-third of worldwide releases, and airborne contaminants from coal-fired electricity generating stations. If in a water-soluble form, the mercury is readily brought back to the ground by precipitation. Once the mercury reaches bodies of water, it is passed up the food chain into fish, notable examples being swordfish and tuna. The global aspects of this problem are demonstrated by the fact that about a fifth of that which deposits in the United States comes from facilities in Asian countries, such as China (Levine, 2004).

A variety of other inorganic chemicals are or can be introduced into foods during processing. One of the most common is sodium, one of the two ingredients (the other being chlorine) in salt. This compound primarily serves as a taste enhancer and preservative. Although the subject is controversial, data from some studies indicate that excessive salt intake is related to hypertension and gastric injury in some individuals. Also added to foods are sulfites and bisulfites, which, in aqueous solutions, form sulfurous acid, an antimicrobial agent. In addition, nitrites and nitrates serve as agents for curing and pickling meats and vegetables. These two substances can also gain access to some foods through uptake into agricultural crops that are produced using nitrogen fertilizers. One of the benefits of nitrates and nitrites is that they inhibit the growth of *C. botulinum* in foods that are vacuum packed. One of the risks is that high concentrations of nitrates in baby food, much of which is converted into nitrites, can cause methemoglobinemia in infants. Another risk is that during cooking, nitrites can react with secondary and tertiary amines to form nitrosamines, a potential carcinogen. Current formulations, however, significantly reduce this risk. To qualify for use, all such additives must be classified as GRAS, that is, they must be "generally recognized as safe" (Marshall and Dickson, 1998).

ORGANIC CHEMICALS

A number of organic acids and their salts are used as preservatives in foods. These include benzoates, which inhibit the growth of yeasts and molds; sorbate salts, which inhibit the growth of yeasts and molds; and propionic acid and propionate salts, which are active against molds (Marshall and Dickson, 1998). Other organic chemicals gain access to food through the use of pesticides and herbicides on agricultural crops. Examples are chlorinated hydrocarbons, polychlorinated biphenyls (PCBs), chlorinated dibenzo-*p*-dioxins, and chlorinated dibenzofurans. Tests in animals show that PCBs can cause reductions in immune system function, behavioral alterations, and impaired reproduction.

Organic contaminants can also be produced in foods, especially meat, through the cooking process. Browned or burned portions of meats that have been charbroiled, whether fried or smoked, contain heterocyclic aromatic amines, many of which have been shown to be highly mutagenic. Examples are benzo-a-pyrene and the polycyclic aromatic hydrocarbons, as well as numerous breakdown products of common dietary amino acids. Measures that have been suggested to avoid the production of these compounds include using alternative processes such as stewing, poaching, or boiling to cook meat and employing a microwave oven to cook fish and poultry.

Another group of compounds that are present in commercially prepared foods are the previously discussed trans fatty acids. Still another is acrylamide, a compound that was identified years ago as a potential industrial hazard from the standpoint of causing certain neurological effects. This compound is present at relatively high concentrations in starch-based foods, such as biscuits, cereals, french fries, and potato chips, that are cooked or baked at temperatures in excess of 120°C (248°F). Laboratory studies show that it is capable of inducing cancer and heritable mutations in rats. The fact that acrylamide has probably been a component of human diets ever since cooking began makes these evaluations particularly interesting. In addition to these sources, abnormal and toxic metabolites are frequently produced when plants are subjected to stress. These include protease inhibitors, hemagglutinins, goitrogens, and allergens (Weiss, 2002).

Antibiotic and Hormone Use in Farm Animals

About 50 years ago, farmers began feeding antibiotics to animals to prevent the spread of infections and to reduce the amount of feed required to fatten them. During the intervening decades, however, studies have increasingly confirmed that such practices have contributed to an alarming increase in the resistance of many human bacterial pathogens to antibiotics. Although the U.S. Congress and governmental agencies have debated the issue, no action to restrict such practices has been taken in this country. In fact, approvals for such use continue to be forthcoming. One example is the use of fluoroquinolones, which the FDA (contrary to the advice of the Centers for Disease Control and Prevention) approved in 1996 for feeding to chickens and turkeys, primarily to prevent mortality associated with infections by *Escherichia coli.* Within three years, more than 15 percent of *Campylobacter jejuni* and 30 percent of *Campylobacter coli* isolated from human patients showed resistance to this antibiotic (Falkow and Kennedy, 2001).

The lack of action to ban such uses in the United States is in distinct contrast to action in other countries of the world. The European Union (EU), for example, banned the use of avoparcin from livestock feed in 1997. The stimulus for this action, which followed two decades of such practices, was to help prevent the further spread of vancomycin-resistant enterococci (VRE) in humans. One of the most prominent of these is *Staphylococcus aureus.* Monitoring showed that there was a subsequent dramatic reduction in the amount of VRE among pigs, chickens, and supermarket chicken meat. In addition, fewer VRE have also been detected in humans, both within the general population and in hospital patients (Ferber, 2002). The EU action is even more significant because *Staphylococcus aureus* is one of the most common causes of hospital- and community-acquired infections in the United States. Increases in resistance among drugs used on patients are estimated to cause perhaps as many as 70 percent of the 90,000 fatal hospital infections that occur in this country each year. The total annual associated cost is estimated to exceed $30 billion (Bright, 1999).

While the EU has been a leader in such actions, it is not alone in these efforts. One of the best examples is the United Kingdom, where such practices were restricted years ago. Denmark, Finland, and Sweden subsequently adopted a similar position. And, as a follow-up to its previous actions, the EU has ordered member countries to end such practices by 2006, and the World Health Organization, which in 1997 initially recommended ending such practices, has recently issued a report of a detailed review of the impacts of the restrictions in Denmark that concluded that eliminating such practices would reduce human health risks without significantly harming animal health or farmers' incomes (Ferber, 2003). Concurrently, one of the major poultry producers in the United States has voluntarily announced that it will no longer use fluoroquinolones in any of its operations. Several leading fast-food chain restaurants have similarly joined in promoting the ban by announcing that they will no longer purchase poultry or products that have been treated with these compounds (Falkow and Kennedy, 2001). Thus while the U.S. government has been slow to act, commercial groups, stimulated by actions in other countries of the world and increasing public recognition of the serious negative impacts of antibiotic use, have taken the lead to ban such uses in this country. Concurrently, however, U.S. pharmaceutical companies are reducing their research on the development of new antimicrobials. Even though this action comes at a time when recent advances in microbial and host genomics have provided a wealth of potentially new antimicrobial targets, those in charge of such operations have concluded that the associated high costs make it impossible for them to continue such activities at this time (Fraser, 2004).

Care in Food Preservation and Handling

A variety of methods are available for safely preserving wholesome food, preventing contamination, and destroying organisms or toxins that may have gained access to or been produced within the food. One of the prerequisites to ensuring that these methods are successful is to seek to maintain the food in a condition that is not favorable for bacterial growth. A major goal in this regard is to avoid conditions that provide warmth, moisture, and a medium that is neither highly acid nor alkaline. On the basis of these and other considerations, the primary methods that have proved to be effective for preserving food may be summarized as follows:

Cooking. Cooking renders food digestible and palatable. Although it also tends to kill many bacteria, this process alone will not preserve food. In fact, partial cooking may render protein foods (meat, eggs, milk, milk products) more susceptible to bacterial growth and permit active increases in the number of harmful organisms or the toxins they may produce. Even when food is heated thoroughly and to a sufficiently high temperature to kill any microorganisms present, it must be eaten promptly or protected from subsequent spoilage.

Canning. The process of canning involves heating food sufficiently to kill any microorganisms present and then sealing it in a container to keep it sterile. The combination of time and temperature required to preserve food by canning varies with the product and its likely contaminants. Acid foods—tomatoes and some fruits—need to be heated to the boiling point for only a few minutes. Nonacid foods—corn and beans—must be heated to higher temperatures (under pressure) for a longer time to prevent undesirable changes in appearance and flavor, as well as, for example, to destroy the anaerobic microorganisms that produce the botulism toxin.

Drying and dehydration. Air drying, one of the most economical and effective ways of preserving food, has been practiced for centuries. Today food can be dried in the sun or by artificial heating processes. Other methods include spray drying, freeze drying, vacuum drying, and hot-air drying. Once the food is reconstituted by the addition of water, bacterial activity resumes, and it is essential that sanitary controls be applied.

Preservatives. As described in the previous section "Inorganic and Organic Chemical Contaminants and Additives," a variety of chemicals are purposefully added to foods to inhibit the growth of microorganisms, to kill them, or to serve as flavor enhancers. These include salt, sugar, sodium nitrate and nitrite, salicylic acid, and sodium benzoate, as well as propionates and sorbic acid. Each carries with it both risks and benefits. One additional method of preserving foods, especially meats, is smoking. This technique is often used since it improves flavor and retards microbial growth.

Refrigeration. Storing food at temperatures lower than 5°C (40°F) will retard the growth of pathogenic organisms and the more important spoilage organisms, but it does not prevent all changes. The level of humidity is also important: too little results in moisture loss; too much promotes the growth of spoilage organisms. Proper air circulation and regular cleaning and sanitizing of chill spaces are mandatory.

Freezing. Bacteria that cause food spoilage do not multiply at freezing temperatures, but once thawing begins, frozen food becomes vulnerable to bacteria and the associated toxins they may produce. Refreezing will not make the food safe, nor will freezing improve the original quality of the product. Thus the selection of appropriate products for freezing is essential. One variation is "dehydrofreezing," in which the food is partially dehydrated (but still perishable) and then frozen. This process provides the space and weight savings of dehydration without depriving the food of its fresh color, flavor, and palatability.

Pasteurization. Pasteurization is an excellent method of preserving food for a short time. Combined with refrigeration, it extends the useful shelf life of dairy products. Milk is generally heated to 63°C (145°F) for 30 minutes—or to 72°C (161°F) for 15 seconds—to kill the pathogenic organisms. Although some heat-resistant organisms will survive, subsequent refrigeration will preserve the milk for up to several weeks.

Irradiation. Through this process, food is exposed to ionizing radiation at sufficiently high doses to kill a large fraction of any microorganisms present. At the doses that are applied, meats and poultry, for example, are not sterilized; they still require refrigeration and proper handling. In this sense, irradiation is directly analogous to pasteurization. It is especially effective in destroying foodborne contaminants such as *Salmonella* and *Escherichia coli.* It also destroys *Trichinella spiralis* in pork (Loaharanu, 2003). In some foods, however, irradiation produces unwanted changes in taste and palatability. Although fears have been expressed about other changes that take place during irradiation, especially the formation of radiolytic compounds, the types and quantities of these compounds in irradiated foods are no different than those in foods processed by other methods of preservation. Irradiation has been approved in the United States for the preservation of pork, chicken, herbs and spices, fresh fruits and vegetables, grains, and seeds used for producing sprouts. In response to legislation passed by Congress in 2002, the USDA now offers irradiated meat as part of its nationwide school lunch program that provides daily meals to more than 25 million children in the United States.

Components of an Effective Sanitation Program

In addition to exercising care in processing, the prevention of foodborne illnesses requires an effective sanitation program. A safe water supply, adequate garbage and refuse disposal, proper wastewater and sewage disposal, and effective insect and rodent control are also essential. Other factors involve equipment and facilities, personnel training and habits, standards and regulations, and enforcement and monitoring.

Equipment and facilities. Equipment used in the preparation or processing of food should be designed to facilitate cleaning. Cutting boards should be made of nonporous materials. Vehicles used to transport food products must be clean and should not carry other products. Refrigerated vehicles must be available for the transport of perishable foods. Facilities should be designed so that all foods, particularly vegetables, can be stored above the floor, where they will remain dry and will not come in contact with powders and sprays used to control insects and rodents.

Personnel training and habits. Personal hygiene is indispensable in the proper handling and preparation of food products. Antimicrobial cleaners should be used on the surfaces on which foods are prepared, and cleaning rags and sponges should be disinfected regularly or replaced. Food handlers must wash their hands after toilet use and before and after work; must avoid contact between open wounds and foodstuffs; must wear clean outer garments, including a cap over the hair; and must avoid using tobacco products while working. Food handlers should be trained in appropriate methods of food storage, garbage disposal, and insect and rodent control. The essential rules for safe food preparation and consumption are summarized in Table 6.2.

Standards and regulations. The basic requirements are that standards and regulations be national in scope and specify the proper methods of processing, preparing, and selling food products; limitations on the types and quantities of chemicals that can be added to foods; restrictions on the quantities, types, and manner in which pesticides can be used on agricultural food crops; and proper labeling requirements for commercial food products. The

Table 6.2 Ten rules for safe food preparation and consumption

1. Choose food processed for safety.
2. Cook food thoroughly.
3. Eat cooked food immediately.
4. Store cooked food immediately.
5. Reheat cooked foods thoroughly.
6. Avoid contact between raw and cooked foods.
7. Wash hands repeatedly.
8. Keep all kitchen surfaces meticulously clean.
9. Protect foods from insects, rodents, and other animals.
10. Use pure water.

principal federal agencies in the United States that have responsibilities related to food safety, with a brief description of their duties, are listed in Table 6.3. In a similar manner, definitions of the safety standards or limits for the control of a range of food contaminants and ingredients in the United States are summarized in Table 6.4. Excellent guidance on the proper preparation of foods to minimize the risks of foodborne illnesses and to ensure the safety of foods served in restaurants, grocery stores, and institu-

Table 6.3 Federal agencies responsible for the safety of the U.S. food supply

Department of Health and Human Services:

Food and Drug Administration, which is responsible for the regulation of food labeling, safety of food and food additives, inspection of food-processing plants, control of food contaminants, and establishment of food standards

Centers for Disease Control and Prevention, which analyzes and reports incidents of foodborne diseases

National Institutes of Health, which conducts research related to diet and health.

Department of Agriculture:

Food Safety and Inspection Service, which is responsible for inspection and labeling of meat, poultry, and egg products, as well as grading of foods

Animal and Plant Health Inspection Service, which inspects food and animal products imported into the United States

Human Nutrition Information Service, which establishes food consumption standard tables for the nutritive value of food and provides educational materials related to food

Other agencies:

Environmental Protection Agency, which develops standards for the use of pesticides on food crops

National Marine Fisheries Service, within the Department of Commerce, which conducts inspections and establishes standards relative to the quality of seafood

Bureau of Alcohol, Tobacco, and Firearms, which regulates alcoholic beverages, and the Customs Service, which inspects food products imported into the United States (both within the Department of the Treasury)

Federal Trade Commission, which regulates food advertising

Table 6.4 Federal regulation of food ingredients

Ingredient	Definition	Safety standard or limit
Unavoidable contaminants	Inherent food substances that cannot be avoided	Adulterated if substance "may render food injurious to health"
GRAS substances	Substances "generally recognized as safe" by the scientific community	Must be "generally recognized as safe"
Food additives	Substances added for specific intended effects, including GRAS substances, color additives, new animal drugs, and pesticides	"Reasonable certainty of no harm"
Substances previously sanctioned	Substances explicitly approved for use by FDA or USDA prior to 1958	Adulterated if substance "may render food injurious to health"
Pesticides	Substances intended for preventing, destroying, repelling, or mitigating any pest or intended for use as plant regulator	Tolerance based on whether substance is "safe for use," considering its benefits
New animal drugs	Substances intended for food-producing animals, excluding antibiotics	"Reasonable certainty of no harm"
Color additives	Dyes, pigments, or other substances capable of imparting color, excluding substances that also have other intended functional effects	"Reasonable certainty of no harm"
Prohibited substances	Substances prohibited from use because they present a potential risk to public health or because the data are inadequate to demonstrate their safety in food	Must not be present in detectable amounts

tions such as nursing homes is given in the *Food Code* published by the Food and Drug Administration (FDA, 1995). The Codex Alimentarius Commission provides guidance on food safety at the international level.

Within the United States, preventive and control measures for foodborne illnesses have been mandated since 1998 through a federal program called Hazard Analysis and Critical Control Points (HACCP). In the case of poultry products, for example, this program requires that processors evaluate and determine the types of contaminants that might gain access during the sequence of steps involved, identify the points that are most vulnerable to their entrance, and institute procedures for their control. Since chickens typically contain bacteria naturally present in their intestines, processors are required to spray carcasses inside and out with an approved disinfectant. To ensure proper implementation of these requirements, USDA inspectors monitor the plants and check production lines to reject any carcasses with evidence of contamination. They also randomly test for the presence of *Salmonella* (Consumers Union, 2003). In the case of frozen eggs, transmission can be prevented by pasteurization, and in the case of contaminated food, it can be prevented by thorough cooking.

Enforcement and monitoring. Within the United States, agencies at the state and local levels have primary responsibility for the inspection of restaurants, retail food establishments, dairies, grain mills, and other food establishments. Their goals are to assure the safe handling, proper labeling, and fair marketing of food products. Methods used to meet these responsibilities include inspection at the point of production or processing, examination of products at the retail or wholesale level of distribution, and licensing of establishments that manufacture or handle foods. Because it is impossible to inspect every food at every site of production, processing, and distribution, the incentives to comply with regulations depend primarily on the probability of detection and the penalties for noncompliance (which can include fines and legal proceedings). In addition, compelling economic and business factors encourage food handlers, processors, and distributors to want to comply with the regulations. No food processor wants to suffer the loss of customer confidence that can accompany a highly publicized foodborne disease outbreak.

Genetically Modified Food

One of the most controversial subjects in the food industry is genetically modified (GM) foods, that is, those that have been produced by plants that have been developed through genetic engineering. The difference in GM plants and/or their products is that they contain either genes from other plant, animal, or bacterial species or modified genes constructed in the laboratory. The goal is to confer on the modified plant one or more advantages. These include the ability to withstand drought, impoverished soils, disease, insects, and fungi, to tolerate a specific herbicide, to exhibit longer shelf life, and/or to provide increased nutritional value. Some estimates indicate that the benefits from reduced pesticide use, alone, may offset the risks (Kennedy, 2003). Such modifications can also be used to reduce the amount of allergenic substances in foods, such as peanuts. GM crops also hold promise of producing higher yields (Eubanks, 2002). Researchers are also identifying genes that can help plants tolerate arid conditions or grow in salty water. This will be of immense help in meeting the world's needs for food, as well as in conserving the ever-decreasing supply of fresh water (Chapter 7). Nonetheless, as with many other modern technological developments, both the risks and benefits of such possibilities need to be carefully assessed and evaluated.

RISKS

The primary risks associated with GM foods include the potential for spreading novel genes to wild plants near areas in which modified crops are being grown, promotion of the development of insects resistant to natural toxins, and the possible introduction of allergens into foods (Ackerman, 2002).

Gene escape. One of the more important considerations is that key genes in GM crops might escape and create genetic pollution and "superweeds." For example, studies in Australia have confirmed that pollen of GM canola plants not only can spread into neighboring fields but, borne by wind and insects, can also be transported as much as three kilometers (two miles) away. These studies followed a Canadian study that showed that several varieties of so-called clean canola seeds contained some genetically modified material. Supporting these observations, the European Union issued a report in 2002 indicating that GM canola is at "high risk" of cross-pollination with other canola crops. This report coincides with other reports of cross-pollination between genetically modified crops and similar varieties of other plants. Although various controls are put in place when the GM seeds are being tested, the USDA currently places no such requirements on the crops once they are approved for use.

Fortunately, there appears to be a solution to this particular problem. The escape of genes to related plants is possible only in cases where they are planted in the vicinity of wild relatives. At the moment, soybeans and corn, the most common GM crops in use in the United States, have no such relatives. For crops such as barley and wheat, where this is not the case, conventional breeding has rendered them "ecologically incompetent," that is, the hybrid forms in common use cannot easily reproduce under natural conditions and are unlikely to cross-pollinate with other species. Another possible approach is to engineer GM crops to be sterile or unable to germinate (Goklany, 2001).

Evolution of genetically resistant pests. Another potential problem is that the use of crops that contain built-in pesticides, such as *Bacillus thuringiensis* (Bt), will lead to the evolution of genetically resistant pests. Buoyed by the fact that it is possible to identify the recessive gene that confers high levels of resistance in agricultural pests, scientists believe that methods can be developed to control this problem. If they can detect the development of resistance during its early stages, hopefully this will provide the time necessary to modify the plants so that they will be defended against the new pest strains. This could, however, lead to a never-ending cycle in which increasingly resistant pests are managed through the development of increasingly complex plants.

Allergenicity. Ultimately, this may prove to be one of the major areas of concern with respect to the human impacts of GM foods because such foods often contain proteins that humans may never have previously ingested. Of particular interest is the possibility that such proteins will elicit potentially harmful immunologic responses, including allergic hypersensitivity (Metcalfe, 2003). Or the proteins may be ones that humans have encountered, for example, as a glycosylated protein in the original plant but as a nonglycosylated protein in the genetically modified plant. This is of particular concern to genetically predisposed people. In their case, exposure may cause an overreaction to an ordinarily innocuous substance. Such reactions range from minor skin rashes, headaches, vomiting, and diarrhea up to, in rare cases, anaphylaxis and death. In fact, there are 150–175 such deaths each year in the United States (Eubanks, 2002).

Resolving these concerns is complicated because the mechanisms involved are highly complex. For example, a person may have an allergic response to a food only when he/she has some type of infection. Another influencing factor is the stage of physical development. Babies and children, for example, are more prone to allergies. Since the only treatment for and/or way to prevent food allergies is to avoid consuming the offending products, it is almost mandatory that GM foods be labeled. Another challenge is to develop tests so that adequate data can be obtained on the potential allergenic effects of various GM foods. Nonetheless, since, as noted earlier, genetic modifications can be used to reduce the amount of allergenic substances in foods, perhaps the technologies that have the potential for creating such problems will provide the mechanisms necessary for solving them (Eubanks, 2002).

BENEFITS

One of the major benefits of GM plants may prove to be their role in increasing the production of food in the less developed countries. Unfortunately, farmers in many such countries are totally dependent on foreign companies for their supplies of genes and seeds. While it might have been anticipated that biotechnology researchers in the developed countries would have focused on GM products specifically designed to meet these needs, this has not been the case. Two of the primary reasons appear to be the increasing costs of creating and commercializing such products and the difficulties in coping with the regulations of countries that discourage trading such products on an international basis (Huang et al., 2002). In response, some leaders have advocated that the U.S. government assist less developed nations of the world by funding research, training their scientists, assisting them in developing appropriate regulations, and encouraging biotech companies to donate technologies and allow free access to patents that are used to produce genetically engineered seeds and animals.

A major exception to this situation is the People's Republic of China, which appears to be developing the largest plant biotechnology capacity outside of North America. Under this program, a host of GM plants are being developed that have been mostly ignored in the laboratories of the industrialized nations. Examples include rice, wheat, potatoes, and peanuts. Stimulated by these activities, small farmers in China are aggressively adopting GM crops whenever they are permitted to do so. One immediately reported benefit has been the increased production efficiency being achieved through the incorporation of *Bacillus thuringiensis* into cotton. As a result of this and other achievements, the amount of land being devoted to the growth of GM plants in China surpasses that anywhere else in the less developed world. Because China has many well-trained scientists, a low-cost research environment, and large collections of germ plasm, it is anticipated that its GM products may in time be exported to both the less developed and the industrialized countries (Huang et al., 2002). Another country that has been active is India. In 2002, its government approved the commercial planting of genetically modified cotton. This approval, which applied to three Bt cotton varieties, was the first granted to any such crop (Kaiser, Holden, and Bagla, 2002). Adding support to these actions is a government panel report issued in the United Kingdom that has given qualified approval to the safety of GM crops (Pickrell, 2003). Additional discussion on GM modified cotton is presented in Chapter 10.

COMMENTARY

In probing the potential risks and benefits of GM foods, it is important to recognize that the growth of agricultural crops utilizes almost 40 percent of the land area of the Earth and results in about two-thirds of the withdrawals of water, a major share of which is "lost" via evaporation and transpiration. Farming also results in releases of nutrients, pesticides, and silt into surface waters. Unless technologies are developed to increase agricultural productivity, these impacts will continue to increase. If, through applications of biotechnological developments, productivity can be increased by 1 to 2 percent annually, this will be sufficient to meet upcoming needs. There will be no need to devote additional land acreage to agricultural use and deal with the accompanying increase in discharges of nutrients and other substances into surface and groundwater. If one combines these attributes with the potential increases in the nutritional value of the world's food supplies, such developments could save millions of lives each year (Goklany, 2001).

Recognizing the need for advice on this important subject, the Codex Alimentarius Commission in late 2003 established guidelines for determining and managing the safety risks from GM foods. The new regulations are designed to standardize and promote risk management and awareness of GM foods, across the 169 Codex-member countries. In response, 35 countries have indicated that they will require safety assessments of foods containing GM components before they can be marketed. The regulations took effect in 2004 (Codex Alimentarius Commission, 2003).

Organic Farming

Stimulated by concerns about the consumption of GM crops and the use of antibiotics in producing agricultural crops and animals for consumption, a dramatic increase in organic farming has occurred throughout the developed countries of the world. One of the advantages cited by advocates is that this approach avoids the pollution caused by pesticides and synthetic fertilizers. How these goals can be achieved is illustrated by some of the steps being taken in the growing of apples. In this case, the eggs of insects that might harm the apples are killed by applying soybean oil to the bark of the trees, placing fake apples coated with glue on the trees to trap apple maggots, releasing sex pheromones to interfere with the mating of moths, and relying on insect predators to eat mites (Marcus, 2001). To avoid the use of synthetic fertilizers, organic farmers initially used so-called biosolids, or sewage sludge, as fertilizer (Chapter 8). In time, however, it was found that biosolids harbor bacteria, such as *E. coli* O157:H7 and *Salmonella,* and may contain pesticides and antibiotics. Since the use of such material posed a risk of contaminating agricultural crops, either through direct contact or through the water with which they were being irrigated, such use has largely been abandoned by such farmers (IFT, 2002).

Even though organic farming is a more benign method of growing crops, it has not been widely adopted except in Europe (Stokstad, 2002). One of the reasons is that the yields per acre are about 20 percent less than those from conventionally farmed plots. As a result, organically grown vegetables generally must command a premium price in order for the process to be profitable. Nonetheless, organically farmed plots clearly have certain advantages in terms of protecting the environment. Experience shows, for example, that they require from a third to a half less fertilizer and more than 95 percent fewer pesticides than the conventional approach. Concurrently, both the fertility and the floral and faunal diversity of the soil are enhanced. This enhancement has been shown to increase the presence of microbes that facilitate the availability of nutrients to the crops being grown (Mader et al., 2002).

In spite of these advantages, several events have damaged the image of certain segments of the industry. These are exemplified by the efforts of certain groups of people to obtain and consume milk that has not been pasteurized, even though this process would not appear to violate the concepts of the operation of an organic dairy farm. Since the sale of such milk is forbidden in many states, some "innovative" organic farm milk producers established "cow-leasing programs" as a method of circumventing the regulations. Under this approach, customers paid an initial fee to "lease" part of a cow. The cow, however, remained on the farm, the farmer milked her, and the milk was provided to the customers. As a result of participating in a cow-leasing program, 75 people in Wisconsin became ill in 2001. Analyses confirmed that the illnesses were due to *Campylobacter jejuni* (CDC, 2002a).

Recognizing the need for guidance, in 2001 the USDA adopted the first national standards for organic food. One of the requirements is that to qualify as being "organic," food cannot be grown using either sewage sludge or synthetic fertilizers. Also banned are the use of most synthetic pesticides on crops and the use of antibiotics in producing organic meat. In addition, organic agricultural crops must not have been genetically engineered or treated with ionizing radiation. For milk to be classified as organic, the cows must have access to pasture. Organic milk, however, can be pasteurized (Marcus, 2001).

The General Outlook

Multitudes of challenges face those who are responsible for ensuring the safety of food. One of the most important in the case of the United States is the fragmented nature of the regulatory system at the national level. Responsibilities for the nation's food-safety program are currently distributed among a dozen federal agencies that, in turn, must interpret some 35 different laws (Chapter 14). Consequently, there is no clear national coordination or oversight of the nation's food-safety program. While the FDA has jurisdiction over about 50,000 food-processing and storage facilities, its budget permits its staff to visit less than one-third of these each year. In contrast, the USDA has a congressional mandate to inspect the carcass of every animal slaughtered and to inspect every meat- and poultry-processing plant on a daily basis. To meet this responsibility, it has some 7,500 inspectors, 10 times the number in the FDA, to check 6,000 meat processors. As is obvious, the allocation of resources in this manner does not maximize the overall ability of the federal government to reduce the associated risks (Taylor and Hoffmann, 2001).

Another challenge is the increasingly global nature of the production and distribution of food. Currently about 60 percent of the seafood, 40 percent of the fruit, and 8 percent of the vegetables consumed in the

United States are imported. This country also exports a significant amount of food to people in other countries (Satcher, 1999). Through such exchanges, new disease agents will almost inevitably be introduced into other countries.

Such challenges, however, are not limited to those of national governments. As noted in Chapter 1, the personal environment within which individuals choose to live has a major impact on their health. This is especially true in terms of the microbiological safety of the food they eat. As a participant in the food-production-to-consumption chain, consumers have obligations no less important than those of food processors (Marshall and Dickson, 1998). If consumers are to fulfill these obligations, they must learn to handle foods properly. Success in instilling this need in the minds of members of the public will require continuing programs of public education.

7

DRINKING WATER

THE QUEST for pure water dates back to prehistoric times. Information on methods for treating water has been found in Sanskrit medical lore, and pictures of apparatus to clarify water have been discovered on Egyptian walls dating back to the fifteenth century B.C. Treatment methods, such as boiling, filtration through porous vessels, and even filtration through sand and gravel, have similarly been prescribed for thousands of years. In his writings on public hygiene, Hippocrates (approximately 460–354 B.C.) directed attention to the importance of water in the maintenance of health (Simon, 2001). The Romans demonstrated a similar awareness of the merits of pure water, as is shown by the extensive aqueduct systems they developed, as well as their use of settling reservoirs to purify water, their rulings that unwholesome water should be used only for irrigation, and the passage of laws prohibiting the malicious polluting of waters (Frontinus, A.D. 97).

The first positive evidence that public water supplies could be a source of infection for humans was based on epidemiological studies of cholera in the city of London by John Snow in 1854 (Chapter 3). His study is particularly impressive when one realizes that at the time he was working, the germ theory of disease had not yet been established. A similar study by Robert Koch in Germany in 1892 provided evidence of the importance of filtration as a mechanism for the removal from water of the bacteria that caused cholera. Subsequent experiments in the United States relative to the control of typhoid fever confirmed his observations and revealed the further benefit of the addition of chemicals to coagulate the water prior to filtration.

One of the most important technological developments in the treatment of water during the twentieth century was the introduction in 1908 of chlorination. This provided a cheap, reproducible method of ensuring the bacteriological quality of water. The dramatic impact of this development, combined with the filtration of water, in reducing deaths from typhoid fever in Philadelphia, Pennsylvania, is shown in Figure 7.1. Prior to that time, typhoid fever had been a major contributor to illness and death in the United States. In fact, during the U.S. Civil War (1861–1865), about twice as many soldiers died from illnesses as did from injuries incurred in battle.

The oceans, which are salty, cover about 70 percent of the Earth's surface and contain an estimated 96.5 percent of its water. Saline or brackish groundwater and saltwater lakes (including large inland seas) make up

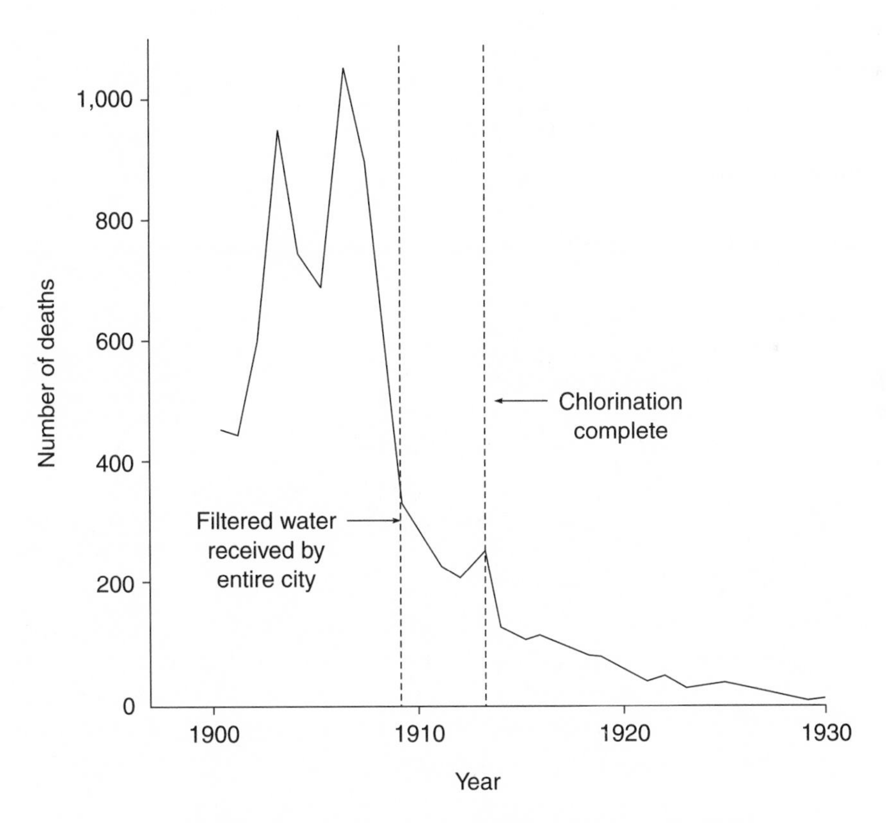

Figure 7.1 Deaths from typhoid fever, Philadelphia, PA, 1900–1930

another 1 percent. The remaining 2.5 percent of the water on Earth is fresh and therefore potentially available for drinking, irrigation, and industrial use. Two-thirds of this, however, is frozen in the polar ice sheets and glaciers. The Antarctic and Greenland ice sheets contain a major portion of this fraction. The remaining 0.8 percent is held in aquifers, soil pores, lakes, swamps, rivers, plant life, and the atmosphere (Montaigne, 2002). Much of this, however, is so deep beneath the Earth's surface that it is not readily accessible. As a result, only about 0.3 percent of the total water on Earth is available for human use. Even so, this small percentage represents a tremendous quantity. The Earth's freshwater lakes, for example, are estimated to contain nearly 125,000 cubic kilometers (30,000 cubic miles) of water, and its rivers and streams contain, on average, an additional 1,250 cubic kilometers (300 cubic miles) or more.

The basic source of all water on Earth is precipitation—rain, snow, and

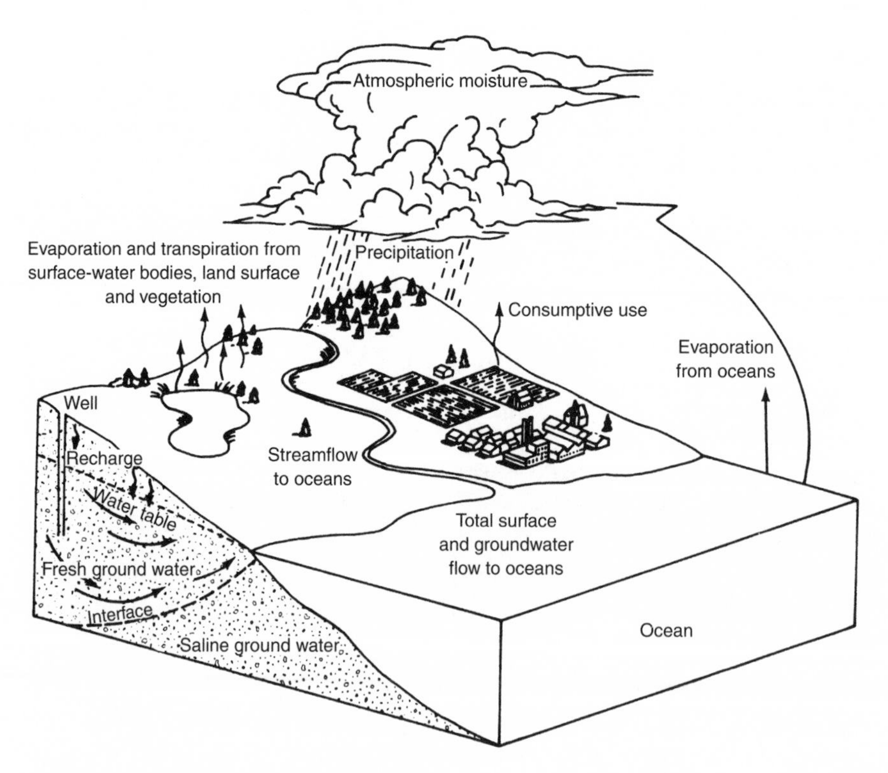

Figure 7.2 The hydrologic cycle

sleet. For the United States, the average annual amount is about 71–76 centimeters (28–30 inches). There are, however, significant geographic and seasonal variations in deposition in this country, as well as the rest of the world. For example, the area east of the Mississippi River typically receives more than twice as much precipitation as the area west of the Rocky Mountains (CEQ, 1998). Because, as noted above, so much of the world is covered by oceans, only about 30 percent of the precipitation falls on land. Of this, about 70 percent is evaporated or transpired (through vegetation) directly back into the atmosphere; 10 percent soaks in and becomes groundwater; and 20 percent runs off into lakes, streams, and rivers. Most of this ultimately flows into the oceans. The overall movement of water from precipitation through various pathways on Earth and back into the atmosphere is called the hydrologic cycle (Figure 7.2).

Sources of Drinking Water

The primary sources of drinking water are groundwater and surface water. In addition, precipitation (rain and snow) can be collected and used. Water within the upper water table can be accessed through dug wells. Such wells generally extend 1.5–6 meters (5–20 feet) beneath the ground surface. Groundwater located in deeper reservoirs or aquifers can be accessed through wells that are driven or drilled. These may penetrate to depths of 450–600 meters (1,500–2,000 feet) (Figure 7.3). Springs, which are outcrops where the underground aquifer intersects the surface of the earth, represent another source of groundwater (Symons, 1992). Sources of surface water include lakes, reservoirs, and rivers. Surface water may also come from protected watersheds. Each of these sources has its advantages and disadvantages.

GROUNDWATER

The widespread use of groundwater stems not only from its general availability but also from economic and public health considerations. Groundwater is commonly available at the point of need at relatively little cost, and reservoirs and long pipelines are not necessary. It is also normally free of suspended solids, bacteria, and other disease-causing organisms unless it contains contaminants introduced by human activities. Unfortunately, as of 1992 it was estimated that more than 10 percent of the community water-supply wells and almost 5 percent of the rural domestic wells in the United States contained detectable concentrations of one or

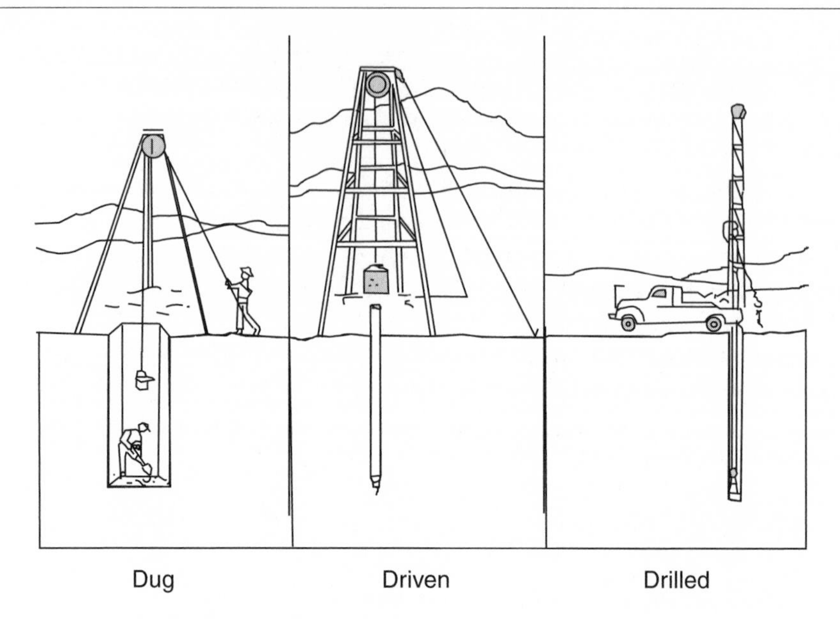

Figure 7.3 The three principal types of wells—dug, driven, and drilled

more contaminants, primarily agricultural pesticides. About 1 percent contain one or more contaminants in excess of health-based limits (Alley et al., 2002).

Accessible groundwater sources are limited in volume and, once depleted, are essentially irreplaceable. Yet farmers and municipalities throughout the world continue to pump water out of the ground faster than it is being replenished. Major portions of the Ogallala aquifer, which underlies the Great Plains section of the United States, have already been depleted. Aquifers in India have been depleted to such an extent that about half of the country now faces groundwater shortages. Similar conditions exist in the People's Republic of China. In fact, it is estimated that lack of water will reduce the production of grain in China and India by 10 to 20 percent within the next several decades (Montaigne, 2002). In many cases, excessive withdrawals are also causing the land to subside. In some areas of Texas, the land has subsided as much as 1 to 2 meters (3–6 feet); in Mexico City, some areas have sunk as much as 10 meters (32 feet). In Florida, where 90 percent of the population depends on groundwater as its source of drinking water, some land areas overlying aquifers have collapsed, and in certain coastal areas, withdrawals have so depleted the volume of freshwater underlying the ground that salty ocean water has moved in to take its place.

PROTECTED RUNOFF

Many homeowners have systems for collecting the rainfall from their roofs, storing it in a cistern, and using it as a source of drinking water. Such sources, however, are almost certain to have some degree of pollution. One step that can be taken to reduce contamination is to delay collecting the water until enough rain has fallen to cleanse the roof. Several types of diversion valves have been developed to accomplish this task. Some systems also incorporate units for filtering the water prior to use. Cisterns in which the water is collected should be watertight, and manholes or other ports of entry should be leakproof.

Rainfall and accompanying runoff can also be collected on a wider scale to provide drinking water to large municipalities. Cities that employ this approach include New York, Boston, and Lisbon, where foresighted planners set aside large land areas for collecting precipitation and runoff in natural and human-made lakes.

SURFACE SUPPLIES

Lakes, streams, and rivers are sources of drinking water for people in many areas. Water from such sources, however, usually requires extensive treatment before use. A further problem is that the adequacy of such supplies is in question in many parts of the world, especially in light of other demands for the water, such as irrigation, fisheries, and habitats for wildlife. Heated debates have ensued—for example, in the western United States—on how the limited surface-water supplies should be managed and allocated. In the final analysis, it will be necessary for all users to learn to accept limitations and to share responsibility for these resources.

Human Uses of Water

Water is absolutely essential to life. From 50 to 65 percent of the human body is composed of water, and variations of as little as 1–2 percent will cause thirst or pain. The loss of 5 percent of body water can cause hallucinations; a loss as large as 10–15 percent can be fatal. Although humans can live several months without food, under hot, dry conditions they can survive only a day or two without water.

In 1995, almost 400 billion gallons of water per day were withdrawn from aquifers and streams in the United States. This demand is equivalent to about 25 percent of the estimated renewable supply (CEQ, 1998). Of this amount, about 75 percent (300 billion gallons) is eventually discharged into rivers and streams, and about 100 billion gallons are consumed and incorporated into manufactured products, agricultural crops, and animal tissue and hence are no longer available for immediate use. Although direct human consumption accounts for only about 150 million gallons per day, water that meets drinking-water standards is routinely used for irrigating lawns, fighting fires, washing cars, cleaning streets, and recreational and aesthetic purposes. With the increasing shortages of water in many areas of the world, however, this approach is rapidly changing. Dual water systems have been constructed in many arid areas, whereby separate plumbing systems deliver high-quality water for human consumption and less pure or reclaimed water for uses such as irrigation and waste disposal.

On a global basis, about 10 percent of the water is applied to household use, and about 70 percent is used for irrigation. The remaining 20 percent is used by industry (Montaigne, 2002). Details on some of the more prominent categories of use are summarized here.

Personal use. Personal use includes drinking, cooking, bathing, laundering, and excreta disposal. On a daily basis, flushing the toilet consumes some 60–90 liters (15–25 gallons); bathing consumes another 60–80 liters (15–20 gallons). Total personal (domestic) water usage depends, of course, on whether a home contains a washing machine and dishwasher, whether it has a swimming pool, the extent to which water is employed to irrigate lawns, and other factors. Only about 2 liters (2 quarts) of the water in this category is actually consumed (for drinking and cooking). The distribution of uses of water in the home is shown in Figure 7.4.

Industrial use. The four largest industrial users are those involved in the manufacture of paper, refinement of petroleum products, and production of chemicals and of primary metals. Within the United States, these groups consume about 30 billion gallons of water per day. An additional 8 billion gallons are consumed by commercial users, including military bases, college campuses, office buildings, and restaurants (Parfit, 1993).

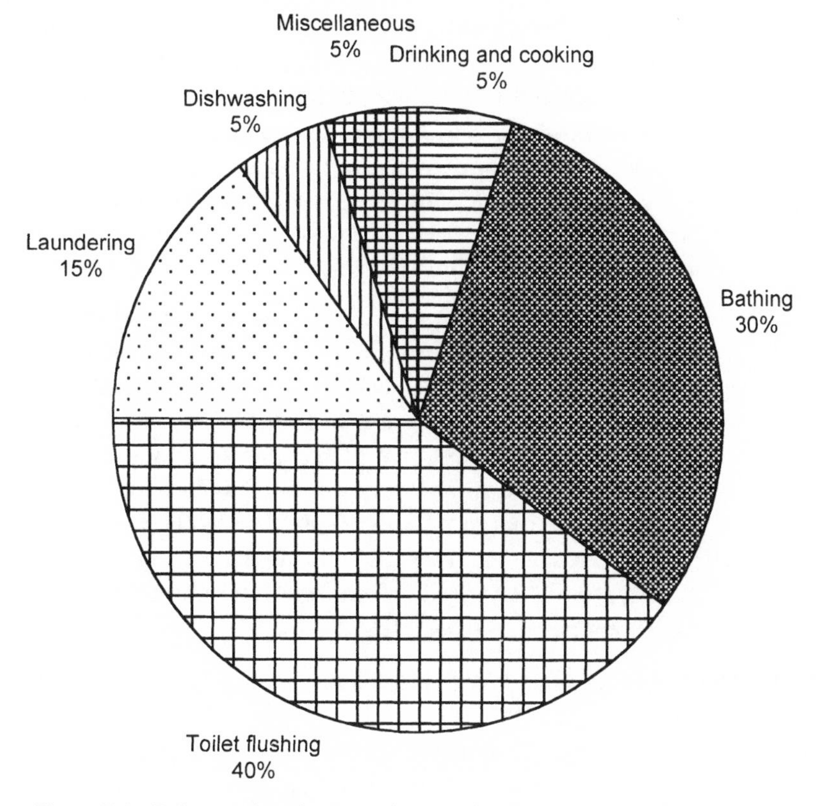

Figure 7.4 Relative distribution of uses of water within the home

Waste disposal. As noted earlier, a major share of the water used by people for personal needs winds up as liquid waste. The water-carriage method of excreta disposal, an outgrowth of the development of the flush toilet (Chapter 8), is particularly wasteful, using almost 250 gallons of purified water to transport a single pound of fecal material to a sewage-treatment plant for disposal.

Recreational and aesthetic use. Boating, sailing, water skiing, spray fountains, and the like fall in this category. Except for discharges of oil and gasoline from powerboats, few of these uses result in significant pollution.

Irrigation. The use of water for irrigation in the United States has increased by a factor of seven since 1900. A total of 55 million acres was being irrigated in 1997, and the amount of water lost through evaporation, transpiration, or incorporation into products or crops was estimated to total about 100 billion gallons per day (CEQ, 1998). About half of this comes from surface-water sources and about half from groundwater. Additional water is used to irrigate golf courses and parks, much of which is reclaimed. In fact, the amount of reclaimed wastewater used in the United States totals more than a billion gallons per day (CEQ, 1998).

Other. Other uses of water include providing supplies for farm animals and aquaculture, transportation (waterways and canals), and the generation of electricity. The last application includes the use of water as a coolant in fossil-fueled and nuclear-fueled electricity-generating stations. For the United States, this accounts for almost 200 billion gallons per day, a portion of which is lost through evaporation. An even larger amount is used to generate electricity in hydroelectric power plants. Although the water is not "consumed," the volumes involved are enormous, amounting to some 3 trillion gallons per day (CEQ, 1998).

Few people in industrialized nations are aware of the many ways in which water is used to support their accepted standard of living. Consider the following requirements: more than 50 glasses of water to grow the oranges to provide one glass of orange juice; 30 liters (8 gallons) to grow a single tomato; 450 liters (120 gallons) to produce one chicken egg; more than 13,000 liters (3,500 gallons) for a steak; and 225,000 liters (60,000 gallons) for one ton of steel, approximately the amount in an automobile (Canby, 1980).

Pathways or Avenues of Human Exposure

Experience has shown that water can have effects on human health through four principal avenues.

Waterborne diseases. These result from the *ingestion* of water that contains the causative organisms for enteric diseases such as typhoid, cholera, and infective hepatitis. Prevention depends on avoiding the contamination of raw water sources by human and animal wastes or removing or destroying the contaminants prior to consumption.

Water-contact diseases. These can be transmitted through direct contact with organisms in water. The most common example is schistosomiasis, which can be transmitted to people who swim or wade in water that contains snails infected with the organism. The larvae, which leave the snail and enter the water, can readily penetrate the skin. Prevention can be achieved through properly disposing human excreta and deterring people from contact with infested waters.

Water-insect-related diseases. Examples are malaria, yellow fever, and West Nile fever, encephalitis or rash, where water serves as a habitat for the disease transmitter, in this case the mosquito. Control requires eliminating mosquito-breeding areas, killing them, and/or preventing their contact with people.

Water-wash diseases. These result from lack of sufficient water for personal hygiene and washing. Shigellosis, trachoma, and conjunctivitis are among the diseases that may ensue.

For purposes of assessing waterborne diseases in the United States, the impacts are divided into two basic avenues of exposure: (1) ingestion and (2) recreational exposures. Ingestion and most recreational exposures would be covered by waterborne diseases, as described earlier. Some recreational exposures, for example, dermatitis, would fall under water-contact diseases.

Impacts of Waterborne Diseases

Although the true magnitude is not known, it is estimated that waterborne bacteria, viruses, and parasites produce about 4 billion cases of diarrhea each year worldwide. In fact, water may serve as the vehicle for the transmission of as much as 80 percent of all illnesses. Groups at highest risk include the approximately 1 billion people who lack access to safe drinking water and the almost 2.5 billion without adequate sanitation facilities. Worldwide, it is estimated that more than 2 million people, most of them infants and children under the age of five, die each year from waterborne diseases (Sawin, 2003). For the United States, the more common waterborne diseases that result from the ingestion of water are caused by bacteria, viruses, and parasites (Table 7.1). Some health officials estimate that these agents may cause up to 1 million illnesses in this country each year.

Table 7.1 Diseases transmitted through contaminated drinking water

Disease	Causative agent	Source
Bacterial infections		
Salmonellosis	*Salmonella* sp.	Animal and human feces
Typhoid fever	*Salmonella typhi*	
Paratyphoid fever	*Salmonella paratyphi-A*	
Shigellosis (bacillic dysentery)	*Shigella* sp.	Human feces
Cholera	*Vibrio cholerae*	Human feces
Leptospirosis	*Leptospira* sp.	Human feces
Gastroenteritis	*Escherichia coli*	Animal and human feces
Diarrhea	*Campylobacter jejuni*	Human feces
Viral infections		
Viral hepatitis	Hepatitis A	Human feces
Acute gastronenteritis	Norwalk-like virus	Human feces
Waterborne parasites		
Amebiasis (amebic dysentery)	*Entamoeba histolytica*	Human feces
Diarrhea	*Cyclospora cayetanensis*	Human feces
Gastronenteritis	*Cryptosporidium parvum*	Animal and human feces
Giardiasis	*Giardia lamblia*	Animal and human feces

BACTERIAL DISEASES

A major outbreak of bacteria-related waterborne disease that occurred among attendees at a county fair in New York in 1999 affected almost 16,000 people, including 10 children who were hospitalized. Although most of the vendors at the fair were supplied with chlorinated water, several food vendors were supplied with water from a shallow well that was not chlorinated. Subsequent investigations indicated that the water was highly contaminated and that the causative agents were *Escherichia coli* O157:H7 and *Campylobacter.* An earlier waterborne bacteria-related disease outbreak, this time due to recreational exposure, occurred in Illinois in 1998. The causative organism was a member of the species Leptospira interrogans and the outbreak involved 375 persons who became ill after swimming in a lake. Twenty-eight of these people were hospitalized, making this the largest outbreak of leptospirosis ever reported in this country (CDC, 2000a).

VIRAL DISEASES

As noted previously (Chapter 6), the Norwalk-like viruses (NLVs) can be readily transmitted through food. The same is true for drinking water, and the impacts are similar. Waters in which NLVs have proved to be a problem include municipal and groundwater supplies, streams, lakes, and swimming pools, as well as commercial ice. Since these viruses are an intestinal organism, the primary mode of transmission is the fecal-oral route (Chin, 2000). As was the case for food, the lack of a readily available analytical method for monitoring NLVs in water has hampered efforts to link NLV strains to specific sources of contaminated water (CDC, 2001b). With the development of molecular diagnostics, this problem as well as the obtaining of data on the frequency of occurrence of waterborne outbreaks caused by NLVs should soon be solved.

PARASITIC DISEASES

As is the case for food (Chapter 6), one of the most common parasitic organisms present in drinking water in the United States is *Giardia.* Persons at highest risk are children in day care, their close contacts, backpackers and campers (via ingestion of unfiltered, untreated water), people who travel to disease-endemic areas, and those who drink water from shallow wells. From the standpoint of recreational exposures, the seasonal peak for young children coincides with the summer recreational season. This may reflect their increased use of communal swimming venues, for example, lakes, rivers, swimming pools, and water parks (CDC, 2000b).

Another parasitic organism for which water serves as a primary mode of transmission is *Cryptosporidium parvum.* A major outbreak involving this organism occurred in Milwaukee, Wisconsin, in 1993. In this case, the organism passed undetected through two water-treatment plants and caused more than 400,000 people to become ill with diarrhea. An estimated 50 to 100 of them died. In more recent years, multiple outbreaks in the United States have been found to be associated with swimming and wading pools, water parks, fountains, hot tubs, and spas. Although any event involving fecal contamination of swimming-pool water increases the probability of the transmission of infectious agents, the probability of the transmission of *Cryptosporidium* is especially high. The reasons are several: its oocysts are extremely resistant to chlorine; they are not efficiently removed by conventional pool filters; and the ingestion of only a few mouthfuls of water from a pool in which only a single fecal accident has occurred can result in infection (CDC, 2001a). For these reasons, the Centers for Disease Control and Prevention has recommended that should a single release of solid fecal matter be observed within a swimming pool, everyone should be directed to leave immediately (CDC, 2001c).

Another parasite, transmitted primarily through the ingestion of contaminated drinking water, is *Dracunculus medinensis,* a filarial worm. This can cause what is called Guinea worm disease, the course of which is as follows. Approximately one year after a person is infected by the consumption of water contaminated by copepods (water fleas) that contain immature forms of the parasite, one or more meter-long adult female worms begin to emerge through the skin. The net impact is that victims of this disease are essentially crippled, unable to work, attend school, care for their children, or harvest crops. Although a variety of methods have been used to control this disease, the approach being used in Sudan is clever, inexpensive, and effective. In this case, every person has been provided with a pipe similar to but larger than a straw that contains a nylon cloth filter capable of removing Guinea worms. Users are instructed to drink by sucking water through the pipe. Through this approach, the number of cases in that country was reduced by 98 percent between 1986 and 2000. The disease has been eliminated in Cameroon, Chad, India, Kenya, Pakistan, Senegal, and Yemen (CDC, 2002b).

Drinking Water and Chemicals

One of the early scientists who studied the role of drinking water in health was H. A. Schroeder of Dartmouth College. One of the results of his studies was what appeared to be a clear correlation between heart disease and the "hardness," or mineral content, of water (Schroeder, 1974). His observations showed that people who drank "soft" water (containing few minerals) had a higher incidence of heart disease—apparently because soft water, being more corrosive, dissolves toxic substances (such as lead and cadmium) from plumbing systems. Although his work was pioneering, scientists in the U.S. Public Health Service had conducted related studies on the effects of fluoride in drinking water in the late 1920s. These studies were an outgrowth of the reports of mottled enamel on the teeth of people who drank water containing relatively high concentrations of this chemical. One of the important ancillary observations was that people who lived in such areas had far less tooth decay than those who lived where mottling was nonexistent. Subsequent studies confirmed that modest intakes of fluoride prevented dental caries, the optimum concentration being about 1 part per million, far less than that which would cause mottling.

Because of this information, one would anticipate that tooth decay would not be a major problem today. On the contrary, it continues to be a major problem throughout the world. In the United States, this disease continues to affect an estimated 50 percent of children aged 5–9 years, 67 percent of adolescents aged 12–17 years, and 94 percent of adults 18 or more years of age. One of the primary reasons for this situation is the opposition of some groups to the addition of fluoride to their drinking-water supplies. In fact, as of 2000 only about 162 million people, less than two-thirds (65.8 percent) of the U.S. population served by public water systems, were receiving fluoridated drinking water. Obviously, policy makers and public health officials at the federal, state, and local levels need to devise new promotion and funding approaches if the required support for this disease-prevention measure is to be achieved (CDC, 2002a).

Although drinking water has not generally been considered a major source of toxic chemical intake, the discovery in the 1990s of high concentrations of arsenic in the groundwater being consumed by an estimated 35 million or more people living in Bangladesh and West India dramatically changed this assumption. Interestingly, the arsenic is of natural origin, occurring through the dissolution of arsenic from the rocks and soils through which the groundwater flows. The urgency of correcting the problem is heightened by the fact that the concentrations of arsenic range up to several thousand parts per billion (ppb), far in excess of the EPA/WHO standard of 10 ppb. In essence, millions of people have been, and are continuing to be, poisoned. Tragically, this situation occurred following a recommendation by international agencies that populations in these countries switch from surface to groundwater sources. The anticipated goal was to provide them a more protected and higher-quality supply. Follow-up studies have shown that an additional million people in Vietnam and Thailand are facing a similar problem (Nordstrom, 2002). A further complication is that the consumption of food products, particularly rice, may represent a larger source of arsenic intake than drinking water.

Trends in Waterborne Disease Outbreaks

Through a collaborative effort, the CDC, the EPA, and the Council of State and Territorial Epidemiologists record and analyze data on the occurrences and causes of waterborne disease outbreaks in the United States. For purposes of analyses, the data are separated into two categories: those outbreaks associated with drinking water and those associated with recreational water exposure. For several reasons, however, there are limitations on the degree to which the reported data reflect the true situation. These include the following factors: the primary responsibility for detecting and investigating such outbreaks rests with state, territorial, and local public health departments; their reports are submitted on a voluntary basis; the unit of analysis is an outbreak, not the number of people affected; and no assessments have been made to quantify the percentage of actual outbreaks that are being reported (Lee et al., 2002). Another factor of interest is that outbreaks due to the ingestion of food are ascribed to that source even in those cases in which the water used in preparing the food is determined to be the source of the problem.

Even with these caveats, the reported data showed that 39 outbreaks, involving drinking water were reported for 1999 and 2000. These represented approximately a 70 percent increase over 1997 and 1998, and they affected 2,068 people, 2 of whom died. Twenty-eight (71.8 percent) of the outbreaks were linked to groundwater sources. During the same two years, there were an additional 59 reported outbreaks involving recreational water. These represented a doubling of the number reported for the previous two years, and they affected 2,093 people, 4 of whom died. Approximately 70 percent of the outbreaks were associated with *Cryptosporidium parvum,* and 25 percent involved dermatitis, 80 percent of which occurred through the use of hot tubs and pools. A major share of the outbreaks in both categories involved gastroenteritis. Reflecting the cautionary comments made earlier, CDC scientists concluded that the increases in the number of outbreaks during the latest two years probably reflected a combination of improved surveillance and reporting at the state and local levels as well as an actual increase in the number of outbreaks (Lee et al., 2002).

Drinking-Water Standards, Implications, and Analyses

The basic U.S. federal law pertaining to potable water is the 1974 Safe Drinking Water Act, which was expanded and strengthened by amendments passed in 1977, 1986, and 1996. Guided by this act, the EPA has developed a series of primary standards, designed to protect human health, and secondary standards, designed to assure that drinking water is aesthetically pleasing in terms of temperature, color, taste, and odor. The primary standards include maximum contaminant levels (MCLs) for selected inorganic contaminants, volatile organic chemicals (including pesticides and certain chlorinated hydrocarbons), and selected radioactive materials, as well as limits for the presence of coliform organisms. The secondary standards include limits for iron, which along with manganese can discolor clothes during laundering; sulfates and dissolved solids, which can have the same effect as a laxative; and minerals that can, for example, interfere with the taste of beverages. They also include limits for suspended solids (turbidity) both for aesthetic reasons and because the efficacy of disinfection is related to the clarity of the water.

To assure that compliance with the primary standards is achievable, the EPA has identified treatment processes that are capable of providing the degrees of removal required. Although no single treatment technique is effective for the removal of all inorganic chemicals, a combination of coagulation, sedimentation and filtration, or lime softening treatment (discussed later) has proved effective for removing many of them. One of the problems with contaminants, such as pesticides and related organic compounds, is that water-purification plant operators must anticipate which contaminant will be present and be ready to remove it. The use of multipurpose removal agents, such as activated carbon, is one approach for addressing these problems. Another very promising and rapidly developing approach is the use of membrane filtration technologies.

The promulgation of limits that require reductions in the concentrations of specific contaminants in drinking water can raise a multitude of contentious issues. One example was the 2001 reduction in the U.S. limit for arsenic in drinking water. The previous limit of 50 ppb was reduced to 10 ppb. Because of the magnitude of the reduction required, the EPA conducted a detailed review of the existing concentrations of arsenic in drinking-water supplies in this country. This revealed that although most large water districts already met the standard, approximately 4,000 water districts did not. Of these, an estimated 97 percent served fewer than 10,000 people. In essence, the primary impacts were on smaller communities that could not benefit from economies of scale in removing such a contaminant, and whose inhabitants are, in many cases, people with lower incomes (Oates, 2002). One possible approach for addressing this problem in smaller communities is to provide assistance from federal and/or state sources.

In the past, the measurement of biological contaminants on an individual basis in water was difficult and tedious. Since coliform organisms originate primarily in the intestinal tracts of warm-blooded animals, including humans, the accepted approach has been to test for these organisms and to use their presence, if confirmed, as an indication of fecal contamination. This situation is now changing. As a result of technological developments, test papers and/or strips, for example, are now available for diagnosing the presence of certain microorganisms on an individual basis. These include *Bacillus brevis* and *Escherichia coli.* A more sophisticated approach is the use of molecular probes, which can not only detect the presence of human feces, but also determine whether an organism, such as *Salmonella,* is present. In a related manner, test strips and/or sticks are available for the rapid determination of the presence of a wide range of individual chemical elements (for example, chromium and lead), as well as chemical compounds (for example, nitrates). To encourage the use of up-to-date methods, the American Public Health Association, the American Water Works Association, and the Water Environmental Federation cooperatively prepare and publish on a periodic basis a book of standard methods for the sampling and analysis of a wide range of physical, chemical, and bacteriological contaminants in drinking water (Clesceri, Greenberg, and Eton, 1998).

Traditional Water-Purification Processes

Preparing water for human consumption is a major industry. There are approximately 170,000 public water systems in the United States, of which about 55,000 are community drinking-water systems. The latter serve about 250 million people, about 90 percent of the U.S. population (CEQ, 1998). As these numbers indicate, most of the community systems are small. In fact, well over half provide water in towns with 500 people or fewer; only about 250 are in towns with populations of 100,000 or more (Symons, 1992). About 11,000 of the larger systems obtain their water from surface supplies and provide drinking water to more than 100 million people. The remainder use groundwater sources. An additional 14 million people (5 percent of the population) obtain their water from private wells; in fact, about 95 percent of rural households depend on groundwater as a source of supply. Including small systems operated by industry and "noncommunity" suppliers—motels, remote restaurants, and similar establishments—that serve the traveling public, the total U.S. output is approximately 200 billion liters (50 billion gallons), or 600–750 liters (160–200 gallons) per person per day.

The capital investment in municipal water-treatment facilities totals about $250 billion, and the annual cost of operating them is roughly $5 billion. Some $2 billion is spent annually in capital improvement of the facilities. Even so, the cost of drinking water remains low, still well under a dollar a ton, or about one cent for more than 150 eight-ounce glasses. As will be discussed later, this does not reflect the true cost of producing the water or the fact that many groundwater sources will be depleted within the next few decades. Another factor that contributes to the unrealistically low prices that are charged for water in the United States is the manner in which existing purification facilities have been permitted to deteriorate. Estimates are that meeting the costs of repairs, upgrades, and the construction of new facilities, coupled with the supporting infrastructure, will approach $140 billion (CEQ, 1998).

The primary purposes of a water-purification or treatment system are to collect water from a source of supply, purify it for drinking if necessary, and distribute it to consumers. About half of the groundwater supplies are distributed untreated. The section that follows focuses on the treatment of drinking water obtained from surface-water supplies and the two principal methods of purifying such supplies, namely, slow and rapid sand filtration.

SLOW SAND FILTRATION

In the relatively simple process of slow sand filtration, the raw water supply is passed slowly through a sand bed 60–90 centimeters (2–3 feet) deep. Soon after a bed becomes operative, a biological growth develops on top of and within the sand that removes and retains particles from the raw water. This process removes most bacteria and disease organisms, including the cysts of *Giardia lamblia.* Because excess turbidity in the raw water supply will rapidly plug the filter-bed, preliminary settling is recommended. A filter-bed area of 185 square meters (2,000 square feet) will provide approximately 100,000 gallons of treated water per day. With proper care, slow sand filter beds can be operated 30–200 days before the top layers of sand have to be scraped, cleaned, or replaced (Leland and Damewood, 1990).

RAPID SAND FILTRATION

Figure 7.5 shows the principal steps in the rapid sand filtration purification process. First, water is pumped or diverted from a river or stream into a raw water storage basin. Such storage provides a carryover or reserve in case the raw water supply becomes unfit for use for several days—for instance, through accidental release of a contaminant upstream of the supply. Storage also removes color and reduces the concentrations of turbidity and bacteria.

The initial step in the treatment process is to add chemicals to the water to create a coagulant. The chemical most commonly used in the United States is $Al_2(SO_4)_3 \cdot 14H_20$, commonly called alum. A less frequently used chemical is ferric chloride ($FeCl_3$). The basic reactions are almost identical:

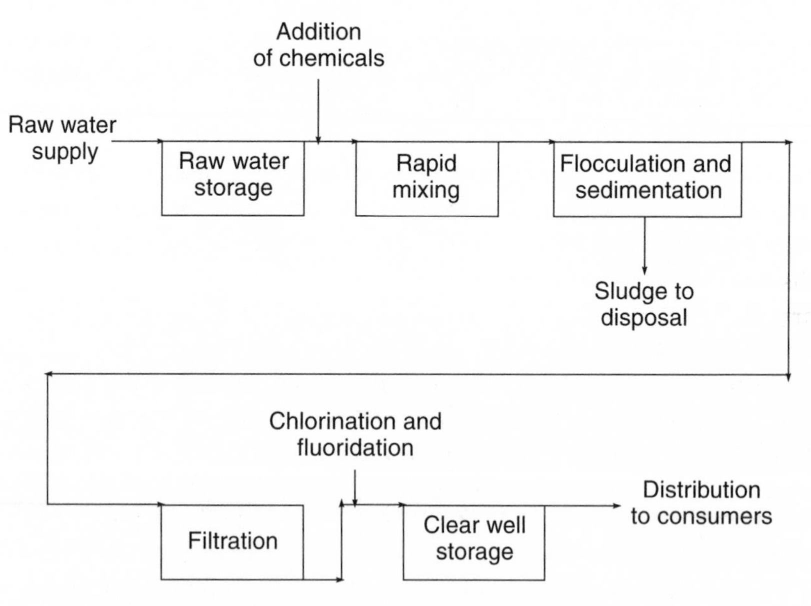

Figure 7.5 Principal steps in the water purification process

$$Al^{+++} + 3HCO_3^- \rightarrow Al(OH)_3 + 3CO_2$$

$$Fe^{+++} + 3HCO_3^- \rightarrow Fe(OH)_3 + 3CO_2$$

The highly positively charged Al^{+++} and Fe^{+++} ions also attract the negatively charged colloidal suspended matter in the water and together with the $Al(OH)_3$ or $Fe(OH)_3$ form a gelatinous mass called floc. Rapid mixing is essential to provide maximum interaction between the positively charged metallic ions and the negatively charged colloidal suspended matter.

Once the water has been rapidly mixed to assure proper coagulation, it is slowly and gently stirred to enable the finely divided floc to agglomerate into larger particles that will rapidly settle. This process, called flocculation, is accomplished by moving large paddles slowly and gently through the water. Since water treatment is performed as a continuous flow-through process, flocculation often takes place as the water enters one end of a large tank, with settling of the floc (sedimentation) occurring at the other end. During flocculation, relatively large particles in the water (including bacteria) are enmeshed in the floc, and ionic, colloidal, and suspended particles are adsorbed on its surface. This process, however, does not remove dissolved contaminants from the water.

Next the water undergoes a period of quiescence. The settled floc or sludge is removed from the bottom of the settling tank and sent to disposal. Originally, such settling was accomplished in a large rectangular tank and required a quiescent period of 2–4 hours. High-rate settling tanks have also been developed in which the water is passed through small-diameter tubes (or between parallel plates) set at an angle within a larger tank (Montgomery, 1985). Because the solids in the water travel a shorter distance before reaching a surface on which to deposit, and because this arrangement provides unique flow conditions, the required detention time for clarifying the water is only about 20 minutes. The space required for the settling tank is also significantly reduced. Because the settled water will still retain some traces of floc, it is next filtered. The filter beds are generally 0.6–0.9 meter (2–3 feet) deep and contain sand or crushed glass as the filter medium. Through a combination of adsorption, additional flocculation and sedimentation, and straining, the filter bed provides a final product of acceptable aesthetic quality.

One advantage of such beds is that they provide an effective method for removing particularly troublesome disease organisms, such as *Giardia* and *Cryptosporidium.* Because *Cryptosporidium parvum* is present in an estimated 65 percent or more of the surface waters in the United States, the EPA now requires that all large drinking-water systems be monitored for this organism and that all surface waters, including those obtained from protected watersheds, be filtered prior to distribution for human consumption (Lee et al., 2002). In time (12–72 hours), the filter bed becomes loaded with floc and must be cleaned by backwashing with purified drinking water. In many cases, the backwash/wastewater is sent to a sewer. The accompanying solids, however, present a formidable disposal problem. A typical water-treatment plant will produce about 250 cubic feet of sludge (three large truckloads) per million gallons of water processed. As will be noted later, new coagulant aids and coagulants have been developed to reduce the quantities of sludge produced.

FINAL STEPS IN PURIFICATION

Although sedimentation and filtration remove a significant portion of the microorganisms from water, these processes alone do not provide adequate protection. Some form of disinfection is required. In the past, the most common disinfecting agent in use in the United States has been chlorine. One of its advantages is that it can be added in sufficient quantity to maintain a small residual throughout the distribution system. Consumers are thereby protected in case bacterial contaminants later gain access to the supply. Unfortunately, the addition of chlorine to water that contains organic contaminants produces chlorinated hydrocarbons, which are carcinogenic. One method that can be used to avoid this situation is to pass the water through an activated carbon bed, as previously mentioned. Another is to avoid the production of chlorinated hydrocarbons by using another type of disinfectant, such as ozone or ultraviolet radiation (Table 7.2). As noted in the section that follows, this is one of the reasons that the use of ultraviolet radiation is receiving increasing attention in the United States. Once the clarified water is disinfected and fluoride is added, it is ready for distribution to the consumers.

Additional steps that can be applied include the removal of iron and manganese, calcium and magnesium, and tastes and odors. Iron and manganese (which, as previously noted, can discolor clothes) are soluble in water only in the reduced chemical state. If they are oxidized, they immediately become insoluble and precipitate. Thus they are readily removed by aeration. Although otherwise harmless to humans, calcium

Table 7.2 Characteristics of various disinfectants

Disinfectant	Characteristics
Chlorine	Widely used in the United States; forms harmful by-products if water contains organic matter; maintains residual in distribution system; requires care in handling as a gas
Hypochlorite	Safer alternative to chlorine gas; can be purchased or produced on-site by electrolysis of sodium chloride, but this process introduces chlorates and bromates as disinfection by-products and adds both sodium and chloride to treated water
Chlorine dioxide	Must be generated on-site since it cannot be transported because of its potential explosiveness; is a strong oxidant that will kill *Cryptosporidium* while chlorine will not; does not provide a persistent residual in treated water; produces its own range of by-products that may be cause for concern
Chloramines	Normally used in conjunction with another disinfectant; do not effectively inactivate viruses or protozoa; do not produce chlorinated by-products; provide a persistent residual in distribution system
Ozone	Must be generated on-site since it is highly reactive; produces no unwanted by-products and will inactivate viruses, bacteria, and protozoa, including *Cryptosporidium;* will also reduce tastes and odors and improve coagulation; does not provide a residual
Ultraviolet radiation	Will inactivate *Cryptosporidium;* effectiveness requires low turbidity; small size of units makes them suitable for installation in existing facilities; overall cost is about double that for chlorine; does not provide a residual

and magnesium give water the undesirable property of being "hard," that is, these chemicals make it difficult to develop a lather when a person uses soap during bathing or washing dishes or clothes. These are also the chemicals that leave a scum or ring in the bathtub. Although hardness is not generally a problem where supplies are derived from surface-water sources, it frequently is where supplies are derived from groundwater sources. The relative amounts of hardness in groundwater supplies in various portions of the conterminous United States are indicated in Figure 7.6.

In the case of a large water-purification facility, the normal procedure for removing calcium is to add calcium hydroxide (lime) and sodium carbonate to the water to interact with the dissolved calcium to form insol-

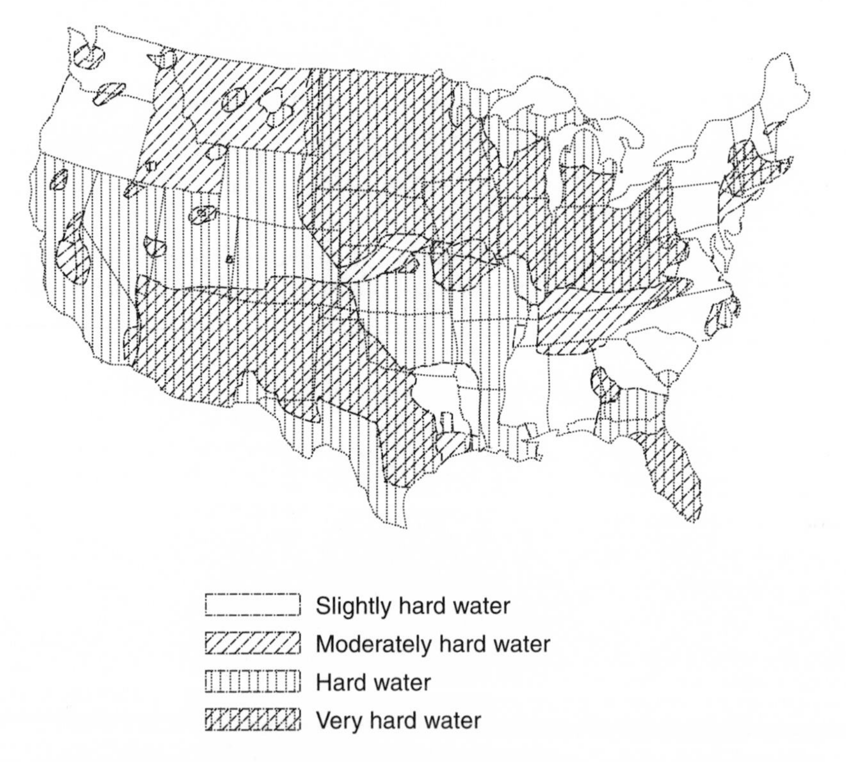

Figure 7.6 Variations of the hardness of groundwaters in different sections of the conterminous United States

uble calcium carbonate, which precipitates and reduces the concentration of calcium. The process also removes some of the magnesium. To remove hardness from small volumes of water, as in systems that serve an individual household, the process normally applied is ion exchange. Unfortunately, this process adds significant quantities of sodium to the treated water. Since sodium is believed to cause high blood pressure in some people, care must be exercised in consuming water treated by this process. One approach that can be used is to connect the softener only to the hot-water line, thus restricting use of the treated water primarily to taking baths and washing clothes and dishes. The cold-water supply, normally used for drinking and cooking, is not connected to the softener and therefore does not contain the added concentrations of sodium. The effects of various steps in the water-purification process on specific characteristics of the raw water supply are summarized in Table 7.3.

Advances and Changes in Water Purification

In the last decade, there have been several advances/changes in the technologies being applied in the purification of water. Three of these are discussed here.

POLYMER COAGULANTS

The success of coagulation depends on how well the floc settles, supplemented by how effectively the remaining particles are removed by the filters. Because suspended and colloidal solids in surface waters possess anionic (negative) charges, they are, in essence, prevented from coalescing into larger particles. One of the primary goals of coagulation is to neutralize these surface charges. Alum not only does this well, but also reacts with the alkalinity in the water to form metal hydroxide precipitates that encapsulate the colloidal particles. At the same time, however, it produces (as previously noted) large volumes of sludge. A major advance in solving these problems has been the development of polymer coagulants. Because these materials possess a cationic charge, they form a dense, rapidly settling floc and do not alter the pH or alkalinity of the water. They also are not as sensitive as alum to temperature and require only minor adjustments in dosage, even if the amounts of turbidity in the raw water vary over a relatively large range. Polymer coagulants, which can be used alone or in combination with alum, also readily enable treatment-plant operators

Table 7.3 Effects of purification processes on specific characteristics of water

Process	Characteristic						
	Bacterial content	Color	Turbidity	Taste and odor	Hardness (calcium and magnesium)	Corrosiveness	Iron and manganese
Raw water storage	+	+	+	±	+	0	+
Aeration	0	0	0	+	0	+	+
Coagulation and sedimentation	+	+	+	0	0	−	+
Lime-soda softening	+	0	+	0	+	+	+
Sand filtration	+	+	+	0	0	0	+
Chlorination or ozonation	+	+	0	±	0	0	+
Carbon adsorption	−	+	+	+	0	0	0

Note: 0 = no effect; + = beneficial effect (aids in alleviating the problem); − = negative effect (adds to the problem); ± = sometimes beneficial, sometimes negative effect.

to meet new, lower limits on the amount of turbidity permitted in the finished water. In some cases, the amount of sludge has been reduced by more than half, and the length of time the filters can be used prior to backwashing has been increased by a third or more (Laughlin, 2001).

ULTRAVIOLET RADIATION

The fact that ultraviolet (UV) radiation is an effective germicidal agent has been known for more than a century. It has for years been used in Europe for disinfecting drinking water. Because of the previously cited problems with chlorinated by-products, combined with tighter restrictions on permissible limits for these contaminants in drinking water, the use of UV radiation is receiving increasing attention in the United States. Its mechanism of action is that it penetrates the cell walls of microorganisms and affects their DNA in such a way that they cannot reproduce. Rather than being killed, the microorganisms are, in essence, inactivated (Tramposch and Fluharty, 2003). Another advantage is that UV radiation will inactivate both *Giardia lamblia* cysts and *Cryptosporidium parvum* oocysts and is capable of effectively treating certain bacteria found to be unaffected by chlorine. In recognition of these and other features, the EPA has designated the use of UV radiation as a best available technology (BAT) for the treatment of drinking water. It can be used alone or in combination with chlorine (Fleming, 2002).

MEMBRANE TECHNOLOGIES

In cases where chlorine is used as a disinfectant, one of the best methods for avoiding the production of chlorinated by-products, as noted earlier, is to remove organic compounds from the water prior to applying the disinfectant. The same is true for hydrogen sulfide and algae growths, both of which can produce bad tastes and odors when subjected to chlorination. Since membrane technologies are effective in removing these types of contaminants, as well as cysts and viruses, they are finding increasing applications in water purification systems in the United States. There are four basic technologies that can be applied. The names used to describe them are based on the sizes of the pores in the membranes and the manner in which they operate.

Microfiltration and Ultrafiltration Microfiltration and ultrafiltration perform their functions through the process of filtration—they physically remove contaminants from water. They are, however, far more effective than conventional sand filters. The sizes of the pores in microfiltration units, for example, range from 0.1 to 10 micrometers; those used in ultrafiltration technologies are even smaller, ranging from 0.001 to 0.1 micrometer. Therefore, they are capable of removing suspended particles as small as bacteria and viruses, providing removals of 99.9999 percent for *Giardia* and *Cryptosporidium* cysts and *Escherichia coli* bacteria. To ensure that the water moves through the membranes at a reasonable rate, a differential pressure of 10 to 50 pounds per square inch is maintained between the intake and discharge sides. As is the case with sand filters, the membranes must be periodically cleaned by backwashing (Johnson, 1999).

Nanofiltration and Reverse Osmosis Nanofiltration and reverse osmosis perform their functions through the mechanism of molecular diffusion. The pore sizes in the membranes used in nanofiltration units are small enough to remove larger molecules and divalent ions; those used in reverse osmosis units are even smaller, having the capability of removing essentially all dissolved ions from water. Because of the extremely small size of the pores (0.5 to 1.5 nanometers), it is necessary to apply much higher differential pressures, ranging from 100 to 1,500 pounds per square inch, to move water through the membranes. In this case, the membranes are kept clean by the use of special cross-flow configurations (Johnson, 1999).

Today, hundreds of water-purification facilities around the world use these technologies. Because such systems are space efficient, their costs are continually being reduced. Because they are so effective in removing biological and chemical contaminants, it is anticipated that their use will increase dramatically in the years ahead. In fact, water-purification facilities that employ membrane technologies with a capacity of up to 100 million liters (approximately 26 million gallons) per day are now in operation in the United States. Plants with capacities of more than 1,200 million liters (approximately 320 million gallons) per day are being developed (Johnson, 1999).

The General Outlook

One of the major challenges facing the world's supplies of water is the matter of sustainability. It is estimated, for example, that 2.3 billion people live in river basins in which surface-water supplies are under stress. This is largely due to the increasing use of water for irrigating agricultural crops. Assuming that consumption patterns continue, by the year 2025 an estimated 3.5 billion people or more, representing almost half of the world's population, will join this group (Johnson, Revenga, and Echeverria, 2001). Similar shortages exist among those who use groundwater as their source of supply. If these challenges are to be effectively addressed, procedures must be adopted to ensure the sustainability of drinking-water resources in much the same way as environmentalists have recognized the need to develop procedures that will provide sustainable fossil-fuel energy supplies, maintain our forests, and preserve biological diversity.

For these reasons, it is imperative that steps be taken to reduce the amounts of water being used. One such step would be to increase the price of water so that it more accurately reflects the associated capital and maintenance costs associated with the purification facilities and distribution systems. Even more important, it should reflect the costs required for the integrated management of the watershed from which surface waters are obtained or the aquifers from which groundwater supplies are being pumped. That people would be willing to pay more for drinking water, particularly if they could be assured that it is wholesome and safe to drink, is demonstrated by the enormous increase in the amount of bottled water now being consumed. In the United States, the amount being paid each year for such water totals about $10 billion.

Another step would be to address the enormous volumes of water being lost through irrigation. While such a practice is commendable from the standpoint of ensuring that food is relatively cheap, present practices are not realistic in terms of sustainability. As previously mentioned, more than half of the water distributed by the more common irrigation systems never reaches the crops. This is primarily the result of leakage in the distribution systems and excess evaporation due to the use of spray systems for delivering the water to the plants (Johnson, Revenga, and Echeverria, 2001). Steps that can be taken to rectify this situation include the installation of underground seepage or drip-irrigation systems, both of which slowly release the water directly onto the plants. Another possible measure would be to introduce salt-tolerant crops that could be irrigated with water from the ocean.

The need for conservation, however, is by no means restricted to irrigation. Each member of a household, particularly in the developed nations of the world, can make sizeable contributions. As an initial step, efforts should be made to ensure that home appliances include water-efficient plumbing components, such as low-flow showerheads, flow restrictors on faucets, and low-water-use toilets (Table 7.4). Other steps include re-

Table 7.4 Potential water savings from using water-efficient instead of conventional household systems

System	Water consumption		Savings
	Liters	Gallons	(%)
Toilets[a]			
Conventional	19	5	
Common low-flush	13	3.5	32
Washdown	4	1	79
Air-assisted	2	0.5	89
Clothes washers[a]			
Conventional	140	37	
Wash recycle	100	26	29
Front-loading	80	21	43
Showerheads[b]			
Conventional	19	5	
Common low-flow	11	3	42
Flow-limiting	7	2	63
Air-assisted	2	0.5	89
Faucets[b]			
Conventional	12	3	
Common low-flow	10	2.5	17
Flow-limiting	6	1.5	50

a. Consumption per use.
b. Consumption per minute.

stricting the size of areas devoted to lawns through the preservation of natural plots that contain native plants that are resistant to dry conditions (Noah, 2002).

A further challenge is that water-purification and distribution systems are potential targets for terrorists. Although a typical large water-treatment and distribution system has several key points of vulnerability, such as the open reservoirs that hold the raw water prior to treatment and the elevated tanks in which the treated water is stored prior to distribution, of special importance is the chlorine gas that is used as a disinfectant at many such plants. Its extremely toxic nature could make it a target of special interest. Switching to other disinfectants to avoid the production of chlorinated hydrocarbons (discussed earlier) would also have the benefit of eliminating the need to store chlorine on the plant site (Kim, 2003). Assessments of vulnerabilities of this type and establishing procedures for their protection are one of the requirements incorporated into the 2002 law that established the U.S. Department of Homeland Security (Consumers Union, 2003).

8

LIQUID WASTE

One of the most common types of liquid waste is human sewage. Basic guidance on its disposal can be found in verses 12 and 13 of the twenty-third chapter of Deuteronomy, where God provided the following instructions to Moses: "You shall have a place outside the camp and you shall go out to it; and you shall have a stick with your weapons and when you sit down outside, you shall dig a hole with it, and turn back and cover up your excrement."

An early and simple method for disposing of human excreta follows this guidance almost to the letter: the pit privy, a hole in the ground with a small closed shelter and toilet built above it. Generally, the hole is approximately 1 meter (3–4 feet) in diameter and about 2 meters (6–7 feet) deep. Privy designs range from the pit privy to those in which excreta are deposited on the surface of the ground and to those in which the excreta are collected in a bucket or tank for later removal and disposal elsewhere. Double-vault pit privies are used by many people in the less developed countries. Alternating the pits each year provides sufficient retention and decomposition to assure the destruction of most pathogenic organisms in the wastes. Improved versions developed later have a screen-covered vent pipe (Figure 8.1), which provides a natural pathway for removing odors and for trapping flies and other insects.

With the development of the water closet or flush toilet, sewage treatment and disposal entered a new era. As far back as the Mesopotamian Empire (3500–2500 B.C.), however, the toilets in some homes were connected to a stormwater drainage system that carried wastes away. In larger homes in Babylon, toilets flushed by hand were connected to vertical

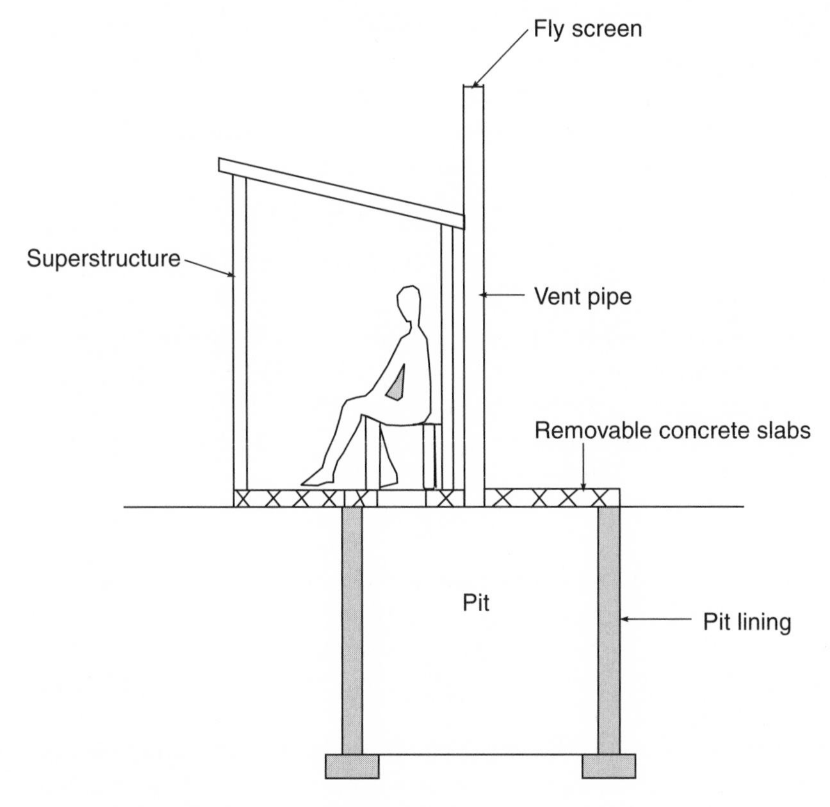

Figure 8.1 Pit privy with ventilation pipe

shafts in the ground that were lined with perforated clay pipe and permitted the liquid to be absorbed in the surrounding soil (Wolfe, 1999). Although as early as 1700 B.C. the royal palace of King Minos in Crete had toilets for which water collected in rain-fed cisterns provided a continuous flow of cleansing water, it was not until 1596 that the modern flush toilet was invented. Even so, the valve that controlled the inflow of water allowed considerable leakage, and this problem was not solved until 1872 when Thomas Crapper, later knighted by the queen of England, invented the first valveless water-waste preventer. The principles of his design continue in use today. Within the next several decades, most wealthy people had at least one indoor water closet that discharged either onto the ground or into a cesspool, an underground pit. In 1855 George Vanderbilt had the first bathroom (consisting of a lavatory, porcelain tub, and flush toilet) built inside an American house. As late as the 1880s, however, only one of every six people in U.S. cities had access to modern bathroom facilities.

Individual Household Disposal Systems

Subsequent widespread use of the flush toilet necessitated methods for disposing of the discharged wastes. Most municipalities constructed systems to transport the effluent to a sewer and then to some form of municipal treatment plant. But even today some 30–35 percent of the U.S. population—85–100 million people—are not served by sewers. They depend instead on some form of on-site subsurface sewage-disposal system. The most common of these is the septic tank.

SEPTIC TANK

A septic tank is usually constructed of concrete or plastic, with an inlet for sewage to enter and an outlet for it to leave (Figure 8.2). As sewage passes through the tank, solids settle to the bottom and are digested through the action of anaerobic bacteria that naturally develop. Although some groups advocate the addition of special types of organisms to enhance digestion, most experts agree that such augmentation is not necessary. Septic tanks in current use have a divider in the bottom and a baffle at the top near the outlet to help prevent carryover of settled solids and

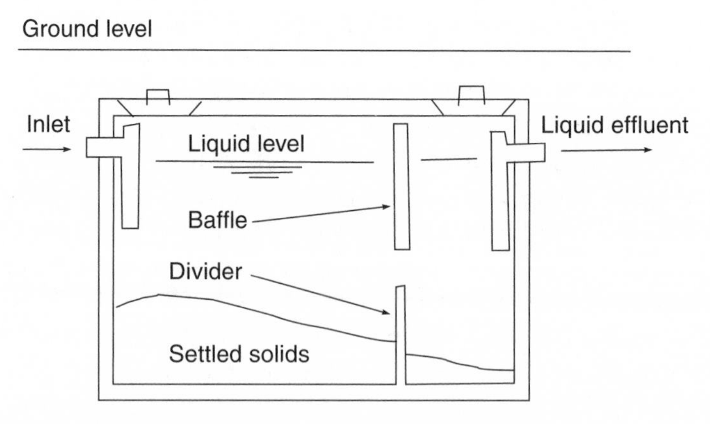

Figure 8.2 Cross section of a typical septic tank

floating material. Under proper operating conditions, the effluent is clear and is discharged into a drain field consisting of open-jointed or perforated pipe buried in the ground so that the liquid can seep into the soil. The purposes of the drain field are several. It acts to disperse the septic-tank effluent over a wide area and thus promotes infiltration of the waste into the soil. Furthermore, natural bacterial populations in the soil continue the digestion of soluble organic materials in the septic-tank effluent. The soil also acts as a filter mechanism to adsorb pathogenic organisms remaining in the waste.

For proper performance, it is generally recommended that (1) the tank hold a volume of at least 2,000 liters (500 gallons), (2) the soil in which the drain field is located be sufficiently porous to absorb the effluent, (3) the land area be adequate for absorption of the volume of flow anticipated, and (4) the tank be cleaned (solids removed) every three to five years. The last recommendation is extremely important because if solids are permitted to build up too long in the tank, they will be carried out with the effluent and will seal the drain field.

Well over half of the land in the United States that is acceptable for the construction of buildings is unsuitable for the installation of septic-tank systems. Nonetheless, almost 40 percent of new housing developments in the United States are being equipped with such systems. It is little wonder, therefore, that about one-quarter of the existing tanks malfunction either periodically or continually. The most common result is that the effluent is not absorbed and breaks through to the ground surface or finds its way into a groundwater source. In either case, the surface soil or a potential drinking-water supply is subject to bacterial and viral contamination. In seeking to solve these problems, many changes have been made in septic-tank systems during the last several decades. These include modifications of the systems themselves, as well as in the systems that feed into them. Among the former is the use of filters to avoid premature plugging of the disposal field with solids (Dix, 2001). Another is the incorporation of a unit for equalizing the flow of wastewater being treated. This improves the settling of the solids prior to release of the effluent. Another change is the use of low-flush toilets, low-water-use dish- and clothes washers, and low-flow showers to reduce the volumes of liquids being discharged into the tanks (Chapter 7).

OTHER TREATMENT SYSTEMS

A variety of alternative treatment systems have also been developed (Hetrick, 2001). These include units in which the sewage is collected in a tank, mixed by a pump to break up the solids, and aerated. Under proper operating conditions, these units are less prone than a septic tank to produce disagreeable odors, and because their effluent contains dissolved oxygen, the probability that the drain field will be plugged by solids is reduced. More sophisticated aerobic systems include features through which the effluent can be recycled and used for flushing the toilet again. Also in use are biological, composting, incinerating, and oil-flushed toilets. One of the most popular of the composting toilets incorporates the Clivus Multrum household excreta and garbage disposal system developed in Sweden. Application of these types of systems in the United States, however, has been extremely limited.

Whether the treatment and disposal system involves a septic tank or an aerobic unit, its design, construction, and location of the drain field are not the only factors that can affect its performance. Particularly troublesome is the discharge of household effluents that contain antibacterial soaps, relatively high concentrations of bleach, pesticides, and strong disinfectants. These can kill the bacteria that stabilize the waste. Also troublesome are grease, fats, oils, and food wastes (from garbage disposal units, for example). These can overload the system (Guy and Catanzaro, 2002).

Advent of Sewer Systems

Between the 1830s and the 1850s, a series of epidemics of cholera and typhoid occurred in London, Paris, Hamburg, and other European cities. These included those during which John Snow conducted his classic epidemiological studies (Chapter 3). Similar events that caused the deaths of thousands of people occurred in the United States between 1832 and 1873. Recognizing that the installation of septic tanks on an individual household basis was not feasible in metropolitan areas, the city of Hamburg, Germany, constructed the first comprehensive sewer system in 1843. This followed by almost 500 years a much simpler system that had been built in Paris, France. Systems similar to the one in Hamburg were subsequently constructed in other cities in Europe and in New York and Chicago. Even so, in all cases these systems served only as a vehicle for transporting the wastes for discharge into a nearby river or lake. Although some of the sewage in the smaller cities in Europe was used to irrigate nearby farmlands, this disposal method proved impractical and unsanitary for all but the smallest cities (Wolfe, 1999).

As would be anticipated, the water bodies into which these wastes were discharged soon became heavily polluted. Recognizing the need to treat such wastes, scientists in England, Europe, and the United States began developing mechanisms for using natural biological stabilization processes for treating such wastes. The Lawrence Experiment Station, which was established in Massachusetts by the State Board of Health in 1887, played a significant role in such activities. By 1890, the staff of this facility had published a report documenting the technical basis for the treatment of municipal wastewater. These and related activities led to the construction and operation of large-scale treatment facilities in the larger cities of the world (Wolfe, 1999). Details of the operation of such facilities are discussed later in this chapter.

Liquid Wastes: A Broader Perspective

In today's society, the sources of liquid wastes extend far beyond those generated in individual households. Within a modern city, such sources include commercial and office buildings, schools, restaurants, and hotels, as well as a wide range of industrial operations. The nature of industrial wastes is often significantly different from that of municipal sewage. Such wastes frequently contain toxic chemicals and other hazardous substances, as well as heated water and various types of suspended materials. If discharged into rivers and lakes without treatment, these wastes, as well as municipal sewage, can be major sources of pollution. If discharged onto the land, they can contaminate the soil and groundwater.

A somewhat oversimplified approach classifies such wastes as *degradable* and *nondegradable.* Domestic sewage is the most common degradable waste, that is, it can be degraded or stabilized by bacteria. Many industrial wastes contain organic residuals that are also degradable. In fact, the quantities of degradable wastes released by industry vastly exceed those in domestic sewage. Primary sources include industrial facilities involved in food processing, meat packing, pulp and paper manufacture, petroleum refining, and chemical production. Notable examples of nondegradable constituents are those that contain inorganic substances, such as ordinary salt and the salts of heavy metals (for example, lead, mercury, and cadmium). A third group of pollutants that do not fall into either of these categories consists of the so-called persistent chemicals, which are best exemplified by the synthetic organic chemicals, such as DDT and the phenols that result from the distillation of petroleum and coal products. Although they can be altered by biological and chemical transformation, these processes are extremely slow.

Many of the sources of liquid wastes, such as industrial and municipal discharges, are readily identifiable and are defined as *point* sources. Severe water-pollution problems, however, are also caused by less obvious and more widespread sources of pollution, the so-called *nonpoint* sources. In fact, these sources, many of which have only recently been brought under regulatory control, may contribute more to water-quality degradation than point sources. Data indicate that one nonpoint source, liquid runoff from agricultural lands, accounts for 39 percent of the pollution being discharged into U.S. rivers (Figure 8.3). In localized cases, it can account for up to 80 percent of the degradation of such waters. The quantities of this type of waste are especially large during spring thaws.

Livestock farms, for example, the so-called hog farms or factories, can be significant nonpoint sources of liquid wastes. Worldwide, upwards of 45 billion animals are maintained on these and other types of animal farms. By 2020, the total is projected to increase to as many as 100 billion. Some 10 billion animals, for example, are now being slaughtered in the

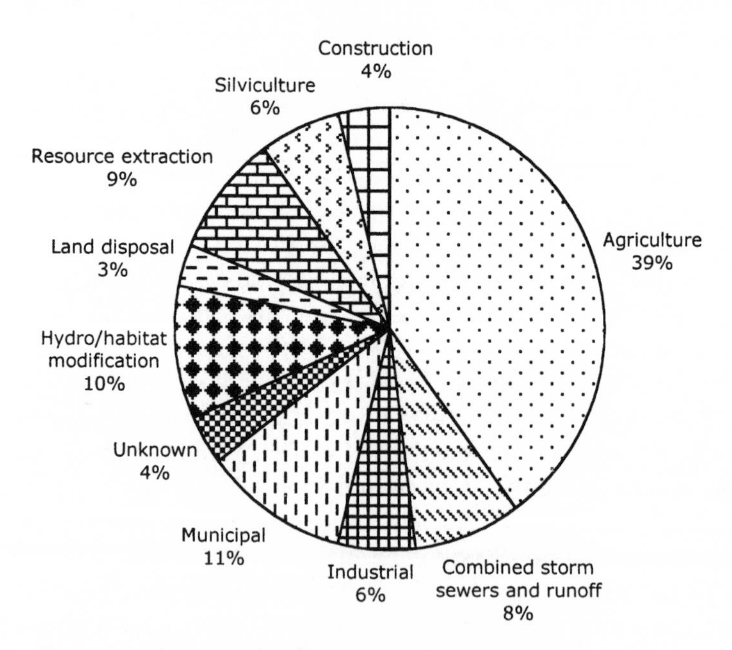

Figure 8.3 Sources of pollution in U.S. rivers

United States each year. A typical hog factory will have 12,000 or more animals that are fed and watered on a mass-production basis. At present, the most common method of handling the resulting wastes is to flush them into giant lagoons, some of which may have capacities in excess of 25 million gallons. Unfortunately, surface overflow and underground seepage from such lagoons have in many cases contaminated nearby surface-water and groundwater supplies. Odors and swarming flies often create noxious conditions for nearby residents. Excess spraying of wastes onto fields has led to runoff that pollutes surface waters. Furthermore, pigs produce nearly twice as much waste as beef cattle and about three and a half times as much waste as chickens (Satchell, 1996). Although efforts have been made to develop improved treatment methods, the lagoon method of disposal appears to be the best process available. In fact, regulations promulgated by the EPA in 2002 designated this approach as the best available technology for handling wastes from such operations.

Another primary contributor to nonpoint source pollution is liquid runoff from urban areas. The cause of this problem is that such areas are dominated by buildings ranging from high-rise offices to suburban single-family dwellings, plus multiple paved areas, such as sidewalks and streets. As a result, a major portion of the rain that falls on such areas collects as runoff and flows into sewers. Prior to reaching the sewers, the water accumulates a host of organic and inorganic contaminants, including animal waste, infectious agents, pesticides, and fertilizers. Urban runoff is a special problem in cities in which the sewers were originally designed to handle only domestic wastes. When a decision is later made to direct runoff into the same sewers, they often overflow, and the runoff, combined with untreated domestic sewage, is released into the environment. Even if a separate storm-drain sewer system is installed, unless the collected runoff is properly treated, its release into the environment, for example, a nearby lake or stream, can create problems.

Impacts of Liquid Wastes

As was done years ago with raw or *untreated* wastes, the most common method for disposing of *treated* liquid wastes is to discharge them into a lake or river. If the wastes contain toxic chemicals and/or pathogenic organisms, and the receiving waters later serve as raw sources for drinking-water supplies, the accompanying contaminants may have direct effects on the health of those who consume the water. If the wastes are applied to land areas for irrigation or other purposes, other avenues of contact with humans may develop. In cases in which the wastes are released into a stream, bacteria within the water will attempt to stabilize the organic matter in the wastes. This process requires oxygen, which must be obtained from that which is dissolved in the water. Fortunately, there are several natural processes that continually replenish the oxygen. These include eddies and other turbulence that serve as aerators, and the production of oxygen by green algae and various plants growing in the water.

So long as these sources replenish the oxygen as rapidly as it is removed, aerobic conditions will be maintained and problems can be avoided. It is when the consumption of oxygen exceeds the supply that problems develop. At this point, the concentration of dissolved oxygen (DO) may become too low to support fish and other forms of aquatic life. Under these conditions, the more desirable varieties of fish will be the first to be affected. If this occurs, one of the initial impacts will be that these varieties will be replaced by pollution-resistant lower orders, such as carp. If all the DO is consumed, anaerobic conditions will result. Instead of releasing carbon dioxide (which occurs under aerobic conditions), anaerobic decomposition produces methane or hydrogen sulfide. The stream or lake will, in turn, become dark and malodorous. After reaching a minimum concentration of DO, the stream will in most cases ultimately recover. This is illustrated in Figure 8.4, which shows what is called the "oxygen sag curve," a schematic plot of the DO concentration in a stream as a function of time or of distance downstream from the point of sewage discharge. In this case, the quantity of pollution was small enough to permit the stream to recover without reaching anaerobic conditions. Additional information is provided in Table 8.1, which summarizes the quality of various waters in the United States and the broader categories and sources of the pollutants they contain. These waters include rivers, streams, and lakes, as well as estuaries and waters near the shorelines of the Great Lakes and the Atlantic and Pacific Oceans.

The harmful effects of liquid wastes on aquatic life in rivers, lakes, and streams are not restricted to oxygen-demanding pollutants. Not only can discharges of suspended solids, toxic chemicals, heavy metals, and other hazardous substances be harmful to aquatic life, but also the receiving waters (even in cases where the wastes have been treated) and fish and shellfish harvested from such waters can be unsafe for human consumption. Analyses conducted in 1999 and 2000 by the U.S. Geological Survey, for example, showed that the range of pollutants in surface waters in the United States is widespread. Most of the samples, which were obtained from 139 sites in 30 states, were collected immediately downstream from

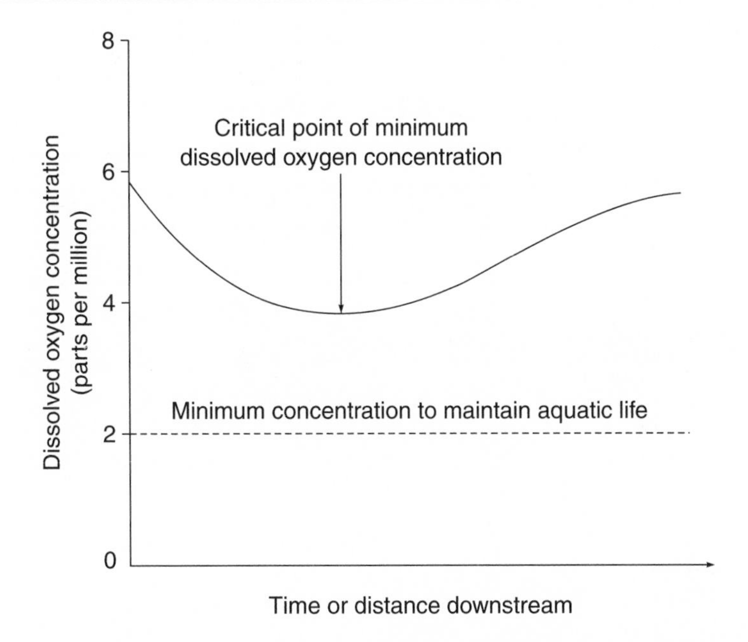

Figure 8.4 An oxygen sag curve, showing dissolved oxygen concentrations as a function of time, distance, or both in a stream into which sewage has been discharged

suspected pollution sources, such as wastewater-treatment plants, urban areas, or agricultural operations. The analyses revealed the presence of a range of antibiotics, other prescription drugs, pesticides, and household chemicals, such as detergents and fragrances. The most commonly observed chemicals were steroids, caffeine, and components of insect repellents, disinfectants, and fire retardants. Although median concentrations were usually relatively low, maximum concentrations occasionally exceeded regulatory limits (Weinhold, 2002).

Another emerging problem is the discharge of excess nutrients into lakes and coastal waters. Notable examples are detergents, fertilizers, and human and animal wastes. One of the most important impacts of such discharges into lakes is a process called eutrophication, through which a lake becomes biologically more productive. This can lead to flourishing blooms of toxic blue-green algae. If a person takes a bath or shower in water that contains these blooms, skin or oral mucosal contact can result in an allergic reaction that resembles hay fever and asthma. Skin, eye, and

Table 8.1 Quality of U.S. waters, 1998

Type	Percent assessed	Rating of those assessed[a]			Primary pollutants	Primary sources
		Good	Good, but threatened	Polluted and/or impaired		
Rivers, streams	23	55	10	35	Silt, nutrients, pathogens	Urban and agricultural runoff, hydroelectric dams, and runoff from wetlands
Lakes, ponds, reservoirs	42	46	9	45	Nutrients, metals, silt	Urban and agricultural runoff, hydroelectric dams, and runoff from wetlands
Estuaries	32	47	10	44	Pathogens, oxygen-depleting substances, metals	Municipal point and nonpoint sources, atmospheric deposition
Shorelines of Great Lakes	90	2	2	96	Organics, pesticides	Atmospheric deposition, releases from shut-down factories, contaminated sediments
Shorelines of oceans	5	80	8	12	Pathogens, turbidity, nutrients	Municipal point and nonpoint sources, land disposal

a. Due to rounding, the sums of the three categories may not equal 100.

ear irritations may also occur. In addition, the ingestion of water that contains the blooms, either through drinking or swimming, may produce gastro- or hepatoenteritis disorders (Pitois, Jackson, and Wood, 2001). Excess nutrients may also be one of the causes of the increased frequency of blooms of the so-called red tide that occurs in coastal waters of many of the world's oceans and can cause paralytic shellfish poisoning (Chapter 6). Another example of the impacts of excess nutrients is the so-called dead zone that has formed in the Gulf of Mexico off the coasts of Louisiana and Texas (Holden, 2002). Concerned about these developments, the National Research Council has recommended that nutrients be considered a high-priority pollutant (Table 8.2).

To combat the impacts of liquid wastes on various ecosystems, major studies are now under way to catalog the genes and proteins of bacteria, such as *Pseudomonas,* a versatile infectious microbe that lives in water, soil, plants, and animals, and *Caulobacter,* a bacterial group common in freshwater streams. Both have the ability to degrade chemical wastes and other pollutants, including aromatic compounds such as benzene and naphthalene. Similar efforts are under way with respect to *Prochlorococcus,* which lives in the ocean and removes carbon dioxide, a major source of global warming (Chapter 20). Armed with such information, scientists hope to better understand their mechanisms of action and to enhance their abilities to stabilize certain of the more important pollutants. Related studies are under way on anaerobic bacteria that live in groundwater sediments. One such organism that reductively dechlorinates trichloroethane, a problem-

Table 8.2 Anticipated national-level priorities for constituents of concern in liquid wastes

Priority	Pollutant group	Example
High	Nutrients	Nitrogen
	Pathogens	Enteric viruses
	Toxic organic chemicals	Polynuclear aromatic hydrocarbons
Intermediate	Selected trace metals	Lead
	Other hazardous materials	Oil, chlorine
	Plastics and floatables	Beach trash, oil, grease
Low	Organic matter	Municipal sewage
	Solids	Urban runoff

atic groundwater contaminant, has recently been discovered. These organisms are readily able to remove trichloroethane from contaminated groundwater sediments (Sun et al., 2002).

Case Study: Polychlorinated Biphenyls

The far-ranging consequences that discharges of wastes can generate have been demonstrated by many past events. A notable example was the discovery in the early 2000s of relatively high concentrations of polychlorinated biphenyls (PCBs) in sediments in the Hudson River upstream from New York City. These were due to discharges of this contaminant during a 30-year period, beginning in 1947, by an electrical equipment manufacturer. Although the releases at the time were within regulatory limits, they were terminated in 1977 when the EPA classified PCBs as "probable human carcinogens" and banned their production. PCBs were subsequently also listed by the EPA as suspected endocrine-disrupting chemicals and ones that may be associated with neurobehavioral alterations in newborn children (ATSDR, 2000).

As is often the case, quantifying the risk associated with the resulting contamination has been difficult. Studies in 1976 showed that rats that were fed large amounts of PCBs developed liver cancer. In contrast, subsequent studies of more than 7,000 people who worked at the manufacturing plant revealed no excess cancers, even though some of them had relatively high levels of PCBs in their blood. At the same time, the permissible concentration for PCBs in fish sold for human consumption, as established by the U.S. Food and Drug Administration, was 2 parts per million (ppm). Tests of fish caught in the upper Hudson River showed concentrations ranging from 2 to 41 ppm, a hundredfold higher in some cases than those in fish caught in cleaner rivers. Studies by environmental scientists from New York University also revealed that 90 percent of a type of codfish caught in the river had developed liver tumors by the time they reached adulthood. As a result of these and related concerns, a ban was placed on the commercial harvesting of all fish except shad, which spends most of its life at sea and is therefore contaminated to a lesser extent (Claudio, 2002).

After considerable debate, the EPA in 2001 mandated that about 40 miles of the river upstream from Albany be dredged, some 2.65 million cubic yards of sediment be removed, the slurry be treated, and the solids be properly disposed of. In all, it is estimated that this will lead to the removal of some 150,000 pounds of the estimated 1.1 million pounds of

PCBs that were discharged. The estimated cost will approach half a billion dollars (Anderson, 2002). As would be anticipated, the manufacturer contested the necessity of the cleanup on the basis of claims that naturally occurring processes have and will continue to rid the ecosystem of the contaminant, and that there is insufficient evidence that PCBs are toxic to humans. Others claimed that dredging the river would resuspend the contaminant and increase its concentrations in the water. Still others pointed out that only one-third of the plant workers involved in the studies had been employed for more than five years, and that no assessments were made of their exposures (Claudio, 2002). These claims, in turn, were challenged, and the situation continues to be a source of controversy.

Water-Pollution Regulations

The principal items of federal legislation pertaining to the control of water pollution are the original Water Pollution Control Act passed in 1948, amendments passed in 1956, and 1972, the Clean Water Act of 1977, and the Water Quality Act of 1987. The 1956 amendments directed primary attention to the establishment of water-quality standards for interstate and navigable waters. The Clean Water Act of 1977 shifted attention to the treatment of point sources of industrial wastes and, through the National Pollution Discharge Elimination System (NPDES), led to the establishment by the EPA of standards for pollutants destined for discharge into public waters or sewer systems. In order to comply, industries that discharged such wastes were required to install the best available control technology, and those that discharged such wastes into municipal sewer systems had to meet secondary sewage-treatment standards. The focus of the regulations was on toxic pollutants that would not be adequately treated by municipal treatment systems (Chapter 14).

With passage of the Water Quality Act of 1987, the objectives were expanded to include the control of nonpoint sources of waste. As part of this effort, the EPA was directed to promulgate regulations requiring that municipal and industrial stormwater dischargers obtain permits to release such wastes into U.S. waters. Included in the permits were the stipulations that those who generated such wastes collect and analyze samples of runoff during initial portions of the rain (when contaminant concentrations are at a maximum) and estimate the quantities of individual contaminants that would be released during a range of anticipated storm events. Those affected also have to describe a stormwater pollution-prevention plan that outlines procedures for preventing releases of hazardous substances and oil into the area within and surrounding the plant or municipality. Congress subsequently brought additional attention to the need for the control of nonpoint sources through passage of the Wet Weather Water Quality Act of 2000.

Another area that has been addressed in recent years is the disposal of the biosolids (or sludge) that is created in the operation of wastewater-treatment plants. In 1979 and 1993, the EPA promulgated regulations to address both the health and aesthetic aspects of using biosolids as a soil amendment, particularly with regard to the use of such materials as a fertilizer for edible crops (Logan, 1999). This will subsequently be discussed in more detail.

Assessing Water-Polluting Potential

A variety of methods are available for determining the quantities of contaminants in a liquid waste. In the main, these methods are generic in nature and provide a broad measure of the polluting potential of the waste, not the identities of the individual contributors. One indicator is the concentration of suspended solids. Another is its nutrient content. Still another is the amount of chlorine required to oxidize the organic matter in the waste. The acidity or alkalinity of the waste may also be used as an indicator of its polluting potential, or "strength." Since the oxidation or stabilization of organic matter requires oxygen, the effective operation of a sewage-treatment plant makes it mandatory that an assessment be made of how much oxygen will be required to accomplish this task. A method for making such an assessment will be described in the next paragraph. Such a method, however, will not necessarily provide information on the quantity of nutrients or toxic chemicals that are present. Since in many cases nutrients and toxic organic chemicals are high-priority pollutants (Table 8.2), tests to evaluate their potential contribution to the impact of a waste may also be necessary.

The method most commonly used for assessing the amount of organic matter in domestic sewage or other nontoxic liquid wastes is what is called the five-day, 20°C biochemical oxygen demand (BOD) test. It is conducted using a sample of the waste that has been inoculated with bacteria and then incubated at the proper temperature in the laboratory. The selection of 20°C (68°F) ensures that the temperature of the incubated sample is representative of the outdoor temperature on a spring or fall day. Under these conditions, the BOD after five days will be approximately 70 percent of that which would be exhibited if the sample were incubated until the bacteria had had sufficient time to stabilize all (100 percent) of the organic matter in it. Measurements of the BOD of the incoming waste and of waste at various stages within a sewage-treatment plant provide an indication of the effectiveness not only of the individual treatment steps but also of the plant as a whole. From the BOD of the effluent from the plant and the rate at which it is being discharged, coupled with the DO content and diluting volume provided by the receiving body, it is possible to estimate the extent to which the DO in, for example, a stream will be depleted. A related chemical test has been developed to assess the oxygen demand of toxic wastes that inhibit bacterial growth and therefore do not permit use of the BOD test. This test, which requires that the sample be chemically digested in the laboratory, yields a measure of the chemical oxygen demand of the waste and is called the COD test.

Treatment of Liquid Wastes

As implied by the previous discussion, methods for treating liquid wastes, particularly domestic sewage, are designed to stabilize or oxidize, through biological processes, the organic matter they contain. This can be most effectively achieved by providing conditions that will optimize the ability of natural biological processes to accomplish this task. This is one of the primary goals in the design and operation of a sewage-treatment plant.

MUNICIPAL WASTES

Overall, the methods for the treatment of municipal sewage and other types of nontoxic liquid wastes are divided into three stages: primary, secondary, and tertiary. Primary treatment consists of holding the wastes undisturbed in a tank for a sufficient period of time to permit the solids within the waste to settle and be removed. Secondary treatment is the use of the previously discussed biological processes for oxidizing the organic matter in the waste. Tertiary treatment involves a variety of processes tailored to the intended uses of the finished product. One of the more common tertiary or advanced methods for treating liquid wastes is very similar to the coagulation, settling, and filtration processes used in treating surface waters to make them acceptable for drinking.

Each of these processes represents a progressive level of purification, and the number of stages applied depends on the degree of treatment required. With modifications, however, higher removals are possible. As shown in Figure 8.5, all municipal sewage-treatment processes begin with the primary stage. Under the 1972 amendments to the Federal Water Pol-

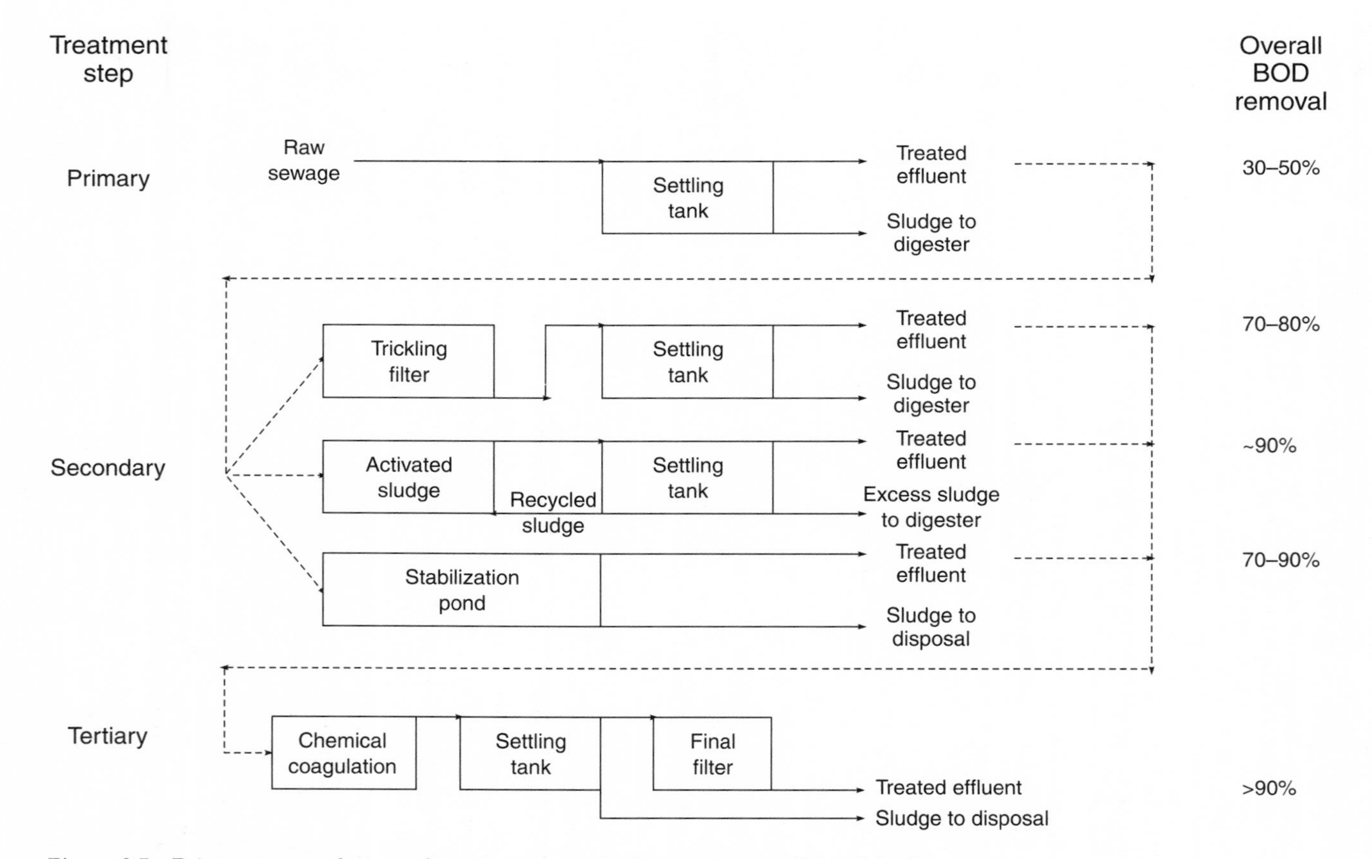

Figure 8.5 Primary, secondary, and tertiary stages in the treatment of municipal sewage

lution Control Act, all wastewater-treatment plants in the United States must also provide secondary treatment.

Primary treatment, as noted earlier, involves holding the sewage in a settling tank to permit the removal of solids by sedimentation. Before the sewage enters the settling tank, it is commonly sent through a chamber or collector to remove sand, grit, and small rocks that might damage pumps or other equipment. The settling tanks are operated on a flow-through basis and are large enough to hold the material for several hours. During that time, approximately half the suspended solids settle out, providing a BOD reduction of 30–50 percent. Grease and light solids that float are removed from the settling tank by a scraper and are pumped along with the settled solids to a large closed tank called a digester, where they are held for anaerobic digestion. Digestion is most effective when the biosolids are heated to 32°C (90°F) or more. At 32°C the biosolids are digested in about 24 days; at 54°C (130°F), in about 12 days. The methane gas produced in the process provides fuel for heating the digester and other applications within the treatment plant.

Secondary (or biological) treatment is accomplished through use of a trickling filter, the activated sludge process, or a waste stabilization pond. The first two methods are aerobic; the last combines aerobic and anaerobic systems. As previously noted, the overall objective is to make conditions ideal for biological stabilization. No special organisms are added; those that are necessary develop and flourish naturally.

The *trickling filter* is a common form of secondary treatment. The term *filter* is a misnomer, since the system does not filter the sewage. Rather, a trickling filter consists of a large tank, roughly 2 meters (6–7 feet) deep, filled with stones 5–10 centimeters (2–4 inches) in diameter over which sewage is intermittently trickled or sprayed from a distributor. The stones rapidly become coated with a biological film or slime. The solids in the sewage percolating through the bed are incorporated into the bacterial growth, where the microorganisms convert the organic matter into cell protoplasm and inorganic matter. When the bacterial growth on the stones becomes too thick and heavy, it sloughs off and is carried away in the liquid effluent leaving the bottom of the filter bed. The effluent is sent to a secondary settling tank, where the bacterial sloughings settle to the bottom as biosolids. The settled effluent represents the treated product. As in the case of primary treatment, the settled biosolids are placed in a digester for anaerobic decomposition. The total reduction in BOD provided by a treatment plant that incorporates a trickling filter is 70–80 percent. For somewhat higher BOD removals, two trickling filters can be used in series, or a single unit can be used and a portion of the settled effluent recycled through the filter bed. Figure 8.6 shows the trickling-filter treatment process.

The *activated sludge process* is another form of aerobic secondary treatment for municipal sewage. Sewage is sent into a large open tank, where it is held for several hours and its oxygen content maintained by means of aerators (air diffusers) or mechanical agitators (paddles or brushes). Rather than growing on the surfaces of stones as in the trickling filter, the microorganisms float as suspended particles in the aerated sewage. The effluent is sent to a secondary settling tank, where the microorganisms settle out, and the settled sewage is the treated product. The overall reduction in BOD is about 90 percent. Some of the microorganisms that have settled out in the secondary tank are pumped back into the aerated tank to maintain an adequate population of microbial growth. The rest of the growth is treated as biosolids and sent to a digester.

Waste stabilization ponds, another method of secondary treatment, have been used in other countries for many years. They were largely ignored in the United States, however, until the 1950s (Gloyna, 1971). Nonetheless, because of their low construction cost, ease of operation, and minimal maintenance requirements, they are now in common use, particularly in the warmer southern states. As the name implies, the basic unit in such a system is an earthen pond having a depth of 1–2 meters (about 3–7 feet), a width of 25 meters (about 80 feet), and a length of 90 meters (about 300 feet). Because effective treatment requires that the waste be retained for 30 to 80 days, a single such pond can serve between 1,000 and 2,000

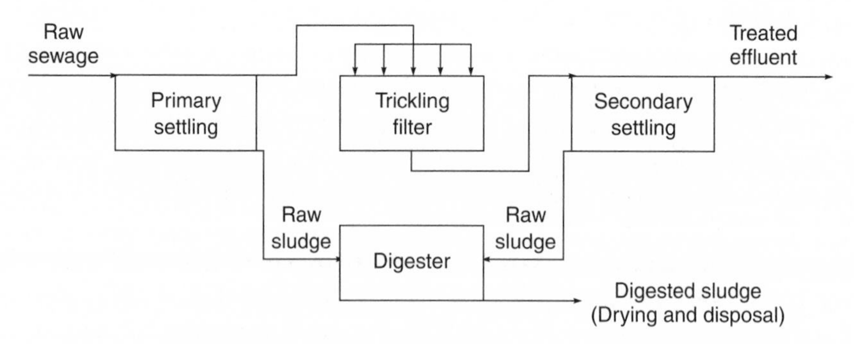

Figure 8.6 Trickling filter sewage treatment

people. The system is operated on a continuous flow-through basis and is effective in treating either raw sewage or sewage that has undergone primary treatment. The principal precaution is not to locate such ponds in soils with fissures that would permit the sewage to move through the ground without filtration, thereby contaminating nearby groundwater supplies.

To serve larger population groups, several ponds can be operated in series. Most ponds operate biologically at two levels: the lower portion is anaerobic, the upper portion is aerobic. In the border area, facultative bacteria (which can live under either aerobic or anaerobic conditions) are active. In some cases, wind-driven mixers are used to increase the amount of oxygen in the upper portions of the pond. Algae growth at the surface also helps assure aerobic conditions. When a pond fills with biosolids, it must be cleaned and the cycle begun anew.

As noted earlier, most methods for *tertiary treatment* of sewage are modeled on those used in the purification of drinking water (Chapter 7). For wastes that contain unusual amounts of organic compounds, or heavy metals and viruses, additional steps may be required. Excess organic compounds are commonly removed by passing the treated waste through two granular carbon beds, each of which provides 30 minutes of contact time. Ozone may be used to disinfect the waste as it passes from the first carbon bed to the second. Heavy metals and viruses can be removed by coagulating the waste, for example, with lime, followed by sedimentation. This process, however, creates large volumes of highly toxic sludge that must be handled and disposed of carefully.

As in essentially all fields of environmental health, there have been significant advancements in recent years in methods for the treatment of municipal wastes. Several of these were developed in Europe. One that relates to the activated sludge process is the addition of solid substrates, such as small polyethylene structures, to the aeration tanks. Such substrates provide a convenient surface on which the bacteria can live. The net result is that the numbers of organisms present are increased by several orders of magnitude. This not only enhances the effectiveness of the treatment system but also enables plant operators to control in a more rigorous manner the age and numbers of bacteria in the aeration tanks (Francisco, 2001).

A second advancement involves changes in the methods for disinfecting the treated waste prior to discharging it into the environment. In the past, the common approach has been to add chlorine. As was the case for drinking water (Chapter 7), reactions of this disinfectant with organic compounds in the waste produce chlorinated hydrocarbons. If chlorine is added in excess and its addition is not followed by some form of dechlorination, the discharged wastewater will be harmful to aquatic organisms. In many cases, its release is a violation of regulations. For example, the International Joint Commission, a U.S.-Canadian advisory group on the control of pollution in the Great Lakes, has recommended banning the application of chlorine to wastewaters discharged into those waters. In many cases, the choice is either to dechlorinate or to use some other means of disinfection (Francisco, 2001). For these reasons and because UV radiation is effective against bacteria, viruses, and parasites, it is now being applied in more than 2,000 wastewater-treatment installations in the United States (Sakamoto, 2000). Since turbidity can severely reduce the effectiveness of UV radiation, it is often necessary to subject wastewaters to an additional filtering step prior to applying UV as a disinfectant. Other advances include methods for removing nitrogen and phosphorus from liquid wastes (Francisco, 2001).

INDUSTRIAL WASTES

As might be anticipated, wastewaters from industrial operations contain a wide range of pollutants. As a result, the treatment of industrial wastes requires not only an expansion in the number of methods applied but also a change in their sophistication. Methods that can be applied, either singly or in combination, include the following:

Physical processes include those designed to remove suspended solids through filtration, centrifugation, or the previously described settling tanks; oils, greases, and emulsified organics through aeration, which causes such materials to float to the surface, where they can be removed by skimming devices; and dissolved materials, such as organic chemicals, which can be accomplished by passing the water through a semipermeable membrane (Furukawa, 1999) or, as noted earlier, through beds of activated carbon.

Chemical processes include the addition of acids to neutralize wastes that are alkaline; bases to neutralize wastes that are acid; and chemicals to coagulate and precipitate suspended solids (as, for example, in tertiary treatment systems applied to the effluents from municipal sewage-treatment plants). Other methods include the use of ion-exchange resins to replace contaminants in the waste with innocuous chemicals and the use of oxidants, such as chlorine, to convert volatile and nonvolatile organic contaminants into nontoxic compounds.

Biological processes include the predigestion of brewery, winery, and meat-packing wastes under anaerobic conditions, often at elevated temperatures to accelerate the process, and the oxidation of certain types of industrial wastes, such as petroleum constituents, under aerobic conditions similar to those applied in the treatment of domestic sewage.

NONPOINT SOURCES

Because of the intermittent flow rate of nonpoint sources and the difficulties in designing facilities to treat them, primary efforts are being directed to controlling the volumes of such releases and the more harmful constituents that they contain. The latter goal is being achieved in agricultural runoff through the optimization of pesticide application rates and timing. Other controls include use of the previously described more effective methods of distributing irrigation water (Chapter 7); application of conservation techniques such as reduced tillage, crop rotation, and winter cover crops; establishing buffer zones, such as vegetative cover along streambanks; and planting strategically placed grass strips and artificial wetlands to intercept or immobilize pollutants.

Among the solutions being proposed for the problem of urban runoff are methods for collecting and "harvesting" the rainwater, rather than sending it to the sewers. One approach is to cover parking lots with permeable surfaces of a honeycomb design that permit rainwater to drain into the soil. In addition to retaining the water, such an approach could also reduce expenditures for storm drains and sewers. Other techniques are designed to slow runoff, allow more water to percolate into the ground, and filter out contaminants. In addition, weirs, movable dams, and detention areas can provide storage capacity in storm and combined sewer systems, thereby reducing the frequency and volume of combined sewer overflows (NRC, 1993). The presence of contaminants in urban runoff can also be reduced by providing convenient disposal sites for used oil and household hazardous waste, collecting leaves and yard trimmings on a frequent basis, and using vacuum equipment for street cleaning.

Disposal of Treated Wastewater

Due to the previously cited problems in releasing treated municipal sewage into rivers, streams, and lakes, increasing attention is being directed to the disposal of treated wastewaters on land surfaces. Advantages of this approach are that it

returns nutrients to the soil, making them available to nourish agricultural crops, golf courses, parks, recreational areas, and forests;

provides a mechanism for reclaiming and preserving open spaces and existing wetlands, as well as for developing new wetlands that, in turn, provide habitats for wildlife;

can create an ideal environment in which natural biological, physical, and chemical processes can stabilize the wastes (wetlands, for example, serve as nutrient sinks and buffering zones to protect streams and other areas);

can provide a ready means, under proper conditions, for recharging groundwater sources; and

frequently results in reductions in wastewater-treatment costs, thus saving funds for addressing other problems.

Another advantage is that a properly developed land disposal system can be operated for 20 or more years. Such a system can also serve as a viable and beneficial alternative to methods commonly employed for the secondary treatment of municipal sewage. The reuse of human waste in aquaculture, in particular, can produce significant benefits and achieve a variety of useful goals. In countries where nutrition requirements exceed food production, aquaculture can assist in closing the gap by using valuable nutrients that would otherwise be squandered. In countries where water quality must be improved, aquaculture can lessen the harmful impacts of excess pollution on watercourses. In arid regions, it can make an important contribution to the conservation of scarce water resources (Edwards, 1992).

Disposal of Biosolids

The magnitude of the biosolids disposal problem is enormous. The EPA estimates, for example, that municipal wastewater-treatment plants in the

United States produce almost 8 million tons annually. Additional biosolids are produced in the treatment of industrial wastes. Although it might seem logical to use such materials as a soil conditioner and fertilizer, questions related to aesthetic and public health concerns immediately arise. Among the most common are questions concerning the possible transmission of disease, especially when such materials are to be used to grow edible crops. Even the use of biosolids on lawns, parks, and golf courses has not been without expressions of concern. If the biosolids are not properly pretreated, any such applications can attract vermin. There have also been multiple objections to the accompanying odors. Another concern is that biosolids tend to concentrate toxic heavy metals from the wastes being treated. This could lead to the uptake of such metals by food crops. If biosolids are incinerated, there is the problem of the release of toxic materials into the air.

Seeking to resolve these problems, the EPA has promulgated regulations that specify the type of treatment biosolids must receive prior to being sent to a landfill, applied to land as a fertilizer, or incinerated. The regulations are designed to reduce the volume of the biosolids, stabilize the organic materials they contain (so as to reduce odors and the attraction of animals), and kill the full range of microorganisms (for example, certain bacteria, viruses, and parasites) they contain (Logan, 1999). For application as fertilizer, the EPA specified three degrees of treatment, depending on the specific use intended. Those for so-called Class A biosolids were the most stringent; those for Class C the least. Under this approach, Class A biosolids must be subjected to rigorous treatment processes such that they contain no detectable pathogens. This requires that they be heated to a minimum of 50°C (122°F) for at least 20 minutes. To ensure that the waste does not attract animals, it must be dried so that it contains at least 75 percent solids (Sims and Bentley, 2001). To confirm that the final product meets the specifications, the treatment-plant operator must perform a suitable sampling and monitoring program. Class A biosolids can be applied as a fertilizer by residents and farmers without site-access restrictions. Materials in this class that also meet stringent trace-element limits have no land-use restrictions (Logan, 1999).

In the meantime, other concerns have developed. One is how to ensure the protection of workers who apply Class A materials, and even more so those involved with handling Classes B and C. As a result, it may become mandatory that all such materials meet the requirements for Class A, regardless of intended use.

Special Problems of Groundwater Contamination

Another concern is that discharges of liquid wastes from industrial, agricultural, and domestic sources can contaminate groundwater. If the biosolids, previously discussed, are placed in a landfill, these types of problems can occur through the leaching of toxic materials and pathogens not only into groundwater but also into surface waters. Once the water in an aquifer is polluted, it is extremely difficult to decontaminate. Even so, there have been multiple events where groundwater sources have been contaminated (Chapter 9).

Many methods have been applied in seeking to remove and/or stabilize groundwater contaminants. These include the full range of previously discussed physical, chemical, and biological agents, applied both in situ and to the water after it has been removed from the ground. One of the earliest approaches is what is called "pump and treat." This, in essence, involves the installation of a series of wells to extract the water and treat it with chemicals, such as alum or ferric chloride, following in general the methodologies used to purify surface waters (Chapter 7). The treated water is then either used or pumped back into the aquifer. Such an approach, however, is frequently both ineffective and expensive. The primary reasons are that (1) the treated water, once pumped back into the ground, is immediately mixed with water that has not been treated; and (2) contaminants that leach out from sediments within the aquifer serve as a continuing source of pollution.

Dissatisfaction with the "pump and treat" approach has led to the development and application of a variety of alternatives. One is to inject chemical reactants into the aquifer either to (1) convert the contaminants into a nontoxic form, (2) precipitate and fix them in place, or (3) mobilize them so they can be effectively extracted and removed by the "pump and treat" approach. Such methods are practical, however, only if the reactants can be injected in a soluble form and the chemistry of the contaminants makes them suitable for reacting as desired. Another approach is to excavate a portion of the aquifer and install subsurface permeable membranes, or reactive barriers, that will remove contaminants from the groundwater as it flows through them. As with the "pump and treat" approach, this method can be very expensive (Lovley, 2001).

Another method that is increasing in favor is bioremediation, using either aerobic or anaerobic organisms. In most cases, the initial approach is to monitor the progress being achieved by the organisms naturally present. The addition of other organisms should be attempted only if the existing rate of progress is not adequate. The success of this approach depends on the nature of the groundwater flow system, the characteristics of the contaminants, and, obviously, their susceptibility to biological degradation. Even if the rate of stabilization is favorable, sufficient monitoring will be required to determine when the groundwater might be acceptable for use.

The General Outlook

There are some 20,000 municipal wastewater-treatment facilities in the United States. These have a total daily treatment capacity of almost 40 billion gallons and represent capital investment of about $4 billion (CEQ, 1998). Due to years of neglect and the changes required by the Clean Water Act of 1977, many of these facilities are in need of major repairs and/or upgrades. Estimates are that the costs associated with improving waste-collection systems, coupled with resolving problems such as the previously discussed sewer overflows, could exceed $180 billion (EPA, 2003). The urgency of meeting these needs is further demonstrated by the fact that surveys of streams, lakes, and estuaries in the United States show that about 40 percent contain water of a quality that is not adequate to support fishing and swimming. Agricultural runoff alone adversely affects some 70 percent of the impaired rivers and almost half of the impaired lakes (Gray, 1999).

One of the basic sources of these and related problems is that all too frequently, the management and control of liquid wastes have been addressed in isolation, rather than holistically. What is needed is the development of a comprehensive plan that addresses wastewater problems on the basis of an entire watershed or drainage basin. An essential part of any such effort is a careful review and evaluation of the interrelationships and relative impacts of each of the contributing polluting sources. Since funds to address all these problems simultaneously are not available, it is essential that a mechanism be established for setting priorities on which problems should be addressed first, backed up by adequate research to provide the scientific information on which to base the associated regulatory programs (Gray, 1999).

The most desirable approach for preventing pollution is to eliminate the production of the waste. Where this is not practical, systems should be designed so that the wastewater can be recycled and reused. Ultimately, such an approach could lead to "closed-loop" systems that produce essentially no liquid discharges. In the automotive industry, not only is wastewater being treated, recycled, and reused, but the sanitary (sewage) effluent from the plants is also being treated and reused for irrigation. In a similar manner, the electronics industry, which requires high-purity water, has found that treating and recycling its wastewater is cost effective. Another approach is to have one industry become the user of the wastewater generated by another. Operators of some electric power plants, for example, are using municipally treated sewage as a source of cooling water.

Finally, it is important to acknowledge that efforts to manage and control wastewater discharges cannot be effectively conducted on a single-country basis. Not only should the developed nations share their expertise with the less developed nations, but they should also recognize that wastewater pollution can move from one country or region to another in much the same way as atmospheric pollutants. Primary examples are the organisms and viruslike particles that accompany the discharge of ballast water at dockside when oceangoing vessels take on cargo. Such activities result in the annual discharge into U.S. ports of an estimated 80 million tons (20 billion gallons) of water, originally taken on at one or more overseas ports (Ruiz et al., 2000).

9

SOLID WASTE

UNTIL the mid-twentieth century, most solid or municipal waste took the form of garbage, yard waste (leaves, grass clippings, tree limbs), newspapers, cans and bottles, coal and wood ashes, street sweepings, and discarded building materials. Most such waste was not considered hazardous and was simply transported to the local land disposal facility or "dump," where it was periodically set on fire to reduce its volume and to discourage the breeding of insects and rodents. Because this practice often led to windblown debris and unsightly disposal facilities, and because people recognized the need for a more technically based method of disposal, it was gradually replaced by the sanitary landfill, where municipal waste was buried in the ground (Figure 9.1). As long as windblown debris and fires were contained, material was covered over and sealed daily (so that breeding and habitation by insects and rodents were controlled), and contamination of nearby groundwater supplies was avoided, the sanitary landfill was considered an acceptable method for disposal.

With the subsequent development of a "throwaway" society and an unprecedented demand for new products, during the next half century the characteristics of municipal solid waste changed dramatically, and its volume increased enormously. While such waste was still classified as nonhazardous, it now contained paint, pesticides, and solvents, as well as construction and demolition debris that included oil and grease, lead, and toxic coatings on wood (Saxe, 2002). Also present were many materials, such as plastics, that are not readily biodegradable. In fact, this component within the waste stream almost tripled between 1980 and 1996. At the

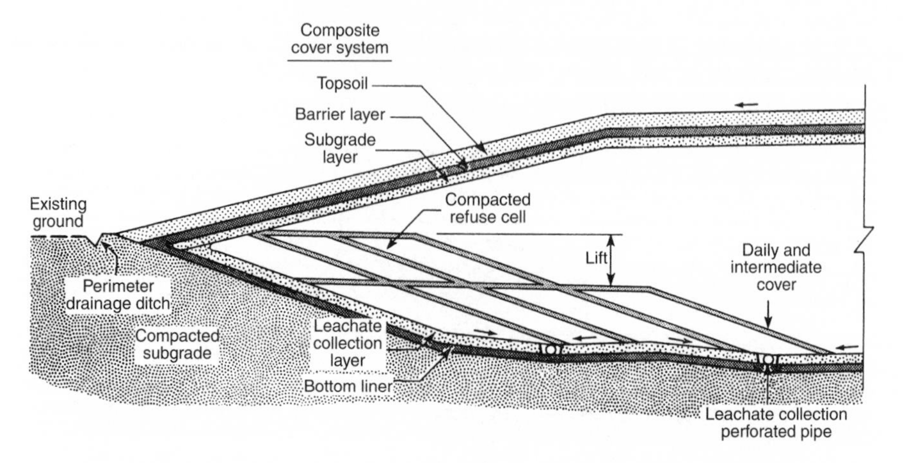

Figure 9.1 Cross section of a typical landfill and leachate collection system

same time, the total per capita quantity of municipal waste generated per day in the United States increased from 1.7 kilograms (3.7 pounds) in 1980 to 2.1 kilograms (4.6 pounds) in 1999, an increase of 24 percent within less than two decades. There was an increase of 3 percent between 1998 and 1999 alone. Today the average person in this country annually produces substantially more than 725 kilograms (1,600 pounds) of municipal solid waste, including almost 70 kilograms (150 pounds) of the previously cited plastics—more municipal solid waste per capita than in any other industrialized nation of the world (CEQ, 1998). At the same time, the amount of hazardous waste being generated had increased in 2000 to some 300 million metric tons per year—more than 1 ton of such waste per person per year. In total, it is estimated that about 6 billion tons of waste are currently produced in the United States each year. Of this, slightly more than 160 million tons are municipal waste.

Another development is the significant reduction in the number of municipal sanitary landfills in operation in the United States. Of the almost 8,000 such facilities that existed in 1988, only about 2,200 remained in 1999. One reason for this is that government regulators, in seeking to make landfills safer, have made them more expensive to own and operate. This has led to the establishment of larger, better-designed and operated disposal facilities on a regional basis. Although it appears that the existing disposal capacity is meeting the need, the reduced number of facilities has led to several major changes. One of the more important is a large increase in the exporting and importing of municipal refuse from one state to another. During 2000, more than 31 million tons of refuse were involved in these types of activities. This was three times the amount imported and exported in 1989, and all but 3 of the 50 states were involved (Wolpin, 2002). The international implications of such practices are discussed later in this chapter.

Types and Classifications

Several terms are used for classifying solid wastes. The definition of solid waste itself was specified by Congress in the Resource Conservation and Recovery Act (RCRA) of 1976. According to RCRA, *solid waste* is defined as "any garbage, refuse, sludge from a waste treatment plant, water supply treatment plant, or air pollution control facility and other discarded material, including solid, liquid, semisolid, or contained gaseous material resulting from industrial, commercial, mining, and agricultural operations and from community activities." Specifically excluded from classification as a solid waste is "solid or dissolved material in domestic sewage," as well as "industrial wastewater discharges regulated under the Clean Water Act" (Chapter 8) (EPA, 1986a).

Having defined solid waste, Congress next defined what is classified as *hazardous waste*. This is any "solid waste, or combination of solid wastes, which because of its quantity, concentration, or physical, chemical, or infectious characteristics may: (1) cause, or significantly contribute to an increase in mortality or an increase in serious irreversible, or incapacitating illness; or (2) pose a substantial present or potential hazard to human health or the environment when improperly treated, stored, transported, or disposed of, or otherwise managed." To implement this definition, the EPA established two basic methods for designating hazardous wastes. Either they are *listed* in accordance with Title 40, Part 261, Subpart D, of the Code of Federal Regulations (CFR), or they have been determined to have certain *characteristics* as specified in Title 40, Part 261, Subpart C, of the CFR. Wastes that are listed are those associated with various manufacturing and industrial processes and with certain commercial chemical products that have been specifically identified by the EPA as consistently posing a hazard to human health and the environment when discarded. Wastes are characterized if they exhibit certain properties, including ignitability, corrosivity, reactivity, or toxicity, based on test results

Table 9.1 Examples of hazardous waste generated by business and industry

Waste generator	Typical wastes
Chemical manufacturers	Strong acids and bases Spent solvents Reactive wastes
Vehicle maintenance shops	Paint wastes containing heavy metals Ignitable wastes Used lead acid batteries Spent solvents
Printing industry	Heavy-metal solutions Waste inks Spent solvents Spent electroplating wastes Ink sludges containing heavy metals
Leather products manufacturers	Waste toluene and benzene
Paper industry	Paint wastes containing heavy metals Ignitable solvents Strong acids and bases
Construction industry	Ignitable paint wastes Spent solvents Strong acids and bases
Cleaning agents and cosmetics manufacturers	Heavy-metal dusts Ignitable wastes Flammable solvents Strong acids and bases
Furniture and wood manufacturers and refinishing	Ignitable wastes Spent solvents
Metal manufacturing	Paint wastes containing heavy metals Strong acids and bases Cyanide wastes Sludges containing heavy metals

or the knowledge of the waste generator. Although these characteristics are important, except for the matter of toxicity they relate only indirectly to the potential health impacts of a waste on the public and the environment. Various industries and the types of hazardous wastes they generate are listed in Table 9.1.

The next category addressed was *nonhazardous waste.* In this case, rather than defining what it is, Congress included in the RCRA certain categories of waste that were excluded from the definition of *hazardous waste.* The most common group excluded was household waste. Also excluded were agricultural wastes used as fertilizer, mining overburden returned to the mine site, and certain wastes produced in the combustion of coal. As a general rule, the waste produced by homeowners is designated as municipal nonhazardous waste, and the chemical waste produced by industry is classified as hazardous. Another type of waste, *mixed waste,* was addressed in the Federal Facility Compliance Act of 1992. This is defined as waste that contains both hazardous chemicals and radioactive materials (U.S. Congress, 1992). The typical content and annual volumes generated of hazardous and nonhazardous, mixed, and radioactive wastes and the federal agencies responsible for their disposal are summarized in Table 9.2. The various categories of radioactive wastes will be discussed later in this chapter.

Table 9.2 Types, regulation, and characteristics of commercially generated waste

Type of waste[a]	Regulating body	Typical content
Nonhazardous	State and local governments	Refuse, garbage, sludge, municipal trash
Hazardous	EPA or authorized states	Solvents, acids, heavy metals, pesticide residues, chemical sludges, incinerator ash, plating solutions
Radioactive	USNRC or agreement states	High- and low-level radioactive waste, naturally occurring and accelerator-produced materials
Mixed	EPA and USNRC or states	Radioactive organic liquids, radioactive heavy metals

a. About 99% of the total waste volume generated each year in the United States is nonhazardous; about 1% is hazardous; the volume of radioactive and mixed waste combined is less than a few ten thousandths of 1 percent of the total.

Wastes of Special Interest

There are multiple wastes within the preceding classifications that pose major challenges in their management and disposal. This is vividly demonstrated to anyone who has observed a junkyard containing masses of rusting cars and trucks. The sheer magnitude of this problem can also be illustrated by the 280 million automobile tires that are discarded in the United States each year, as well as similar numbers of home appliances, such as refrigerators, stoves, washing machines, and clothes driers. Added to these are millions of discarded electronic products, such as television sets and computers. The more than 3 billion household and industrial batteries that are sold annually in the United States pose another special management and disposal problem. Prominent among the devices that require batteries are cell phones, notebook computers, and power tools.

In addition to being a waste that must be disposed of, tires can readily provide breeding grounds for mosquitoes (Chapter 10). Equally important, they frequently serve as fuel for multitudes of fires. Once ignited, the resulting fires are extremely difficult to extinguish. The magnitude and nature of the disposal problems associated with computers, television sets, and other electronic devices are equally challenging. The annual volume sent to disposal numbers in the millions of tons. A further problem is that cathode-ray tubes (CRTs), a common component of television sets and computers, contain relatively large amounts of lead. In fact, such tubes have been classified by the EPA as a hazardous waste. The associated printed circuit boards and batteries also contain lead, plus smaller amounts of cadmium and mercury (Table 9.3). Had the new television and computer screens that incorporate liquid crystal technology not been developed, it had been estimated that the CRTs destined for disposal in the United States during the next decade could have contained as much as 1 billion pounds of lead (Chapter 17).

Health and Environmental Impacts

Numerous epidemiological studies have been conducted to evaluate whether the health of people living near hazardous waste disposal sites is being adversely affected, particularly through an increase in cancer rates. Most such studies have been inconclusive. Even at Love Canal, such studies revealed no evidence of higher cancer incidence than in the rest of New York State (Golaine, 1991). Although other studies have shown

Table 9.3 Toxic materials in desktop computers

Material	Components	Chronic health effects
Arsenic	Doping agents in transistors, printed wiring boards	Skin sores, hypertension, peripheral vascular disease, skin and bladder cancer (ingestion), lung cancer (inhalation and ingestion)
Beryllium	Printed wiring boards, connectors	Lung damage, allergic reactions, chronic beryllium disease, reasonably anticipated to be a human carcinogen
Cadmium	Batteries, blue-green phosphor emitters, cathode-ray tubes, printed wiring boards	Pulmonary damage, kidney disease, bone fragility, reasonably anticipated to be a human carcinogen
Chromium	Housings, hardeners	Lung cancer (inhalation), liver and kidney disease, strong allergic reactions, may cause DNA damage
Cobalt	Batteries	Respiratory irritation, reduced pulmonary function, asthma, pneumonia, and lung cancer (inhalation)
Gallium	Semiconductors, printed wiring boards	Evidence of carcinogenesis in laboratory studies of animals
Lead	Radiation shielding, metal joints, printed wiring boards	Damage to kidneys and nervous, endocrine, and reproductive systems, serious adverse effects on brain development
Mercury	Batteries, switches, printed wiring boards	Chronic brain, kidney, lung, and fetal damage, increases in blood pressure and heart rate, allergic reactions
Nickel	Cathode-ray tubes, printed wiring boards, structural components	Allergic reactions, asthma, chronic bronchitis, impaired lung function, reasonably expected to be a human carcinogen

apparent associations between living near a hazardous waste disposal site and increased risks of certain types of cancer, as well as birth defects, investigators are careful to point out that because of limitations in the data, it is too early to reach any definitive conclusions. For several reasons (Chapter 2), such an outcome is not unexpected. First of all, the earliest recognizable effects of low-level chemical exposures (headache, malaise, minor skin irritation, and respiratory tract complaints) tend to be common to many conditions. In addition, many of the illnesses (such as cancer) that might be anticipated have latency periods of 10 to 40 years. Under these conditions, it is difficult to establish patterns of exposure and equally difficult to gather data on a sufficiently large population group to verify a definitive relationship.

For these reasons, most assessments of the risks associated with toxic materials, particularly those present in disposal facilities, have been derived from evaluations based on various hypothesized exposure scenarios (Chapter 17). Key factors in such assessments are the nature of the toxic materials in the wastes and estimates of how much of each might become airborne or be leached out and gain access to aquifers and other sources that may later be consumed by humans or farm animals or used for irrigating agricultural crops. Account must also be made of the possibility that toxic materials can leach from solid wastes and enter a nearby stream from which fish are caught and consumed. Once the exposures have been estimated, the associated health risks are calculated using information on the toxicology of each of the materials involved. Although assessments of this type are extremely difficult, even when only a single material is involved, they are far more complicated when the potential exposures involve a mixture of agents, as is frequently the case. It is essential in any such evaluation to ensure that all possible pathways of significant exposure have been identified. These types of challenges are exemplified by the case study that follows.

While the impacts of solid waste on human health are important, attention needs also to be directed to its impacts on other types of animals in other segments of the environment. Certain particular types of solid waste, namely, plastic fishing gear, six-pack beverage yokes, sandwich bags, and certain types of plastic cups, that have been discarded into the ocean entrap and kill an estimated more than 1 million seabirds and 100,000 marine mammals every year. In fact, plastics may be as much a source of mortality to marine mammals as oil spills, heavy metals, and other toxic materials combined (Shea, 1988). Recent surveys in the far reaches of the Pacific

Ocean have revealed the presence of extensive areas that contain up to a million pieces of plastic per square mile. Some of these result from materials illegally jettisoned from ships; other portions occur as a result of accidental spills. Because plastic is lighter than seawater, it floats on the surface for years, gradually breaking up into smaller and smaller particles that end up in filter-feeding animals, such as jellyfish. Due to their nature, plastics adsorb toxic chemicals and become part of the food web when they are eaten by turtles. In a similar manner, birds take in larger pieces of plastic when they mistake them for fish. Although an international convention called MARPOL bans the dumping of plastics at sea, the agreement is not enforced on the open ocean. Newer biodegradable plastics that have been developed offer hope for ultimately solving the problem (Hayden, 2002).

Case Study: Chromated Copper Arsenate Wood

In the early 1970s, wood treated with chromated copper arsenate (CCA) began to be widely used in the United States for the construction of structures to be used in outdoor land, aquatic, and marine environments. The treatment process, which involves applying the chemical under pressure so that it enters the pore spaces of the wood, was designed to prevent fungal and microbial decay. By the late 1990s, such wood was being used in almost 80 percent of the preserved wood market in the United States. In fact, by the early part of the twenty-first century, it was estimated that nearly 450 million cubic feet had been sold in this country. Because fungal and microbial decay are especially troublesome in Florida, the use of CCA-treated wood there was quite extensive. In fact, it was used to construct the boardwalks and decks in essentially all of the 150 state parks in that state.

Studies initiated in the 1990s showed that arsenic from the treated wood was leaching into the soil beneath such structures, and that the concentrations in some soils in Florida were in excess of federal and state limits. This led to concerns that children playing on such equipment might ingest arsenic through, for example, licking their hands. Tests showed even higher rates of leaching from similarly treated wood that was used to construct docks and marinas (Tom, 2001). Although tests of workers who were regularly exposed to the raw materials used in the wood found levels of arsenic that were deemed to be insignificant, many members of the public continued to be concerned. On the basis of the evaluations of the associated risks, including those to children, the EPA concluded that it was not necessary for homeowners to remove their backyard decks and picnic tables or to dismantle swing sets and jungle gyms. Questions continued to be raised, however, and the accompanying concerns were subsequently heightened by studies that showed that the level of arsenic on the surfaces of such products does not significantly diminish with time. Indeed, the risk of being exposed to relatively large amounts of arsenic persists for as much as two decades (Lavelle, 2002).

As a result of these concerns and those of state regulatory agencies, several major retailers voluntarily agreed to discontinue selling such wood, and four U.S. manufacturers agreed to withdraw the chemical from the treatment of wood for residential use by 31 December 2003 (Lavelle, 2002). In the future, all CCA-treated wood sold in the United States will be required to be accompanied by detailed safety handling information. Its use, however, will be restricted to certain industrial applications, such as pier marine pilings, highway barriers, and plywood used in the roofs of homes. Concurrently, the Consumer Product Safety Commission has agreed to request public comments on petitions that could lead to an outright ban on the use of CCA-treated wood (Tom, 2001).

But the problem does not end there. Once facilities made of CCA-treated wood have reached the end of their useful life, they are dismantled and frequently disposed of in so-called construction and demolition landfills. Analyses of groundwater samples collected near several unlined landfills of this type in Florida revealed arsenic concentrations more than double the EPA limit for drinking water (O'Connell, 2003). In other cases, it was found that the CCA-treated wood from construction and demolition projects was being burned to generate electricity, and the ash was being applied to agricultural fields, such as those used to grow sugarcane. In still other cases, the wood was being ground into a mulch that was applied to the soil. Tests showed that some of the ash contained arsenic concentrations of several hundred parts per million. Chromium concentrations were also high (Tom, 2001).

Trends in Waste Management

For many years, agencies and organizations responsible for protecting the environment accepted the wastes that were generated and tried to develop satisfactory methods for their treatment and/or disposal. This is referred today as the "end-of-pipe" approach. With the coming of the previously mentioned throwaway society and the rapid expansion of industrial activities, environmentalists and the U.S. Congress soon realized that the generation of waste was becoming overwhelming, and that new approaches had to be developed. With passage of the Resource Conservation and Recovery Act of 1976, Congress mandated that the reduction or elimination of the generation of hazardous wastes at their source (that is, pollution prevention) should take priority over the management of such wastes after they have been produced. The principal programs and goals of this act are summarized in Table 9.4. Congress expanded the approach by passing the Pollution Prevention Act of 1990, which stated that it was the policy of the United States that, whenever feasible, pollutants that cannot be prevented should be recycled, and those that cannot be prevented or recycled should be treated and disposed of in an environmentally safe manner (U.S. Congress, 1990).

Today the generally accepted philosophy is that waste management and disposal should not and cannot be regarded as "freestanding" practices that require their own justification. They must be made an integral part

Table 9.4. Principal programs and goals of the Resource Conservation and Recovery Act (1976)

Solid-waste program (directed primarily at management and control of nonhazardous solid wastes)
Primary goals:
To encourage environmentally sound solid-waste management practices
To maximize reuse of recoverable resources
To foster resource conservation

Hazardous waste program ("cradle-to-grave" system for managing hazardous waste)
Primary goals:
To identify hazardous waste
To regulate generators and transporters of hazardous waste
To regulate owners and operators of facilities that treat, store, or dispose of hazardous waste

Underground storage tank program
Primary goals:
To provide performance standards for new tanks
To prohibit installation of unprotected new tanks
To provide regulations concerning leak detection, prevention, and corrective action

of the processes that generate them. In accord with this view, the challenges and potential difficulties of waste management and disposal must be addressed at the time the decision is made to initiate a given process or operation. If, upon assessment and review, it is determined that a proposed activity will generate wastes that have no available option for disposal, it should not be approved (DOE, 1999). The EPA took further action in 1993 by announcing that it was committed to a policy that places the highest priority on waste minimization (EPA, 1993b). Under this approach, the agency requires generators of hazardous waste to certify on their shipping manifests that they have a waste-minimization program in place. The same certification is required for owners and operators of facilities that receive a permit for the treatment, storage, or disposal of hazardous waste on the premises where such waste was generated. Methods for minimizing the production of hazardous waste include the following:

Separating or segregating waste at its source to prevent hazardous materials from contaminating nonhazardous waste and thereby making the entire mixture hazardous;

Eliminating raw materials that generate a large amount of hazardous waste or *substituting* raw materials that generate little or no hazardous waste for those that generate a large amount (for example, the use of nonhazardous materials);

Changing manufacturing processes to eliminate steps that generate hazardous waste, or altering processes so that the waste is no longer produced (for example, using more effective and efficient methods of applying paints).

Table 9.5 summarizes some of the techniques that can be used to minimize the production of specific types of solid waste.

WASTE MINIMIZATION

Although waste minimization was initially viewed as one more regulatory burden, many industrial leaders now acknowledge that it has at least three major advantages. First, it makes disposal inherently safer because of the reduced risk and volume of the wastes being generated. This, in turn, makes disposal of the resulting waste more acceptable to the public. Second, minimization reduces the overall cost of waste management and disposal. The savings arise not only from the reduction in the volume and risk of the waste, but also from a reduction in the indirect costs, for ex-

Table 9.5 Techniques for minimizing the production of hazardous wastes

Inventory management and improved operations
Inventory and trace all raw materials
Emphasize use of nontoxic production materials
Provide waste-minimization or reduction training for employees
Improve receiving, storage, and handling of materials
Modification of equipment
Install equipment that produces minimal or no waste
Modify equipment to enhance recovery or recycling options
Redesign equipment or production lines to produce less waste
Improve operating efficiency of equipment
Maintain strict preventive maintenance program
Production process changes
Substitute nonhazardous for hazardous raw materials
Segregate wastes by type for recovery
Eliminate sources of leaks or spills
Separate hazardous from nonhazardous and radioactive from nonradioactive wastes
Redesign or reformulate end products to be less hazardous
Optimize reactions and raw material use
Recycling and reuse
Install closed-loop systems
Recycle on-site for reuse
Recycle off-site for reuse
Exchange wastes
Treatment to reduce toxicity and volume
Evaporation
Incineration
Compaction
Chemical conversion

ample, insurance and long-term liability (NCRP, 2003). Third, waste minimization is often facilitated through reduced consumption and more efficient use of raw materials. This, in turn, can provide a significant cost savings.

WASTE RECYCLING

One step that can facilitate the recycling of solid wastes is to reuse them within the process (for example, recycling lead storage batteries) or to transfer the waste to another industry that can use it as input to its production process. If reuse requires prior treatment of the waste, this should be considered. It is mandatory for the success of any such efforts that they receive the unequivocal support of all levels of corporate management. One way to achieve this is to make recycling a part of the culture of the organization. The development of a recycling culture, however, should not be restricted to industrial organizations. It can take place at the national level, as exemplified by Executive Order 12873, issued by the U.S. President in 1993, that required all federal agencies to purchase only recycled copier paper. The U.S. Congress took similar action in 1996 through passage of the Mercury-Containing and Rechargeable Battery Management Act. This act required operators of stores that sell batteries to accept them back at the end of their life for possible recycling of their toxic components.

Governmental organizations at the state and local levels can also play a key role in assuring the success of a recycling program. More than 40 states in the United States have established recycling goals. One of the most ambitious is Rhode Island, which seeks to attain the recycling of 70 percent of its garbage within the next few years. In a similar manner, the state legislature in Oregon has mandated that its Department of Environmental Quality conduct an annual comprehensive survey of progress in the management of solid waste. At the local level, nearly 4,000 communities now levy user charges, often called "pay-as-you-throw," or unit-based pricing, on municipal solid waste (Portney and Stavins, 2000). Through the stimuli of these types of incentives, the amount of waste being generated per person is decreasing and the percentage being recycled is increasing. In some states, overall recycling is now at almost 40 percent (Padgett, 2001). The net effect is a reduction of more than 25 percent over the past 20 years in the percentage of municipal waste that is sent to landfills for disposal.

Similar actions to promote recycling can be initiated by members of the public. The success of such efforts is demonstrated by the fact that millions of people in the United States routinely sort their trash, fill recycling bins, demand to be able to purchase products made of recycled materials, and avoid products with wasteful packaging. More than 80 percent of the cities in the United States now have operational curbside recycling programs, and more than 60 percent have programs for collecting recyclables from multifamily buildings. Such recycling, however, is not without its limitations. A major problem is the required initial investment in additional collection vehicles and sorting equipment. Whereas normal municipal waste can be loaded onto a truck and compacted for efficient transport, recycled materials cannot. As a result, the amounts of waste that can be hauled by the trucks used for collecting and transporting recyclable materials are far below normal. This leads to increased energy consumption and air pollution. Since the recyclable materials must subsequently be carefully sorted, often by hand, the system is labor intensive, further increasing the costs. In addition, it is important to note that the success of any recycling program is dependent on a demand for products in which the recycled materials can be incorporated. Unfortunately, balancing demand with supply has frequently been a problem, especially during the early phases of a recycling program.

Waste Treatment

Treatment is defined as any method, technique, or process, including neutralization, that is designed to change the physical, chemical, or biological character or composition of a hazardous waste so as to neutralize it, recover energy or material resources from it, render it nonhazardous or less hazardous, or make it safer to transport, store, or dispose of, more amenable to recovery or storage, or smaller in volume (EPA, 1993b). Treatment may be either thermal (for example, incineration), chemical, or biological (especially for hazardous wastes that contain organic materials). Where methods for neutralizing a waste or rendering it nonhazardous are not available or are ineffective, immobilization (stabilization) can often be effective, especially for inorganic hazardous wastes.

The general goal is to convert hazardous waste into a solid form for disposal. Treatment may be initiated at any stage prior to or following solidification, for example, in tanks, surface impoundments, incinerators, or land treatment facilities. Because many of these processes are waste specific, the EPA has not attempted to develop detailed regulations for any particular type of process or equipment; instead, it has established general requirements to assure safe containment (EPA, 1986a, 1986b).

In general, four processes (incineration, heat treatment, solidification and/or stabilization, and chemical treatment) are being used to treat solid wastes. Each of these is described here.

INCINERATION

Incineration deserves special mention because it is one of several processes available both for reducing the volume of solid and hazardous waste and for destroying certain toxic chemicals within it. The increased use of plastics in packaging, however, has created a corresponding increase in the amount of polyvinyl chloride in solid waste. When such plastics are burned, they produce hydrochloric acid. This extremely corrosive compound can destroy incinerator components such as metal heat exchangers and flue-gas scrubbers and can threaten human health if it is released into the atmosphere. Hydrochloric acid can also be produced in incinerators by the combustion of foods and wastes that contain chloride salts. A further problem is that incomplete combustion of some organic materials in the presence of chlorides can produce dioxins, a toxic group of compounds.

These and other potential threats to human health have led to stringent regulations on emissions from incinerator facilities, particularly in light of the realization of the health effects of extremely small airborne particulates (Chapter 5). Although modern technology will provide almost any degree of cleanup required, the economic costs can be high. One response has been to construct and operate centrally located incinerators to serve a group of waste producers. In many communities that (for environmental, political, economic, and other reasons) have a limited capacity for direct disposal of solid waste in landfills, incineration has become the principal method of intermediate treatment. One reason is that the resulting ash is generally in a physical and chemical form that is more readily disposed of than the original waste, and it is biologically and structurally more stable. In addition, many of the compounds it contains are insoluble, so their long-term leaching by rain and groundwater is minimized.

Heat Treatment

Heat, applied at moderate temperatures, is effective in treating soils, particularly those that are contaminated with volatile solvents such as creosote and diesel and gasoline fuels. This approach has been used for years to enhance the removal of oil from the ground. The heat can be applied either through submerged electric heaters or steam-injection wells. Electric heating is especially effective in clay soils, which are not very permeable and thus tend to have higher moisture content. The presence of water not only enhances the conductance of the electricity but also produces steam, which expands and dries the clay matrix. Steam treatment, in contrast, is more effective in soils that are more permeable. The net result is that the heat either volatilizes or immobilizes the contaminants. In the latter case, the contaminants can either be removed or destroyed in place. An interesting by-product of this approach is that in some cases, the presence of heat has attracted thermophilic bacteria that have assisted in stabilizing the contaminants, including some that other bacteria could not. This approach also has the benefit of not requiring workers to handle the contaminated soil, and the soil, in turn, does not have to be removed and transported elsewhere. Since 70 percent of the Superfund sites are contaminated with solvents, heat-treatment methodologies may have widespread application (Black, 2002).

SOLIDIFICATION AND/OR STABILIZATION

Solidification and/or stabilization of solid wastes can be accomplished by several techniques. Plasma power is one of the newer technologies that is being applied in the treatment of such wastes. The temperatures that this technology is capable of producing (in excess of 7,000°C) can melt or vaporize contaminated soil and a full range of typical wastes and garbage and produce a glass- or sandlike residue. Through this process, hazardous and toxic chemicals and biological agents are reduced to their elemental components. This technology, which was developed for laboratory studies of heat shields designed to protect spacecraft during reentry, is being used in Japan for the treatment of municipal solid waste and automobile shredder residue. One plant, which has a capacity of 20 tons per day, went into operation in 2002. The hot gases that are produced, which consist primarily of hydrogen and carbon monoxide, are sent to a secondary combustion system where they are mixed with water to form steam, which can be used to run a turbine and generate electricity. Most of the sandlike residue is currently mixed with cement to form bricks that are used in pavements. Much larger plants are under construction. A similar plant has been constructed in France, and plans are under way to construct a plant in the state of Georgia that will have the capacity to process upwards of 100 tons of tires per day. The steel in the tires will be drawn off as ingots. Since tires also contain sulfur, the off-gases will be treated to remove this contaminant (Link-Wills, 2002).

CHEMICAL TREATMENT

One of the common applications of chemical treatment is in the treatment of corrosive solids, such as lime or cement kiln dust. These can be neutralized by using either chemicals or acidic wastes from other operations within a plant. Specially formulated solutions are being used to leach organic or inorganic contaminants from soils either in situ or ex situ.

Through this process, some compounds can be chemically converted to related but much less mobile or less toxic versions; for example, chromium VI can be converted to less toxic chromium III. In a similar manner, some chlorinated organics, such as polychlorinated biphenyls, can be degraded in soils or other solids by using various sodium-based reagents.

Waste Disposal

Disposal, by definition, means the discharge, deposition, injection, dumping, spilling, leaking, or placing of any solid or liquid waste into or onto the land or water. It has crucial ramifications for environmental health because disposal may permit the waste and/or its constituents to enter the terrestrial environment, be emitted into the air, or be discharged into surface waters. The potential contamination of groundwater is also a concern.

As previously mentioned, the primary method for managing and disposing of municipal and hazardous waste is burial in the ground. Such disposal includes a range of options (EPA, 1986b):

Landfills. Disposal facilities in which the waste is placed into or onto the land. In most landfills, the wastes are isolated in discrete cells within trenches. To prevent leakage, landfills must be lined and have systems to collect any leachate or surface runoff. In this regard, it is important to distinguish between landfills and surface impoundments. The latter are typically considered storage units; they are not an effective method for disposal. A typical hazardous waste disposal facility is shown in Figure 9.2.

Underground injection wells. Steel- and concrete-encased shafts placed deep in the earth into which wastes are injected under pressure. Although this method was used in the past on a regular basis, it is being applied in the United States today only in the case of oil and gas wells that are exempted from hazardous waste regulations. In general, underground injection of hazardous wastes is no longer permitted in the United States.

Waste piles. Noncontainerized accumulations of insoluble solid, nonflowing hazardous waste. Some waste piles serve as final disposal, many as temporary storage pending transfer of the waste to its final disposal site.

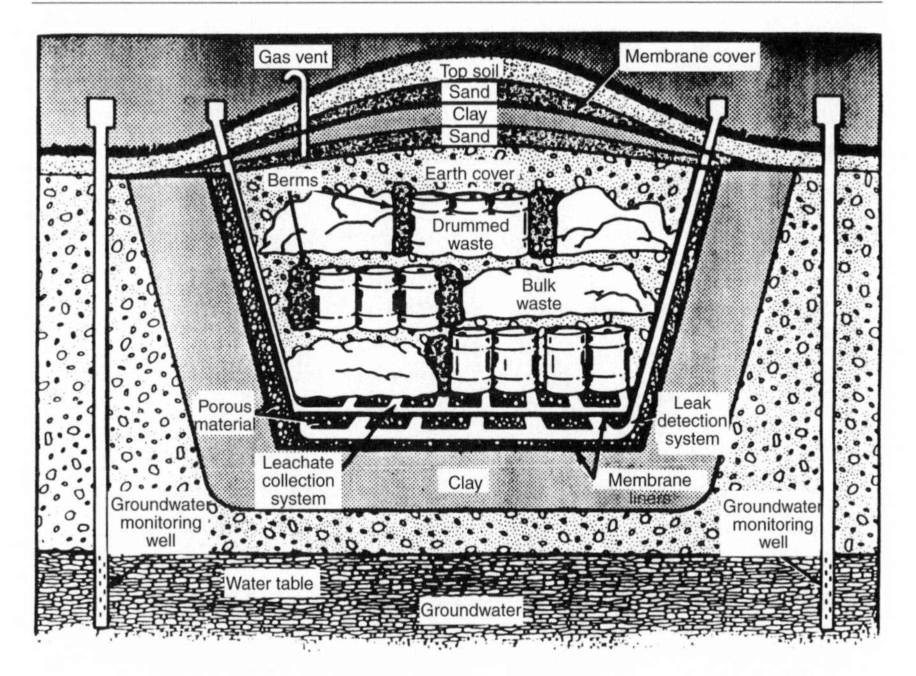

Figure 9.2 Land burial facility for hazardous waste

Land treatment. A disposal process in which solid waste, such as sludge from municipal sewage-treatment plants, is applied onto or incorporated into the soil surface (Chapter 8). Under proper conditions, microbes occurring naturally in the soil break down or immobilize the hazardous constituents.

More than 200,000 sites in the United States are now being used or were formerly used as sites for the disposal of municipal wastes. Although the wastes as buried were classified as nonhazardous, about 35,000 of these sites are known to have received hazardous chemicals and other materials from small-quantity industrial generators. In addition, a certain amount of the waste from most households contains hazardous materials. Through the 1984 amendments to the RCRA, Congress mandated that the EPA develop new criteria to provide better protection of the public from the potential health risks associated with these facilities (EPA, 1986b). Responding to this mandate, the EPA requires disposal practices for municipal waste that closely parallel those for industrial (hazardous) waste. These requirements, which apply to all aspects of the siting, design, construction, operation, and monitoring of such facilities, can be summarized as follows (EPA, 1993a):

Location. Landfills must not be located on a floodplain. They must also not be built on wetlands unless the proposed operator can show that the landfill will not lead to pollution. They also cannot be located in areas subject to landslides, mudslides, sinkholes, or major disruptive events such as earthquakes, which could lead to pollution. In addition, they cannot be located near airports, where birds that are frequently attracted to such facilities might constitute a danger to aircraft.

Design. Landfills must be designed to avoid contaminating groundwater. Ancillary requirements include lining the bottom of the landfill with clay, covered by an impervious synthetic-material liner, coupled with a system to collect and treat any leachate (liquids) that may collect within the liner.

Operation. No hazardous waste should be disposed of in a municipal landfill, and the waste that is disposed of must be covered daily with dirt to prevent the spread of disease by rats, flies, mosquitoes, birds, and other animals. In addition, access to the landfill must be restricted to prevent illegal dumping and other unauthorized activities; the site must be protected by ditches and levees to prevent stormwater flooding; and any runoff that occurs must be collected and controlled.

Monitoring. Generally, landfill owners or operators must install monitoring systems to detect groundwater contamination. Monitoring for changes, such as subsidence, that may be indicative of possible problems is encouraged. If contamination is observed, the concentrations must be reduced to assure compliance with federal limits for drinking water. Methane gas that is generated through decomposition of the waste must also be monitored and controlled, if necessary.

Closure and postclosure care. Upon ceasing operation, landfills must be closed in a way that will prevent subsequent problems. The final cover must be designed to keep liquid away from the buried waste, and for 30 years after closure, the operator must continue to maintain the cover, monitor the groundwater to be sure the landfill is not leaking, and collect and monitor any subsequent generation of gases. As will be noted in Chapter 18, methods are now being applied to collect the methane gas released from landfills and use it as a source of energy.

As shown by the data presented in Table 9.6, the relative use of landfills for the disposal of municipal solid waste has been decreasing during the past two decades. Concurrently, the percentage of the waste that is being composted and recycled has been dramatically increasing. Such an increase is both logical and beneficial. It is logical because upwards of a third of municipal waste (for example, food products, yard trimmings, and wood) is organic in nature. It is beneficial because any such wastes that are composted do not have to be collected and transported to a disposal facility. At the same time, these changes reflect the increasing numbers of residents who have initiated home and backyard composting operations. Adding to this movement is the ready availability of lawnmowers that mulch the cut grass and leave it on the ground. To promote this approach, some municipalities provide subsidies to homeowners who purchase this type of mower. Wood chippers that cut brush debris into small particles and make it readily suitable for composting are also increasingly being used. These are now commercially available for use on an individual household basis. Another contributing factor is the promotion by community leaders of the establishment of centralized composting facilities (Stuckey and Hudak, 2002).

Table 9.6 Trends in the disposal of municipal solid waste

	Percent		
Disposition	1980	1990	1996
Landfills	81.3	66.6	55.4
Combustion			
Waste to energy	1.8	15.2	16.1
Without energy	7.3	1.1	1.1
Recycled			
Composted	0.0	2.2	5.4
Other	9.6	14.9	21.9

Superfund and Associated Sites

Thousands of waste disposal sites established in the United States during the past 30 to 50 years were improperly designed or operated and have leaked, or have the potential to leak, hazardous waste into the environment. Recognizing the severity of this problem and the urgent need for cleanup of these sites, Congress in 1980 passed the Comprehensive Environmental Response, Compensation, and Liability Act (CERCLA).

SUPERFUND SITES

CERCLA, more commonly referred to as the Superfund Act, authorized the EPA to investigate various waste disposal sites and to identify them as potential Superfund sites. By 1994, more than 40,000 such sites had been so identified. Those sites with the highest levels of contamination and deemed to present the most serious threats to health are placed on what is called the National Priorities List (NPL). By September 1995, almost 1,400 sites had been so designated. For these, the EPA identifies the potentially responsible parties and gives them an opportunity to implement cleanup. If they fail to do so, the EPA arranges for the cleanup, using Superfund money, and then seeks to recover the costs from the responsible parties. As of 1995, work was under way at more than 90 percent of the NPL sites, final cleanup activities were in progress at about 35 percent, and such activities had been completed at another 25 percent (CEQ, 1997).

As part of what might be called a "streamlining" effort, the EPA concluded in late 1995 that approximately 24,500 of the potential Superfund sites were of such low priority that they could be removed from the list. Through that effort, the number of sites that remained in the Superfund inventory was reduced to about 15,500. The sites that were removed are now covered by the Brownfields Development Initiative (discussed later). The net impact of this change is that slightly more than 1,300 sites remain on the NPL or have been proposed for listing (CEQ, 1997). About 200 of these sites are former municipal landfills; many others were contaminated by operations of the U.S. Departments of Defense and Energy.

One approach that can be used for the cleanup of these sites is to excavate the contaminated material and transport it to a new burial site. Because in many cases the quantities involved are enormous, various methods for on-site treatment are being developed. These involve one or more of the previously described physical and chemical processes. Which treatment option is selected depends on the types of contaminants and the relevant properties of the soil—for example, its clay and humus content. In the case of soils that contain organic chemicals, the most proven separation technology is volatilization, using vapor extraction and/or forcing air through the soil. Many sites are also being remediated by solidification/stabilization. One approach is to render the contaminated soil inert by mixing it with additives such as cement (Fox, 1996).

Another method that is being extensively evaluated is biological treatment, which offers two distinct advantages: it is inexpensive and has the unique potential for rendering hazardous constituents nontoxic. In contrast to the more conventional applications of such processes (for example, in the treatment of domestic sewage), biological treatment of contaminated soils is an immature field that offers high expectations but is confronted with numerous scientific and engineering challenges (Hughes, 1996). Treatment can be pursued through the introduction of new organisms or through depending on attenuation by the organisms already present. The latter approach is deemed acceptable if public health is not at risk, natural mechanisms will degrade or decrease contaminant concentrations over a reasonable period of time, and monitoring can be used to assure that the concentrations of the contaminants of concern are indeed being reduced.

THE BROWNFIELDS REDEVELOPMENT INITIATIVE

In addition to the previously cited low-priority facilities that were removed from the list of Superfund sites in 1995, there are up to 600,000 abandoned, idled, or underused industrial and commercial facilities in the United States, many of which have low levels of contamination but are in need of cleanup and restoration. In many cases, these sites are located in economically depressed areas. The Brownfields Redevelopment Initiative is designed to stimulate their cleanup, revitalize the properties, and restore their usefulness. Recognizing the benefits of this program, many state and local governments have agreed to provide economic incentives to private-sector companies that redevelop such areas. Such incentives include grants, tax exemptions or abatements, low-interest loans, waiver of impact and permit fees, expedited development approvals, and marketing and promotional assistance (Verbit, 2001). The EPA is also providing strong support. Its staff has agreed that if an enforceable arrangement can be entered into by a responsible state/local agency and a willing developer, its primary role will be to observe and ensure that progress is being made.

Buoyed by the success of the program, Congress passed the Small Business Liability Relief and Brownfields Revitalization Act, which became law in 2002. This act increased the incentives provided for the cleanup and reuse of brownfields (Isler and Lee, 2002). As a result, long-neglected parcels of land in many areas of the country, particularly ones in which the degree of contamination is low to moderate, are now being converted into valuable new property. Common uses are to convert the land into parks and/or sites for industrial buildings. For those sites in which the extent of the contamination is very low, only minimal cleanup may be required. Where the concentrations of contaminants are relatively high, it may be necessary (as in the case of some Superfund sites) to excavate the contaminated soil. This can be very expensive, both because of the equipment and transportation involved and the necessity to locate an acceptable site for disposal of the contaminated soil.

Initially, the public was almost universally opposed to the construction of houses and schools or the creation of open spaces on reclaimed brownfields sites. In the late 1990s, however, this view changed for at least two reasons: such sites provided a readily available supply of land on which to build new housing units, which were in high demand; and regulators were careful to ensure that the degree of cleanup was acceptable to the local community.

GREYFIELDS

Although not as well known as brownfields, the redevelopment of greyfields represents another major effort under way to revitalize certain areas within cities. Greyfields are exemplified by failing malls and strip malls that do little if any business. A key characteristic of such areas is that they are large tracts of land accompanied by both empty retail space and parking lots. Outclassed by newer, more modern malls and shopping centers, these dying business districts have simply not generated sufficient revenue to sustain their use. The land they occupy, however, can be very useful and economically profitable to local communities. The concept of mixed-use redevelopment has demonstrated that greyfields can be converted into vibrant city centers that will be both profitable and sustainable (Chen, 2002).

Management of Radioactive Wastes

As is true of hazardous chemical wastes, the management and disposal of radioactive wastes are receiving extensive governmental attention. Groups involved at the federal level include Congress, the EPA, the U.S. Nuclear Regulatory Commission (USNRC), and the Department of Energy (DOE). In general, Congress passes relevant legislation (Table 9.7), the EPA sets

Table 9.7 Principal federal laws related to management and disposal of radioactive waste

Year	Law	Public Law number
1954	Atomic Energy Act	85–703
1978	Uranium Mill Tailings Radiation Control Act	95–604
1980	Low-Level Radioactive Waste Policy Act	96–573
1983	Nuclear Waste Policy Act of 1982	97–425
1986	Low-Level Radioactive Waste Policy Amendments Act of 1985	99–240
1987	Nuclear Waste Policy Amendments Act	100–203
1992	Energy Policy Act of 1992	102–486

applicable environmental standards, and the USNRC develops regulations to implement the standards. As the ensuing discussion shows, such wastes have been separated into four separate groups. The first group, low-level radioactive wastes, has been subdivided into Class A, B, and C wastes, depending on the types and quantities of radioactive materials they contain. Of the three, Class A is the least hazardous and Class C the most.

LOW-LEVEL RADIOACTIVE WASTES

Low-level radioactive wastes, which represent by volume more than 80 percent of the total radioactive waste generated by the commercial sector, include those produced through the operation of nuclear power plants and related industrial facilities, the decommissioning and decontamination of nuclear facilities, and the use of radioactive materials in medicine, research, and industry. While this amounts on average to about 100,000 cubic meters (3.5 million cubic feet) per year, it contains much less than 1 percent of the total quantity of radionuclides that will ultimately need to be sent to disposal. The vast majority of the radionuclides are contained in high-level radioactive wastes (discussed later). By volume, the major share of low-level waste is generated by industrial organizations; only 8 percent is produced through the operation of nuclear power plants.

At present, low-level radioactive wastes produced in the United States are being disposed of in one of three facilities: the Barnwell facility in

South Carolina, the US Ecology facility in Washington, and the Envirocare facility in Utah (Zacha, 2003). Each of these facilities, however, has restrictions on either the types of wastes it can accept or the states in which the wastes can originate. Although the Barnwell facility currently accepts Class A, B, and C wastes, its acceptance of wastes will be restricted, beginning in 2008, to generators located in states that are members of the Atlantic Compact. As of 2003, wastes being accepted by the US Ecology facility were restricted to generators in the State of Washington. While the Envirocare facility is open to waste generators throughout the United States, it is licensed to accept Class A wastes only. The relative volumes and activities of the radionuclides disposed of in each of these facilities are shown in Figure 9.3.

In earlier days, the approach commonly used in the disposal of low-level radioactive wastes was shallow land burial, very similar in principle to the approach used in the disposal of municipal wastes. As better methods were developed, there was a gradual but steady shift in what was considered to be acceptable (Table 9.8). A major stimulus for this change was the increasing involvement of citizen groups in planning such

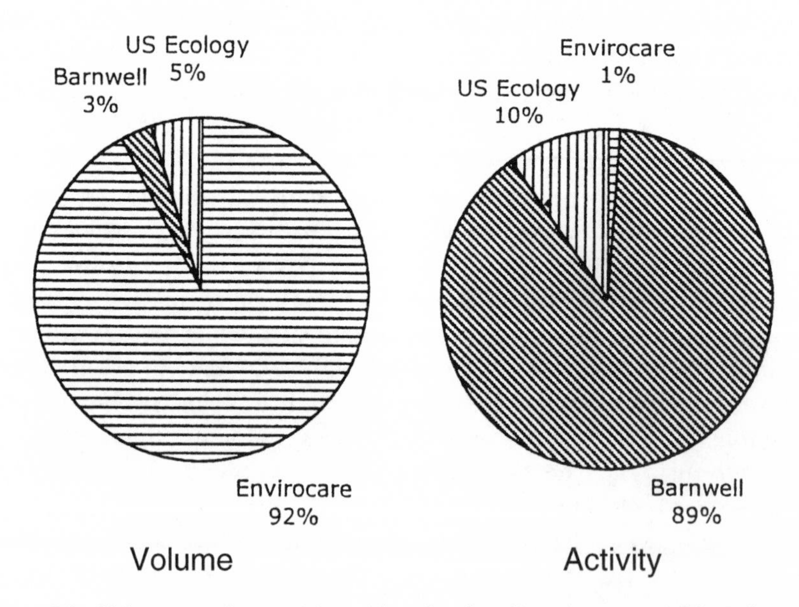

Figure 9.3 Volumes and quantities of low-level radioactive waste shipped to U.S. disposal facilities, 2000

Table 9.8 Trends in low-level radioactive waste disposal facility requirements

Factor	Early approach	Intermediate approach	Latest approach
Technology	Simple landfill	Advanced landfill with liners	Multiple engineered barriers
Waste containers	Wooden boxes and 55-gallon drums	Metal and high-integrity containers	Metal and high-integrity containers placed within a vault
Container handling	Open facility: random dumping	Open facility: individual placement	Covered facility: individual placement
Record keeping	Simple records	Detailed records with computer storage	Detailed records using on-line computer

activities. As a result, low-level radioactive wastes are now being more securely packaged and placed in more robust facilities (Figure 9.4).

URANIUM MILL TAILINGS

Uranium mill tailings are low-level radioactive wastes that were produced primarily as a result of activities pertaining to national defense. They are the materials (so-called tailings) that remain after the uranium metal has been separated from the original ore. On a relative basis, the volumes (measuring in the millions of cubic meters) are significantly larger than those of the low-level radioactive wastes generated by the commercial sector. For this reason, mill tailings are handled "in place," that is, they are stabilized and provided with a cover to protect them from wind and water erosion. The Army Corps of Engineers has jurisdiction over the management and disposal of these wastes.

TRANSURANIC WASTES

Transuranic wastes are those that contain, as the name implies, elements heavier than uranium that also have half-lives and concentrations in ex-

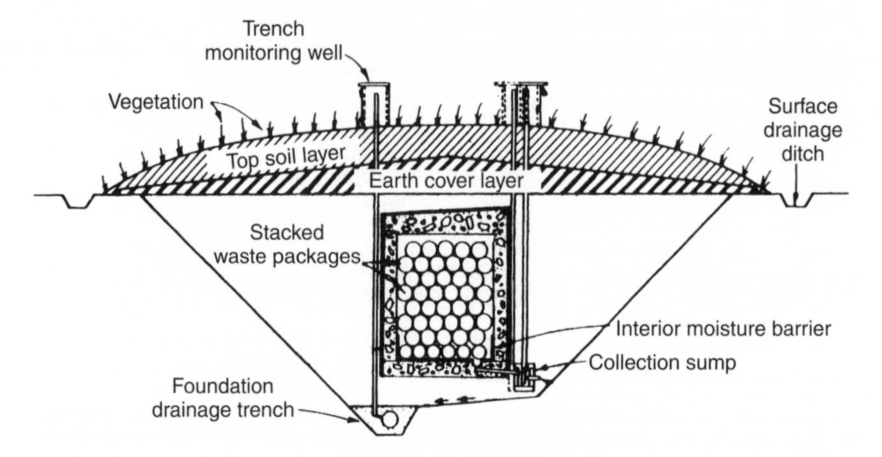

Figure 9.4 Earth mounded concrete bunker for disposal of low-level radioactive waste

cess of certain stipulated limits. Because the radionuclides in transuranic wastes tend to be long lived and highly toxic, they are being disposed of in the Waste Isolation Pilot Plant, a deep underground repository that has been constructed in southern New Mexico. Plans for this facility were reviewed by the EPA, and the facility was approved for operation in March 1999. The first storage room, which now contains more than 10,000 barrels/drums of waste, was filled in 2002. Additional shipments are being made on a continuing basis.

HIGH-LEVEL RADIOACTIVE WASTES

High-level radioactive wastes include spent (used) fuel removed from commercial nuclear power plants and fission-product wastes that were produced in the process of manufacturing plutonium for nuclear weapons. The latter activities are no longer being conducted in the United States. The spent fuel wastes now being stored at the individual nuclear power-plant sites, are tentatively scheduled for disposal in an underground geological repository proposed for construction in Yucca Mountain in the southwest region of the state of Nevada. Prior to disposal, the fuel will be enclosed in containers specifically designed to assure long-term retention of the associated radioactive materials. Liquid wastes generated from the chemical processing of spent fuel in past years in the United States will be vitrified and sealed in robust containers prior to disposal. Long-term confinement of the high-level waste will be dependent on engineered barriers (including the solidified form of the waste and the container in which it is encased), combined with the natural geological features of the site.

In accord with legislation passed by the U.S. Congress, standards for the proposed repository were established by the EPA; its construction and operation were assigned to the U.S. Department of Energy; and the review and ruling on the acceptability of the facility were assigned to the USNRC. If the proposed facility is deemed acceptable, the first waste is tentatively scheduled for disposal in 2010.

Other countries of the world are moving ahead with similar plans for the disposal of their high-level radioactive waste. Sweden and Finland, for example, are making progress in their respective plans for selecting a site and initiating construction of an underground repository. In the former case, two possible sites are under consideration, with the specific site to be confirmed by 2007. The schedule calls for the proposed repository, which is to be located in bedrock (500 meters, about 1,650 feet) beneath the surface, to be ready for commercial operation in 2015. In the case of Finland, the goal is to have the proposed repository, which is to be located on the island of Olkiluoto, ready to accept high-level waste in 2020. Other European countries that are conducting research and exploring possible repository sites include Switzerland and the Czech Republic (Sperber, 2002).

International Waste Transport

As noted in the introduction to this chapter, the reduction in the number of solid-waste disposal sites in the United States has led to an increase in the export of such wastes from one state to another. Similar shipments are taking place internationally, some of which have resulted in dramatically unfavorable situations. In addition, investigations have revealed that in several cases such exports, involving shipments from developed countries to less developed countries for disposal, were arranged through what are called "silent trades," negotiated in secret. Others, however, had been arranged under contracts signed with governments of the importing countries. Unfortunately, such contracts were assessed by officials in the receiving countries solely in terms of the economic benefits. One such arrangement led to the disposal of several thousand tons of hazardous wastes at inland and coastal sites in Lebanon in 1987. As would be anticipated, this action produced a critical environmental situation that was exacerbated by the fact that the human, technical, and financial resources to manage the associated environmental and public health impacts were not available.

Recognizing the need for action, in 1989 the United Nations Environmental Programme convened a meeting in Basel to review these matters and develop recommendations for avoiding the repetition of such events. Three primary recommendations were an outgrowth of these deliberations: (1) before any such waste is shipped, appropriate officials within the recipient country must be notified and indicate their consent; (2) officials in the countries through which the waste would travel must similarly be notified; and (3) officials in the transient and importing countries must provide their written consent to the arrangements. Participants in the convention also stipulated that they considered participation in "silent trades" to be a criminal act (Jurdi, 2002a). These recommendations were expanded in scope at a follow-up conference held in Geneva in 1994 that involved the countries within the Organization for Economic Cooperation and Development (OECD). One recommendation was that all transboundary movements of hazardous waste from OECD to non-OECD countries be banned (Jurdi, 2002b).

The General Outlook

The generally accepted philosophy today is that waste management and disposal must be recognized as an integral part of any type of industrial operation. Effective implementation of this philosophy requires that the challenges and potential difficulties of waste management and disposal be considered at the time the decision is made to initiate the industrial operation that will generate them. This is particularly true for the multitudes of consumer products that enter the solid-waste stream on a daily basis. In this regard, one might readily ask whether an industry that claims to be environmentally responsible should not be held accountable if the products it sells are purposefully designed to fail (even under routine use) and/or rapidly become obsolete. Examples range from cell phones, television sets, and computers to automobiles. Why must a person purchase a new car simply because the old one is no longer in style? Perhaps the world is in need of a change in culture so that style or some relatively minor modification is not a primary motivation for discarding a product that could continue to be used. Many economists have called for the establishment of a system that would require all manufacturers to take back their products at the end of life. This might well be one approach to reduce these types of practices.

Under the present classification system, some wastes are being managed more stringently than necessary, leading to higher costs than warranted, while others are being managed less stringently than necessary, with the potential of adverse effects. Efforts are under way within the United States to develop a common risk-based approach for classifying all types of solid waste. One of the goals of these efforts is to eliminate these types of problems. Another benefit would be to increase public confidence in waste-management and disposal activities. As envisioned, such a system might also make it possible to establish an exempt class of solid wastes. This designation would be restricted to those wastes that pose a risk sufficiently low that they do not impose an unacceptable risk on any member of the public or the environment. Such wastes could be managed as if they contained nonhazardous materials, that is, they could either be disposed of in a municipal/industrial landfill or recycled and reused (NCRP, 2002).

10

RODENTS AND INSECTS

SCIENTISTS estimate that there are more than 3 million insect species in the world. Of these, nearly 1 million have been identified, including more than 100,000 species of butterflies and moths, more than 100,000 species of ants, bees, and wasps, and almost 300,000 species of beetles. At any one time, the total number of insects on Earth is believed to be about 1 million trillion (Wilson, 2002). Even so, about 4,000 new varieties are discovered each year. Some, such as the honeybee and the silkworm, bring financial benefits; in fact, honeybees are responsible for the pollination of some $10 billion worth of agricultural products in the United States each year. Other insects (such as the butterfly and lightning bug) are aesthetically pleasing. Still others (such as flies, mosquitoes, boll weevils, corn borers, termites, and locusts) are destructive and may even be dangerous to humans. The mosquito, in particular, is the vector (transmitter) of a wide range of disease agents. So far as is known, it serves no useful purpose. Mosquitoes do not, for example, serve as an important food for any other creature, nor do they pollinate flowers (Shaw, 2001). That is not to say, however, that mosquitoes have not played a major role in the history of the world. On the contrary, mosquitoes have dealt fatal blows to armies, they delayed the construction of the Panama Canal for years, and, as will be noted in the subsequent discussion, they continue to kill millions of people each year through the transmission of diseases such as dengue fever, yellow fever, encephalitis, West Nile fever, filariasis, and malaria (Spielman and D'Antonio, 2001).

Rodents too are known transmitters of disease agents and represent a major challenge to environmental health. It is estimated that in the United

States there are 140 million rats, or one for every two people. Table 10.1 lists rodents, as well as various insect and noninsect vectors, and the diseases they can transmit. These vectors and vector hosts have major public health, social, and economic impacts throughout the world.

The descriptions that follow of various rodents and insects will feature a vast array of diseases, some of which can be perpetuated solely in nonhuman hosts. Such infections are known as zoonoses. These include bubonic plague, in which rats serve as the nonhuman host; rabies, for which bats, raccoons, foxes, and skunks serve as the nonhuman host; and various hantaviruses, for which deer mice serve this function. Also featured will be emerging zoonoses, such as West Nile fever, for which mosquitoes serve as the vector but for which birds serve as the major reservoir, and Lyme disease, for which the deer tick serves as the vector but for which deer serve as the nonhuman host.

Table 10.1 Public health impact of various disease vectors and hosts

Vector	Impact
Flies	Diarrhea, dysentery, conjunctivitis, typhoid, cholera, fly larvae infestations, annoyance
Mosquitoes	Encephalitis, malaria, yellow fever, dengue, West Nile virus, filariasis, annoyance, bites
Lice	Epidemic typhus, louse-borne relapsing fever, trench fever, bites, annoyance
Fleas	Plague, endemic typhus, bites, annoyance
Mites	Scabies, rickettsial pox, scrub typhus, bites, allergic reactions, annoyance
Ticks	Lyme disease, tick paralysis, tick-borne relapsing fever, Rocky Mountain spotted fever, tularemia, bites, annoyance
Bedbugs, kissing bugs	Bites, annoyance, Chagas' disease
Ants	Bites, annoyance
Rodents	Rat-bite fever, leptospirosis, hantaviruses, salmonellosis

Rodent-Related Zoonoses

Rodents have been a public health problem for centuries. Most famous is the role of rats in the successive epidemics of bubonic plague, collectively known as the Black Death, that swept Europe in the fourteenth century. One of the earliest recorded epidemics was launched in 1347 in Genoa, when ships arriving from Black Sea ports brought with them rats that were carrying infected fleas. The subsequent spread of bubonic plague depopulated some 200,000 towns and in three years killed 25 million people, or a quarter of the population of Europe. The Black Death remains the greatest calamity in human history. In 1665 another epidemic of plague killed 100,000 people in London. A major outbreak in India in 1994 resulted in at least 1,000 cases, with almost 100 deaths. Worldwide, up to 3,000 cases of bubonic and pneumonic plague occur each year, including from 10 to 15 sporadic cases in the United States. The causative agent is the bacterium *Yersinia pestis.*

Four species of rodents are of environmental concern in the United States today: the Norway rat, the roof rat, the house mouse, and bats. Species of importance in other parts of the world include the Polynesian rat (*Rattus exulans*), which has spread from its native Southeast Asia to New Zealand and Hawaii, and the lesser bandicoot (*Bandicota bengalensis*). which is predominant in southern Asia, especially India. The main impact of these rodents is their widespread destruction of food, particularly grains (Canby, 1977).

Knowing the characteristics of the rodents that pose an environmental problem is essential to their control. Figure 10.1 highlights the distinguishing characteristics of the most common domestic rodents in the United States.

The Norway rat (*Rattus norvegicus*) is characterized by its relatively large size and short tail. Norway rats frequent the lower parts of buildings and inhabit woodpiles, rubbish, and debris. They also burrow under floors, concrete slabs, and footings and live around residences, warehouses, and chicken yards and in sewers. They nest in the ground and have a range of 100–150 feet.

The roof rat *(Rattus rattus)* is characterized by its smaller size and longer tail. Roof rats live in grain mills, dense growth in willows, and old residential neighborhoods. They are excellent climbers and frequently occupy shrubbery, trees, and upper parts of buildings. They usually nest in buildings and have a range similar to that of the Norway rat.

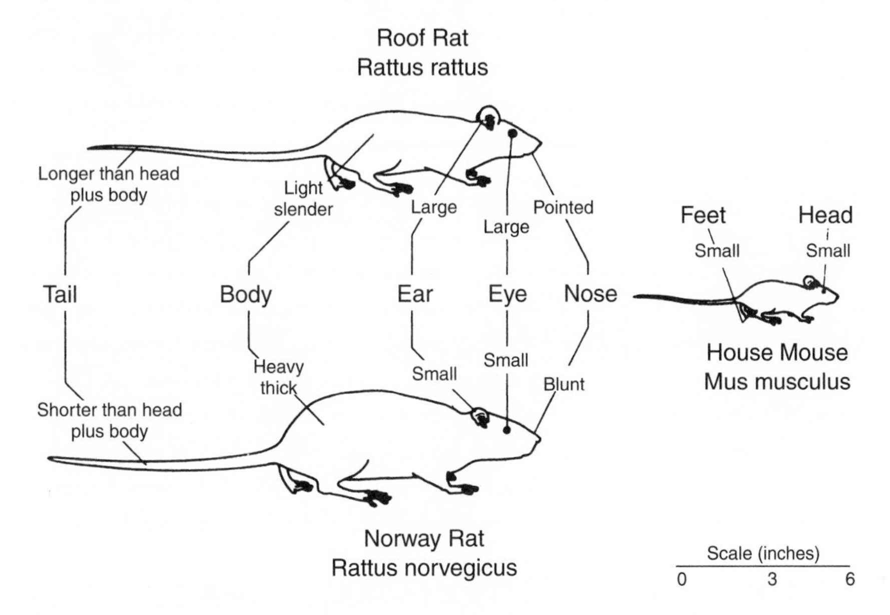

Figure 10.1 Distinguishing characteristics of three species of rodents

> The house mouse (*Mus musculus*) is characterized by its small size, including small feet and eyes, and long tail. Mice live in buildings and in fields, and their range is limited (3–10 meters, 10–30 feet).

As will be noted in the discussions that follow, bats can also be an important transmitter of diseases, such as rabies.

RATS

Among the fastest-reproducing mammals, rats have a gestation period of 21–25 days and can reproduce every 60–90 days. A typical litter ranges in size from five to nine. Their life span is 9–12 months. Rats are very intelligent and survive in a hostile human environment by means of complex social mechanisms.

Characteristics and Impacts Although their vision is poor, rats have a keen sense of smell; they like the same food as people and prefer it fresh. Yet they can eat decayed food and consume contaminated water with no apparent ill effects. Rats also have a well-developed sense of touch via their nose, whiskers, and hair. All rats are accomplished swimmers.

At the same time, rats have a strict social structure. Although those in neighboring colonies will tolerate one another, those in nonadjacent colonies are openly hostile. Rats are rarely seen in the daytime. They seldom pick a fight with people. They follow established paths, which are readily identified by the presence of droppings (feces), grease marks (where the rats have rubbed), urine stains (located by using ultraviolet light), and their characteristic odor. Rats are nonetheless extremely adaptable and can survive under adverse conditions (Canby, 1977). Scientists who returned to Pacific islands that were virtually destroyed in nuclear weapons tests in the early 1950s found flourishing colonies of rats. The same species that lives in burrows in the United States and in attics in Europe can live in palm trees in the South Pacific. Other species, finding shortages of food on land, have learned to dive into lakes and ponds to catch fish.

Rats affect human life in many ways. In addition to bubonic plague, rats can transmit typhus fever through infected fleas, salmonellosis through food contaminated by their urine, and rat-bite fever through a spirochete in their blood. In poor housing conditions, infants, paraplegics, and people under the influence of alcohol or drugs are especially vulnerable to rat bites. On babies, the targets for mutilation are often the nose, ears, lips, fingers, and toes. The impacts do not end there. For many inner-city residents, the presence of rats is a vivid and gruesome symptom of community environmental degradation, a token of the larger pattern of social and economic breakdown and disorder in the real world of the urban poor. The appalling quality of life in such conditions often becomes clear to others only during urban renewal, when rats from buildings that are being torn down or renovated stream into adjacent neighborhoods.

Rats can also have significant economic impacts. One rat can eat 10 bushels of grain or 40 pounds of food per year. Rats are estimated to destroy 20 percent of the world's crops annually. In locations such as India, they compete seriously with humans for food. In the United States, rats are estimated to cause about $1 billion in losses per year. This estimate includes fires caused by rodents chewing on electrical wiring. In fact, about 25 percent of the fires in rural areas are caused by rodents (Canby, 1977).

Control The control of rats is complex because it is so closely tied to human behavior and to large-scale social and economic factors. Effective control includes the following measures:

Eliminating food sources. Rats cannot live and reproduce without food. Making food unavailable to them, however, requires control of garbage and refuse, which in turn requires comprehensive public education. Garbage should be stored in metal cans with tight-fitting lids and collected twice a week; otherwise, storage containers will be filled, and residents will switch to plastic bags or cardboard boxes, which rats can easily tear open.

Ratproofing. Basic to the long-term control of rats is the enactment of strict codes to ensure that all new buildings are ratproof. That is, they should be designed and constructed not to have any openings large enough for rats to enter or leave. In addition, all existing buildings should be surveyed to confirm that they are similarly ratproof. Where deficiencies are found, all openings should be sealed, and a concrete floor or underground shields should be installed outside the building walls to prevent rats from burrowing underneath. (A young rat can squeeze through a ½-inch opening; a young mouse, through a ¼-inch hole.) Buildings that cannot be ratproofed should be demolished.

Traps, fumigants, or poisons. Once all buildings have been ratproofed and food has been made unavailable, the rat population can be reduced by trapping, fumigation, or poisoning programs both inside buildings and in adjacent outdoor areas. Traps avoid the use of poisons, but the rats that are caught must be collected and buried or incinerated. In the process, any fleas that survive may transfer to people. Fumigants consist of gases, such as calcium cyanide and methyl bromide, that are released inside buildings. They provide a quick kill but require care to prevent dangerous exposures of humans during use. Poisons are generally placed in food (baits) for the rats. Examples include warfarin, a slow-acting anticoagulant rodenticide; red squill, a bitter-tasting red powder that causes heart paralysis; zinc phosphide, a fast-acting black powder with a garlic odor that reacts with acid in the rat's stomach to produce phosphine gas; and norbormide, a fast-acting poison that causes shock impairment of blood circulation.

Effective use of rat poisons is difficult. Some rats, for example, have become immune to warfarin and, as a group, they have developed efficient feeding strategies that enable the members of a colony to avoid poisoned baits and to adjust to sudden changes in the food supply. Both laboratory and wild rats tend to avoid any contact with novel objects in their environment. Typically they avoid a new food for several days, and they may never sample it if their existing diet is nutritionally adequate. Eventually, small sublethal quantities may be ingested. If feeding animals become sick, the entire colony thereafter avoids the new food. Other approaches being developed or considered for rat control include single-dose chemosterilants that could sterilize both male and female rats and new rodenticides that are rat specific and thus not hazardous to nontarget animals.

MICE

Mice can serve either as a direct source of disease or as a host for insects that transmit certain diseases.

Characteristics The house mouse can live in any structure to which it can gain entrance. Mice that live outside during the summer tend to move into buildings with the onset of cold weather and heavy rains. Nondomesticated varieties live outdoors essentially all the time; in this case, it is virtually impossible to control their populations.

Disease Transmission Although the effects of most mice are primarily of a nuisance nature, some varieties play a major role, either directly or indirectly, in the transmission of serious diseases. One example of the latter is the previously cited Lyme disease, in which the white-footed mouse serves as a host for the ticks that transmit the disease. For this reason, this disease is covered later in the discussion on ticks. Examples of the diseases in which mice play a dominant role are those caused by the hantaviruses, which for many years have caused episodes of pulmonary disease and killed thousands of people in East Asia. In 1993, an outbreak occurred in the southwestern United States that was subsequently shown to be due to the Sin Nombre virus. Investigations confirmed that the deer mouse *(Peromyscus maniculatus)* was the predominant carrier, and that the disease can be transmitted to humans either through direct contact with infected rodents, rodent droppings, or nests or through inhalation of aerosolized virus particles from mouse urine and feces. As of mid-1999, a total of more than 200 cases had been confirmed in 30 states in this country. Another example of these types of zoonoses is the previously unknown Nipah viral disease, which caused more than 100 deaths in Malaysia during 1999. It closely resembles the Hendra viral disease, which killed two people and more than a dozen horses in Australia in 1994 and 1995 (Enserink, 2000). Still another is the Whitewater Arroyo viral disease, which led to the deaths of several people in California during 1999 and 2000.

Control The most effective way to reduce the risk of disease from the Sin Nombre virus is to limit the exposure of humans to rodents and their excreta. Specific measures to reduce exposures include eliminating food sources available to rodents, limiting possible nesting sites, sealing holes and other possible entrances for rodents, and using "snaptraps" to catch the rodents or rodenticides to kill them. Since brooms and vacuum cleaners can spread the virus, they should not be used to clean contaminated areas.

BATS

As indicated earlier, bats are being increasingly recognized as an emerging factor in the transmission of disease, particularly in the United States. One such disease is rabies, a viral infection transmitted in the saliva of infected mammals. Worldwide, more than 50,000 cases of human rabies occur each year. In contrast to the United States, the vast majority of these are due to dog bites. During 2000, five people in the United States were diagnosed with rabies. Although they were hospitalized and treated, all five died (CDC, 2000). A further problem is that unlike the situation with dogs, people can be bitten by bats and not realize it.

Although it is tempting to condemn bats for their role in the transmission of rabies, they perform a major role in the control of insects. The Mexican free-tailed bat, for example, consumes up to 70 percent of its body weight in insects each night. On average, it is estimated that the 100 million such bats that migrate from Mexico to caves in south central Texas each summer consume as much as 2 million pounds of insects, that is, about 1,000 tons of insects, each night. The overall reduction in the numbers of corn earthworm and tobacco budworm moths is conservatively estimated to save farmers tens of millions of dollars in damages each year, not to mention significantly reducing the amounts of insecticides necessary to protect their crops (McCracken and Dickman, 2002).

Insects

Insects are highly specialized. Houseflies, for example, have hundreds of eyes mounted in such a way as to provide them with wide-range vision, coupled with unusual visual powers. Some insects can detect sex attractants more than 15 miles away. One of the unique characteristics of certain insects is their ability to protect themselves from cold weather. Those with dark colors survive by absorbing sunlight; others gain heat by basking on dark surfaces or have heavy layers of hair or scales that retard heat loss. Some survive subfreezing temperatures by lowering the freezing point of their body fluids, producing compounds that function in a manner similar to the antifreeze used in automobile engine cooling systems (Conniff, 1977).

Insects infect multitudes of people with diverse agents of disease (Table 10.2). Mosquitoes alone cause millions of new cases of malaria worldwide that result in the deaths of about 2.7 million people each year, predominantly children in sub-Saharan Africa. Its economic impacts in that region of the world are vast, amounting to half a billion dollars annually (Satchell, 2000). In addition, an estimated 120 million people throughout the world have lymphatic filariasis, an infection caused by a parasitic worm transmitted by mosquitoes. In fact, filariasis is one of the most rapidly spreading diseases.

Insects also have an enormous economic impact on agricultural production. They attack all stages of plant life, eating seeds, seedlings, roots, stems, leaves, flowers, and fruit; after the harvest, they eat the stored product. Flies and other insects can reduce the yield of milk from dairy cows and eggs from chickens and can cause cattle to lose weight. In many parts of the world, the persistent biting of mosquitoes, black flies, and other bloodsucking insects seriously impairs the productive capacity of workers and sometimes even brings their activities to a standstill.

Table 10.2 Global impacts of tropical disease infections

Disease	Insect Vector	Number of countries affected	Number of people infected (millions)[a]	Total population at risk (millions)[a]
Chagas' disease	Triatomines (kissing bugs)	21	16–18	90
Leishmaniasis	Sandflies	80	12	350
Lymphatic filariasis	Mosquitoes	76	120	900
Malaria	Mosquitoes	>100	300–500[b]	>2,400
Onchocerciasis (river blindness)	Black flies	37	17–18	100

a. Numbers are approximate.
b. Number of cases occurring annually.

MOSQUITOES

Essentially every person in the world has heard the buzzing of mosquitoes and suffered their bites. Their characteristics, especially through interactions with humans, often provide the key to their control.

Nature and Characteristics There are more than 2,500 species of mosquitoes (Shaw, 2001). Though seemingly frail, they show remarkable abilities in flight: those that fly during the day navigate by polarized light from the sun; those that fly at night navigate by the stars. Their wings move even faster than those of a hummingbird—an estimated 250–600 strokes per second (Conniff, 1977)—and produce the familiar whine that is their mating call.

Only female mosquitoes bite people. Some bite only during daylight; others bite only at dusk or at night. They subsequently land on water and lay hundreds of eggs in a raft-shaped mass smaller than a grain of rice. Larvae hatch two days later and swim and feed in the water, breathing through a tube at the surface. After 12 days, the larvae give rise to the pupal stage. Two days later, the young mosquito emerges. Fortunately, less than 5 percent of the eggs become mature adults, and each fall the initial frost kills most of the adult mosquitoes. Nonetheless, mosquitoes can thrive almost anywhere, from the heat of the arid wastes to the frigid Arctic (Shaw, 2001).

Disease Transmission Mosquitoes are transmitters of diverse agents of disease. *Anopheles quadrimaculatus,* which breeds in swamps, is the principal vector of malaria; and *Aedes aegypti,* which breeds predominantly in artificial containers (cans, bottles, old tires), is the urban vector of yellow fever and dengue fever, a debilitating viral disease common in parts of Asia, West Africa, and the Americas. Another disease, the previously cited West Nile fever, has proved to be a major problem in the United States. First observed in New York City in 1999, within three years it had spread to California, and by the fall of 2002, more than 1,400 cases had been detected in humans, 66 of which were fatal. A host of factors add to the complexity of controlling this disease. Since West Nile fever primarily circulates between birds and mosquitoes and probably only incidentally infects humans, horses, and other mammals, control of this disease is complex. A further problem is that very little is known about *Culex pipiens,* the species that transmits the virus among birds in the northern portions of the United States.

Control The control of mosquitoes and mosquito-transmitted infections involves two basic steps: (1) reducing the mosquito population by eliminating their breeding habitats—draining land areas in the case of *Anopheles quadrimaculatus* or applying insecticides or other agents to kill the adult mosquitoes or their larvae; and (2) preventing mosquitoes from biting people and providing medical treatment to individuals who have been, or are subject to being, infected.

As with rodents, effective control of mosquitoes requires extensive knowledge of their life cycles and breeding habits, as well as their role as vectors of disease. Even with this knowledge, mosquito control remains complex because of the large number of species involved and their widely different breeding places, biting habits, flight ranges, and relations to disease. Shoreline towns may be troubled by salt-marsh mosquitoes, inland towns by freshwater mosquitoes. Control is also complicated by the mosquito's rapid development of new behavior patterns, such as the shift from indoor to outdoor blood feeding and resting, in response to insecticide control programs.

Whereas earlier programs to kill mosquitoes provided temporary relief at best, the discovery and exploitation of *Bacillus thuringiensis israeliensis* (Bti), a natural enemy of mosquito larvae, appears to be changing this situation dramatically. The bacterium was discovered in the gut of dead mosquito larvae in an oasis in the Negev Desert. It appears to kill only mosquito and black-fly larvae; it has shown no toxicity to humans or other nontarget organisms. More recently a second larvicide that incorporates the *Bacillus sphaericus* bacterium, commonly present in the soil, has been developed and approved for use in the United States to control mosquitoes, such as the *Culex,* which breed in municipal wastewater lagoons and stormwater basins.

Chemical insecticides continue to be widely used to control adult mosquitoes. In the past, the most commonly used insecticide was DDT [1,1,1-trichloro-2,2-bis(*p*-chlorophenyl)ethane]. Unfortunately, as was the case with the other organochlorine compounds, such as aldrin and dieldrin, DDT proved to be persistent. This fact, combined with the fact that DDT is bioaccumulated within the environment, led the EPA to ban its use in the United States in 1972. A similar ban was placed in effect in Europe. Later, the use of certain other organochlorine compounds (such as aldrin and dieldrin) was also prohibited. As a result, malathion, a far less persistent insecticide, is now the most commonly used insecticide for the control of mosquitoes in this country. While the use of DDT remains controversial, it offers many advantages in the battle against malaria. Two are that it is cheap and its effectiveness is intertwined with the behavioral characteristics of the mosquitoes that transmit malaria. Since *Anopheles* mosquitoes typically bite people indoors at night and the added weight of the blood meal makes it difficult for them to fly, they immediately fly to a nearby wall to rest and excrete the excess fluid from their bodies. If the wall has previously been sprayed with DDT, it will be absorbed into their waxy body coating, they will be killed, and the transmission cycle will be interrupted. Since infectious mosquitoes are those that are being killed, the net effect is far more beneficial than that of reducing their absolute numbers (Shaw, 2001).

Following the bans on DDT, malaria rates, which had been reduced dramatically in many areas of the world through its use, immediately began to increase. Concurrently, the Agency for Toxic Substances and Disease Registry (ATSDR, 2002) concluded, on the basis of an exhaustive review of the literature, that there was no evidence that exposure to DDT at concentrations present in the environment causes birth defects or other developmental effects in people. In a similar manner, the World Health Organization has repeatedly stated its opposition to banning its use, particularly in view of the large number of deaths caused by malaria in the less developed countries. Nonetheless, the bans in the United States and Europe have continued.

Other controls include elimination of breeding zones by digging drainage canals, preventing construction and other practices that lead to the creation of stagnant water, changing the salinity of existing waters, and raising and lowering the water level in lakes, such as those created by dams, to disrupt the life cycle of the mosquito. People can stay indoors except on breezy days or hot afternoons; install screens on doors, windows, and porches; wear protective clothing; cover the sleeping areas for babies and children with nets; and apply mosquito repellents to the skin. The most effective repellents are those that contain diethyltoluamide (DEET). To ensure that children are not harmed, only products containing less than 35 percent of this compound should be used.

FLIES

Various kinds of flies can also be major transmitters of disease agents. River blindness (onchocerciasis, for example, is caused by the bite of small black flies that breed only in rapidly flowing streams with high dissolved oxygen content. Although the disease occurs primarily in Africa, it also threatens more than half a million people in Mexico, Guatemala, Venezuela, Colombia, Ecuador, and Brazil. Worldwide, it is estimated that there are 17–18 million cases of river blindness and that at least 300,000 people have lost their vision as a result. In addition to its direct impact on people, the disease has also had detrimental effects on agricultural production in the affected countries because people no longer are willing to cultivate fertile bottomland near rivers (Carter, 2002).

Principal Species Among the members of the fly family, three have been selected for discussion here. One is the housefly *(Musca domestica),* which is present in many of the temperate parts of the world and may be a carrier of the agents for several diseases. The other two are the screwworm fly *(Cochliomyia hominivorax),* because of its potentially devastating impact on livestock, and the Mediterranean fruitfly *(Ceretitis capitata),* because of its destructive effects on citrus and other fruits.

The housefly. Gray and about a quarter-inch long, the housefly breeds in a variety of decaying animal and vegetable matter, and its larval stage is the maggot. In rural areas, horse, pig, cow, or chicken manure frequently serves as a breeding habitat; human excreta can also be involved where proper disposal methods are not observed.

The housefly's larval stage lasts 4–8 days; the pupal stage, 3–6 days. In warm weather the average time from the laying of eggs to the emergence of the adult is 10–16 days. Flies live 2–8 weeks in midsummer; in cooler weather, up to 10 weeks. Although flies have been reported to travel several miles in one day, most flies present in a given area probably originated nearby.

Although their role in transmitting disease is difficult to document, houseflies pick up and carry a wide range of pathogens (including viruses, bacteria, protozoa, and eggs and cysts of worms) both externally (on their mouth parts, body and leg hairs, and the sticky pads of their feet) and internally (in their intestinal tract). As a rule, pathogens picked up by the larvae are not transmitted to the adult fly, and most pathogens picked up by adult flies do not multiply in them. The germs on the surface of a fly often survive only a few hours, especially if they are exposed to the sun. In contrast, pathogens can live in the intestinal tract and be transmitted to humans when the fly vomits or defecates. In order to eat, the housefly regurgitates a fluid that dissolves its food. Part of this effluent may remain behind on the food when the fly departs and may contain pathogenic organisms. Specific diseases in which houseflies may play a role include typhoid, dysentery, diarrhea, cholera, yaws, and trachoma.

The screwworm fly. The adult has a metallic blue body and three vertical black stripes on its back between its wings. It is about twice as large as the housefly. In contrast to the housefly, it lays its eggs in fresh wounds of warm-blooded animals. Any accidental or surgical wound, a fresh brand mark, or the navel of a newborn animal can serve as the site for initial invasion by screwworm maggots. In warm areas populated by screwworm flies, few newborn calves, lambs, kids or pigs or the young of larger game species escape attack.

The maggots hatch in 12–24 hours and begin feeding on the flesh head down, soon invading the sound tissue. They become full grown in about five days, drop out of the wound, burrow into the ground, and change to the pupal or resting stage. The adult flies emerge from the pupal case after about eight days during warm weather, live for two to three weeks, and range for many miles. The larvae that feed in the wound cause a straw-colored and often bloody discharge that attracts more flies, resulting in multiple infestations by hundreds to thousands of maggots of all sizes. Death is inevitable in the case of an intense infestation unless the animal is found and treated.

Early in the twentieth century, screwworm flies were present in southern Texas and northern Mexico and annually migrated northward into Louisiana and Arkansas. In 1933, screwworm flies appeared in Georgia, presumably introduced through shipment of infested cattle from the Southwest. During that summer, screwworm flies spread southward into Florida, where they were a problem in the 1930s and early 1940s. Subsequent outbreaks occurred in the United States in the late 1950s and the 1970s. As is obvious, screwworm flies can have devastating effects on livestock growers. Losses from screwworm infestations along the Atlantic seaboard in 1958 were estimated at $20 million (Richardson, Ellison, and Averhoff, 1982). Today losses from a major outbreak would be many times that amount. For this reason, ranchers in states such as Texas and Florida gladly pay a tax per head of cattle to finance control programs. Although the screwworm fly continues to be a problem in Central and South America and in Libya, methods (discussed later) have been developed to bring it under control.

The Mediterranean fruitfly. Also known as the Medfly, this fly is slightly smaller than the common housefly, has yellowish orange spots on its wings, and thrives in warm climates. Scientists believe that it originated in West Africa. By 1850 it had spread throughout the Mediterranean region; it was found in Australia in the late 1800s and in Brazil and Hawaii in the early 1900s. In 1929 it was discovered in Florida.

A Medfly typically lays her eggs in a ripe, preferably acidic fruit by drilling tiny holes in the skin or rind while the fruit is still on the tree. Choice targets are oranges, grapefruit, peaches, nectarines, plums, apples, and quinces. In 2–20 days the eggs hatch into larvae, which eat their way through the fruit, causing it to drop to the ground. The larvae later burrow into the ground, where they pupate. Adult flies emerge after some 10–50 days.

Although quarantines of fruit and other measures have brought the Medfly under control, infestations recurred in Florida and Texas between 1930 and 1979 and again in the 1980s. The Medfly also appeared in California in 1975, 1980, 1987, and 1990. Because the export of fruit is prohibited from any areas where the Medfly has been detected, the economic impact is tremendous and could, if infestations were left unchecked, approach a billion dollars a year in crop damage in California alone.

Control Several approaches can be used to control flies. The specific technique depends on the habits of the species in question.

Although installing screens in buildings helps reduce contact between houseflies and people, it does not reduce the fly population. That objective calls for other approaches, one of the most important of which is a careful sanitation program. Keeping garbage and excreta covered and disposing of them promptly and properly will eliminate a primary breeding ground. Timely disposal of garbage, especially decaying fruit, and prompt removal and disposal of infested fruit that has fallen from trees have proved effective in controlling the Medfly in Israel and Italy. These measures, however, have essentially no effect on the screwworm fly. One method of control in this case is to restrict the breeding of cattle so that births occur only during the winter months, when the screwworm fly population is at a minimum.

Chemical insecticides are widely used for killing flies inside buildings. The two basic approaches are to wipe or spray the insecticides on indoor surfaces or to hang insecticide impregnated tapes from indoor ceilings. In all cases, however, care must be taken to avoid contaminating foodstuffs. Outdoor control measures include the application of larvicides to breeding areas and the use of bait stations and sprays. As was the case with mosquitoes, the principal insecticides initially used for the control of flies were the organochlorine compounds, such as DDT, dieldrin, and chlordane. As a result of the previously described controversies surrounding the use of DDT, there has been a similar shift to the use of less persistent organophosphorus compounds (such as malathion, used to control the screwworm fly and Medfly), the carbamates (such as Sevin), and the pyrethroids (such as permethrin).

In some cases, unique approaches have been developed in the use of insecticides for the control of specific fly species that transmit certain diseases. In the case of the tsetse fly, one method that has proved effective is to impregnate cow-sized rectangular sheets of cloth with synthetic pyrethroids, an insecticide, plus a mixture of chemicals, such as acetone and octenol, that are exhaled by cattle as they breathe and phenols that are present in cattle urine. In essence, the sheets of cloth are designed to represent a "fake" or "artificial" cow. Once the flies land on the impregnated cloth and come into contact with the insecticide, they are killed. More than 60,000 such cloth cows have now been deployed on ranches in Zimbabwe, and the technique is also being applied successfully in parts of Zambia, South Africa, and the Ivory Coast. This has led to the virtual elimination of human deaths, and nagana infections in cattle were reduced from some 10,000 in 1984 to about 50 per year today. A further benefit is that this approach is much less expensive than the procedures employed in the past (Lecrubier, 2002).

Another approach that is being increasingly applied is radiation sterilization. Developed in the late 1950s, this technique is relatively straightforward. The first step is to artificially breed and grow millions of adult male flies. The flies are then sterilized with radiation and released. The result is that eggs of the indigenous female flies with which the sterile males mate do not hatch. The technique benefits from the fact that insects generally mate only once. It is applicable, however, only where the density of the fly population is low. Otherwise, sufficient numbers of sterile males to have an impact cannot be bred. Nonetheless, the technique has been successfully applied for the eradication of the screwworm fly in the United States, as well as in parts of Mexico, Central America, and Libya. It has also become a standard tool for the control of the Medfly throughout the Western Hemisphere. It is likewise being used to control the melon fly in Japan and is supplementing the "fake-cow" approach in the attempt to eradicate the tsetse fly in sub-Saharan African countries. This follows the successful application of the technique in 1997 to eliminate the presence of this fly on the Tanzanian island of Zanzibar.

TICKS

Ticks are a good example of noninsect pests that can be important vectors of disease. For many years the primary disease of concern relative to this vector was Rocky Mountain spotted fever. The recent upsurge in this disease and, most especially, Lyme disease has caused renewed interest in these pests.

Characteristics and Disease Transmission Ticks are leathery-bodied, eight-legged arthropods with mouthparts that enable them to penetrate and hold fast in the skin and withdraw blood from animals. The female mates while attached to a host and usually feeds for 8–12 days. The tick that plays a major role in the transmission of Lyme disease in the United States differs from one part of the country to another. On the East Coast and in the northern Great Plains, the disease is spread by the bite of *Ixodes dammini;* in the western states it is transmitted by *Ixodes pacificus.* Both are common parasites of deer and mice. Because young stages (nymphs) of these ticks are only as large as a poppy seed, they often go undetected on humans.

The first cluster of cases of Lyme disease was reported in Connecticut in 1975; it is now present in almost all of the states in the United States. In fact, Lyme disease is the most common arthropod-borne disease in this country; almost 18,000 cases were reported in 2000. This is more than double the number of cases reported in 1990. A contributing factor to this increase was a shift of major farming activities from eastern to western regions of the United States, the accompanying abandonment of many of the farms in the Northeast, and the subsequent regrowth of trees that provide an ideal habitat for both the deer and its ticks. Most cases occur in the northeastern, mid-Atlantic, and north central regions of this country, and the highest numbers occur during June and July, reflecting the May and June peak months of the host-seeking activities of infective ticks (Matuschka et al., 1997).

Control As with mosquitoes and malaria, the control of ticks and Lyme disease can be complex. On a long-term basis, there is a need to avoid changes that increase outdoor environments that facilitate transmission of this disease. On a shorter-term basis, the disease can be limited by reducing tick populations through area control and vegetation management with insecticides, avoiding tick-infested areas, applying chemical repellents (for example, DEET), and promptly removing any ticks that become attached (CDC, 2002). Prompt removal is essential because the Lyme disease spirochete, *Borrelia burgdorferi,* is not likely to infect the patient before the vector ticks begin to engorge. Also of importance is the control of ticks on pets and in buildings.

As is the case with other such pests, effective control of ticks requires a fundamental understanding of their biology. Recognizing that one of the hosts for the ticks that transmit Lyme disease is the white-footed mouse, scientists developed a system for distributing cotton balls impregnated with permethrin in areas foraged by these animals. Once this is done, mice collect the cotton to use as a liner for their nests, and the ticks they normally carry are killed (Spielman, 1995). A similar approach has been used for years in the control of typhus fever, which is transmitted by rat fleas. In this case, the insecticide is distributed along the rat "runs," or paths.

Trends in Pest Control

With the surge in the development of artificial pesticides, beginning in the late 1940s, multiple chemical compounds were marketed and applied throughout the world for the control of insects. Although such applications continued for years without much concern about their potential health implications, the publication of Rachel Carson's now-classic *Silent Spring* in 1962 stimulated a dramatic change. More than anything else, this book alerted people to the potential toxicity of pesticides in nontarget species, including many wild animal species, as well as humans. As previously discussed, these concerns were based on a host of factors, all related to specific characteristics of the pesticides being used—their bioaccumulation and persistence in the environment; their presence in groundwater and surface-water supplies as a result of seepage and runoff from agricultural lands (Chapter 8); and demonstrations, in many cases, of their toxicity in humans. For these reasons, other control methods, such as the application of technologies involving biological control and genetic engineering, are being developed. The integrated approach, which involves a combination of methodologies, is also being increasingly applied.

BIOLOGICAL CONTROL

As noted earlier, several varieties of *Bacillus thuringiensis,* a naturally occurring bacterium, are deadly to mosquito larvae. Another variety of Bti, *Bacillus thuringiensis kurstaki* (Btk), has proved effective in killing the gypsy moth and spruce budworm larvae. Still other varieties, in combination with natural predators and fungi, are controlling certain agricultural pests such as the potato beetle.

Unfortunately, experience has shown that the potential damage caused by the introduced agent may be as extensive as that caused by the target pest species. One example was the introduction of parasitic wasps from Texas and China into Hawaii 50 years ago to prey on sugarcane pests. These wasps are now dominant players in the food web of remote native forests (Stokstad, 2001b). Another example was the introduction into the

United States of the weevil *Rhinocyllus conicus* for the control of exotic thistle species. It soon proved to be a problem since it also attacked native thistle. As a result, nonindigenous weeds are now spreading and invading some 700,000 hectares of wildlife habitat in the United States each year (Pimentel, 2000).

One of the major pests in the United States today is the fire ant. A nonnative pest, it was reportedly introduced into this country from South America in soil used as ballast on ships that docked in Mobile, Alabama, in 1918. They subsequently migrated northward and today are present in almost all of the southern states in the eastern United States. Once entrenched, they dominate and regularly constitute up to 99 percent of the total ant population. In the course of their activities, they damage electrical equipment, air conditioners, and farms, as well as domestic lawns and gardens. They also bite and sting people, releasing a venom comparable to that of a bee. In addition, they have significant ecological ramifications. In the Galapagos Islands, they eat the hatchlings of tortoises and have also attacked the eyes and cloacae of the adult reptiles. In the Solomon Islands, they have reportedly occupied the areas where incubator birds lay their eggs, and their stings have reportedly blinded dogs. In Gabon in West Africa, they have reportedly had the same effects on cats and elephants (Hayashi, 1999).

Although the presence of fire ants can be limited, to some extent, by treating the individual mounds with over-the-counter insecticides, reinfestation rapidly occurs since individual mounds are often interconnected via underground tunnels. One promising longer-range approach is the introduction of the phorid fly, also called the humpback fly, which kills fire ants through its reproductive process. This fly lays its larvae into the back of the neck of the fire ant. As the larvae pupate, they feed on the fire ant's brain, release an enzyme that destroys the muscles holding the ant's head in place, and kill the ant. Eventually, an adult fly emerges from the ant's head. Since the fire ant appears to be the only host that provides this service, evaluations indicate that the introduction of the phorid fly should not have any effect on any other part of the environment. Field studies conducted in North Carolina in 2000 showed that phorid flies reproduced through five generations within the first year. No flies, however, were found to be present the following year. Although the reason is not known, it may be that the flies were not able to survive the intervening winter (Huepel, 2002).

GENETIC ENGINEERING

The capabilities that scientists have developed to engineer the genetics of plants and animals could have profound impacts on the control of insects and insect-related diseases. As the following discussion will confirm, there are multitudes of ways in which these technologies might be applied.

Completion of the mapping of the genes of *Plasmodium falciparum,* the parasite that causes malaria, and *Anopheles gambiae,* the mosquito that transmits it in many parts of the world, will hopefully lead to the development of improved insecticides and repellents, as well as new approaches for combating this disease. If, for example, researchers can pinpoint the genes that enable *A. gambiae* to have the finely honed smell and taste receptors to seek out humans for its blood meal, they may be able to develop better repellents. Similar studies of the genes of mosquitoes may make it possible to target the proteins that enable mosquitoes to develop resistance to pesticides, or to alter the genes that enable them to serve as an intermediate host for the parasite that causes malaria. Another possibility would be to introduce into the mosquito population individuals that have been genetically altered to carry genes that interfere with reproduction. Still another would be to add a gene to the mosquito that prohibits the malaria parasite that it ingested during its blood meal from moving from its gut into its saliva. Control would then be accomplished by releasing into the environment millions of mosquitoes that have been altered in this manner. As in the case of the methods for biological control, questions immediately arise, such as whether the changes incorporated into the mosquitoes might make them more efficient in spreading other diseases.

Because of the multitudes of unexpected developments in the application of biological controls, potential problems of this type obviously need to be carefully addressed. Extensive testing, for example, would need to be conducted prior to the release of modified mosquitoes into the environment. What types of testing should be done, how extensively, and for how long are questions that remain to be answered. A decade ago, for example, scientists genetically engineered the cotton plant to enable it to make insecticidal proteins from the bacterium *Bacillus thuringiensis* (Bt). So-called Bt-cotton plants, altered in this manner, were able to control several major pests, including the cotton and pink bollworm and the tobacco budworm. An additional benefit was that the modified cotton required about half as many treatments with chemical pesticides as ordinary cotton. As often appears to be the case, it soon became apparent that insects can adjust to these natural toxins, just as they do to synthetic chemical pesticides. This has raised concerns that extensive use of modified crops will lead to widespread resistance that could render both the crops and Bt sprays useless (Stokstad, 2001a).

INTEGRATED PEST MANAGEMENT

One of the strategies for the control of pests that has gained widespread acceptance is what is called an integrated approach. Successful implementation of this concept, commonly termed *integrated pest management,* involves acquiring detailed information about a given pest (including its physiology, predators, and life cycle); becoming thoroughly familiar with the technical measures available for its control and the related political, industrial, and environmental factors; and then applying the most effective combination of control strategies and techniques. This approach involves consideration of the full range of available educational, cultural, biological, chemical, and legal controls, including the use of pest parasites, pathogens, pheromones, predators, and resistant crops. One of the goals is to reduce the need for the application of pesticides. Another is to minimize unnecessary health and environmental side effects of vector-control activities while assuring maximum protection of the public and the environment.

Planning the control program should take place at the local level, with full input from the community. Strategies may include rotating crops to interrupt the cycle of specific pests, interspersing one crop with another to confuse insects, carefully timing control efforts (that is, applying pesticides only when insects appear and using them in carefully controlled amounts), introducing natural predators to combat specific pests, and applying an insecticide developed specifically for a given pest. Mechanical methods of pest control, such as soil aeration, tillage or no-till, tractor-mounted flaming devices, vacuuming machines, and pest barriers, are often an integral part of such a program. As with any such development, there are disadvantages. Implementation of the integrated approach is labor intensive and must be structured around the growing cycle. Nonetheless, farmers increasingly are adopting one or more aspects, and the method is increasing in popularity (Leslie, 2004).

The General Outlook

One message generated by the discussions in this chapter is the increasing importance of zoonotic diseases in the world today. Of the estimated 1,700 diseases that plague humankind, almost 50 percent are believed to be of this variety, that is, as noted earlier, they represent an infection that, while having the potential of causing a disease in human hosts, can be perpetuated solely in nonhuman animal hosts (Spielman and Kimsey, 1997). In fact, of the slightly more than 150 so-called emerging diseases, almost 75 percent are zoonoses. In response to this trend, U.S. government officials have developed programs for increased monitoring of emerging infections and promoted the enhancement of international efforts to control the spread of such pathogens (Pennisi, 1996). One factor that makes such programs necessary is the extensive amount of international travel taking place in the world today. Each day an estimated 2 million people cross international borders.

A second message is the need for the developed nations of the world to assume far more responsibility in assisting the less developed countries in controlling all types of diseases, most especially those that are transmitted by insects. Malaria is a prime example. Although mosquitoes serve as the vector, humans serve as the reservoir. Reducing the incidence of malaria anywhere in the world helps protect people everywhere. An equally important benefit of such efforts is that in multitudes of cases, the people who live in the less developed countries are poor because they are sick. They simply do not have the energy to work and earn income. By reducing their burden of disease, not only would the developed countries be reducing disease and suffering, but they would also reap a multiple return on their investment through increased trade and economic gains.

It must also be recognized that in most instances, the increase in infectious diseases can be directly tied to one or more environmental changes that have facilitated contact between the vectors of disease and their human hosts. The presence of Lyme disease in the northeastern United States, for example, can be directly attributed to reductions in the spaces that formerly separated the living areas of people and deer herds. These and other environmental trends demonstrate the dynamic relationships that link an ever-changing landscape, the vectors that exploit these instabilities, and the pathogens that may thereby affect human health. Similarly, the continued proliferation of rodents is almost totally a result of urbanization, the deterioration of many of our inner cities, and the lack of proper garbage and refuse disposal. Some scientists predict even more dramatic changes; such as the transmission of insect-related diseases within areas previously not affected, if the predicted global warming materializes (Chapter 20).

A third message is the continuing evolution of insecticide resistance by agricultural pests. Just as drug resistance in microbes (Chapter 6) threatens the use of antibiotics to cure human patients, this development threatens human welfare through its impact on disease transmission among agricultural crops. From 1970 to 1980, for example, the number of resistant arthropod species increased from an estimated 200 to almost 450. Today they number more than 500. Although some of these responses are evolutionary in nature, others have occurred at a relatively rapid pace. While the time delay for insecticides, such as the organophosphates (introduced in 1950) was 15 years, the time delay for the pyrethroids (introduced in 1972) and the neonicotinoids (introduced in 1989) was only 6 years. Although this was initially considered a regional, and perhaps a national, problem, it has now been found that through natural migration or human-mediated transport, resistant pests have the capacity to disperse and transfer genes over large areas within a very short period of time (Denholm, Devine, and Williamson, 2002).

As in all aspects of life, there is much that scientists, public health officials, and farmers can learn from nature relative to developing more effective methods for using pesticides. Studies of the natural environment can provide useful information on possible methods for avoiding the evolution of resistance to human-made toxic chemicals. It has been known for some time, for example, that certain long-lived trees, such as the mountain birch, generate chemical weapons to deter caterpillars from eating their leaves. Yet experience shows that the caterpillars never seem to develop resistance to these weapons. The explanation appears to be the ability of the mountain birch to generate a multitude of resistant chemicals on a fluctuating basis at various times within the growing season. While caterpillars might be able to evolve countermeasures against one of these, they are not able to overcome them all (Kaiser, 2000). Similar approaches in the application of pesticides might reap equivalent dividends.

11

INJURY CONTROL

FOR PURPOSES of evaluation, injuries are generally classified as *unintentional* (e.g., those that occur in motor-vehicle accidents) or *intentional* (e.g., acts of violence). In the first category, accidents of various types in the United States account for more than 20 million disabling injuries and almost 100,000 deaths and result in approximately 500,000 hospitalizations and 4 million emergency department visits annually. Worldwide, unintentional injuries annually account for about one-third of all hospitalizations and about 3 million deaths. In many countries, the problems of injury and injury control have assumed an importance equal to that of infectious diseases. At the same time, more than 50,000 deaths occur in the United States each year through intentional injuries, two-thirds from suicides and one-third from acts of violence.

In either category, the magnitude of the challenge is enormous. During 2000, for example, about 12 percent (almost 35 million) of the people in the United States were treated for nonfatal injuries associated with the full range of such events (NSC, 2001). Of these, about 2 million were treated for violence-related injuries. While falls accounted for the most nonfatal injuries overall, 7 million, compared to about 3 million who were treated for disabling or nondisabling injuries resulting from motor-vehicle accidents, the importance of the latter group should not be underestimated. Within the United States, motor-vehicle crashes kill more children and young adults between the ages of 1 and 24 years than any other single cause. In fact, such events are the leading cause of death in this country from unintentional injuries for persons of all ages.

The significance of motor-vehicle accidents is further demonstrated by

Table 11.1 Unintentional injuries by industry, United States, 2000

Group	Deaths per year		Disablity injuries
	Total	Per 100,000	
Agriculture[a]	780	22.6	130,000
Mining and quarrying	110	21.2	20,000
Construction	1,220	13.6	470,000
Transportation and public utilities	930	11.5	380,000
Manufacturing	660	3.3	630,000
Government	450	2.2	580,000
Trade	420	1.5	750,000
Services[b]	630	1.3	940,000
All industries	5,200	3.8	3,900,000

a. Includes forestry and fishing.
b. Includes finance, insurance, and real estate.

the fact that they are the source of about 40 percent of the deaths involving unintentional injuries in the United States; the remainder result from community and recreational activities, fires, and work-related accidents. Even in the case of work-related injuries and deaths, motor-vehicle accidents play a significant role, accounting for 2.5 percent of the four million unintentional injuries and about 25 percent of the more than 5,000 work-related deaths in the United States during 2000 (NIOSH, 2003). Nonetheless, progress is being made. Overall, the death rate among workers in 2000 was less than 20 percent of that in 1960; it was less than half of that in 1990. Fatality rates for various industries are shown in Table 11.1 (NSC, 2001).

Addressing the Problem

According to Julian Waller (1994), injury control as a public health endeavor began in Germany in 1780, when Johann Peter Frank urged that injury and its prevention be addressed not only by individuals but also by nationwide public health programs. In the mid-1900s, several state and local health departments in the United States initiated modest data-collection efforts and child-safety, burn-prevention, and other programs. The effects of these programs on behavior, morbidity, or mortality were never fully evaluated, however.

In 1942 Hugh De Haven, an engineer at Cornell University, published an article (De Haven, 1942) that began a conceptual revolution in injury control. He showed how people successfully survived falls of 15–45 meters (50–150 feet), in some cases with only minor injuries, through proper dispersion of kinetic energy in amounts as high as 200 times the force of gravity. Through this process he demonstrated that it is possible to disconnect the linkage between accidents and the resultant injuries. His studies in turn led to the development and introduction of seat belts and other occupant restraints as an effective method of reducing injuries in automobile accidents (Waller, 1994).

In 1961 J. J. Gibson observed that injury events have only five agents, namely, the five forms of physical energy: kinetic or mechanical energy, chemical energy, thermal energy, electricity, and radiation (Gibson, 1961). Shortly thereafter, William Haddon expanded on this concept through the initiation of a movement to incorporate sounder scientific and public health concepts into the development of accident- and injury-prevention programs. Instead of relying primarily on attempts to change human behavior, he applied an environmental approach to injury control. In so doing, he followed the lead of De Haven by concentrating on the prevention of injuries, not accidents. He developed a generic approach to the analysis, management, and control of such injuries, which he treated as fundamentally a result of the rapid and uncontrolled transfer of energy (Haddon, 1970). His approach can be applied to all types of occupational and environmental hazards, ranging from automobile accidents to oil spills to major accidents in nuclear power plants. It can also be applied to controlling acts of violence.

To facilitate an analytic approach, Haddon divided accidents into three phases: the pre-event phase (the factors that determine whether an accident occurs), the event itself, and the post-event phase (everything that determines the consequences of the injuries received). The factors that operate in all three phases are the humans involved, the equipment they are using or with which they come in contact, and the environment in which the equipment is operated. Combining the three accident phases and the three factors yields a nine-cell matrix (Figure 11.1) that public health workers can use to determine where best to apply strategies to prevent or control injuries.

Because vehicular accidents account for almost half the deaths resulting

Phases	Factors		
	Human	Equipment	Physical and socioeconomic environment
Pre-event	(1)	(2)	(3)
Event	(4)	(5)	(6)
Post-event	(7)	(8)	(9)

Figure 11.1 Matrix for the analysis of accidents

from unintentional injuries in the United States, they are used as examples in the discussion that follows.

PRE-EVENT PHASE

The goal in the pre-event-phase is to reduce the likelihood of a vehicular collision. Factors that should be considered include the following:

1. Humans involved: driver impairment by alcohol or other drugs; the thoroughness of testing procedures for licensure; the degree of enforcement of traffic rules and regulations, including mandatory use of seat belts; and the availability of mass transportation as an alternative to the use of private vehicles

2. Equipment: the condition of headlights, tire treads, and brakes (and whether they include antilock features); the size and visibility of brake lights; the speed the vehicle can attain; and vehicular crash tests

3. Environment: the presence of barriers and traffic lights to protect pedestrians; the design, placement, and maintenance of road signs for ready comprehension; and the design of bridge abutments to prevent or reduce impact damage

EVENT PHASE

The goal in the event phase is to reduce the severity of the "second collision," for example, when the victim hits the windshield or steering column. Factors that can reduce the extent of injuries include the following:

4. Humans involved: proper use of seat belts and child-resistant systems; and driver abstention from alcohol (which affects cell membrane permeability, so that even in low-impact collisions people who have consumed alcohol are more likely to sustain severe or even fatal neurological damage)

5. Equipment: whether the vehicle is equipped with an airbag, collapsible steering column, high-penetration-resistant windshield, interior padding (for example, on the dashboard), recessed door handles and control knobs, and structural beams in doors; low bumpers with square fronts to reduce the likelihood of pelvic and leg fractures in pedestrians who are hit; and, on large trucks, a bar under the rear end to prevent cars from "submarining" beneath them

6. Environment: breakaway sign posts, open space along the sides of the road, wide multiple lanes, guardrails to steer vehicles back onto the road, and road surfaces that permit rapid stopping

POST-EVENT PHASE

The goal in this phase is to reduce the disabilities due to the injuries. Factors that can reduce or limit the effects of injuries include the following:

7. Humans involved: rapid and appropriate emergency medical care, followed by adequate rehabilitation; properly trained rescue personnel; and injury severity scores to help medical personnel evaluate multiple traumas and predict outcomes

8. Equipment: fireproof gasoline tanks to prevent fires after an accident

9. Environment: "jaws of life" to extract victims from vehicles; helicopters for rapid transport of victims to medical-care facilities; trauma centers equipped to handle injured victims; ramps and

other environmental changes to reduce the real "cost" to the victims of being disabled; and rehabilitation of the victims

Vehicular Accidents

Through application of these and other strategies, deaths in the United States caused by most categories of vehicle-related accidents have shown a continuing decrease over the past several decades. Since 1980 the number of deaths of pedestrians, for example, has decreased by more than 45 percent; those for pedalcyclists (bicyclists) by almost 35 percent; those for motorcyclists by almost 25 percent. Concurrently, deaths of people in passenger vehicles—cars, pickup trucks, utility vehicles, and cargo vans—have been reduced by almost 20 percent (NSC, 2001). Nonetheless, the challenges continue. In 2003, for example, 43,220 people were killed in the United States in motor vehicular accidents, an increase of more than 400 deaths compared to 2002. Fifty-eight percent of the victims were not wearing their seat belts (Durbin, 2004).

Total deaths, however, do not reflect the true story of what has been accomplished, particularly during a period in which the number of vehicles on the road and the distances they were being driven were undergoing enormous increases. In 1970, the estimated number of vehicles in the United States was about 111 million, the number of vehicle miles traveled was 1,120 billion, and the death rate per 100 million miles was 5.21. In 2000, the estimated number of vehicles was about 224 million (more than double), the number of vehicle miles traveled was 2,688 billion (almost 2.5 times as many), and yet the death rate per 100 million miles was 1.60 (a decrease by a factor of more than 3) (Figure 11.2). As a result, the total number of deaths in 2000 (43,000) was 27 percent less than in 1970 (54,633). Had these improvements not been achieved, the number of people dying each year in the United States would today have approached 120,000. Nonetheless, motor-vehicle accidents continue to account for more than 11 percent of all disabling injuries in this country, and the associated annual economic costs exceed $200 billion (NSC, 2001). It should also be noted that the number of deaths has increased in recent years.

The development of a program for preventing or reducing injuries suffered in vehicular accidents has political, social, behavioral, and economic aspects. It therefore requires a multifaceted approach that involves new technical advances as well as new policies and strategies.

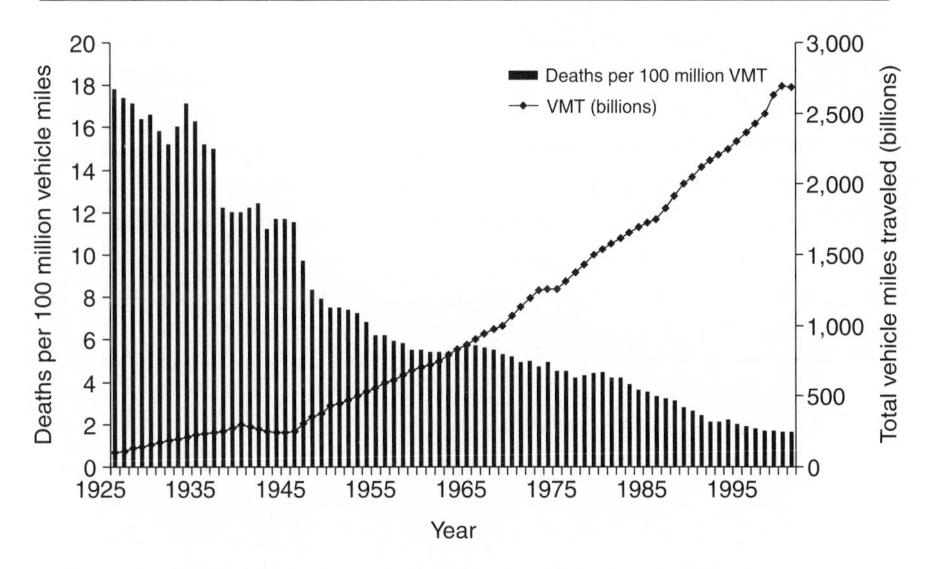

Figure 11.2 Trends in total vehicle miles traveled (VMT) and death rates per 100 million VMT, United States, 1925–2000

TECHNICAL ADVANCES

A number of safety-enhancing technical advances have been incorporated into motor vehicles in recent years. A primary stimulus has been the increased safety consciousness on the part of the public. Several of these advances are discussed here. In one case—the use of speed-monitoring devices by motorists—the advance is being used to circumvent measures that are being taken to improve safety.

Air bags. Studies show that a combination of lap/shoulder belts and air bags offers the best available protection for motor-vehicle occupants. This combination, however, is not a cure-all. In fact, the added fatality-reducing effectiveness of the air bags is estimated to be about 11 percent over and above the benefits of using safety belts alone. Nonetheless, they are saving an estimated 1,300 lives each year in the United States. To increase the protection, a number of manufacturers now install air bags in the doors and outer walls of cars to protect against side impacts. Unfortunately, it was soon found that children and persons of short stature, who were sitting in the front seat and were too close to either the steering wheel or the dashboard at the time of deployment of an air bag could be killed. To resolve this problem, the U.S. Department of Transportation issued a rule, effective in January 1998, that permits vehicle owners who meet certain qualifying criteria to have air-bag on-off switches installed in their vehicles.

Improved head restraints. These devices can be extremely effective in preventing whiplash injuries in rear-end collisions. In order to do so, however, the restraint must be behind and close to the heads of the occupants. Unfortunately, even as late as 1999 a third of the head restraints installed in new cars did not meet these basic requirements (IIHS, 1999). Until these problems are corrected, the full benefits of these devices will not be realized.

Antilock brakes. Early evaluations on the test track indicated that antilock braking systems would provide many benefits in emergency braking situations, especially on road surfaces that are wet and slippery. While this has been confirmed in the case of large trucks, especially tractor-trailer units, such systems have reduced neither the frequency nor the cost of automobile accidents. Although there may be several explanations, one is that emergencies on the road often involve complicated scenarios that differ significantly from test situations. In addition, it may be that individuals who drive cars with antilock brakes place too much confidence in the system and take more risks.

Daytime running lights. Daytime running lights (reduced-intensity headlamps) have proved to be particularly beneficial in reducing automobile accidents in urban areas, where traffic congestion is heavy and demands on driver attention are numerous. Although they have been approved for use in the United States, not all manufacturers have chosen to install them. This is in contrast to Sweden, Norway, Finland, and Canada, where they are mandatory. Such lights reduce collision damage by several percent.

Truck-trailer visibility. Studies have shown that collisions of other vehicles with the trailers of large trucks can be reduced, particularly at night and in bad weather, simply by making them more visible. One low-cost approach (about $100 per unit) is to add reflective material to their sides and rear. Recognizing this fact, the National Highway Traffic Safety Administration now requires that all new truck trailers in the United States be equipped in this manner. Estimates are that this reduces accidents involving such units by about 15 percent.

Radar detectors. Many drivers, including operators of commercial trucks, use radar detectors to alert them to the presence of police speed-monitoring units. Because the devices have only one purpose—to alert speeding drivers to slow down at that particular time—they have been banned in many states. They have also been banned nationwide by the Federal Highway Administration on all commercial vehicles, primarily trucks, used in interstate commerce. This action reflects the fact that while medium and heavy-duty trucks account for about 8 percent of the vehicles involved in fatal accidents, the occupants of such trucks account for only about 2 percent of the fatalities (NSC, 2001).

POLICY AND ETHICAL ISSUES

A number of policies and strategies have been developed and applied in the past 35 years to improve vehicular safety. In at least one case (monitoring driver behavior), a policy or action developed primarily with other goals in mind has been found to influence vehicular safety.

Alcohol. In spite of major efforts by groups such as Mothers against Drunk Driving (MADD), alcohol continues to be a causative factor in about 40 percent of all traffic fatalities in the United States (Durbin, 2004). Even more disturbing is that the number of deaths in which alcohol played a role has been increasing in recent years, after having remained steady during most of the 1990s. In addressing this problem, MADD recommends that state agencies establish more well-publicized sobriety checkpoints and enact tougher penalties against drivers who refuse alcohol tests when stopped, are found to be driving with a license that has been suspended due to an earlier drunken driving arrest, or have a blood-alcohol level of 0.15 percent or more.

Speed limits. During the mid-1970s, when a nationwide speed limit of 88 kilometers (55 miles) per hour was imposed to conserve fuel, vehicular deaths in the United States were reduced dramatically. This is not surprising because vehicles traveling at high speeds allow the driver less time to react to an emergency and require longer distances to stop. Later, when fuel became more readily available and this restriction was withdrawn, many states increased the speed limit on interstate highways to as much as 110 kilometers (70 miles) per hour or more. As would be anticipated, this led to an immediate increase in the number of deaths. Another factor that has recently encouraged high speeds is the additional horsepower of the engines being installed in many cars and trucks. Rather than promoting safety, many automobile manufacturers now emphasize how fast their vehicles can accelerate from 0 to 95 kilometers (60 miles) per hour.

Ticketing traffic violators. One of the most common sources of accidents on urban and suburban streets is the motorist who "runs" a red traffic light. Research shows that this problem can be effectively reduced by the installation of cameras that photograph the license plate of the offending motorist, who is subsequently ticketed by mail. Although such cameras have been challenged on the basis of invasion of privacy and related ethical issues, they are being installed in an increasing number of cities (IIHS, 1999). Knowing that such a violation will be recorded, regardless of whether a police officer is present, can obviously serve as a major deterrent for drivers who may be tempted to commit such violations.

Cellular telephones. Concerned about distractions to drivers from the use of handheld cellular telephones, the New York state legislature passed legislation in 2001 that banned the use of such devices in cars. Multiple communities in various states followed suit on the basis of cumulating evidence that such use can seriously impair a driver, regardless of the type of telephone (handheld or remote) being used. For similar reasons, the use of hand-held cellular phones by people operating a motor vehicle has been outlawed by regulators in Great Britain. This includes making it illegal to pull off the road and leave the engine operating while using such a phone (Black, 2004). Other tests, using driving simulators, demonstrated that drivers who are talking on any type of telephone tend to be slow to react, particularly to unexpected events; it takes them longer to brake; and they are more likely to miss traffic signals. At the same time, those responsible for weighing this issue need to recognize that cellular telephones have multiple benefits. These include providing (1) a ready means for vehicular occupants to contact emergency services; (2) "peace of mind" to travelers who fear a vehicle breakdown; and (3) tremendous flexibility in contacting people, which, in turn, leads to significant reductions in the number and duration of automobile trips that must be made (Lissy et al., 2000). This is not to say, however, that such activities need take place while a person is driving a vehicle.

Teenage drivers. While young drivers, 15 through 20 years of age, represent only about 7 percent of all licensed drivers in the United States, they are involved in 15 percent of all fatal accidents and 18 percent of all police-reported accidents. At the same time, data clearly show that teenagers who live in states with fewer steps in the licensing procedure have higher crash rates. To counteract this problem, the legislators in a number of states have established what is called a graded approach for granting driving licenses to teenagers (NSC, 2001). Although specific requirements vary, a three-level system is quite common. In general, the three levels, each of which encompasses a one- to two-year period, include (1) a learner's permit, (2) a provisional license, and (3) a full license. Each of the first two levels has a set of conditions that must be fulfilled before the candidate can move up to the next. Any violations within either of the first two levels essentially require the applicant to restart the process. After such a system was introduced in North Carolina, fatal accidents among this age group were reduced by more than half. A supplement to this approach has been adopted in the state of Texas that makes the avoidance of tobacco an essential component of the right of a teenager to hold a driver's license.

Older drivers. Elderly drivers, particularly those more than 85 years old, have accident and fatality rates in excess of those for the 16- through 19-year-old group. Furthermore, the percentage of the U.S. population in the older age range is rapidly increasing—from an estimated 16 million people over the age of 65 in 1990, representing 10 percent of all drivers, to an estimated 50 million in 2025. Anticipating this situation, some safety experts have advocated that older drivers be required to attend special senior driver education programs and to undergo periodic retesting. Others have suggested that one or more types of restrictions be placed on such drivers, including, for example, that at a certain age their licenses be revoked. As in the case of cellular telephones, however, addressing these issues will not be easy. For many older persons, the inability to drive would seriously impair their quality of life, for example, through social isolation and adverse effects on their physical and mental health. Other considerations are that compelling older people to walk may increase their frequency of injuries from falls, many will have difficulties in using public transportation, and walking home with goods purchased at local stores will increase their vulnerability to crime.

Safety (seat) belts. Safety belts are a well-proven method for protecting passengers in motor vehicles. They saved the lives of an estimated 11,000 people more than four years old in the United States during 1999 (NSC, 2001). Nonetheless, even though essentially all motor vehicles in the United States are equipped with these devices, 27 percent of the people did not use their safety belts in 2000. Disturbingly, the percentage is higher among younger age groups; in fact, of the teenagers who were killed either as drivers or passengers in this country during 2000, only one-third were wearing safety belts (CDC, 2002). Although there are several ways to address this issue, experience has clearly demonstrated that law-enforcement groups can be very effective. As a general rule, those states that have achieved the highest rates of use are those that have comprehensive programs that are supported by laws mandating safety-belt use; aggressive law enforcement; and vigorous educational programs. Unfortunately, the movie industry, among others, is not supporting such activities as well as it could. While, as noted above about 70 percent of the vehicle occupants in this country now routinely wear safety belts, the percentage of use by actors in movies produced during the last decade has ranged from 10 to 30 percent. Since movies are later shown on television, much the same situation exists in that industry (Jacobsen et al., 2001).

Monitoring driver behavior. Shortly after air bags came into use, some automobile owners in the United States submitted claims stating that the air bags in their cars had deployed without cause. In response, several automobile manufacturers quietly installed electronic units in their vehicles to record key performance factors that could be extracted and made available for use in their defense. In many respects, these units are similar to the "black boxes" installed in airplanes. Soon thereafter, safety officials recognized that the recorded data would be a source of extremely useful information in analyzing the causes of motor-vehicle accidents. Such information would include how fast the vehicle was moving just prior to the event, the exact time when the brakes were applied, and how fast it was moving at the time of impact. Stimulated by these benefits, U.S. manufacturers have indicated that they now plan to install such units in their new vehicles on a routine basis. At the same time, other people recognized that the presence of such units, particularly without the owners being informed and granting their consent, raised ethical and policy questions similar to the previously discussed use of cameras to identify and ticket traffic-light violators.

Crime prevention. Scientists in the United Kingdom have developed a system that calls the owner on the telephone if his/her car is stolen, tracks the location of the vehicle using a global positioning satellite, and turns the engine off when the police are ready to make an arrest. A similar service is available in the United States for trucks and turns the engine off if the drivers stray from their approved route. Other services in this country and Canada track the location of rental cars and deactivate their engines if they are stolen. Similar devices could be used to turn off the engines of vehicles that are being pursued by the police (Law, 2002). Applications of these technologies for these and related situations could reduce not only the number of accidents but also the number of injuries and deaths of the people involved.

Vehicle size and body style. To reduce the consumption of oil in this country, in 1978 the U.S. Congress established for automobiles what are called corporate average fuel economy (CAFE) goals (Chapter 18). One of the easiest methods for achieving compliance with these requirements was to reduce the size and weight of the cars being manufactured. Unfortunately, smaller vehicles involved in accidents can be a source of increased injuries and deaths. In fact, among the 11 existing vehicles in the United States with driver death rates at least twice as high as the average, 10 are small, and 1 is midsize; none is large. Conversely, 8 of the 12 vehicles with the lowest driver death rates are large, and the other 4 are midsize; none is small (IIHS, 1994). The policy issues in this case are difficult to resolve. It is not surprising, therefore, that when congressional leaders revisited the CAFE goals in 2002, they decided not to make them more stringent. They did, however, broaden these requirements to include minivans, sport utility vehicles (SUVs) and pickup trucks, all of which had previously been exempted.

If, ultimately, everyone is driving a smaller car, perhaps the number of deaths might become acceptable. Even then, however, occupants in smaller cars would still be vulnerable to possible collisions with large trucks that will not have been downsized. Another but publicly contentious alternative would be to increase the taxes on gasoline. One of the added benefits of this approach is that it would discourage people from driving as much.

Daylight saving time. Because it adds an hour of sunlight to the afternoon commuting time and increases the visibility of both vehicles and pedestrians, the adoption of daylight saving time is a proven method of reducing vehicular accidents. Although this step also eliminates an hour of sunlight in the morning, the increase in accidents at that time is not enough to outweigh the lives saved in the afternoon, when many more pedestrians and vehicles are on the road.

OTHER CONSIDERATIONS

Other vehicle-related issues that need to be addressed are unintentional injuries and deaths of motorcyclists and pedestrians and collisions of motor vehicles with animals.

Motorcyclists. Motorcycle registrations (4.1 million) represent less than 2 percent of the total number of vehicle registrations (224 million) in the United States. Even so, during 2003 motorcycle accidents accounted for 8 percent of the occupant deaths and about 2 percent of the injuries in motor-vehicle accidents in this country (Durbin, 2004). As these numbers imply, the ratio of deaths to injuries in the case of motorcycle accidents is far higher than that for other types of vehicular accidents. In fact, the death rate per vehicular mile for motorcycle riders is 24 times that for occupants of passenger cars, trucks, and buses (NSC, 2001). One protective measure that has been promulgated in all but 3 of the 50 states is to require that motorcyclists wear helmets. A decision in 2000 to discontinue this requirement in Florida, following a period of enforced use, led to an increase of more than 20 percent in deaths during the following year (Mullet, 2004).

Pedestrians. During 2001, there were an estimated 4,882 pedestrian deaths and 85,000 pedestrian injuries in motor-vehicle-related accidents in the United States. About 50 percent of these occurred when pedestrians crossed or entered streets; about 8 percent occurred when people were walking along the roadway. In 38 percent of the accidents that involved the death of a pedestrian, the driver, the pedestrian, or both were intoxicated (NSC, 2001). As might be anticipated, pedestrian deaths are a special problem among the elderly (Figure 11.3). In fact, the death rate for people 75 years of age or older is more than twice that for people 65 to

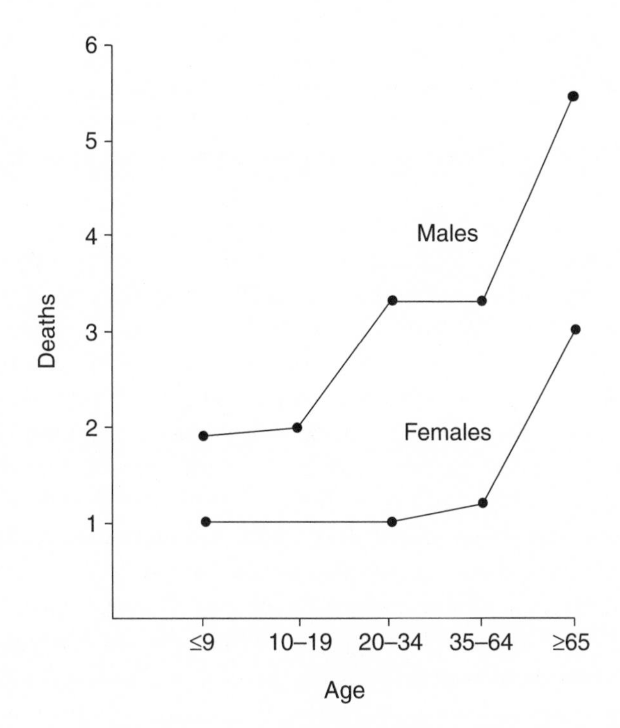

Figure 11.3 Pedestrian deaths per 100,000 people as a function of age

74 years of age; it is far higher than that for people younger than 65. Even so, it is equally important to recognize that about 15 percent of all motor-vehicle deaths sustained by children and young people, 0–19 years of age, occur to them as pedestrians. Another group of "pedestrians" who require special consideration are highway and street construction workers, about 1,000 of whom were killed in the United States during the 1990s. Most of these deaths involved vehicles and moving equipment, and more than a third involved workers on foot who were struck by a vehicle. One of the most important measures for reducing such deaths is to require that highway and street workers wear high-visibility wearing apparel. Another step is to use fluorescent and retroreflective material on headgear and on the gloves of people who wave flags to give directions (NIOSH, 2001). Steps that can be taken to protect pedestrians in general include providing separate pathways for walkers, placing sidewalks well back from the road, restricting on-street parking, and requiring that the exterior of motor vehicles have no sharp edges or protrusions (CDC, 1999c). Noting that front-end collisions account for up to 30 percent of those involving pedestrians, the European Union has established standards to address this issue. Special attention is being paid to knife-edged front-engine sports cars and SUVs (Rendell, 2004).

Collisions with animals. In addition to increasing the spread of various insect and animal-related diseases (Chapter 10), the encroachment of society upon wildlife habitats has led to a dramatic increase in the number of collisions between motor vehicles and wild animals, particularly deer. Nationwide, such collisions annually cause the deaths of 100 to 150 people. States with high deer populations, such as Michigan, North Carolina, Pennsylvania, and Wisconsin, each experience from 40,000 to 50,000 deer-vehicle collisions each year. In each state, the cost for repairing the associated damages ranges from $50 to $80 million; nationwide, it exceeds hundreds of millions of dollars. One countermeasure that is being applied is a deer whistle that is mounted on a vehicle and is activated by the onrushing air. Although the resulting ultrasonic sound is supposed to repel deer, the effectiveness of such devices has not been confirmed. Another approach is to install specially designed roadside reflectors to try to prevent animals from crossing in front of vehicles. Presumably, the reflector, illuminated by the headlights of the oncoming vehicle, frightens the deer and causes them to stop.

NEW DEVELOPMENTS/TECHNOLOGIES

In the early 1990s, Congress appropriated about $650 million to be spent on stimulating the development of what would become known as intelligent vehicle highway systems (IVHS). Often referred to as "smart cars" or "smart highways," IVHS encompassed a range of high-technology approaches specifically designed to reduce vehicular injuries and deaths. Products developed through this program include the previously discussed antilock braking systems; electronic message boards that alert drivers to upcoming road conditions; sensors in the rear bumpers of cars that sound an alarm when the driver is backing up and about to collide with an obstacle; related alternate or backup small television-like screens that show the driver what is behind the car; and onboard navigation systems that enable drivers to determine exactly where they are and to select the best route to reach their destination. Another product is a pulse-transmitting system that can determine the location, size, and distance of objects in front of a car and transmit that information back to sensors that will alert the driver. In case of an impending collision, the system will even apply the brakes.

The design of improved safety belts is also being explored. These include four-point X-shaped devices based on the full harnesses worn by race-car drivers. Since such harnesses will move the mounting points for the belt outward, their use should eliminate the chafing inherent in the designs being used today. Another device, now being installed in new cars in the United States, is a sensor that will inform drivers when one or more of the tires are underinflated. Additional benefits should be forthcoming when manufacturers complete the shift from 12- to 36-volt batteries. This change will enable the installation of features such as the "steer-by-wire" approach, in which instructions to the front wheels will be electronically transmitted by the steering wheel, rather than through a direct mechanical or hydraulic linkage. Removing the steel shaft through which such instructions are now transmitted will eliminate what has been a major source of chest injuries to drivers in the event of a crash (McCosh, 2001). Another system that has been developed in Germany will activate certain safety features just prior to a crash. Through detectors designed to sense the probability of a crash, the system will not only activate the air bags, move the steering wheel forward, and pretension the safety belt, but will also extend the front bumper to help dissipate the energy of the crash (Phillips, 2001).

Sports and Recreational Injuries

Sports and recreational activities are a major source of unintentional injuries and deaths. Basketball and bicycle riding together accounted for more than a million hospital emergency department visits in the United States during 1999 (Table 11.2). In all cases except for basketball and exercising, the percentage of injuries in the 5- to 14-year-old age group exceeded that in the 15- to 24-year-old group. Of special concern is the use of playground equipment that leads to injuries to more than 200,000 chil-

Table 11.2 Sports related injuries treated in hospital emergency departments in the United States, 1999

Sport	Number of Injuries[a]
Baseball and softball	340,000
Basketball	600,000
Bicycle riding	600,000
Exercise	160,000
Fishing	70,000
Football	375,000[b]
Horseback riding	70,000
Roller skating	140,000
Skateboarding	60,000
Soccer	175,000
Swimming	100,000[c]
Volleyball	65,000
Weight lifting	65,000

a. All numbers are approximate.
b. Includes both touch and tackle football.
c. Includes injuries associated with pool slides, diving boards, and related equipment.

dren in this country each year. One of the most prominent sources of such injuries is the unyielding nature of the surfaces on which such equipment is often installed. Because 70 percent of these types of injuries occur due to falls to the ground, all such equipment should be installed on shock-absorbent surfaces, such as sand, wood chips, small round gravel, or rubber. Other preventive measures include providing adequate spacing around separate items of equipment, ensuring that the equipment is appropriate for the age group using it, and maintaining it in a safe condition (CDC, 1999b).

At the same time, the nature of the sources of injuries to children is constantly changing. A good example is the small foot-propelled scooter that was introduced into the U.S. market in the late 1990s. These devices, which had small, low-friction wheels similar to those on in-line skates, proved to be extremely popular and led to a dramatic increase in the number of scooter-related injuries. The problem was exacerbated by the introduction in the spring of 2000 of a new aluminum version that weighed less than 10 pounds and could be folded for easy portability and storage (CDC, 2000). By 2000 the number of children being treated for scooter-related injuries had increased to more than 40,000. About 85 percent involved children less than 15 years old, and almost 25 percent were less than 8 years old. Many of the injuries could have been prevented or reduced in severity if protective equipment (helmets, elbow pads, and knee pads) had been worn (NSC, 2001).

When all age groups are considered, one of the major recreational sources of death in the United States is drowning, which accounted for some 3,900 fatalities during 2000. Some 900 of these occurred at home and involved either swimming pools or bathtubs. More than 730 occurred during recreational boating, and alcohol consumption was reported to be a contributing factor in about 25 percent. About 90 percent could have been avoided if the victims had been wearing life jackets. Another source of deaths was pedalcycling (particularly bicycling), which accounted for slightly more than 800 deaths in 1998. About 85 percent of these occurred as the result of a collision with a motor vehicle. Although only 32 percent of the pedalcyclist deaths in 1983 involved riders 21 years of age and older, by 1985 this had increased to 50 percent, and by 1998 it had increased to 75 percent, perhaps reflecting the growth of bicycling as a form of adult exercise. As is true of many other activities, males are much more involved in fatal bicycle accidents than females, accounting for 87 percent of the deaths during 1998 (NSC, 2001). The risk of serious head injury in such events can be reduced by as much as 90 percent if the cyclist wears a protective helmet; in fact, if all cyclists wore helmets, perhaps 500 lives could be saved and 135,000 head injuries prevented in this country each year.

Intentional Injuries (Violence)

Although the public health community was slow in acknowledging the importance of unintentional injuries as a public health problem, it has been even slower in recognizing acts of violence (suicides, homicides, and assaults) as sources of intentional injuries. While the modern age of international terrorism has tremendously expanded both the nature of such acts and the number of people who can be affected by a single event, the discussion here will concentrate on suicides, homicides, and assaults. In this regard, certain facts are apparent. One is that the personal environment plays a prominent role in determining the extent and nature of violence in a community. Neighborhoods that inspire people to befriend one another, that are protective of local children, and that share resources appear to provide the kind of support that fosters healthy development. These characteristics may explain why some poor urban neighborhoods escape the violence that takes an enormous toll only a few blocks away. In contrast, neighborhoods that are socially and politically disorganized create conditions that contribute to antisocial behavior. When such behavior begins to dominate, there is an exodus of the small businesses that typically provide the glue that holds a neighborhood together. The way is then paved for illegal economies, such as drug dealing and gambling. This is one of the reasons that the "brownfields" program for restoring contaminated sites in abandoned urban areas (Chapter 9) is receiving such widespread support.

Another relevant factor is the situation within the home, particularly the relationship between husband and wife or other heads of households. Studies repeatedly demonstrate that more violence is caused by family and former friends than by strangers. In fact, 29 percent of the women who were murdered in the United States in 1992 were killed by a husband, ex-husband, lover, or suitor. At the same time, many such assaults are not reported. Contributing factors include the economic dependence of the wife on her husband, her desire to preserve the home, her concern about being separated from her children, and fear for her own safety should she try to leave. Two examples of violence-related events are discussed here. The first emphasizes that acts of violence are not restricted to the community and the home; the second emphasizes the increasing recognition that too little effort is being directed to seeking out the causes and developing programs for prevention of such events.

WORKPLACE HOMICIDE

In the United States during 1999, more than 20,000 people were physically assaulted while at work, and almost 1,000 were killed (NSC, 2001). Workplaces with the highest number of deaths are grocery stores, eating and drinking places, taxicab services, and justice or public order establishments. Occupations with the highest rates are taxicab drivers/chauffeurs, law-enforcement officers, gas-station or garage workers, and security guards. Factors that increase the risk for homicide among workers include the exchange of money with the public, working alone or in small numbers, and working late at night or in the early morning hours (NSC, 2001). Control measures include the installation of physical barriers, such as bullet-resistant enclosures with pass-through windows on critical service counters; alarm systems and panic buttons; video surveillance with closed-circuit television; bright and effective lighting; and training of employees in the identification of hazardous situations and appropriate responses (Mandelblit, 2001).

FIREARMS

Firearms are a major source of violence in the United States, accounting for the deaths of almost 65 percent of the people who are murdered and almost 60 percent of those who commit suicide. If all types of firearm-related deaths in this country are considered, the total number of people killed each year equals about 70 percent of those killed in motor-vehicle accidents. In some of the more populated states, gun-related deaths represent more than 25 percent of the total number of injury-related deaths. Although this may seem surprising, in reality it is not. The estimated number of guns owned by civilians in this country exceeds 200 million; in fact, handguns are present in about 25 percent of our households (Miller, 2002). While there is no denying that there are other major contributing factors to violence, such as poverty and lack of higher education, these data clearly show the need to gather more background and understanding of the relationship of guns to these types of events, as well as the need to increase the amount of funds that are directed to prevention of violence versus its control. At present, the former represents less than 6 percent of the latter.

The role of guns in violence-related deaths is illustrated in other ways.

A recent ten-year nationwide study of suicide deaths among children in a group of five states with the highest rates of household gun ownership (so-called high-gun states) versus those in five states with the lowest rates of such ownership (so-called low-gun states) found that while the number of nongun suicides in the two groups of states were similar, seven times as many children killed themselves with guns in the five high-gun states as in the five low-gun states. Likewise, the number of children murdered with guns in the high-gun states was more than three times that in the low-gun states. It is little wonder, then, that gun-related suicides or murders of children aged 5 to 14 in the United States now rank as the third-leading cause of mortality in this age group, being exceeded only by motor-vehicle accidents and cancer. The overwhelming contribution of the prevalence of firearms as a factor in gun-related suicides, homicides, and other types of violence among this age group is vividly illustrated by the data presented in Figure 11.4 (Miller et al., 2002).

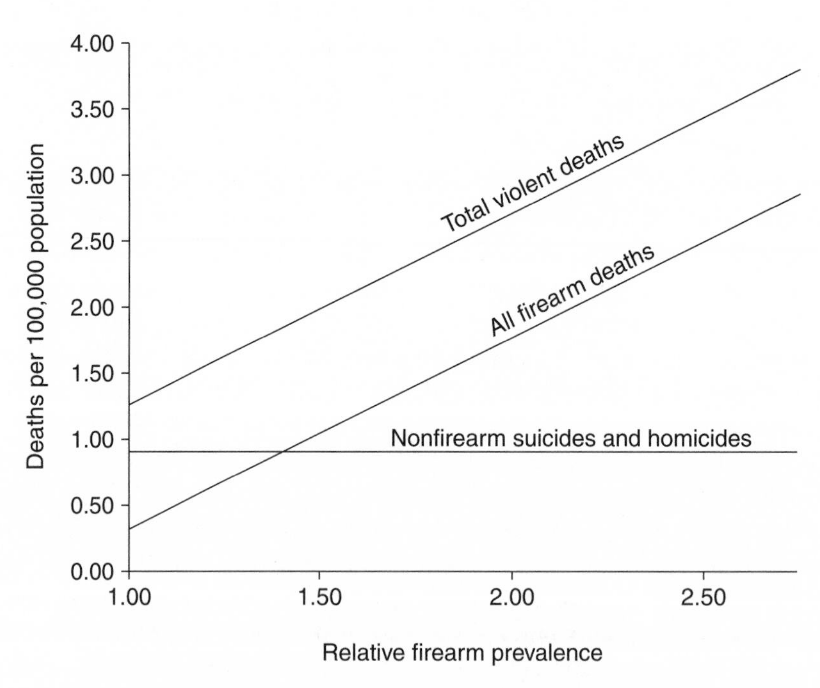

Figure 11.4 Rates of firearms prevalence and violent deaths among 5– to 14–year olds, United States, 1988–1997

Through the newly developing National Firearms Injury Statistical System, a major effort is under way to collect background data on gun-related acts of violence in the United States. Data being collected include whether an event occurred indoors or outdoors; the type of gun used; its make, model, and source; whether it was equipped with safety features; and whether drugs and/or alcohol were involved. Initially, the system is limited to nine sites. In the long term, it is envisioned that the program, which is directed by the Harvard Injury Control Research Center, will be expanded to include all 50 states and will be coordinated and financially supported by the federal government. Another step that is needed is to assign the regulation of the manufacture of guns and bullets to a specific federal agency, such as the Consumer Product Safety Commission. The development of guns that tranquilize rather than kill the person being shot should also be considered. Such guns could then be made available to people who desire a gun solely for protection (Dwortzan, 2000).

Other Sources of Injuries

Unintentional injuries arise from many sources, some of which have been covered in the preceding sections. Three other types are briefly discussed here.

ACCIDENTS INVOLVING CHILDREN

Although the discussion on recreational injuries shows that there are many sources of unintentional deaths and injuries among children, one that deserves special attention is the family farm. From 1992 through 1999, almost 350 deaths and more than 1,300 nonfatal injuries were reported among the approximately 300,000 young people, 19 years of age or under, engaged in such work in the United States. Some of these were in the 10- to 13-year-old range (NSC, 2001). The prevention and control of such events is hampered by the fact that the safety requirements of the Occupational Safety and Health Act (Chapter 14) are not enforceable on the vast majority of U.S. farms. The same is true for the Fair Labor Standards Act. Although the latter prohibits youth aged 16 years of age or younger from performing hazardous agricultural tasks, it does not apply to children employed on family farms. As a result, many children perform tasks that are both prohibited in other industries and inappropriate for their age (CDC, 1999a).

FALLS

Overall, falls are the leading cause of unintentional nonfatal injuries in the U.S., accounting for more than five million such injuries, as well as more than 6,000 deaths, in 2000. More than 70 percent of these events involved people 75 years of age or older; more than 80 percent involved people 65 years of age and older. More than half of these occurred in the home or on the premises, and 30 percent involved leisure activities. In the case of children less than 12 years of age, more than two-thirds of these types of injuries occurred at home (NSC, 2001). As with many other sources of injuries, increased effort is needed to identify and evaluate the origins and causes of these events, particularly among the elderly. Wheelchair users represent a specific group in need of attention. A recent study showed that almost 40 percent of such users had suffered at least one fall during the previous year, and almost half had been injured. In many cases, a contributing factor was that the home in which the person resided had not been modified to accommodate the use of a wheelchair. While about 2 million community-dwelling people in the United States use wheelchairs, only 10 percent of them live in home environments properly suited for such use (Berg, Hines, and Allen, 2002).

FIRES

Fires in residences were estimated at 383,000, almost three-quarters of all structure-related fires in the United States, during 1999. Almost three-quarters of these occurred in one- or two-family dwellings (NSC, 2001). Death rates were highest during December through February, reflecting the seasonal use of heating devices (e.g., portable space heaters and wood-burning stoves). The most vulnerable age groups were children aged less than 5 years and adults aged 65 years or over. Their death rates were two to six times the average for all ages (CDC, 1998). In all, these events accounted for 2,920 deaths and 16,425 injuries in the United States that year. The leading source of residential fire-related deaths is cigarette smoking (26 percent). Such fires result in the deaths of more than 1,000 people (including about 100 children) each year. About 150 of these are caused by children playing with cigarette lighters. Arson is also an important factor and is the leading cause of fire-related injury (19 percent) and economic loss (18 percent) (NSC, 1990).

As with most environmental and public health problems, the control of deaths and injuries from fires requires a systems approach. Increased fire-fighting capabilities, stricter enforcement of building and housing codes, and intensified pursuit of arsonists are all helpful, but these approaches alone will not control the problem. These activities must be supplemented by the installation and continued maintenance of smoke detectors and sprinkler systems in buildings and increased attention to the design, installation, operation, and maintenance of heating systems. It is also important that sleeping garments, especially those worn by children, be fire resistant, and that bedding and upholstered furniture be not only fire resistant but also incapable of releasing toxic gases when exposed to heat and flame. Another helpful measure would be the marketing of fire-safe cigarettes that would not burn hot enough to ignite upholstery and of a childproof cigarette lighter.

The General Outlook

As noted earlier, there has been a dramatic reduction in the deaths per mile due to motor-vehicle accidents. How this was accomplished is an outstanding example of applying the systems approach to this type of problem. Motor vehicles were modified to include headrests, energy-absorbing steering wheels, shatter-resistant windshields, safety belts, and air bags. Roads were improved by better delineation of curves, adding center-line stripes and reflectors, converting to breakaway sign and utility poles, illuminating many key sections of roadways at night, installing barriers to separate oncoming traffic lanes, and designing guardrails to guide vehicles back onto the road should the driver lose control. Other measures adopted included improved driver licensing and testing and vehicle inspections, the enactment and enforcement of traffic safety laws, and reinforced public education. Further contributing factors were the previously discussed better enforcement of safety-belt and child-safety-seat requirements, motorcycle helmet laws, and the establishment of the graduated system for licensing young drivers.

If similar approaches are designed and applied to other sources of injuries, similar progress can be made. At the same time, however, this is not to imply that the problem of motor-vehicle injuries and deaths has been solved. Many challenges remain. While annual motor-vehicle crash-related fatalities involving alcohol were reduced from about 52 percent in 1990 to as noted earlier, about 40 percent in 2003 (Durbin, 2004), alcohol remains a major contributor to the associated injuries and deaths and carried with it an estimated economic burden of almost $30 billion for the year 2000 (NSC, 2001). Another continuing challenge is the relatively high rate of deaths among young drivers and passengers. Since these deaths, many of which could readily be avoided, occur at such a young age, the years of life lost are not only a tragedy for the families concerned, but also a cost in terms of the contributions these people could have made to society had they been able to lead full and productive lives.

Interestingly, there never seems to be an end to the challenges that continue to emerge. Eleven children in this country died during 1998 after being entrapped in the trunks of cars (CDC, 1998b). To avoid such events, some manufacturers are installing emergency releases that are readily visible to anyone inside a trunk and will enable him/her to escape. Even more serious is the number of children who are dying after being left inside a vehicle parked in the summer sun (CDC, 2002b). Under such conditions, temperatures inside a car, with the windows closed, can reach 55°C–78°C (131°F–172°F) within as short a time as 15 minutes. From 1998 to 2002, the number who died each year in the United States ranged from 25 to 33; data for 2003 showed that the total may have reached 40 (Associated Press, 2003).

These and many injury-related problems emphasize once again how important it is that members of society recognize their obligations in helping to ensure their own safety as well as that of their families. Parents, for example, need to ensure that doors to cabinets for the storage of household cleansers and other toxic agents are child resistant; that stairs are equipped with handrails and padding; that play yards are fenced; and that, as noted previously, the ground beneath swings, slides, and other playground equipment is covered with soft dirt or other shock-absorbent material. Obviously also of importance is close parental supervision of small children at all times.

At the same time, legislated codes and standards can also contribute to the reduction in the numbers of childhood injuries and deaths. Examples include the requirement that all toxic materials be sold in containers with childproof caps; that hot-water heaters have temperature limits to prevent scalds and burns; that barriers be installed on upstairs windows of residential buildings; that fences be erected around swimming pools; that electrical outlets near the floor be covered; that paint used on indoor walls, furniture, and equipment for children be lead free; and that control knobs on stoves be located out of reach of children.

12

ELECTROMAGNETIC RADIATION

All human beings are constantly exposed to natural radiation, artificial radiation, or both. What is formally known as electromagnetic radiation is propagated through space in the form of packets of energy called photons, which travel at the speed of light (3×10^{10} centimeters per second). Each photon has an associated frequency and wavelength. Its energy is directly proportional to its frequency and is expressed in units of electron volts (eV)—the energy that an electron would acquire in being accelerated across an electrical potential difference of one volt. Higher-energy photons, such as cosmic rays, have frequencies of 10^{21} hertz (Hz, or cycles per second) or more and energies of 10^{7} eV or more; lower-energy photons, such as those associated with electric and magnetic fields, have frequencies of 1–10^{3} Hz and energies only a tiny fraction of an eV. Photons in the intermediate-energy range (10^{-2}–10 eV), such as those associated with infrared and visible light, have frequencies of 10^{12}–10^{15} Hz. High-energy photons are extremely penetrating and can have effects far from their source; the effects of lower-energy photons are concentrated near the source. Only intermediate-range electromagnetic radiation can be detected by the human senses. The energy ranges for the various types of radiation (Figure 12.1) have not been precisely defined in every case; overlaps are common.

As it moves through space, electromagnetic radiation interacts with the atoms of which matter is composed. Only photons in the higher-energy ranges, such as cosmic rays, x rays, and gamma rays, have sufficient energy to ionize these atoms by interacting with the orbital electrons and stripping them away. These are referred to as *ionizing* radiation. Electro-

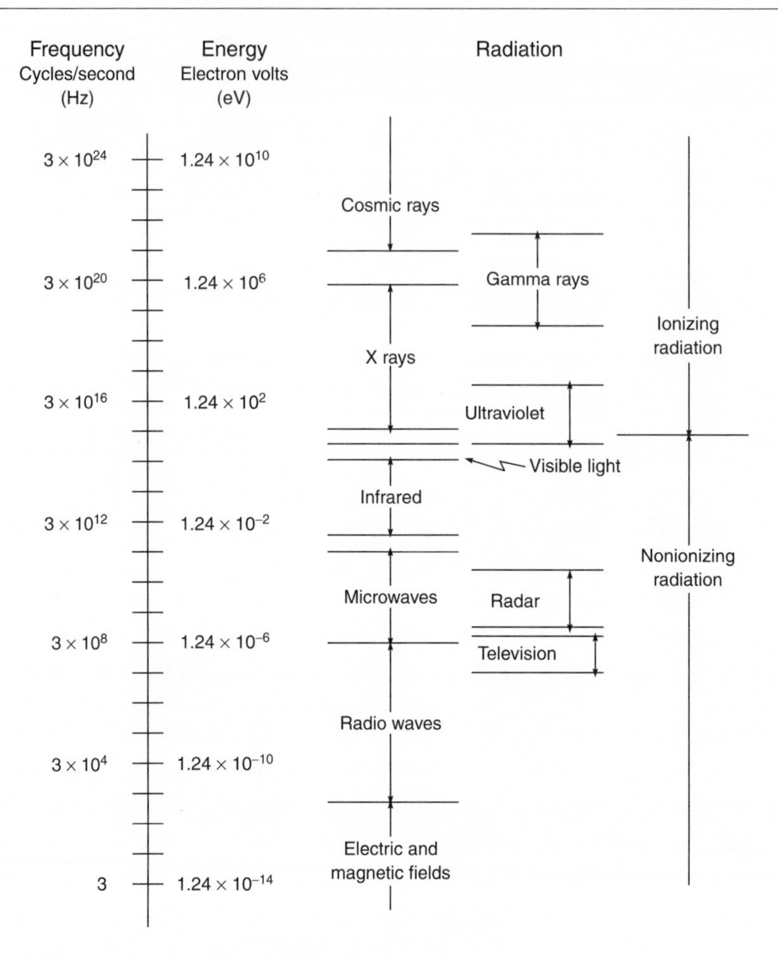

Figure 12.1 The electromagnetic spectrum

magnetic radiation in the lower-energy ranges, such as the lower-frequency range of ultraviolet waves, as well as infrared waves, microwaves, and radio waves, do not possess sufficient energy to be ionizing. These are referred to as *nonionizing* radiation. Once an electron is removed, it exhibits a unit negative charge, and the residual atom shows a net unit positive charge. The two products are known as an ion pair (Figure 12.2). This transfer of energy to atoms can result in chemical and biological changes that are harmful to health.

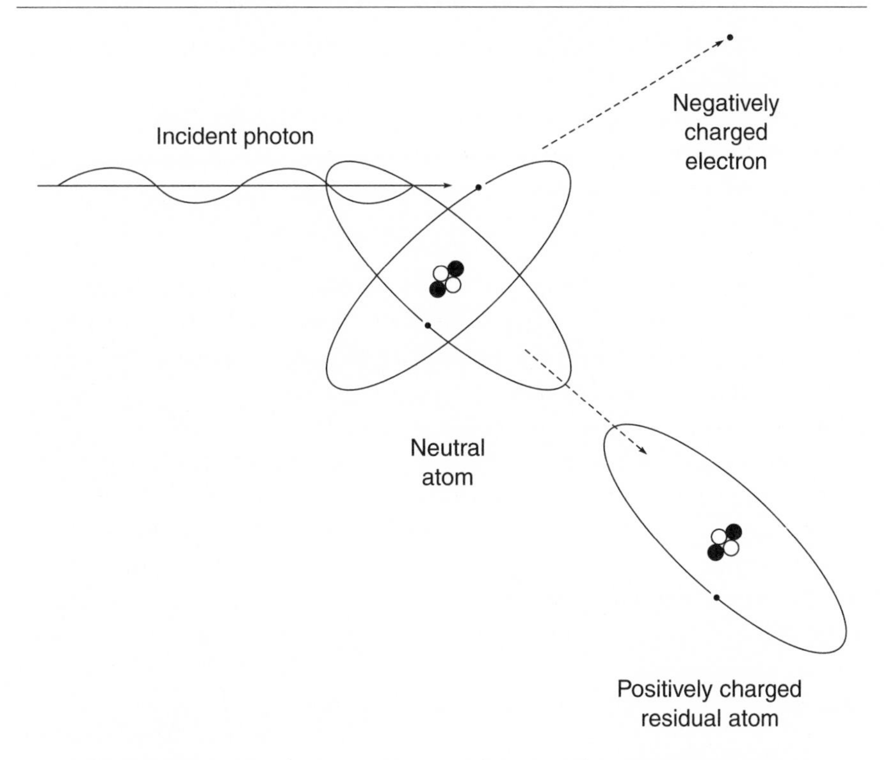

Figure 12.2 Interaction of an x or gamma photon with a neutron atom to produce an ion pair

Nonionizing Radiation

Although the biological effects of ionizing radiation have been recognized and reasonably well understood for some time, questions remain concerning the nature and effects of certain types of nonionizing radiation, in particular the photons associated with lower-energy electric and magnetic fields. Consequently, techniques for assessing such radiations, especially in terms of the specific parameters that need to be measured to evaluate their potential impacts, are still evolving. Each of the major sources of nonionizing radiation is discussed in the following sections.

ULTRAVIOLET RADIATION

The principal natural source of *ultraviolet radiation* (UVR) is the sun. Since the atmosphere serves as an absorbent, the amount of UVR reaching the

Earth at a particular location increases with altitude. Since the stratospheric ozone layer also serves as a protective barrier, its depletion is a matter of considerable interest. Today, as a result of technological developments, there are many artificial sources of UVR. These include the electric arcs used in lights, welding torches, plasma jets, germicidal lamps, and tanning lamps.

High short-term (acute) exposures to UVR can produce marked systemic effects, including fever, nausea, and malaise. Cumulative (chronic) effects include aging of the skin and premalignant or malignant changes. The effect of most concern is the development of malignant melanoma, the most serious form of skin cancer, the incidence of which doubled in the United Kingdom between 1979 and 1994. In seeking to provide guidance on avoiding such cancers, detailed studies have been conducted on the protection afforded against UVR by various types of clothing. Although ordinary window glass will remove most of the higher frequencies, the protection afforded by clothing depends on its composition and color. Whereas cream-colored cloth woven of 100 percent cotton provides a protection factor of less than 10, the same cloth in bright pink or turquoise provides a protection factor of more than 30. This is important because a protection factor of less than 10 is not considered adequate for exposures occurring during the summer. The protection factor also depends on whether the material is stretched, which reduces its effectiveness, and whether it is wet, which can either increase or reduce the effectiveness (Agnew et al., 1998). Because studies show that exposure to tanning lamps, especially by young people, can lead to a doubling of the risk of several types of skin cancer, including basal-cell and squamous-cell cancers, some health officials have recommended that tanning salons be closed to minors.

VISIBLE LIGHT, INCLUDING LASERS

The health effects of visible light may be direct or indirect. An example of the former is a retinal burn caused by looking at the sun during an eclipse without adequate filtration; an example of the latter is injury from an accident caused by insufficient or excessive lighting. Insufficient lighting can result in a fall; excessive lighting, such as the bright headlights of an oncoming car, can cause a crash.

Other health problems can result from the use of devices in which beams of visible light can be focused both temporally and spatially. One of these, the laser, has found a wide range of applications, which include the alignment of tunnels, distance measurement, welding, cutting, drilling, heat treatment, entertainment (laser light shows), and surgery. Lasers are also used in videodisc players, supermarket scanners, and facsimile and printing equipment. In most cases, laser units are totally or partially enclosed to prevent exposure to direct or scattered radiation. A common exception, however, is the use of laser pointers, which permit lecturers to focus audience attention on relevant parts of projected slides. Unless care is exercised, such use can readily lead to harm because even minute quantities of laser light can burn a small hole in the retina and permanently impair the vision of any person whose eyes are subjected to the direct beam.

Another concern is the increasing incorporation of lasers into a wide range of novelty products. For this reason, regulations have been promulgated in the United Kingdom requiring that laser pointers be classified according to their output. Only those for which the power of the beam is less than the level believed to be damaging to the eye or is less than the level at which the involuntary blink or closing of the eyelid will provide protection are permitted for use in unsupervised areas. Another protective measure is to require that the size of the beam be expanded so that it produces a spot on the screen measuring between one and two centimeters in diameter. Not only is this considered to be easier to detect by viewers, but it also reduces the incentive that such devices be used as a novelty product (O'Hagan and Hill, 1998). Nonetheless, the same property that makes higher-power lasers damaging if they are improperly used makes them uniquely beneficial for use in correcting problems within the eyes. One common example is to change the curvature of the cornea and correct the problem of nearsightedness. A summary of the mechanisms of interaction and examples of adverse effects of exposures to radiation from different parts of the optical spectrum, including visible light and lasers, is presented in Table 12.1.

INFRARED RADIATION

All objects emit infrared radiation to other objects that have a lower surface temperature. One example is the heat that reaches the Earth from the sun; another is the heat produced by a stove or by the radiant heating units used in many dwellings. Fortunately, the sensation of heat quickly provides adequate warning of extreme conditions. Infrared radiation does not penetrate deeply into tissues, but if it is not controlled, it can cause burns on the skin surface, cataracts in the lens of the eye (which has poor

Table 12.1 Mechanisms of interaction and examples of adverse effects of exposures to radiation within different portions of the optical spectrum

Part of spectrum	Mechanisms of interaction	Adverse effects
Ultraviolet radiation (180–400 nm)	Photochemical alterations of biologically active molecules, such as DNA, lipids, and proteins	Acute erythema, keratitis, conjunctivitis, cataracts, photoretinitis, accelerated skin aging, skin cancers
Visible radiation (380–600 nm)	Photochemical alterations of biologically active molecules in the retina	Photoretinitis ("blue-light hazard")
Visible and near-infrared radiation (400–1,400 nm)	Thermal activation or inactivation; photocoagulation	Thermal injury, skin burns and retinal burns, thermal denaturation of proteins, tissue coagulation/ necrosis
Middle and far-infrared radiation (3 μm–1 mm)	Thermal activation or inactivation; coagulation	Thermal injury, skin and corneal burns, cataracts, thermal denaturation of proteins, tissue coagulation/ necrosis
Laser radiation (180 nm–1 mm)	Photochemical, photothermal, photoacoustic, exposure duration <1 μs; photoablative, exposure duration <1 ns; bubble or plasma formation (change of phase); nonlinear optical effects	Tissue damage, skin burns, ocular burns, tissue vaporization

heat-dissipating mechanisms), or retinal damage. Cataracts can readily be prevented by wearing protective glasses. In more generalized situations, such as the interior of buildings, excess heat is commonly controlled through the use of air-conditioning and ventilation systems, with the excess being dissipated into the outdoor environment.

MICROWAVE RADIATION

Sources of microwave radiation include radar, radio and television transmitters, satellite telecommunication systems, and microwave ovens. Microwaves are used in industry to dry and cure plywood, paint, inks, and synthetic rubber and to control insects in stored grain. They are used in medicine to provide deep-heat therapy for the relief of aching joints and sore muscles.

The human body is largely transparent to the lower frequencies of microwaves, and those in this energy range (Figure 12.1) produce no biological effects. As the frequency increases, however, the energy is increasingly absorbed, reaching a maximum at about 3×10^8 Hz, the ultrahigh-frequency (UHF) television range. At still higher frequencies ($>10^9$ Hz), less of the energy is absorbed, and above 10^{10} Hz the skin acts as a reflector. Potentially the most hazardous microwaves are those in the range 10^8–10^9 Hz since at these frequencies there is little or no heating of the skin and the thermal receptors are not stimulated.

From the standpoint of the public, one of the most common sources of microwaves is the microwave oven. In fact, such appliances are being used in an estimated 9 of every 10 homes in the United States. To perform their function, microwaves agitate the water molecules and cause them to vibrate millions of times each second and rub against one another. The accompanying friction manifests itself as heat. Although the window in the oven door is relatively clear, it contains a thin metal wire mesh with holes large enough for visibility but too small for the microwaves to escape. Although earlier it was feared that microwave ovens could interfere with pacemakers and other implanted medical devices, such devices now contain shields to protect them from most ambient radiation (Ropeik and Gray, 2002).

During the 1990s, increasing concern was expressed about two widely used devices that are possible sources of microwave radiation—traffic radar devices and handheld cellular telephones. Although epidemiological data did not confirm the association, the former was cited as a possible source of testicular cancer among police officers. In the case of cellular telephones, the issue was whether the moderately low levels of radiofrequency energy (in the low microwave range) emitted could contribute to brain cancers or other adverse health effects. Again, however, there is no scientific evidence to support such a concern (Ropeik and Gray, 2002). Even so, the director general of the World Health Organization has recommended that parents limit the use of cellular telephones by children. One of the apparent reasons is a report that laboratory studies had shown that cultured human cells shrank after one hour of exposure to a cellular telephone (Kirschner, 2002).

ELECTRIC AND MAGNETIC FIELDS

All atoms contain positively and negatively charged ions. Since most objects contain a balance of ions with such charges, they are electrically neutral. When this balance is upset, an *electric field* is produced that leads to effects that can be readily observed. Common examples are the attraction between a comb and a person's hair or the charge experienced by a person who walks on a synthetic rug during cold weather. When electric charges (electricity) flow through a wire, a *magnetic field* is generated. In both cases, however, the field is confined to the vicinity of the source. Any home appliance that has an electric motor can be a source of a magnetic field. Typical examples are refrigerators, clothes washers, and vacuum cleaners. Also of importance are electric mixers and can openers, as well as personal items such as electric shavers, hair dryers, electric toothbrushes, electric blankets, and the magnets used to hold notes and pictures on refrigerators (Ropeik and Gray, 2002). Interestingly, the magnetic fields produced by many of these sources are comparable to, or far in excess of, those present under electric power transmission lines. Even the Earth itself produces a magnetic field (Valberg, 2001).

While several investigators have claimed to have observed a link between childhood cancer and electric and magnetic fields, particularly those associated with transmission lines, laboratory evidence in support of these findings has not been confirmed, and an accepted mechanism by which such fields can cause disease has not been identified. Since electric fields are too weak to penetrate the skin, scientists believe that only magnetic fields can have any biological effects (Valberg, 2001). A summary of the mechanisms of interaction and adverse effects of exposures to electric and magnetic fields is presented in Table 12.2.

STANDARDS FOR CONTROL

Worldwide, the primary source of guidelines for limiting exposures from sources of nonionizing radiation is the International Commission on Non-Ionizing Radiation Protection (ICNIRP). In developing its recommendations, the ICNIRP conducts detailed reviews of the relevant epidemiological and laboratory studies. Where there are large uncertainties in knowledge about the health effects of a particular agent, the exposure limits are reduced in proportion to the magnitude of the uncertainties. In all cases, the limits are sufficiently restrictive to avoid immediate effects for those cases in which thresholds are known to exist. Specific sources addressed to date include optical radiation, for example, ultraviolet, visible, and infrared, including lasers; the nonionizing portions of the electromagnetic spectrum, for example, microwaves; and other radiofrequency fields, including those down to the range that encompasses static electric and magnetic fields (ICNIRP, 1998). Plans call for the possible development of guidelines for ultrasound and infrasound exposures.

Similar guidance has been provided by the American Conference of Governmental Industrial Hygienists (ACGIH, 2003). This includes recommended limits for occupational exposures to lasers, static magnetic fields, sub-radio-frequency magnetic fields, sub-radio-frequency and static electric fields, radio-frequency and microwave radiation, light and near-infrared radiation, and ultraviolet radiation.

Ionizing Radiation

Ionizing radiation (Figure 12.1) includes machine-produced x rays, discovered by Wilhelm Roentgen in 1895, and alpha, beta, and gamma rays, first observed when Antoine-Henri Becquerel discovered naturally radioactive materials in 1896. Since the development of the nuclear reactor in late 1942, multitudes of artificially produced radioactive materials have been added to these sources.

BIOLOGICAL EFFECTS

Biological effects in living organisms exposed to ionizing radiation involve a series of events. The first is the previously discussed ionization, which ejects electrons from the atoms in the molecules. The residual molecule, left with a positive charge, is highly unstable and will rapidly undergo chemical changes. One such change is the production of "free radicals," which are extremely reactive chemically. The ensuing reactions may in turn lead to permanent damage of the affected molecule, or the energy may be transferred to another molecule and the free radicals may recombine. The time required for this chain of physical and chemical events to take place is on the order of a microsecond or less. The subsequent development of biochemical and physiological changes, however,

Table 12.2 Mechanisms of interaction and examples of adverse effects of exposures to electric and magnetic fields

Part of the spectrum	Mechanisms of interaction	Adverse effects
Static electric fields	Surface electric charges	Annoyance of surface effects, shock
Static magnetic fields	Induction of electric fields in moving fluids and tissues	Effects on cardiovascular and central nervous systems
Time-varying electric fields (<10 MHz)	Surface electric charges	Annoyance of surface effects, electric shock, burns
	Induction of electric fields and currents	Stimulation of nerve and muscle cells, effects on nervous system functions
Time-varying magnetic fields (<10 MHz)	Induction of electric fields and currents	Stimulation of nerve and muscle cells, effects on nervous system functions
Electromagnetic fields (100 kHz–300 GHz)	Induction of electric fields and currents, absorption of energy within the body	Excessive heating, electric shock, burns
	>10 GHz: surface absorption of energy	Excessive surface heating
	Pulses <30 μs, 300 MHz–6 GHz: thermoacoustic wave propagation	Annoyance from microhearing effect

may require hours; in the case of latent cancers, it may require years (Little, 1993).

All cells are susceptible to damage by ionizing radiation, and only a very small amount of energy needs to be deposited to produce significant biological change. For example, if all the deposited energy were converted to heat, a dose of radiation sufficient to be lethal to human beings would raise the temperature of the body by only 0.001°C. Fortunately, ionizing radiation can be accurately assessed, using other methods of measurement, at exposure levels several orders of magnitude below those required to produce measurable biological effects.

Although radiation is similar in some respects to other carcinogenic agents, it is unique in that it has the ability to penetrate cells and to deposit energy within them in a random manner, unaffected, for example, by the cellular barriers presented to chemical agents. As a result, all cells in the body are susceptible to damage by ionizing radiation (Little, 1993). For many years, the critical irreversible chemical change responsible for the biological effects that subsequently follow appeared to be a direct consequence of damage to deoxyribonucleic acid (DNA), the basic material that controls the structure and function of the cells that make up the human body. Now, however, it is recognized that the traversal of an ionizing particle through the nucleus of a cell is not a prerequisite for producing genetic damage or a biological response. In fact, studies in which as few as 1 percent of the cells in a population were traversed by alpha particles have demonstrated that there is intercellular communication in the transmission of damage signals to nonirradiated cells. This type of interchange is what is known as the "bystander" effect, namely, that cells in close proximity to those that are traversed by radiation can undergo similar biological changes. Since the bystander cells suffer point mutations, versus the double-stranded breaks in the cells that are directly hit, this suggests that the biological effects of radiation are more complex than previously thought.

The potential effects of radiation on cells can be divided into three categories: (1) at high doses, it can cause death; (2) at lesser doses it can inhibit mitosis; and (3) at any dose it can cause alterations in the genetic material of the cell. Because of the effects of radiation on mitosis (category 2), the most sensitive tissues are those in which the cells frequently divide—for example, the precursor cells in the bone marrow that give rise to white blood cells and platelets, and the cells that line the stomach and small intestine. Muscular and brain tissues, where cell division is less pronounced, are far less sensitive. At still lower doses, even those insufficient to affect the ability of the cell to divide, radiation may produce mutations or other heritable alterations in DNA metabolism in the genetic material of cells. Presumably, such changes are responsible for the long-term somatic effects of radiation, such as cancer induction. When mutations involve germinal cells in the gonads, they may be passed on to the offspring of the irradiated individual and cause heritable genetic effects (Little, 1993). To date, however, ionizing radiation has not been identified as a cause of such effects in humans (ICRP, 1991).

UNITS OF DOSE

On the basis of knowledge about the deposition of energy and its associated biological effects, units have been developed for expressing the doses that result from exposures to ionizing radiation (Table 12.3). The most common is that for the *equivalent dose,* the *sievert* (Sv). Since this unit is far in excess of the doses usually encountered in the workplace and the ambient environment, subunits have been developed, such as the millisievert (mSv), which, as the name indicates, is one thousandth of the Sv. As commonly applied, the Sv and mSv express the dose to an individual. For certain purposes, such as comparing the relative societal impacts of several sources of ionizing radiation, the International Commission on Radiological Protection (ICRP, 1991) has developed what is called the *collective dose.* As noted (Table 12.3), it is calculated by multiplying the total number of people exposed (expressed in units of persons) by their average individual dose (expressed in units of the sievert). As noted later (Chapter 15), however, care must be exercised in the application of the concept of collective dose, particularly in terms of assessing the risk to large population groups who have received very small doses.

On the basis of total dose and dose rate, the effects of radiation exposure can be classified as either *deterministic* or *stochastic.* Deterministic effects are those for which the severity of the effect varies with the dose, and for which a threshold may therefore exist. Such effects are generally associated with *acute* exposures involving doses in the range of a Sv or more, delivered to part or all of the body within a short period of time. From the standpoint of exposures to individual portions of the body, deterministic effects may include cataracts, sterility, and tissue damage (for example, erythema). In terms of acute doses to the whole body, such effects range from nausea, vomiting, and diarrhea in the lower dose ranges to severe effects, including death, at higher levels (Table 12.4). Acute doses

Table 12.3 Units of dose for ionizing radiation

Unit	Description
Roentgen	The roentgen, now obsolete, was first introduced at the Radiological Congress held in Stockholm in 1928 as the special unit for expressing exposure to ionizing radiation. It was based on the quantity of electrical charge produced in air by x or gamma radiation. One roentgen (r) of exposure will produce about 2 billion ion pairs per cubic centimeter of air. Later it was noted that the exposure of soft tissue or similar material to 1 r resulted in the absorption of about 100 ergs of energy per gram. By multiplying the amount of energy absorbed by what is called a radiation-weighting factor (which takes into consideration the nature of the radiation, plus other factors), it is possible to estimate the accompanying biological effects. For x, gamma, and beta radiation, the radiation-weighting factor is 1. For alpha radiation, it is 20.
Gray	The gray is the unit of *absorbed dose* and is equivalent to the absorption of 10^4 ergs per gram. If soft tissue or similar material is exposed to 100 roentgens, the amount of energy absorbed is equivalent to about 1 gray.
Sievert	The sievert is the unit of *equivalent dose* (often simply called the *dose*). One sievert (Sv) is equal to 1,000 millisievert (mSv). It was designed to provide a means of expressing the biological effects of all types of ionizing radiation on an equivalent basis and is commonly used for expressing the dose to all or a portion of the body of an individual.
Person-Sv	After the development of the sievert, the need for a unit for expressing the societal risk associated with doses to more than one person was recognized. The resulting unit is the *collective dose,* the product of the number of people exposed and their average dose. Basic to the use of this unit is the assumption that the relationship between the dose and the accompanying health effects is linear. Accordingly, a dose of 0.10 Sv to 100,000 people (yielding a collective dose of 10,000 person-Sv) would, by definition, have the same societal impact as a dose of 0.05 Sv to 200,000 people.

Table 12.4 Biological effects in humans of acute whole body external doses of ionizing radiation

Dose (Sv)	Effects
0–1	Long-term effects possible, but serious immediate effects on average individual highly improbable
1–2	Minimal symptoms; nausea, fatigue, and possible vomiting
2–3	Transient mild to moderate nausea and vomiting in 20–70% of exposed individuals with onset at 2–3 hours and continuing for first 2 days; after latent period of up to 2 weeks, loss of appetite and general malaise appear but are not severe; recovery in about 3 months unless complicated by previous poor health
3–5	Transient moderate nausea and vomiting in 50–90% of exposed individuals within first few hours, followed by latent period of up to 1 week without definite symptoms; then steady decline in health with eventual death to 50% of those exposed, some as early as 2–6 weeks after exposure
5–8	Onset of moderate to severe nausea and vomiting in 50–90% of exposed individuals within 1 hour; diarrhea, hemorrhage, fever, puapura, inflammation of mouth and throat toward end of first week; rapid emaciation; death may occur within 6 weeks for more than 50% of those exposed at low end of dose range, within 3–5 weeks in 90% of those exposed at high end of dose range
>8	Severe nausea and vomiting within minutes; death to probably all exposed individuals within 2–3 weeks

to specific body organs, for example, the testes or ovaries, can produce sterility (Table 12.5). Stochastic effects, in contrast, are those for which the probability that an effect will occur, rather than its severity, is regarded as a function of the dose, without threshold. These types of effects, for example, solid tumors and leukemia, are anticipated to follow some years after the receipt of low doses over a long period of time (ICRP, 1991).

DOSE-RESPONSE RELATIONSHIPS

Ionizing radiation has sometimes been termed a universal carcinogen in that it induces cancer in most tissues and most species at all ages (including the fetus). In reality, radiation has proved to be relatively weak in terms of both its carcinogenicity and its mutagenicity. As a consequence, few human data exist on the harmful biological effects of ionizing radiation at low doses. Much of the information derives from observations involving relatively high doses received over short periods of time. A prime example is the ongoing epidemiological study of survivors of the World War II atomic bombings in Japan. Similar studies are ongoing on the evaluation of health effects among populations exposed as a consequence of the accident at the Chernobyl nuclear power plant (Chapter 19). Other studies that have provided valuable data include evaluations of lung cancers in uranium miners exposed underground to airborne radon and its decay products, bone cancers in young women who ingested radium and thorium while painting radioactive luminous markings on the faces of clocks and watches, and breast cancers in women with tuberculosis who had multiple fluoroscopic chest examinations.

Difficulties in interpreting data on the health effects of ionizing radiation are exemplified by the epidemiological studies of the survivors of the atomic bombings in Japan. By the year 2000, the total number of cancer deaths in this population was about 7,500. Of these, less than about 500 are estimated to be attributable to radiation. Further complicating the studies was that even as late as 2004, more than 40 percent of the survivors were still alive (Malakoff and Normile, 2004). Until a higher percentage have died, estimates of the full extent of the latent effects cannot be com-

Table 12.5 Estimated threshold doses for deterministic effects of acute radiation exposures of specific body organs

Health effect	Organ	Dose (Sv)
Temporary sterility	Testes	0.15
Depression of blood-cell-forming process	Bone marrow	0.50
Reversible skin effects (e.g., erythema)	Skin	1.0–2.0
Permanent sterility	Ovaries	2.5–6.0
Temporary hair loss	Skin	3.0–5.0
Permanent sterility	Testis	3.5
Cataract	Lens of eye	5.0

pleted. Even then, quantification of the effects will be difficult. A further complication in the application of the results of these studies to evaluations of the biological effects due to protracted exposures (as in the case of radiation workers) is that the health effects per unit of dose are generally significantly less under chronic exposure conditions. For these and other reasons, the uncertainties in applying risks estimated on the basis of one set of conditions to another are enormous.

Closely interwoven with these challenges is whether there is a threshold for the health effects that may be induced by low-level radiation exposures (Chapter 15) (NCRP, 2001). Nonetheless, scientists believe that as an outgrowth of the existing data, a reasonably reliable basis exists for expressing the relationship between the biological effects of ionizing radiation and chronic exposures received at low dose rates. In fact, several models have been developed for expressing this relationship (Figure 12.3). In general, mutations induced by radiation in cultured human cells follow a linear model (graph *a*). In contrast, cell death is related exponentially to dose (graph *b*). A further complication is that the shape of the

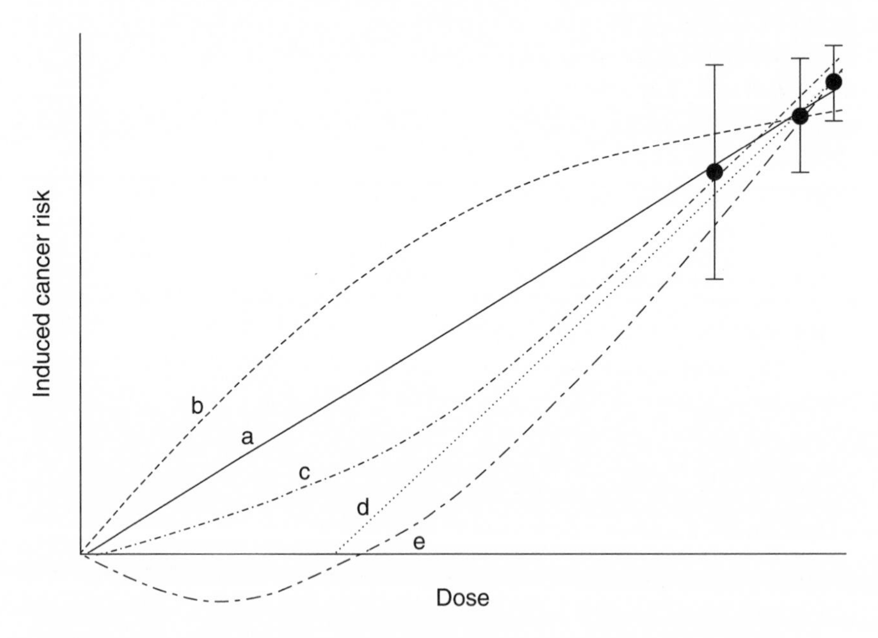

Figure 12.3 Dose response models for quantifying the effects of ionizing radiation

curve appears to vary with the nature of the ionizing radiation. After considerable debate, the Committee on the Biological Effects of Ionizing Radiation (BEIR) of the National Research Council (BEIR, 1980) suggested that for x rays and gamma rays, the data appeared to support a relationship that combined the linear and quadratic models (graph *c*). Although newer data appear to support this conclusion with regard to leukemia, it now appears that the induction of solid tumors (cancers of the thyroid, breast, and bone) follows a linear model (NCRP, 2001). The relationship between dose and risk, if there were a threshold, would be as depicted in graph *d*. Finally, the possibility that small doses of ionizing radiation are beneficial is depicted by graph *e*. This is similar to the J-shaped dose-response curve discussed in Chapter 2.

TYPES OF EXPOSURES

Both external and internal exposures from ionizing radiation represent potential risks to human health in the workplace and general environment. Regardless of the assumptions regarding the dose-response model, for sources of ionizing radiation *external* to the body, the resulting biological effects will depend on (1) the total dose, (2) the dose rate, and (3) the percentage and region of the body exposed. In general, the potential for harmful effects increases in response to increases in each of these three factors. *Internal* exposures result from the presence or deposition of radioactive materials in the body through ingestion or inhalation. The potential for harm in this case depends on the types and quantities of material taken in and the length of time they remain in the body. Radionuclides that emit alpha particles present a higher risk than those that emit beta particles. In general, the larger the quantity of radioactive material consumed and the longer it remains in the body, the higher the risk.

The ICRP has developed annual limits on intake for a large number of radioactive materials. These include guidelines for determining intake limits for workers (ICRP, 1994), as well as members of the public (ICRP, 1996). For radionuclides that are distributed rather uniformly within the body, the permissible intake is calculated on the basis of the equivalent dose-rate limit for the whole body. For radionuclides that concentrate predominantly in a single organ, the annual limit on intake is based on the concept of effective dose (Chapter 15). To protect people from airborne radionuclides, the annual limits on intake have been converted into derived air concentrations. In a similar manner, permissible concentrations have been calculated for specific radionuclides in drinking water. Both types of limits have been incorporated into the regulations of the U.S. Nuclear Regulatory Commission (USNRC) and the Occupational Safety and Health Administration (OSHA). Techniques for assuring compliance with the regulations for occupational exposures include analyses of intakes via inhalation and possible ingestion, as well as bioassays and whole-body counting (Chapter 4). Techniques for monitoring the movement of radioactive materials within the environment and for estimating doses to members of the public are described in Chapter 16.

Natural Background Radiation

Since the beginning of their existence, humans have been exposed to significant levels of ionizing radiation from two sources: naturally occurring radioactive materials in the Earth and cosmic radiation from outer space. While both of these expose the human body externally, radioactive materials can also cause internal exposures if they are ingested or inhaled. People in high-flying aircraft and those who participate in space missions are also externally exposed to rapidly moving charged particles.

COSMIC RADIATION

The annual dose from *cosmic radiation* at sea level is about 0.3 mSv. Since, as was the case with UVR, the atmosphere between the Earth and outer space serves as a shield, the accompanying dose rate increases with altitude. At an altitude of about 1,600 meters (1 mile), it is about 0.5 mSv per year; at about 3,660 meters (12,000 feet), about 1 mSv per year; at about 9,150—12,200 meters (30,000–40,000 feet), where commercial subsonic aircraft operate, the range is 45–70 mSv per year; at about 15,240 meters (50,000 feet), where commercial supersonic aircraft operated, until flights were terminated in 2003, the range is 80–90 mSv per year; and at about 19,800 meters (65,000 feet) and higher, where future supersonic aircraft may operate, it could be in the range of 100–175 mSv per year or more. Passengers, in general, however, do not remain at these altitudes for long periods of time. Another factor that mitigates the problems of exposure to people at the higher altitudes flown by supersonic aircraft is that they fly much faster and complete their flights in less time. As a result, the total dose for a given trip is about the same whether a trip is made in a supersonic or subsonic aircraft.

Taking into account the fact that passengers on flights across the United States in subsonic aircraft are airborne only a few hours, the actual increase in dose per trip is only a few hundredths of a mSv. Nonetheless, because they fly so much more frequently, the doses to aircraft crews can be substantial. Those who fly on the nonstop North Pole route between New York and Hong Kong encounter unusually high levels of cosmic and solar radiation, primarily because the north magnetic pole attracts charged particles from outer space. Crew members involved in these flights, as well as others, receive annual doses ranging as high as 5 mSv (Waters, Bloom, and Grajewski, 2000). As will be noted later, this is substantially higher than the doses received by medical x-ray technicians and commercial nuclear power-plant workers in the United States. Cosmic radiation doses received by people flying between various cities of the world are summarized in Table 12.6.

TERRESTRIAL RADIATION

As in the case of cosmic radiation, the external dose rates that individual members of the population of the world receive from naturally occurring radionuclides in the soil vary widely. In some areas of the United States, they may be as low as a few tenths of a mSv per year; in other areas, they may be as high as one mSv per year or more (Figure 12.4). The regions with the highest dose rates are those associated with uranium deposits in the Colorado plateau, granitic deposits in New England, and phosphate deposits in Florida; those with the lowest rates are the sandy soils of the Atlantic and Gulf coastal plain. In contrast, the coastal, monazite-bearing areas in the states of Rio de Janeiro and Espirito Santo, Brazil, yield ex-

Table 12.6 Cosmic radiation dose to airline passengers on subsonic flights between various cities of the world

Trip	Duration (hours)	Altitude		Dose (μSv)
		(Meters)	(Feet)	
Los Angeles to Honolulu	5	10,700	35,000	14
London to New York	7	11,300	37,000	37
Athens to New York	>9	12,500	41,000	62
Tokyo to New York	12	12,500	41,000	70
Hong Kong to New York Transpolar route	17–20	12,500	41,000	100

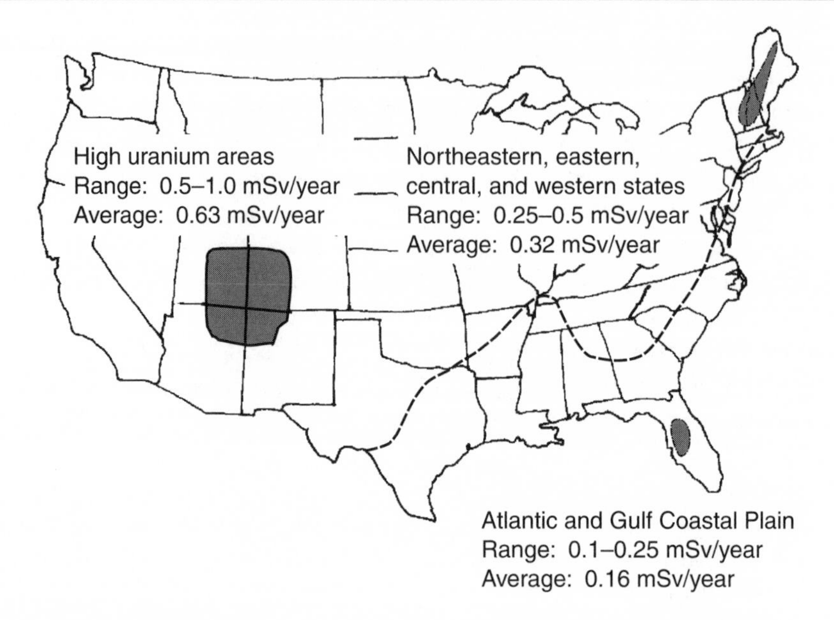

Figure 12.4 Terrestrial dose rates from natural background radiation in the conterminous United States

ternal dose rates up to 30 mSv per year; the same is true for coastal areas in Kerala and Tamil Nadu, India, and certain areas in the People's Republic of China (ICRP, 1999).

INTERNALLY DEPOSITED RADIONUCLIDES

As noted earlier, humans are exposed to ionizing radiation due to naturally occurring radioactive materials within the body. The primary radionuclide of consequence, from the standpoint of *ingestion,* is ^{40}K, an extremely long-lived radioisotope of potassium. This radionuclide contributes an average whole-body dose rate ranging from 0.15 mSv per year for women to 0.20 mSv per year for men. Since the dose rate is influenced by the amount of muscular tissue in the body, it tends to decrease with age. Although one might be inclined to want to avoid foods containing potassium so as to reduce these dose rates, the amount of potassium in the body is under homeostatic control. Maintenance of a proper balance of this element is essential to health.

A comparable dose rate is contributed through the ingestion of other naturally occurring radionuclides. The primary sources are radioactive isotopes of radium (primarily ^{226}Ra), followed by those of hydrogen (^{3}H, commonly known as tritium) and carbon (^{14}C). While ^{226}Ra is continually produced through the decay of naturally occurring radionuclides, beginning with uranium, ^{3}H and ^{14}C are continually replenished through the interaction of cosmic rays with various atoms in the atmosphere. Of this group, the highest contributor to the dose is ^{226}Ra. In contrast to ^{40}K, however, this radionuclide is not essential to the body, and the amounts ingested can be significantly affected by what, and how much, a person eats and drinks (NCRP, 1987a).

Another very important avenue of intake is *inhalation*. The major contributor in this case is radon (^{222}Rn). Although radon itself is a short-lived gas, it decays into a solid radioactive radionuclide that, in turn, decays in sequence into other solid radioactive materials. Since radon is continuously released into the air from the ground (Chapter 5), it and its decay products are present in the air in many homes and buildings. While the radon that is released outdoors is diluted by copious quantities of air, that which is released into the home is not. As a result, it is estimated that dose rates to the lungs of people throughout the world average about 16 mSv per year. The associated risk is estimated to be equivalent to that resulting from a whole-body dose rate of 1.3 mSv per year (ICRP, 1999).

TOTAL DOSE FROM NATURAL SOURCES

For the world as a whole, the average total dose rate per person from natural background radiation sources is estimated to be 2.2 to 2.4 mSv per year (Table 12.7). About one-third of this is due to external exposures to cosmic radiation and terrestrial sources, and about two-thirds to exposures to radionuclides deposited within the body. Overall, about half of the dose is a result of exposures to radon and its decay products. Although the dose rates in various parts of the world cover a wide range, more than 90 percent of the people on Earth incur dose rates less than about 5 mSv per year; about 99 percent receive dose rates less than 7 mSv per year (ICRP, 1999). For people living in the 48 conterminous states within the United States, the total dose rates from all natural background radiation sources are estimated to range from less than 1 mSv for people living in radon-free houses at sea level on the Atlantic and Gulf coastal plain to 5 to 10 times this value or more for those living at high elevations on the uranium-bearing lands of the Colorado plateau. On average, the estimated dose rate from natural radiation sources to members of the U.S. public is

Table 12.7 Effective dose rates from natural background radiation

Source of exposure	Average dose rates worldwide (mSv/year)	Dose rates in high-background areas (mSv/year)
Cosmic radiation	0.39	2.0
Terrestial gamma radiation	0.46	4.3
Radionuclides in the body (excluding radon)	0.23	0.6
Radon and its decay products	1.3	10
Total (rounded)	2.2–2.4	>10

about 3 mSv per year. The primary reason for the higher value than the worldwide average is a much higher estimate (2.0 mSv per year) for the contribution from radon and its decay products.

ASTRONAUTS: DOSES IN OUTER SPACE

With the continued interest in space exploration, as exemplified by the establishment of the International Space Station and initial planning for human missions to planets, such as Mars, far more attention is being directed to the radiation doses that people taking part in such activities will receive (Chapter 15). In fact, concern about radiation exposures is becoming one of the dominating factors in developing plans for such missions. While the dose to crew members who spent 90 days on the former USSR *Mir* station was about 70 mSv, estimates are that the dose to crew members in a 1,000-day round-trip mission to Mars will be about 1,000 mSv (1 Sv). As part of the strategy for reducing doses on the *Mir* station, USSR scientists provided shielded enclosures for protecting the astronauts while they slept. A similar approach is being considered for protecting crew members who take part in long-term deep-space missions (Long, 2001).

Technologically Enhanced Natural Exposures

Dose rates from certain radiation sources of natural origin can be increased by human activities. When this occurs, the doses are said to have been *technologically enhanced.* One example is the increase in concentrations of radon inside buildings that have been constructed to be more energy efficient and therefore have low air-exchange rates with the outdoors. Another example is the use of tobacco. This will be discussed later in this chapter in the section on "Consumer Products." For most people, the classic example of technologically enhanced exposures is that which occurred in conjunction with the production of uranium mill tailings in the course of separating uranium from the original ore as mined so that it could be used for the development of nuclear weapons and as fuel in nuclear power plants. In the ground, the uranium ore did not constitute a significant source of exposure. Once it was processed to remove the uranium, and the tailings (which contained the radium and other long-lived naturally occurring radioactive decay products that were in the original ore) were left exposed on the surface of the ground, they represented a readily available source not only for the release of radon into the atmosphere, but also for the transport of radionuclides into nearby surface waters. The problem was exacerbated when several thousand tons of the tailings, which appeared to be a good source of clean sand, were subsequently used as fill under and around nearby houses and in public projects, such as road construction. Once the problem was discovered, extensive remediation programs were required to remove the tailings from the homes and to stabilize those at the uranium mill sites.

Artificial Sources: Radiation Machines

The principal artificial sources of ionizing radiation are radiation machines, primarily x-ray generators, and radioactive materials produced in nuclear reactors and particle accelerators.

MEDICAL AND DENTAL APPLICATIONS

Some 2 billion diagnostic medical x-ray examinations, including more than 500 million dental x-ray examinations, are performed worldwide each year (ICRP, 2000). In the United States, substantially more than half of the population visit their doctor or dentist annually and undergo some type of x-ray procedure. These procedures are accomplished through the use of about 400,000 x-ray units in medical and dental practices in this country (Figure 12.5). While such procedures are designed to gain information to improve the health of patients, they expose both the patients and the x-ray machine operators to ionizing radiation. Procedures for protecting the operators include limiting the time of exposure, maintaining an ade-

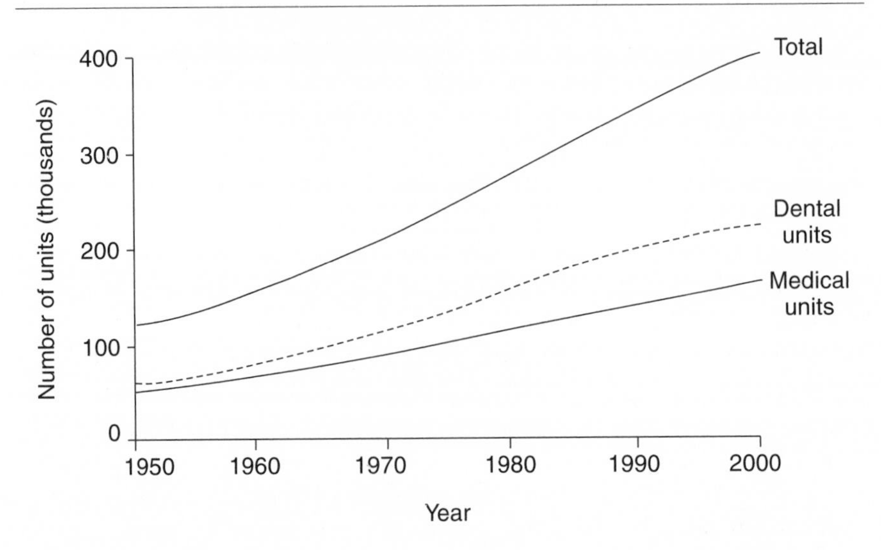

Figure 12.5 Increase in the number of medical and dental x-ray units in the United States, 1950–2000

quate distance between the x-ray beam and the operator, providing adequate shielding, and improving the designs of the machines, themselves. Through the effective application of various combinations of these and other control measures, the average dose rate to medical and dental x-ray personnel in the United States is well below 5 mSv per year, and essentially all of them receive less than 10 mSv per year. Similar procedures have been followed to ensure protection of the patients. These include regular inspections by federal, state, and local regulatory agencies to confirm that x-ray units are operated at the proper voltage, that x-ray beams are adequately filtered and collimated, and that physicians and dentists use faster (lower-dose) films and up-to-date processing techniques and equipment (NCRP, 2003a).

Several major new developments, however, pose the potential for increases in these dose rates. One is digital technology, through which the recording of transmitted photons on an image intensifier or other receptor, rather than on film, permits the images to be manipulated by a computer. The accompanying advantages have led to widespread use of this technology in vascular radiology, as well as other examinations. A second is interventional technology that incorporates techniques in which x-ray fluoroscopic images provide guidance to radiologists and other physicians in conducting semisurgical diagnostic and therapeutic procedures.

One of the outgrowths of these developments is a much wider application of computed tomography (CT), or computer-assisted tomography (the so-called CAT scan), which was perfected in the 1970s. Initial applications of this technique were primarily in examinations of the head and nervous system. Soon thereafter, this technology revolutionized the practice of neurology and neurosurgery. By the 1980s, the process had expanded to techniques for guiding the placement of needles, probes, and catheters into various body cavities during interventional treatments. In the 1990s, emphasis shifted to examinations of disorders of the chest and acute abdominal emergencies. Today new techniques and applications, such as virtual endoscopy, CT angiography, and fusion imaging, especially positron emission tomography/computed tomography (PET/CT) and CT screening, have evolved (Ferrucci, 2002).

Concurrent with these developments, applications of CAT scans have increased tremendously, primarily because they have two distinct advantages over previously available techniques. First, they produce the diagnosis without invading the body and avoid traditional surgical exploration and other invasive testing. Second, they enable surgeons to conduct their preoperative planning in a more leisurely and detailed manner (Ferruci, 2002). At the same time, however, the accompanying doses to the patients are relatively high (Table 12.8) (ICRP, 2000). In some cases, the accompanying doses to the skin of patients in procedures such as cardiac catheterization have been sufficiently high to cause erythema as well as radiation-induced skin injuries. One technique that holds promise for solving this problem is automatic tube current modulation, which ensures that the current to the x-ray machine is at all times the minimum required to retain the quality of the image being produced. This has been supported by the development of computer software that reminds operators of the dose consequences of technique selection (Westerman, 2002). Even so, it should be recognized that other technological advances have led to the development of methodologies that are permitting what were formerly high-dose examinations to be conducted with much reduced, sometimes even zero, radiation exposure. Examples are examinations of the upper gastrointestinal tract and the large intestine, which today are being conducted by techniques such as gastroscopy and sigmoidoscopy or colonoscopy, which were made possible through advancements in fiber optics.

Table 12.8 Effective doses for typical diagnostic x-ray examinations, 2000–2001

Examination	Effective Dose (mSv)
Chest (single film)	0.01
CT chest	6–8
Skull	0.1
CT head	2
Abdomen	1
CT abdomen	6–10
Thoracic spine	0.7
Lumbar spine	1.3
Pelvis	1
CT pelvis	6–10
CT abdomen-pelvis	12
CT chest-abdomen-pelvis	15
Intravenous urography	2.5
Barium enema (including fluoroscopy)	7
Mammography (screen film)	0.1

INDUSTRIAL APPLICATIONS

Industrial x-ray devices primarily include radiographic and fluoroscopic units used to detect defects in castings, fabricated structures, and welds and fluoroscopic units used to detect foreign material in items such as food products. Today there are about 16,000 active industrial radiographic installations in the United States; some 40,000–50,000 people are occupationally exposed in their operation. The primary concern is the control of exposures of the x-ray machine operators. The same techniques of filtration, coning, shielding, and limiting the time of exposure apply here as in applications of medical x rays.

COMMERCIAL APPLICATIONS

As noted in Chapter 6, various types of x-ray devices are being used to irradiate food products to extend their shelf life and/or to destroy disease-causing organisms. Such devices, as well as sealed sources of radioactive materials, are also used to sterilize medical supplies and equipment.

SECURITY APPLICATIONS

Since the 1970s, x-ray machines have been used increasingly to inspect luggage at airports as a security measure against aircraft hijackings and bombings; tens of thousands of such units are in operation at U.S. airports today. Although travelers often pass close to these units when entering the boarding area, their advanced design keeps the doses extremely low, in the range of a few thousandths of a millisievert per inspection. The metal detectors used for checking passengers are not a source of radiation exposure.

At the same time, there is an increasing trend on the part of U.S. government agencies and other institutions to use various types of x-ray devices to screen members of the public for purposes of national security. One particular application is the detection of contraband. Such events may involve the screening of a large number of members of the public or a small number of suspected individuals. The effective dose per screening ranges from about 0.1 to as much as 10 μSv. The magnitude depends on the nature of the equipment used and the purpose of the screening (NCRP, 2003b).

RESEARCH APPLICATIONS

High-voltage x-ray machines and particle accelerators are common equipment in the laboratories of universities and research organizations. More than 1,000 cyclotrons, synchrotrons, Van de Graaff generators, and betatrons are in operation in the United States, plus about 3,000 electron microscopes and some 20,000 or more x-ray diffraction units. Modern electron microscopes are shielded to protect the operators, but diffraction units still account for a significant number of radiation injuries (primarily burns on the hands).

Radioactive Materials

More than 20,000 hospitals and academic, industrial, and research organizations in the United States have been licensed to use radioactive materials. Such materials, for example, are used in about 10 million medical diagnostic procedures each year. Worldwide, the number is in excess of 35 million. Radioactive materials are also used in the United States each year to perform some 200,000 medical treatments and about 100 million laboratory procedures. The latter, while obtaining diagnostic information, do not require the administration of radioactive materials to the patient. Prominent examples are radioimmunoassay tests on blood and bodily fluids from patients. As is the case with medical x rays, joint actions by federal and state agencies have been effective in controlling the accompanying exposures. All users of significant quantities of such materials must be licensed; part of the licensing procedure involves a demonstration that applicants have the training, equipment, and facilities to handle such materials safely. As a result, doses associated with diagnostic uses in patients are low: a typical patient receives an estimated whole-body equivalent dose of 4.4 mSv. The estimated dose rate to nuclear medicine technical personnel who prepare and administer these materials is 4.0 mSv per year, less than 10 percent of the limit; that to nonnuclear medicine hospital personnel is less than 0.1 mSv per year. The average dose rate to a member of the U.S. public due to associated radionuclide releases into the environment is very low, about 0.4 μSv per year (NCRP, 1996).

Artificially produced radioactive materials are also widely utilized in universities and other institutions for teaching and research. One of the best-known research applications is the use of ^{14}C and other radionuclides for dating artifacts. Radionuclides are also routinely used as tracers in chemistry experiments and in chemical and polymer synthesis. Another such application is in the gas chromatograph. In addition, both portable and fixed devices, such as thickness, level, and moisture-density gauges and static eliminators are used in a wide range of industrial operations. Sealed capsules that contain radionuclides that emit gamma radiation and combinations of radionuclides that produce neutron radiation are used to log wells during explorations for oil and gas. In addition, as noted earlier, sealed radionuclide sources that emit gamma radiation are being used for sterilizing medical supplies and equipment, disinfesting food products, and extending the shelf life of poultry and other perishable products (Chapter 6).

Nuclear Power Operations

As of 2004, more than 100 electricity-generating units powered by nuclear reactors were operating in the United States (Figure 12.6). These units had a generating capacity in excess of 100,000 megawatts and were producing more than 20 percent of the nation's electricity (USNRC, 2001). This

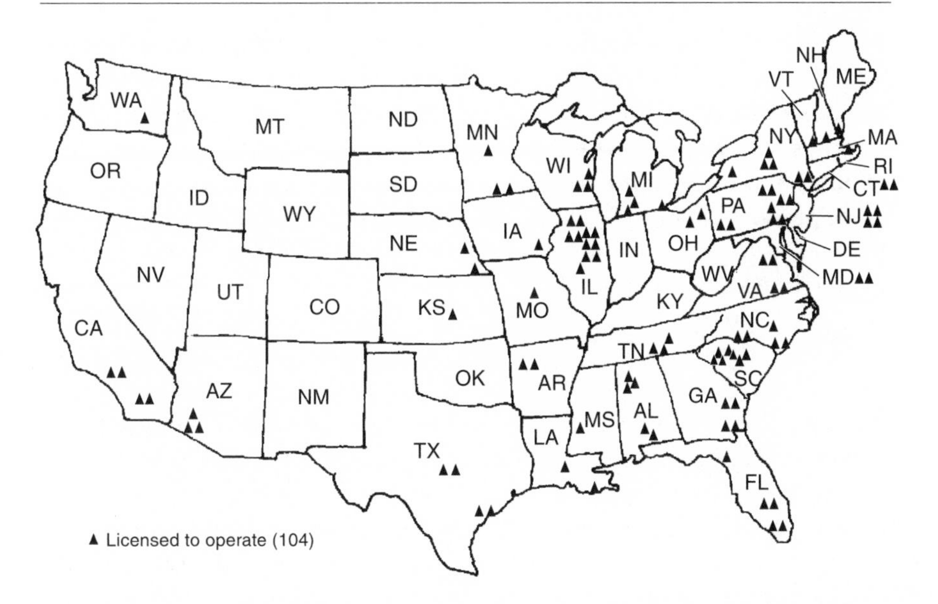

Figure 12.6 Nuclear power plants in the United States, 2000 (none in Alaska or Hawaii)

amounts to about 31 percent of the world's net nuclear-generated electricity. In addition, about 35 nonpower reactors were being used in the United States for training and research. More than 150 additional reactors are being used by the military services as propulsion units in submarines, cruisers, and aircraft carriers. Worldwide, 438 nuclear power plants, with a total generating capacity of more than 350,000 megawatts, are producing about 17 percent of the global electricity supply. Eight new units are scheduled to begin operation in 2004 (Tompkins, 2004). Thirty-one nations produce electricity using nuclear power, in several cases as a major source, ranging from France at 75 percent to Sweden, Ukraine, and South Korea at about 40 percent and to Germany, Japan, and the United Kingdom at about 30 percent (USNRC, 2001).

Data for calendar year 2002 show that no worker at any of the U.S. plants exceeded the annual dose limit of 50 mSv; in fact, the average dose to each of the approximately 100,000 people who worked at these plants and received measurable doses was far less than 5 percent of the limit. In addition, their industrial safety accident rate was 0.22 per 200,000 work hours, which makes this one of the safest industrial work environments

(INPO, 2003). These results reflect the regulatory efforts of the USNRC and supporting activities of the Institute of Nuclear Power Operations (INPO), a self-evaluation organization supported by the nuclear industry. A major reason for achieving such low doses is the fact that the operators of all such facilities apply the "as low as reasonably achievable" (ALARA) principle in terms of the dose rates that they permit their personnel to receive, as well as the amounts of radionuclides that can be released into the environment (Chapter 18).

Nuclear Weapons Testing

From the late 1940s through the early 1960s, the United States and the former Soviet Union conducted hundreds of nuclear weapons tests in the atmosphere. Lesser numbers were conducted by the United Kingdom, France, and the People's Republic of China. In accordance with the test ban treaty, tests after 1962 were, in the main, conducted underground, and radionuclide releases to the atmosphere were minimized.

Nonetheless, during earlier periods of active testing, large quantities of radioactive materials were released into the atmosphere and subsequently deposited throughout the world. Because of continuing interest in the associated possible long-term biological effects, detailed studies have been performed to estimate the accompanying doses. On the basis of these efforts, it is estimated that the average effective (whole-body equivalent) dose to the world's population through the year 2000 was about 0.4 mSv from external radiation and about 0.6 mSv from internal sources, yielding a total of about 1 mSv. The main contributor to these doses was ^{137}Cs. The total effective doses (Chapter 15) beyond 2000, summed over present and future generations, are projected to be about 2.5 mSv. The addition in dose will primarily be due to long-lived ^{14}C (Bouville et al., 2002).

Similar studies have been performed on the doses to the U.S. public from the tests conducted at the Nevada Test Site (NTS), as well as those due to global fallout from tests in other parts of the world. In the former case, the estimated average dose received by the U.S. population from external radiation was about 0.5 mGy (a unit for expressing the amount of energy absorbed, Table 12.3). An additional 0.7 mGy was estimated to have been received from fallout from high-yield tests conducted at sites outside this country. As would be anticipated, there was considerable variation in the doses to individuals in the United States, particularly in the case of fallout from the NTS. The most exposed individuals were those

Table 12.9 Average doses to U.S. population from weapons-testing fallout

		Thyroid dose (mGy)		Bone marrow dose (mGy)	
Source	Age group	External	Internal	External	Internal
Nevada Test Site	Adult	0.5	5	0.5	0.1
	Child[a]	0.5	30	0.5	0.1
Global sources	Adult	0.7	0.7	0.7	0.6
	Child[a]	0.7	2	0.7	0.9

a. For a child born 1 January 1951; external doses were assumed to be age independent.

who lived in areas immediately downwind. For these and other reasons, external doses to members of the U.S. public are estimated to range from a low of perhaps one-fourth of the average to a maximum of about four times the average (Bouville et al., 2002).

In the case of fallout from the NTS, the doses to various body organs from external sources of fallout were rather uniform. In contrast, the internal doses to individual organs through the ingestion and inhalation of radioactive materials varied markedly. This was especially true for doses to the thyroid. As was the case for external sources, the internal doses were influenced both by the distance and direction of the exposed individuals from the NTS and, most especially, the consumption of milk. The estimated average external and internal doses to the thyroid and bone marrow for an adult and child member of the U.S. population are summarized in Table 12.9. For the reasons cited, it is estimated that the doses to the thyroids of some children could have been as high as 10 times the values shown. In the case of bone marrow, the estimated internal dose for children, as a result of global fallout, is about a third higher than that for adults. As the data indicate, the estimated external dose for children is about the same as that for adults (Bouville et al., 2002).

Consumer Products

Natural and artificially produced radioactive materials are present in a variety of consumer products that are in common use in the home and at work (NCRP, 1987b). Among the devices used in the home, the smoke detector deserves special mention. Records of fire-related deaths show that such detectors save several thousand lives in the United States each year. The associated radiation exposure is minimal. Prominent among the consumer products used in the workplace are static eliminators and thickness gauges. The former, which contain radioactive materials, are widely applied to reduce the electrical charge buildup on materials, such as those in newspaper printing and photography. The 30,000–50,000 people occupationally exposed receive an estimated annual whole-body dose of 3 to 4 μSv. Thickness gauges that incorporate radioactive materials are similarly widely applied in industry. Typical applications include ensuring that the thickness of sheet steel, aluminum, and other metal products is maintained within standards. The 15,000–20,000 people exposed receive an estimated average annual whole-body dose of less than 1 mSv.

On a personal basis, people in the United States use a host of luminous watches and clocks that contain radioactive materials. While the units sold a half century ago contained ^{226}Ra and produced relatively high localized dose rates, manufacturers long ago switched to tritium (^{3}H), which does not emit penetrating radiation. In a similar manner, the materials used in making false teeth, which were formerly glazed with uranium, have been replaced with nonradioactive acrylics. However, one consumer product that is a major contributor to radiation dose remains in widespread use. That is tobacco, a product that contains relatively high quantities of ^{210}Po, a naturally occurring radionuclide. In fact, assessments by the ICRP indicate that it is one of the most hazardous such materials on Earth. When tobacco is used in manufacturing cigarettes, and smokers light up, the ^{210}Po is volatilized and taken into the lungs. An estimate of the annual dose to the lungs of the 50 million smokers in the United States, assuming an average of 1.5 packs of cigarettes per day, is 160 mSv (NCRP, 1987b). The whole-body equivalent dose is estimated to be more than 10 times the recommended limit (1 mSv per year) for an individual member of the public (Chapter 15).

Summary

The contributions of various ionizing radiation sources to the annual dose to the average member of the U.S. public are summarized in Figure 12.7. Because the dose rate from cigarettes is so high and is primarily limited to smokers, it is not included. As may be noted, the dose rates from various components of natural background radiation are far higher than those from any other sources, making an overall contribution of 82 percent to the total dose to the average nonsmoking member of the U.S. public.

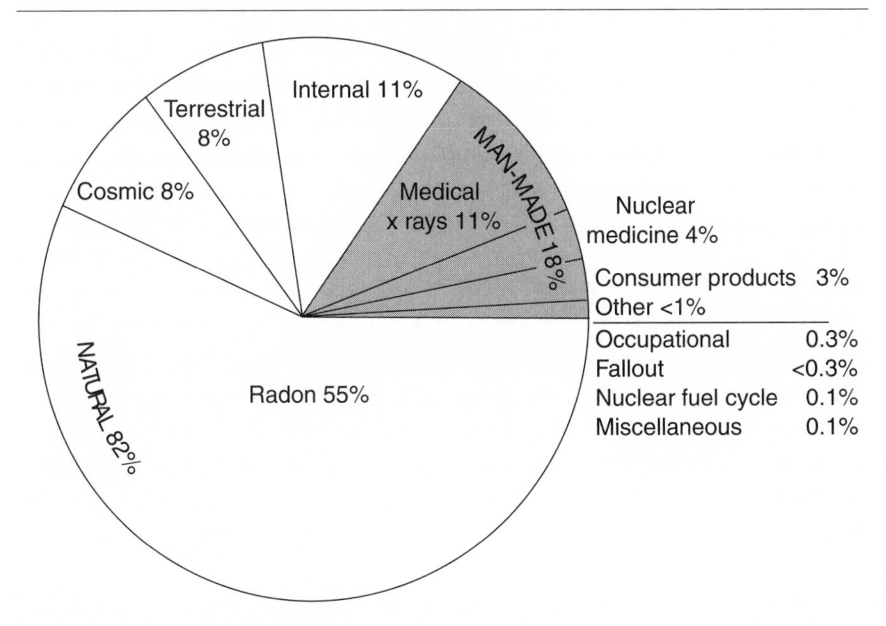

Figure 12.7 Relative annual dose contributions to the U.S. population of the principal sources of ionizing radiation

The General Outlook

The public urgently needs a more realistic understanding of the radiation sources that are important in our daily lives. All too often, the potential benefits of certain radiation sources have been denied or not fully attained owing to misconceptions about their importance. A notable example is the use of radiation for the preservation of food; many of the same misconceptions exist with respect to applications of nuclear power. While it is true that radiation is to be respected, it need not be feared.

Perhaps one of the most significant messages in this chapter is that in spite of the many environmental and occupational sources that can potentially cause exposures to radiation, the primary concerns arise through the personal environment. This is true for both ionizing and nonionizing sources. On the basis of the information presented, those people who desire to keep their dose rates to a minimum should avoid overexposure to the sun, ensure that the doctors who examine them use x-ray equipment that is up-to-date and well maintained, and, above all, avoid smoking cigarettes. Another message is the multitude of ways in which applications of radiation are helping to solve various types of problems in our daily lives. As noted, smoke detectors, which contain radioactive material, have reduced fire-related deaths significantly. One reason is that their cost is very low. Similar benefits have resulted from applications of nonionizing sources, for example, the widespread benefits of microwave ovens and the improvements in airline safety made possible by radar. Applications of radionuclides in agricultural research have provided similar benefits. Through the use of radionuclides to identify the best methods for applying nitrogen fertilizers in rice production, the quantities required have been reduced by up to half. Nuclear techniques such as gamma or neutron irradiation have induced mutations in seeds and thus created new varieties of food crops. As previously described (Chapter 10), radiation is being used to control insects such as the screwworm fly, the Mediterranean fruit fly, and the tsetse fly. It is also being proposed for use in locating hidden land mines, which continue to kill people in war-torn countries, such as Afghanistan.

13

ENVIRONMENTAL ECONOMICS

As early as the 1930s, studies by economists had shown that free access by industry to scarce environmental resources and their consequent overuse would lead to environmental problems (Pigou, 1932). Even so, during passage of the landmark federal legislation that was enacted in the United States during the early 1960s as part of the "environmental revolution" and the follow-up Clean Air Act Amendments of 1970 and the Federal Water Pollution Control Act Amendments of 1972, attention was directed solely to the traditional "command-and-control" approach (Oates, 1999). That the suggestions of economists were ignored can be ascribed to two factors (Oates, 1999):

> First, both environmentalists and industrial officials were opposed to levying fees or taxes of any kind. The former objected because they opposed placing a "price" on the environment, and they had concluded that the very existence of pollution demonstrated a failure of the effectiveness of a market-system approach. Industrial officials were nonsympathetic because they believed that adoption of a market-system approach would, in many cases, be more restrictive. They also favored the status quo because under it, the controls imposed on new, possibly competitive industries were often more restrictive than those that were applied to ongoing operations.
>
> Second, the field of environmental economics was in its infancy. Although a limited number of economists had explored their possible role in environmental protection, their knowledge of the environmental field did not extend beyond a general conceptual level. Little thought had been given as to how their ideas could be implemented. The situation was exacerbated by a similar ignorance on the part of those in the environmental field of the approaches being proposed by the economists.

Challenges to the Economists

Recognizing these realities, economists have sought to provide a framework through which regulators can set environmental standards that represent a balance between protecting health and other values and, at the same time, taking into account the costs and possible adverse consequences of the regulations being imposed. One approach has been to assign economic (dollar) values to all the relevant factors that must be considered. Despite progress, however, economists have not been able to provide a mechanism for assigning such values in even a reasonably precise manner. For this reason, the approaches being developed involve a host of decisions that are qualitative and judgmental in nature (Portney, 2000b). As will be noted later, one decision that has proved to be particularly difficult is the assignment of a minimum economic value to the actions that should be required to save a human life. Such considerations reflect the moral and ethical nature of the economic approach.

There are multiple illustrations of these types of challenges. One is the issue of how much timber should be harvested in the Pacific Northwest region of the United States. Reducing the number of trees that can be cut could well mean that harvesting activities will simply be shifted to countries such as Malaysia and Indonesia. In fact, some economists estimate that the resulting environmental damage will be substantially higher in these countries than it would have been if the same amount of timber were harvested in the United States (Sedjo, 1999). Another is the arrangement through which U.S. companies contract with industrial groups in the less developed nations to manufacture products for sale here. In some cases, one of the reasons such products can be manufactured at lower cost overseas is that less attention is paid to controlling releases of liquid and airborne contaminants and to protecting the health and safety of workers. Stimulated to some degree by public pressure, some U.S. companies have voluntarily adopted a code of standards for human rights and environmental protection and are assisting their contractors in implementing these standards, regardless of the location of the manufacturing facilities.

Challenges to National Leaders

National leaders must also make moral decisions in confronting problems that lead to damage of the global environment (Chapter 20). Examples are activities that contribute to the depletion of the ozone layer, the deposition of acid rain, and increases in the releases of atmospheric gases that cause global warming. In many ways, leaders of individual countries have little incentive to reduce the global damage that is being caused by their activities. To do so unilaterally could easily place them at a competitive disadvantage, particularly if it means reducing the consumption of fossil fuels. Addressing these types of problems, as well as those just cited, requires collective action on a global basis. Such actions include international agreements to impose order on the world's nations that are analogous to property rights. At the same time, the fact that the moral implications of environmental protection also apply to the actions of individual members of society must not be overlooked. It must also be recognized that the "morality" of such issues is often dictated by the various, often conflicting, cultural norms of the peoples involved. The bumper sticker of the 1980s, "Think Globally and Act Locally," is as relevant today as it was then.

Terms and Concepts

As background for a discussion of the approaches that are being developed by environmental economists, it is important that certain terms and concepts be understood. Several of these are described here.

EXTERNALITIES

Externalities are costs associated with the manufacture of a product that frequently are not considered in establishing its selling price. A prime example is the economic value provided to various manufacturers through the discharge into the environment of wastes generated by their operations. Since it is difficult to place a value on these types of benefits, some economists advocate the use of surrogate prices that are expressed in the form of pollution taxes and effluent fees.

UTILITY

Utility refers to the benefits or improvements in the welfare of given members of a community that result from the goods and services that are being provided. In dealing with environmental problems, the goal of most economists is to achieve a situation in which there is no practical way to increase the utility of some member or members of the community, and thus make them better off, without reducing the utility of other members of that same community. At this point, the economy is said to be *Pareto optimal* or *Pareto efficient* (Dorfman and Dorfman, 1977).

MARGINAL BENEFITS AND COSTS

It is generally assumed that to be acceptable, the benefits of controlling a specific pollutant must exceed the costs imposed if control were not implemented. Nonetheless, when a certain level of pollution control is achieved, economic analyses will show that any additional expansion in control activities will increase the costs more than the benefits. This is illustrated in Figure 13.1, which depicts the increase in the total costs and total benefits from the burning of increasing amounts of coal. During the initial stages, both the costs and benefits increase as the consumption of

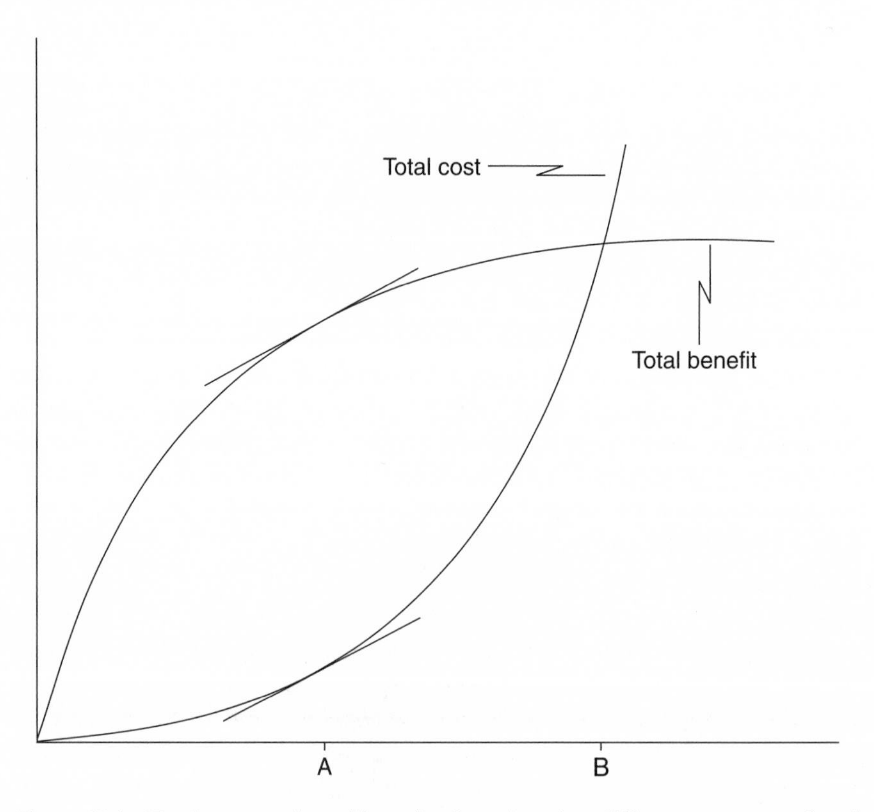

Figure 13.1 Total costs and total benefits from burning different amounts of coal

coal increases. When the consumption of coal is low, the benefits increase far more rapidly than the costs, due to the fact that the energy produced permits people to heat their homes and travel, while, at the same time, the air is not overburdened with pollution. Up to point A (in terms of coal consumption), the rate of increase in total benefits exceeds the rate of increase in total costs. Beyond that point, the rate of increase in total costs exceeds that of total benefits. In essence, any increase in the consumption of coal beyond point A costs more than it is worth. In fact, at point B, there is no increase in total benefits, while the total costs continue to increase. The net gain from the activity is said to be maximized at this point. This principle is the cornerstone of economics (Ruff, 1977).

COST-BENEFIT ANALYSIS

Cost-benefit analysis is basically a process for comparing the benefits and costs of a specific policy or other action to society and the environment. The conventional approach, in the environmental context, is to define benefits as the reduction in the detrimental impacts of pollution brought about by its control. The costs are defined as the value of the resources that are consumed in order to reduce pollution. As would be anticipated, assessments of the costs and benefits are closely intertwined. An improvement in the quality of the environment, for example, may be viewed as a benefit or negative cost. An increase in the cost of manufacturing a product or providing a given service that occurs through substituting an environmentally superior input may be viewed as a cost or negative benefit. When attempts are made to take factors such as the social aspects into account, such assessments become even more difficult. As a result, the outcome of a cost-benefit analysis is accompanied by large uncertainties. At best, it identifies cases in which the differences between marginal costs and benefits are sufficiently large to be indicative of grossly inefficient regulations.

COST-EFFECTIVENESS ANALYSIS

Cost-effectiveness analysis is a process that is closely related to cost-benefit analysis. It is particularly useful for evaluating situations in which information on the benefits of the control of a particular pollutant is not available or the benefits are especially difficult to assess. In a basic sense, this process is designed to answer the question "Which of the available policies or strategies, all of which are capable of achieving a desired goal, can accomplish the task at least cost?" Once the analysis has been completed, the various control policies or strategies can be ranked in terms of their relative monetary costs. Although this approach avoids the need to assign a monetary value to the benefits of a given policy, it also provides no information on the fundamental question as to what level of control should be sought.

COMMAND-AND-CONTROL (C & C) MEASURES

For years, *command-and-control (C & C) measures* were the primary mechanism for the control of environmental pollution. They continue to be widely applied today. This approach depends primarily on directing environmental agencies to establish air- and water-quality standards. Once this is done, these agencies are told to require that industrial companies limit their waste emissions to comply with the legislated standards. Such measures are very prescriptive, often specifying the control technologies that are to be applied. Little regard is given to the economic impacts associated with the regulations.

MARKET-BASED MEASURES

In contrast to C & C measures, *market-* or *incentive-based measures* are designed to stimulate and/or encourage industrial operators to limit their discharges of pollutants. The most common examples of such incentives are (1) effluent fees and (2) tradable emission permits, both of which are performance based. Both of these, which are discussed in more detail in what follows, require a process of trial and error for their implementation. The goal is to establish an optimum level for the effluent fees or for the volume of tradable emission permits that are being made available. The two approaches will lead to the same results only if the regulatory agency is successful in establishing a value for either the effluent fee or the volume of permits, as the case may be, that corresponds to the equilibrium or crossing point of the community's demand and supply curves for environmental improvement.

Effluent fees. The basis of affluent fees is to place a fee or tax on each unit quantity of pollution that is discharged. This serves as a major incentive for industries to reduce the amount of pollution they release and thereby to reduce the fees they must pay. Since plant managers have the option of either treating their wastes or releasing them and paying the fee, they are encouraged to search not only for manufacturing methods that produce less waste, but also for more economical methods for treating the wastes that are generated. Effluent fees, however, also have negative aspects. For example, it is often difficult for regulators to quantify the economic cost caused by the pollution, and it is equally difficult for them to set fees that will achieve the desired Pareto-optimal level. In addition, this approach does not limit the total amount of pollution that can be released; rather, it sets the cost for control. In fact, critics argue that the system of effluent fees implies that the government endorses pollution (Watkins, 2000).

Tradable emission permits. The concept of tradable emission permits was originally established by the Clean Air Act Amendments of 1990. It is now being applied in the regulation of discharges of liquid wastes as well. Although such permits are applied in several ways, one of the more common is for regulators to set limits on the total amount of pollution that can be discharged within a given area during a specified period of time and then to permit industry to buy and sell these rights. A new operator who desires to establish a facility in an area where all the permits have been claimed has two options. He/she must either purchase the pollution rights of an existing company or assist one or more of them in reducing their discharges by a sufficient amount to "make room" for the releases from the proposed new facility. An operator who reduces emissions below the allotted limit or ahead of the timetable set by regulations can earn credits that can be applied to future emissions or sold to the operator of another facility. This may even include trades between point and nonpoint sources (Chapter 8). In fact, industrial organizations that are point sources are now compensating nonpoint sources, such as farms, for reducing their discharges of liquid wastes. If such programs are properly executed, the cost savings can be substantial. Examples of the range of environmental applications of tradable emission permits are shown in Table 13.1.

Bubble policy. The bubble policy is a variation of the tradable-permits policy. It is directed to existing, as opposed to new, sources. Its application can be illustrated as follows. Assume that airborne particulate emissions from a steel mill are exceeding the regulatory limits. Analyses performed by the operator show that the cost of additional reductions in the emissions from some sources within the mill would be relatively high, while reductions for other sources would be relatively low. Under the "bubble policy," the U.S. Environmental Protection Agency provides the operator the option of reducing emissions from those sources within the mill that can be accomplished at lowest cost. The only requirement is that the emis-

Table 13.1 Examples of major federal tradable-permit systems

Program	Traded commodity	Dates in Operation	Effects	
			Economic	Environmental
Emissions trading program	Criteria air pollutants under Clean Air Act	1974 to present	Status not affected	Savings of $5–$12 billion
Lead emission reductions	Rights for lead in gasoline traded among refineries	1982 to 1987	Dramatic reductions in airborne levels	$250 million annual savings
CFC trading for ozone protection	Production rights for some CFCs, based on depletion potential	1987 to present	Targets achieved ahead of schedule	Effects not readily quantifiable
Acid rain reduction	SO_2 emission-reduction credits, primarily among electric utilities	1995 to present	Targets achieved ahead of schedule	Annual savings of up to $1 billion
Clear skies initiative	Multipollutant cap for NO_2, SO_2, and Hg	Proposed in 2002	Yet to be implemented	Proponents project high savings

sions, as a totality, comply with the regulatory limits. This, in essence, represents an application of tradable emission permits within a single large facility.

Societal Implications of Pollution Control

Pollution is almost a "natural" by-product of living. While sewers provide an effective vehicle for transporting domestic liquid wastes, unless these wastes are adequately treated prior to discharge into a lake or river, environmental degradation will occur. While pesticides are essential to averting famines in the less developed countries of the world, the improper and excessive use of these materials can lead to contamination of groundwater aquifers. Both of these examples illustrate the problems that will arise anytime the natural dilution and/or recovery capabilities of an environmental system are overwhelmed. A classic example of such a problem is described in the article "The Tragedy of the Commons" (Hardin, 1968). Hardin discussed a time in the distant past in which a parcel of land in a local community, and the associated pasture that it provided, were open to everyone. That is, everyone had free and unmanaged access to the land. Under these conditions, each of the local herdsmen sought to keep as many cattle as possible on the land, referred to by Hardin as the "commons." The goal of each herdsman was to maximize his gain. In so doing, each user considered himself free to add more animals to his herd. So long as tribal wars, poaching, and disease kept the numbers of both the herdsmen and the grazing cattle well below the carrying capacity of the land, this arrangement was satisfactory. When that condition ceased, however, the land was destroyed and what had been a common resource was lost. In essence, freedom of access to the commons and self-interest on the part of its users brought ruin to all. The lesson to be learned is that in a society without clearly defined property rights, those who pursue their own interests can destroy the public good.

There are numerous examples of related situations today. All have underlying economic and social implications. Ranchers who lease national lands for grazing their animals exemplify this situation. Without strict controls, the net result will be the same as with the herdsmen on the commons. Overfishing of lobsters has similarly led to depletion of this source of food, and excessive cutting of timber can lead to the destruction of forests. The value or utility that visitors seek in national parks can similarly be lost if they are overused and abused. In essence, parks are today exactly what the commons were in the past. If they are not protected, they can become of no value to anyone (Hardin, 1968). The concept of the commons, if it is justifiable at all, is acceptable only under conditions of low population density.

Whereas the tragedy of the commons can be averted by granting individuals the right to ownership of property, it is typically not practical to exclude people from using, or to prevent them from benefiting from, environmental resources such as rivers, streams, and the atmosphere. The reason is that either it is not physically possible, or the placement of restrictions on their access would not be socially acceptable. This *free-rider* phenomenon is prevalent in many environmental processes. Economists describe this by stating that the air and water lack the property of *excludability* (Dorfman and Dorfman, 1977). Nonetheless, there are likely to be enormous economies in their joint consumption or use. If such economies are to be achieved, however, methods must be developed to ensure that such use can be applied in a manner that ensures not only that these externalities are considered in the cost of the products being manufactured, but also that the salient components of the environment will be adequately protected from unacceptable pollution. It is also essential that all components and systems within the environment be considered. These include not only the air and water and the soil and minerals, but also the ecological processes that tie these systems together. Otherwise, there will be what economists call "market failure."

The Challenge of Environmental Protection

To set the stage for the discussion that follows, one may well ask, "What is the purpose of environmental protection?" If it is to ensure that the air and water are clean, then the question arises, "What is the purpose of clean air and water?" If the answer is to further human enjoyment of life, then the next question is, "Given that resources are finite, but the desires and wants of people are insatiable, how can we best promote such enjoyment?" The answer lies in making intelligent marginal decisions on the basis of costs and benefits. The objective as noted earlier, is to determine the level of pollution control at which the costs of additional abatement begin to exceed the benefits. Environmental protection and pollution control are designed to achieve a variety of objectives. These include enabling people to breathe comfortably, enjoy the mountains, and swim in the water, as well as to enjoy other variables or components of healthy lives.

Many other aspects of life, such as good food, comfortable habitats, and rapid transportation, also contribute to our enjoyment. The question is not which of these is most desirable, but rather what combination is most desirable (Ruff, 1977).

Assigning Economic Values to Natural Resources and Human Life

From the foregoing discussion, it is apparent that if the policies recommended by economists are to be applied, mechanisms must be developed for assigning economic values to the benefits and costs of various human practices. As might be expected, the assignment of such values in either case is not straightforward. Yet the failure to quantify the benefits, for example, of a natural resource such as fresh water can have far-reaching implications (Frederick, 2001). In fact, a major reason for the increasing water scarcity and freshwater ecosystem decline is that water is undervalued throughout the world (Johnson, Revenga, and Echeverria, 2001). This is exemplified by the policies that many governments are pursuing to ensure that agricultural crops can be produced at prices that are politically acceptable. As a result, the cost for irrigation water generally represents only a small fraction of the associated capital and management costs (Chapter 7). The application of equitable market-oriented mechanisms is essential if this problem is to be solved. If the prices being charged for water reflect the costs of supply and distribution, then agricultural producers would be expected to respond by improving the efficiency with which they use the water.

This example is relatively simple compared to many others. To illustrate this point, consider the steps required and the factors to be considered in evaluating the benefits of what might appear to be simple control measures, such as those for the prevention of water pollution (Figure 13.2). What, for example, is the benefit of draining a swamp to reduce mosquito populations? Even more complicated are assessments of the benefits of reducing the releases of greenhouse gases to avert global warming. In fact, the environment is often seen as a prime example of the failure of economists to establish the "right" prices for costs and benefits. The difficulties arise, as noted earlier, because no one owns the environment and no one can therefore charge for its use. Although many people are affected by the quality of the environment, it would be both difficult and expensive to organize them to pay for what they enjoy, or to prevent them from continuing to enjoy it if they refused to pay. Recognizing this void, economists

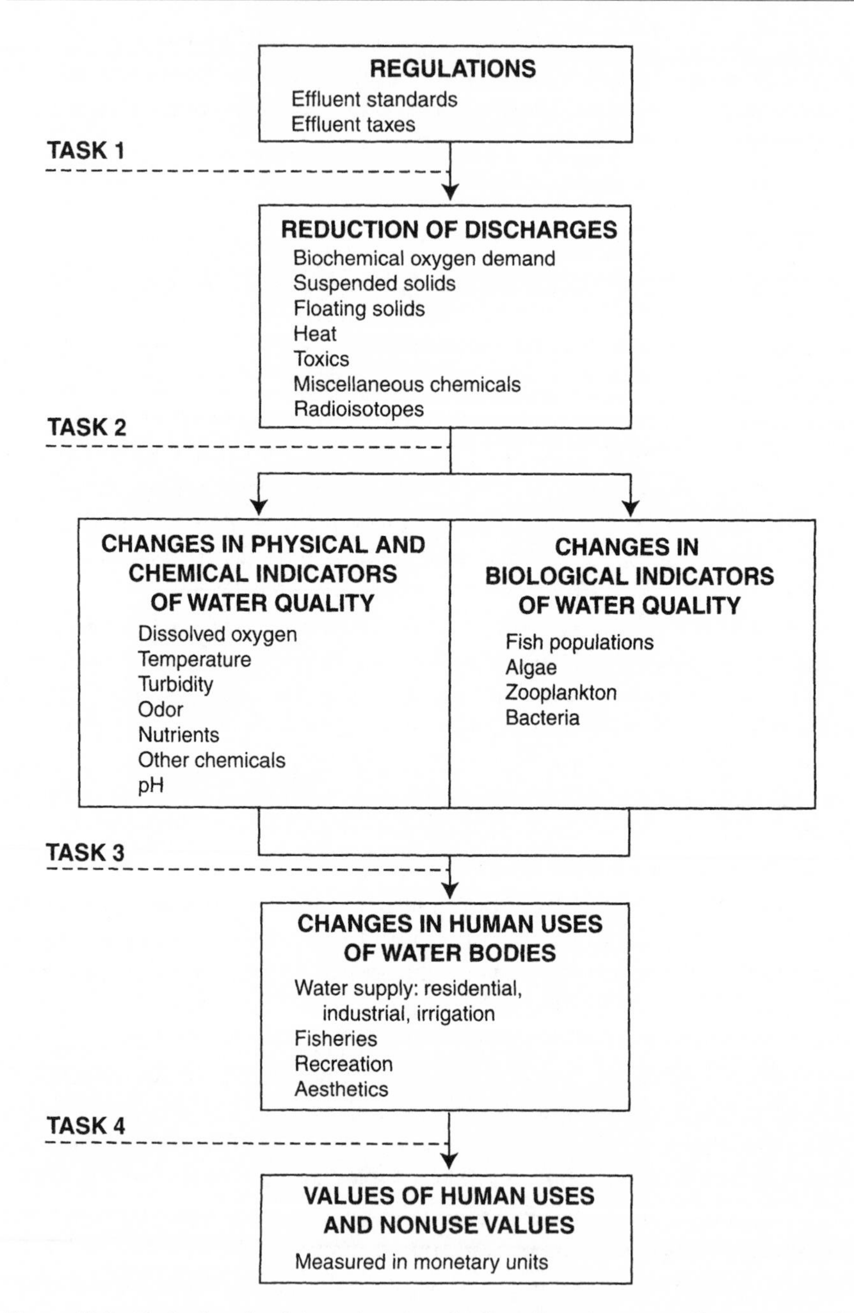

Figure 13.2 Tasks involved in estimating the benefits of water pollution control policies

have proposed that they intervene and assign proper values to these benefits (Roberts, 1982). Examples of efforts to meet these challenges are described here.

VALUE OF NATURAL RESOURCES

Although it is widely recognized that natural resources are the source of fossil fuels, such as coal and oil, few people realize the contributions made by other aspects of our environment, most especially the associated ecosystems. Even though freshwater ecosystems occupy less than 1 percent of the Earth's surface, they deliver goods and services of enormous global value (Johnson, Revenga, and Echeverria, 2001). The former include food, timber, genetic resources, and medicines; the latter include water purification, flood control, shoreline stabilization, biodiversity, conservation, disease regulation, and maintenance of air quality, as well as aesthetic and cultural benefits. Unfortunately, relative to other forms of capital, these resources are often poorly understood and inadequately monitored. To correct this deficiency, efforts are being made to undertake a global assessment of the condition and future prospects of these extremely important natural assets (Chapter 20; Ayensu et al., 1999).

There are several methods through which the values of natural resources can be estimated. One is to assess them, as described earlier, on the basis of the inherent benefits they provide. Another is to base the assessments on the value that people place on their associated benefits. Under this approach, the costs of pollution can be inferred from the higher prices people are willing to pay for property in nonpolluted areas. Another approach is to ask people how much they are willing to pay to have pollution reduced. One difficulty in this case is that there is no procedure for guaranteeing the accuracy of the responses. People's estimates of the cost of pollution will obviously depend on how they think the estimates will be used. If they anticipate being compensated for their losses, their estimates will be high. If they anticipate being charged for controlling the pollution that caused the problems, their estimates will be low. Another important consideration is the extent of the area being affected. Assessments of the costs of pollution that has global implications are far more difficult than those that involve pollution that is limited to a local or regional level (Ruff, 1977).

In spite of these difficulties, data on the negative impacts of human activities on various natural resources are gradually being developed. Several examples follow (Mock and Vanassect, 2001).

Forests. Forests can serve as a major factor in absorbing CO_2 and helping control global warming. The likely cost to plant a sufficient number of trees to offset 1 million tons of carbon emitted annually from a medium-sized coal-fired plant is $7 million. The value of the trees that have resulted from the regrowth of forests in Russia, for example, has been estimated to range into the tens of billions of dollars (Webster, 2002). Recognizing the benefits of these types of activities, the Kyoto Protocol (Chapter 20) permits governments to achieve their carbon reduction targets through reforestation and improved forest management. Additional stimuli are being provided by the World Bank, which has established a $145 million fund to support such trade-offs. To encourage such activities in its member countries, the European Union is developing a regional market for carbon dioxide emissions (Henriquez, 2004).

Freshwater systems. The amount allocated to New York City to protect its watershed against organisms such as *Giardia* and *Cryptosporidium* was $1.5 billion. The projected cost of building a water-filtration plant to accomplish the same objective was $4 billion (Johnson, Revenga and Echeverria, 2001).

Topsoil. Because of improper farming practices and an accompanying drought, some 200 million tons of topsoil were lost from the U.S. Great Plains during one dust storm in 1934. The current value of this topsoil is estimated to be $9 trillion.

There are other natural resources whose economic values need to be evaluated. One illustration is the philosophical and ethical issues surrounding the maintenance and preservation of biodiversity (Ferraro and Simpson, 2001). For example, what is the economic loss when a fish or plant becomes extinct and society has not established its value either to humans or to various ecosystems? Might a potential wonder drug never be discovered if it is made extinct? This is not an unreasonable scenario.

VALUE OF HUMAN LIVES

Assessments that involve policies that directly relate to, or can be expressed in terms of, saving human lives are complex. Although many people believe that the value of a human life cannot and should not be expressed in monetary terms, this approach has nonetheless become routine practice in some industries, notably the life insurance sector. It is important to recognize, however, that the value assigned is, in reality, a shorthand expression of the willingness of society to pay, divided by the reduction in risk expressed in units of the number of lives being saved. As such, the "individual" whose life is in question is viewed in abstract or statistical terms. In this approach, the expenditures that are justified to save a life, as expressed in various environmental regulations, cover a wide range. There are several reasons for this. A primary one is that saving lives at least cost is not one of the considerations taken into account in developing regulations. A second, but far less important, reason is that all such assessments are accompanied by a large range of uncertainties. In reality, most environmental regulations are formulated with little, if any, regard for their marginal cost-effectiveness.

Two of the most common approaches for assigning values to the saving of lives are called *revealed preference* and *contingent valuation* (Hammit, 2000). Revealed preference is based on the assumption that people prefer the choices they make to the available alternatives. For example, the value of reducing the risk of a fatality can be inferred by comparing wages and workplace-fatality rates among several types of occupations. In general, people engaged in occupations with higher workplace-fatality risks are offered higher wages. Workers who elect to work in such occupations are assumed to prefer the extra compensation more than they dislike the extra risk. Workers who choose lower-wage, lower-risk jobs are assumed to have the opposite preferences. Studies show that workers in the second group are willing to sacrifice $300 to $700 in extra annual pay for a job that carries with it a reduced annual fatality risk of 1 in 10,000. This means that 10,000 such workers would be willing collectively to sacrifice from $3 million to $7 million to prevent one "sure" death. This can be assumed to be the value they place on a statistical life (Portney, 2000a).

The second approach, contingent valuation, is based on the results of a survey in which people are asked which of several hypothetical situations appears to be the most desirable. Because this approach does not require that people face real choices, it is a much more arbitrary method than revealed preference. Since the respondents do not have as strong an incentive to evaluate their choices carefully, the results of such a survey are generally considered less credible. Standard economic theory implies that the amount that members of society are willing to pay to reduce the risk of mortality increases with income and, to a lesser degree, with the magnitude of the risk involved. Very little is known about the effects of other economic and cultural differences. Other examples of the values placed on human life are illustrated by the data presented in Table 13.2. While none of the listed estimates is exact, they do provide perspective. In prin-

Table 13.2 Amount of money spent to save a "statistical life" as implied by various societal activities in the United States (expressed in 2000 dollars)

Activity	Cost per fatality averted[a]
Environmental protection	
Protection of workers in underground mines	$85,000,000 (maximum)
Avoidance of thyroid cancer death due to emissions from nuclear power plants	$40,000,000 (minimum)
Disposal of radioactive waste	$30,000,000
Compliance with air-pollution standards	$5,800,000 (central estimate); $700,000–$16,300,000 (range)
Motor-vehicle safety[b]	
Air bags (driver only)	$1,000,000
Passive three-point safety belt	$750,000
Median barrier improvements	$700,000
Breakaway sign posts	$350,000
Driver education	$300,000
Rescue helicopters	$200,000
Medical care[b]	
Kidney dialysis	$600,000
Breast cancer screening	$250,000
Colorectal cancer x-ray screening	$90,000
Cervical cancer screening	$75,000
Provision of medical x-ray equipment	$12,000

a. All values are approximate and are presented to provide perspective.
b. The values in this category were tripled to account for inflation.

ciple, it appears that the values of improvements in health depend on a number of economic, social, and cultural factors (Evans et al., 2001).

Assigning Economic Values in the Industrial Sector

Experience shows that the assignment of economic values within the industrial sector is complicated in that it requires a similar need to examine all facets of a given operation. One approach that is being used is what is called the life-cycle assessment. In this procedure, analysts take into account the full range of the environmental impacts of manufactured materials and products, beginning with their fabrication, continuing with the impacts resulting from their use, and concluding with an evaluation of the options available for their disposal at their end of life. Such assessments are proving to be important tools in pollution prevention and the promotion of so-called green design efforts. Under this process, the selection of product design, materials, processes, reuse or recycle strategies, and final disposal options is subjected to a careful examination of energy and resource consumption, as well as those environmental discharges associated with each prevention or design alternative.

Uncertainties and Voids in Information

As is implied by the preceding discussion, assessing the benefits of our natural resources and the costs of reducing and/or eliminating the stresses being placed upon them cannot be precisely estimated. It is therefore important that the associated uncertainties be quantified, and the voids in information be identified, so that informed decisions can be made in deciding among the various available alternatives. These uncertainties derive from a host of factors. In the area of pollution control, it is often difficult to predict which of several available pollution-control technologies will be selected by the operators of the facilities being regulated. It is also difficult to estimate the cost of applying a given technology, especially when it is new and/or under development (Freeman, 1982). The problem is compounded by major uncertainties not only about the basic data on the identities, quantities, and chemical nature of the pollutants being discharged, but also about the degree to which the neighboring population is being or will be exposed. Even if the doses are well known, there will always be uncertainties (Chapter 2) about the relationship between a given dose and the associated health effects, as well as the risks these effects imply.

Because of these difficulties, some analysts have recommended that costs and benefits be expressed in terms of risk. In this approach, the risks, for example, might be expressed in terms of number of lives saved, the number of years of life saved, or the net change in the amount of pollution discharged into the environment. Since these units are physical rather than monetary, the hope is that they will be easier to estimate and apply (Graham, 2001). Another tool that has proved useful in rendering decisions in the face of uncertainties is decision analysis. In this approach, probabilities are assigned to the possible magnitudes of the relevant variables, and the laws of probability are used to compute expected values and confidence intervals. Such an approach is designed to quantify terms such as a statement that a given chemical is "highly toxic" or that there is a "small chance" that a given event will occur. The resulting probabilities provide a means for reducing the ambiguity of statements about uncertainty (Freeman, 1982).

Under the conditions outlined, the only basis for making a choice between the two approaches for environmental protection would be the ease of enforcement. Because effluent fees and taxes are generally easy to administer, they have usually proved to have an advantage. Compliance with prescriptive regulations can, in contrast, prove difficult for regulated industries to achieve. This is especially true if the regulations require what appear to be excessive capital expenditures and/or conflict with the self-interests of the companies involved. Under these conditions, experience has shown that the regulated entities often resort to protracted lawsuits and other delaying actions in seeking to avoid the large financial burdens that the regulations would impose. In contrast, there are fewer pretexts available, and less inducement, for seeking to avoid charges and effluent fees (Dorfman and Dorfman, 1977).

Other Market/Incentive-Based Approaches

The various social and economic control strategies and policies discussed thus far have applied to the discharge of liquid and airborne pollutants. Examples of other types of "pollutants" to which market-based controls can be applied are described here. All such applications have essentially the same goal. Rather than penalize industrial production, they are designed to penalize environmental destruction through the imposition of taxes and fees on pollution and resource depletion. Benefits of the application of these types of controls include not only behavioral changes on the part of both manufacturers and consumers, but also significant reductions in the costs of pollution control (Ruffing, 1999).

FEES OR TAXES ON PRODUCTS

Common products to which fees are applied include batteries, television sets, refrigerators, and air conditioners, all of which contain toxic materials. The fees serve as a stimulus to manufacturers to establish readily accessible methods for recycling such products. Essentially all companies that sell automobile batteries in the United States, for example, take back the used battery for recycling at the time a new one is purchased. This approach can also be applied to products that generate large volumes of waste. One example is used automobile tires, which represent a major contribution to the solid-waste problem (Chapter 9). In this case, the primary purpose of the fee is to assist in covering the costs for treating, recycling, and/or disposing of the worn-out tires.

The control of lead emissions from automobiles is a classic example of the successful application of fees or taxes. For years, this toxic material was added to gasoline in the United States to enhance the performance of automobile engines. In time, it was recognized that inhalation of automobile emissions and the consumption of agricultural crops grown on contaminated farmlands adjacent to highways were leading to significant intakes of lead by humans. To reduce the accompanying risks, legislators in the United States and many other countries placed a tax on leaded gasoline as a stimulus to refiners and distributors to develop safer additives. In some countries, such as Thailand, the proceeds from the tax were used to reduce the price of gasoline that did not contain lead, thus providing an incentive for people to switch to a safer product. In the case of the United States, another stimulus for the production of nonleaded gasoline was the requirement that automobiles be equipped with catalytic converters to control various other types of toxic airborne emissions. Such converters would not function on cars whose engines used leaded gasoline. The net result of these and other efforts was that the amount of lead released into the atmosphere in this country was reduced by more than 98 percent between 1970 and 1994 (Chapter 5).

Whether a fee or tax on a toxic-material-containing product will enhance environmental protection depends on several factors. For such a fee to be successful, it is first necessary that a less harmful substitute for the toxic material be readily available. Fortunately, this was the case for leaded gasoline. It was also the case when the refrigerator-manufacturing industry was required to replace chlorofluorocarbons that were being used as a coolant. Second, it is necessary that the substitute be available at a comparable or lower cost. A third necessity is that the fee or tax imposed be high enough to cause an increase in the price of the product that is sufficient to reduce the associated demand. Closely intertwined with the last requirement is that the fee or tax represent a relatively large portion of the total cost of the product. Otherwise, the manufacturer might simply absorb/hide the imposed tax within the overall costs of production (Ruffing, 1999).

DEPOSIT-REFUND SYSTEMS

The primary example of a deposit-refund system is a refundable charge that is applied to beverage containers. Another common application is to products such as automobile batteries, television sets, and air conditioners that, as noted earlier, contain toxic materials. Benefits of this approach are that it rewards people for handling such items in an environmentally responsible manner; the deposit-refund system is generally applied at the point of sale in retail stores and is therefore simple to implement and manage; the associated administrative costs are low; and it conserves natural resources. This is definitely the approach preferred by regulators, and it has also received wide public support. As might be anticipated, some groups, such as the beverage container industry, have indicated that they prefer the promotion of community recycling programs (Chapter 9). Nonetheless, the effectiveness of the deposit-refund system has proven vastly superior, indicating once again the benefits of applying economic incentives in solving such problems. The return rates in some countries, for example, Norway, Sweden, Finland, and the Netherlands, have approached 100 percent (Ruffing, 1999).

The General Outlook

Although for many years people in the environmental field tended to ignore the role that economics plays in environmental degradation, this is no longer the case. This change has been due to several factors. One of the most important has been the realization on the part of essentially all groups associated with environmental protection that many of these types of problems of the world are basically economic in nature. Indeed, their origins can be directly related to the necessity for humans to buy and sell goods to raise money, to acquire land for subsistence agriculture, and to feed, clothe, and house their families. On a broader scale, it can be similarly shown that global warming is promoted by the fact that fossil fuels are the most economical and most convenient of our energy sources. In like manner, the rapid increase in species extinction can be shown to be largely due to the economic forces that drive humans to hunt wildlife, log forests, and build in fields, marshes, and wetlands (Powell, 2002).

A further stimulus to this change is that the performance/incentive-based environmental control measures developed by economists have proved to have far more appeal than the C & C measures developed by regulators. This was almost immediately seen to be the case from the standpoint of industrial organizations and environmentalists. Today regulators also acknowledge this. In the case of industrial organizations, the application of performance/incentive-based measures, as exemplified by tradable emission permits, enables them not only to avoid paying a tax, but also to have the opportunity of receiving (often under some form of a grandfather provision) a valuable permit. In addition, many industrial organizations take pride in the goodwill generated by a perception of being good stewards of the community's natural resources. Environmentalists favor such permits since, by restricting the number available, the regulatory authority can directly and unambiguously limit the amount of emissions. Environmentalists will thus be able to quantify the success of their efforts. Regulators, in turn, favor this approach because it eliminates the necessity of having to perform the often difficult task of setting, and then having to adjust, tax rates to induce the needed reductions in waste discharges. Their constituencies appreciate the "do-something" approach on the part of the government.

Another factor that has contributed to this change is that as their field has matured, economists have been able to provide a compelling conceptual rationale for the tradable-permit approach. In addition, they have been able to conduct a substantial number of empirical studies that have documented the large cost savings for the producers, and ultimately the consumers, that have resulted from the application of these concepts (Oates, 1999). Not stopping here, they have demonstrated a host of other environmental concerns to which these concepts can be applied in an effective manner. This is typified by the use of these concepts to limit catches of certain types of fish, for example, in New Zealand, where fishermen are allocated shares in the total allowable catch within a given fishery. As in the case of tradable emission permits, the fishermen have the right to catch, transfer, sell, or lease their quotas (Day, 2001). A similar approach is being used in the U.S. Pacific Northwest to protect salmon and steelhead, an oceangoing rainbow trout. In this case, it is being applied under the Endangered Species Act.

Nevertheless, much work remains. This is exemplified by the fact that as previously noted, the C & C approach continues to be widely applied in the United States. The situation is exacerbated by the fact that few of the C & C regulations have sunset clauses, and there is little incentive for Congress to review and modify the basis on which these regulations were established. The C & C approach is even more common practice in Europe. In fact, tradable emission permits and other market-based approaches have yet to be adopted in that part of the world. Although there are several strategies to consider for overcoming this resistance, another approach would be for economists to develop methods for making C & C measures more effective while at the same time reducing the costs of compliance (Oates, 1999). In the end, the most important contribution that environmental economists have made to protect health is through providing a clear objective for evaluating pollution decisions. By developing mechanisms for weighing the costs of more pollution control against the associated benefits, they have provided society with a balanced approach. They have also championed the wise use of scientific information by involving all relevant disciplines in building integrated environmental assessment models. Finally, economists have diligently pursued the goal of providing regulators the tools necessary to promote efficient abatement programs. Through these efforts, regulators should increasingly be able to design pollution-control programs that are more effective and thus able to deliver an adequately safe environment at the lowest possible cost (Mendelsohn, 2002).

14

ENVIRONMENTAL LAW

ALTHOUGH laws providing for protection of the environment date back to the late 1800s, modern U.S. environmental law had its birth with the enactment of the National Environmental Policy Act (NEPA) in 1969 (P.L. 91-190). This law declared that it was national policy to (1) encourage productive and enjoyable harmony between people and their environment, (2) promote efforts that will prevent or eliminate damage to the environment and the biosphere and stimulate the health and welfare of man, and (3) enrich our understanding of ecological systems and natural resources important to the nation. In enacting this legislation, Congress recognized the profound impact of human activities—particularly of population growth, high-density urbanization, industrial expansion, resource exploitation, and new and expanding technological advances—on the interrelation of all components of the natural environment.

The basic goal of the NEPA was to administer federal programs in the most environmentally sound manner consistent with other national priorities. As such, the act required that federal agencies prepare an environmental impact statement (EIS) for all major federal projects that are anticipated to have a significant effect on the environment. The statement must include an evaluation of the repercussions of alternative courses of action. Examples of projects or activities covered were the construction of new facilities (bridges, highways, airports, hydroelectric and/or nuclear power plants), the dredging or channeling of rivers or harbors, large-scale aerial spraying of pesticides, and the disposal of hazardous materials. A significant initiative of the NEPA was the establishment of the Council on Environmental Quality (CEQ) within the Executive Office of the President.

Responsibilities assigned to the CEQ included developing new environmental programs and policies, coordinating the wide array of federal environmental efforts, ensuring that officials responsible for federal activities take environmental considerations into account, and assisting the president in assessing and solving environmental problems.

Although the NEPA was far reaching and all-encompassing, it focused primarily on process rather than substance. Nonetheless, in many cases proposed projects were halted after initiation of NEPA litigation. In other cases, federal agencies have refrained from taking controversial actions to avoid the expense and delay of litigation. Although the NEPA has undoubtedly been beneficial, the unanswered question is whether the benefits have been sufficient to justify the expense and accompanying delays (Findley and Farber, 1992; 2000). For these and other reasons, attempts are frequently made to modify the NEPA. In 2002, for example, the CEQ undertook a review to determine whether increased logging should be permitted to reduce the frequency of wildfires in our national forests. While such a change would please the logging industry, it was opposed by many environmental organizations. One recent decision by the Justice Department that could have far-reaching implications is that the NEPA and other environmental laws do not apply to waters more than three miles beyond U.S. shores. This could reduce the protection being provided to whales, dolphins, and other marine life (Associated Press, 2002).

Interestingly, the NEPA contains no requirements on how clean the air and water must be or how much pollution can be discharged into the environment. To meet these and other needs, Congress has passed a wide range of laws that address air and water pollution, solid-waste disposal, the purity of food and drinking water, and problems of the occupational environment (Table 14.1). The increasing pace of such legislation in recent years is illustrated in Figure 14.1. This is supplemented by Table 14a which includes a chronological listing of some of the major items of environmental legislation passed by the U.S. Congress from the late 1800s to the present. One of the objectives is to illustrate the variety of subjects that have been addressed. For the sake of brevity, in most cases subsequent amendments to the laws have not been included. As may be noted from Figure 14.1, there was a burst of activity between 1965 and 1995. Prominent among the laws passed during this period were those that established the Environmental Protection Agency, the Occupational Safety and Health Administration, and the Nuclear Regulatory Commission. Although the pace has slowed since 1995, Congress is still active in the field of environ-

Table 14.1 Purpose and/or scope of various environmental and occupational health laws

Law	Date	Purpose and/or scope
Environment		
Solid Waste Disposal Act (Resource Conservation and Recovery Act)	1965 1976 1984	"Cradle-to-grave" coverage of wastes; prohibition of land disposal of nontreated hazardous waste
National Environmental Policy Act	1969	General protection of the environment; requirement of environmental impact statements
Clean Air Act and Amendments	1970 1990	Consideration of public exposures to airborne contaminants; specification of required treatment and ambient standards
Toxic Substances Control Act	1976	Requirements for EPA notification on use, testing, and restriction of certain chemical substances
Clean Water Act and Amendments	1977	Restrictions on pollution discharges into rivers and streams
Comprehensive Environmental Response, Compensation and Liability Act (Superfund)	1980 1986	Requires cleanup of existing disposal sites; establishes financial responsibility for sites
Emergency Planning and Community Right-to-Know Act	1986	Establishes requirements that sources must assess and annually report to EPA their emissions to air, land, and water; establishes emergency release notification system
Pollution Prevention Act	1990	Requirements for prevention, reduction, or treatment of pollution at its source, with disposal to the environment being a last resort
Radioactive Waste Disposal		
Uranium Mill Tailings Radiation Control Act	1978	Requirements for remediation of former uranium mill processing and disposal sites
Nuclear Waste Policy Act	1982	Requirements for disposal of high-level radioactive wastes

Table 14.1 Purpose and/or scope of various environmental and occupational health laws (continued)

Law	Date	Purpose and/or scope
Low-Level Radioactive Waste Policy Amendments Act	1985	Requirements for disposal of low-level radioactive wastes
Food and Water		
Federal Food, Drug, and Cosmetic Act	1938	Protection of foods against pesticides and harmful additives
Federal Insecticide, Fungicide, and Rodenticide Act	1972	Registration of chemicals used for pest control
Safe Drinking Water Act and Amendments	1974 1986	Provision of maximum limits for contaminants in public drinking water and techniques for their removal
	1996	Requirements for specific analyses for emerging contaminants, such as *Cryptosporidium;* provides a revolving fund to help state and local authorities improve drinking-water systems
Food Quality Protection Act	1996	Repeal of Delaney clause; eases regulation of processed foods and tightens regulation of raw foods
Worker Safety		
Occupational Safety and Health Act	1970	Assurance of safe and healthful conditions for U.S. workers
Energy		
Atomic Energy Act	1954	Assurance of safe handling of radioactive materials and safe management of nuclear facilities
Energy Policy Act of 1992	1992	Requirements for improvements in the energy efficiency of a wide range of items, including transportation vehicles and industrial and home appliances, and promotion of the use of renewable resources; also stipulated procedures for the establishment of radiation protection standards for the disposal of high-level radioactive wastes

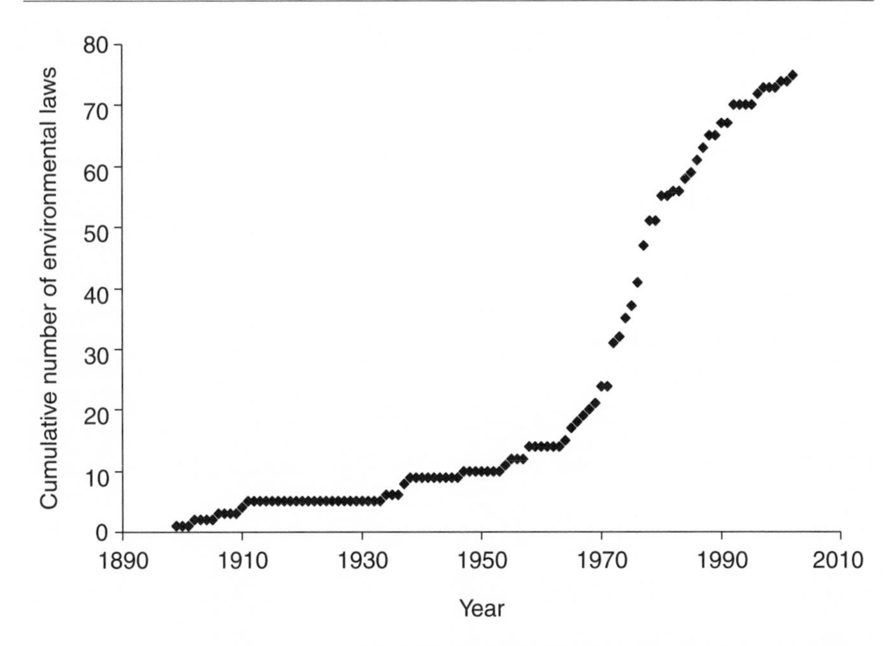

Figure 14.1 The growth of environmental laws in the United States, 1890–2002

mental protection. Some of the more important laws are discussed in the sections that follow.

The Clean Air Act and Amendments

Although air-pollution ordinances were passed in Pittsburgh in 1895 and in Boston in 1911, it was not until 1952, with the passage of an air-pollution law by Oregon, that such action was taken on a statewide basis. While other states soon followed, it was not until 1955 that such a law, the Clean Air Pollution Control Act (P.L. 84-159), was passed at the federal level. Although this law recognized that the individual states have primary responsibility for ensuring the quality of the air within their borders, funds provided in support of such activities were limited to air-pollution *research* and *training*. Direct federal involvement in the *control* of air pollution had to await the passage of the Clean Air Act in 1963. This was followed two years later by an amendment that provided for the establishment of the first federal emission standards for automobiles. This, in

Table 14a Chronological listing of major environmental legislation passed by the U.S. Congress

Year	Legislation
1899	Rivers and Harbors Act
1902	Reclamation Act
1906	Pure Food and Drug Act
1910	Insecticide Act
1934	Taylor Grazing Act
1937	Flood Control Act
1937	Wildlife Restoration Act
1938	Federal Food, Drug, and Cosmetic Act
1947	Federal Insecticide, Fungicide, and Rodenticide Act
1948	Water Pollution Control Act
1954	Atomic Energy Act
1955	Air Pollution Control Act
1958	Fish and Wildlife Coordination Act
1958	Food Additives Amendment
1964	Wilderness Act
1965	Solid Waste Disposal Act
1966	National Historical Preservation Act
1968	Wild and Scenic Rivers Act
1969	National Environmental Policy Act
1970	Occupational Safety and Health Act
1970	Resource Recovery Act
1972	Marine Protection, Research, and Sanctuaries Act
1972	Coastal Zone Management Act
1972	Parks and Waterways Safety Act
1972	Marine Mammal Protection Act
1973	Endangered Species Act
1974	Deepwater Port Act
1974	Safe Drinking Water Act
1975	Energy Policy and Conservation Act

Table 14a (continued)	
1976	Energy Conservation Act
1976	Toxic Substances Control Act
1976	Federal Land Policy and Management Act
1976	Resource Conservation and Recovery Act
1977	Clean Water Act
1977	Saccharin Study and Labeling Act
1977	Surface Mining Control and Reclamation Act
1977	Soil and Water Resources Conservation Act
1978	Energy Tax Act
1980	Environmental Education Act
1980	Comprehensive Environmental Response, Compensation, and Liability Act
1980	Federal Insecticide, Fungicide and Rodenticide Act Amendments
1980	Acid Precipitation Act of 1980, Clean Air Act Amendment
1982	Nuclear Waste Policy Act
1984	Environmental Programs and Assistance Act
1986	Emergency Planning and Community Right-to-Know Act, Title III of the Superfund Amendments and Reauthorization Act
1988	Ocean Dumping Ban Act
1990	Clean Air Act Amendments
1990	Pollution Prevention Act
1992	Energy Policy Act
1992	Federal Facility Compliance Act of 1992
1992	Residential Lead-Based Paint Hazard Reduction Act of 1992
1996	Food Quality Protection Act of 1996
2000	Wet Weather Water Quality Act
2000	Clean Water Act Amendments
2002	Small Business Liability Relief and Brownfields Revitalization Act

turn, was followed by legislation in 1967 that required that the federal government support research on the effects of air pollution, operate a monitoring network, and promulgate criteria to serve as the basis for setting emission standards.

A major turning point in progress on the control of air pollution in the United States occurred in 1970 when Congress amended the Clean Air Act. Two of the most significant requirements of these amendments were that the national ambient air-quality standards (NAAQS), set by the Environmental Protection Agency (EPA), provide "an ample margin of safety to protect the public health," and that limits be established for controlling emissions from both stationary and mobile sources. These changes were followed by additional amendments in 1977 and 1990 (P.L. 91-604; 101-549). Among the most significant of the 1977 amendments was that different levels of stringency were to be applied to airborne emissions in areas that met the existing air-quality standards, the so-called attainment areas, versus those that did not, the so-called nonattainment areas.

To ensure that levels of pollution in attainment areas would not increase, the 1977 amendments incorporated a "prevention of significant deterioration" requirement. New stationary sources in such areas were required to use the "best available control technology" (BACT). At the same time, the amendments required the EPA to take into account the costs of compliance. Existing sources in nonattainment areas were required to use "reasonably available control technology" (RACT), which, as the wording implies, represents a lesser level of control that can be achieved at lower cost. In essence, the objective of RACT was to provide a minimum level of control in nonattainment areas. To "drive" industry to achieve better controls, the EPA in some cases specified the percentage reductions to be achieved.

Among the emissions that received special attention were the "toxic air pollutants," or "air toxics." These are primarily substances that tests have shown to be carcinogenic. The establishment of standards for these pollutants has been much more difficult than anticipated. One of the primary problems is whether they have threshold concentrations below which they have no biological effects and, if not, how much risk to the public should be considered acceptable. This challenge is one that is faced in all aspects of the control of environmental and occupational stresses.

One of the important contributions of the previously cited 1990 amendments to the Clean Air Act was the separation of air pollution standards into several classes. These extended the 1977 amendments by including harm-based standards designed to protect public health, technology-based standards requiring application of various levels of control technology, and technology-forcing standards designed to ensure that industry develop and apply the very best control technology. The 1990 amendments also required that specific attention be directed to airborne particulates 10 micrometers in size or less, those in the respirable range. As noted in Chapter 3, these requirements have now been expanded to include concentration limits for airborne particles 2.5 micrometers in size or less, as well as those 10 micrometers in size or less (Table 5.1, Chapter 5). Another aspect of the Clean Air Act amendments of 1990 is that they permit the buying and selling of air-pollution emission allowances (Chapter 13). In the case of pollutants that have worldwide implications, such as those that can cause depletion of the ozone layer or lead to global warming, a suggested approach for implementing such exchanges was outlined in the Kyoto Protocol of 2000.

The Water Pollution Control and the Clean Water Acts

Federal laws pertaining to maintaining the quality of U.S. rivers and streams predate those for all other components of the environment. The Rivers' and Harbors Act of 1899, for example, barred the discharge or deposition of refuse in navigable waters without a permit. The first federal legislation, however, that was designed to address the conventional aspects of maintaining the quality of surface waters was not enacted until 1948, when Congress passed the Water Pollution Control Act. This act provided funds for relevant federal research and associated investigations. Amendments to this act in 1956 authorized the states to establish water-quality criteria and sponsored enforcement conferences to negotiate cleanup plans for bodies of water whose maintenance required the involvement of several states. More important, these amendments provided federal support for the construction of new municipal sewage-treatment plants.

The Water Pollution Control Act was followed by the Water Quality Act of 1965, which was designed to assist the states in meeting the water-quality standards that they had established. Next came the Water Quality Act Amendments of 1972 and the Clean Water Act Amendments of 1977. The ultimate goal of these laws was to provide "fishable and swimmable" rivers and streams on a nationwide basis. As with the amendments to the Water Pollution Control Act, these laws continued to provide federal funds for the construction of municipal sewage-treatment plants (Freeman, 2000). One of the provisions of the Clean Water Act that was fortuitous in terms of responding to acts of terrorism (Chapter 19) was a requirement that operators of facilities that have the potential for major accidental liquid releases develop and maintain a plan describing the actions to be taken in case such an event occurs. The plan must identify the people who will handle the incident and outline their specific duties and responsibilities, as well as the associated reporting and record-keeping requirements. The act further requires that effective application of the plan be guaranteed by regular training of the personnel who will be involved.

One of the most far-reaching provisions of the Clean Water Act was the institution of a permitting mechanism, the National Pollution Discharge Elimination System (NPDES), through which limits on the discharge of pollutants could be established. As part of this effort, the EPA can delegate to individual states the authority to establish the standards and to issue NPDES permits. To qualify, each state must demonstrate that its control program is substantially the same and at least as stringent as that of the EPA. Some 40 states have met these requirements and currently issue such permits. The first preparatory step is to designate the intended use of each body of water, that is, whether for drinking, swimming, fishing, or boating. This is followed by the establishment of discharge standards. Although one would assume that these standards would be harm based to ensure compliance with the intended use of the receiving body of water, this has not generally been the case. Instead, most of the discharge standards have been set on the basis of what can be achieved using available cleanup technology. That is, the effluent release standards are predominantly technology based, as contrasted to harm or risk based (Freeman, 2000).

The 1987 amendments to the Clean Water Act modified the NPDES program by expanding the number of chemical constituents that must be controlled. The amendments also significantly changed the thrust of enforcement by requiring increased monitoring and control of toxic constituents in wastewater and in discharges of polluted runoff from city streets, farmland, mining sites, and other nonpoint sources. With passage of the Wet Weather Water Quality Act of 2000, Congress directed specific attention to the problems of overflowing sewers, which have been occurring on an increasingly frequent basis in many U.S. cities, particularly during heavy rains (Chapter 8).

Safe Drinking Water Act

Closely tied to the control of the discharge of liquid wastes into rivers and streams is the provision of a safe drinking-water supply. The Clean Water Act, however, has been generally interpreted as applying only to surface waters. From this perspective, it did not provide protection for one of the most important drinking-water sources—groundwater. This situation was partially remedied through passage in 1974 of the Safe Drinking Water Act (P.L. 93-523), whose general objective was to ensure that public drinking-water supplies are free of potentially harmful materials and that they meet minimum national standards for protection of public health. The act includes regulations pertaining to the underground injection of liquid wastes.

Under the Safe Drinking Water Act, the EPA was authorized to establish national standards, including the specification of maximum contaminant levels (MCLs) for specific substances in drinking water. Under this mandate, the MCLs must be set at a level that allows no known or anticipated adverse health effects. As was the case with airborne contaminants, the EPA has adopted the premise that there is no safe level of exposure to a carcinogen. At the same time, the agency has recognized that the costs of controlling human exposure must be reasonable. As in the case of the Clean Water Act, the EPA may either establish an MCL for a specific substance or prescribe a technique for its removal. Since the feasibility of achieving a specified MCL or of implementing a given treatment will change with advancing technology, the act requires the EPA to revise and update the regulations on a continuing basis.

Through passage of the 1977 amendments to the Safe Drinking Water Act (P.L. 95-190), Congress recognized the finite nature of the nation's water supplies and the need to assess present and future supplies and demands. This act included requirements for an analysis of the projected demand for drinking water, the extent to which other uses would compete with drinking water needs, the availability and use of methods to conserve water or reduce demand, the adequacy of present measures to assure adequate and dependable supplies, and the problems (financial, legal, or other) that must be resolved in order to assure the availability of adequate quantities of safe drinking water for the future. Of particular interest is the emphasis on conserving water and reducing demand; this approach is part of the increasing recognition of the need to manage the nation's limited natural resources in a sustainable manner. The Safe Drinking Water Act also reflected concerns about the potential health effects of by-products that occur in drinking water through use of disinfectants such as chlorine (Chapter 7).

The Safe Drinking Water Act, as amended in 1986 and 1996, specified additional contaminants to be regulated and the development of acceptable treatment techniques for each. The amendments further required that all drinking-water supplies be disinfected, prohibited the use of lead products in drinking-water conveyances, and renewed the emphasis on the need for protection of groundwater sources. A major stimulus for the 1996 amendments (P.L. 104-182) was the recognition that many water supplies were not being analyzed for some of the emerging biological contaminants, such as *Cryptosporidium.* The new amendments require that such analyses be performed, and that consumers be provided with data on the concentrations of these and other contaminants in their water supplies. The 1996 amendments also required the EPA to publish every five years a list of contaminants that were either known or anticipated to occur in public water systems and might need to be regulated. The first such list, called the "Drinking-Water Contaminant Candidate List," was published in March 1998. It included 10 microbial contaminants (EPA, 1998). In addition, the 1996 amendments established a revolving fund to help state and local authorities pay for water-system improvements; directed the EPA to conduct cost-benefit analyses in establishing new regulations; and permitted the agency to consider overall risk reduction when establishing standards, as well as the outcome of these analyses in adjusting MCLs, if warranted.

Pollution Prevention Act

Recognizing that the Clean Air Act and the Clean Water Act emphasized pollution treatment or remediation rather than prevention, Congress passed the Pollution Prevention Act of 1990 (P.L. 101-508). This act formally made it national policy to assure that pollution is prevented or reduced at the source, recycled or treated in an environmentally safe manner, and disposed of or released into the environment only as a last resort. It also created a clearinghouse to encourage the sharing and transfer of source reduction technology and provided financial assistance to states to promote pollution prevention. Through advocating the substitution of less hazardous substances for those used in a wide variety of industrial processes, the law has led to significant reductions in worker exposure (Chapter 4).

Solid Waste Disposal Act

The original law passed by Congress to address the problem of solid waste was the Solid Waste Disposal Act of 1965 (P.L. 89-272). It was supplemented by the Resource Recovery Act of 1970 (P.L. 91-512), which was designed to promote the demonstration, construction, and application of waste management and resource-recovery systems. Finding these efforts inadequate, Congress passed the Resource Conservation and Recovery Act (RCRA) of 1976 (P.L. 94-580). Although in essence RCRA was designed to regulate the storage, transfer, transport, and disposal of hazardous substances, its goals were quite sweeping: (1) to protect human health and the environment; (2) to reduce waste and conserve energy and natural resources; and (3) to reduce or eliminate the generation of hazardous waste as expeditiously as possible. As with the air- and water-pollution laws, the requirements imposed by the solid-waste laws were to be implemented at the state level.

To achieve these goals, Congress incorporated requirements into this law for implementing three distinct yet interrelated programs. The first, outlined under subtitle D, encouraged states to develop comprehensive plans for managing solid wastes, with emphasis on those of a "nonhazardous" nature, for instance, household wastes. The second, outlined under subtitle C, established a system for controlling hazardous waste from its generation to ultimate disposal (the cradle-to-grave approach). The third, subtitle I, was designed to regulate certain underground storage tanks through the establishment of performance standards for new tanks and the development of methods for detecting, preventing, and mitigating leaks in existing facilities.

Recognizing the need for increased control over the types and forms of solid waste being sent for disposal, Congress passed the Hazardous and Solid Waste Amendments Act of 1984 (P.L. 98-616). One of its stipulations was that hazardous wastes must be treated in a prescribed manner prior to disposal in order to limit the releases of hazardous chemicals should water inadvertently interact with the wastes. The act also closed loopholes in previous rules that allowed toxic wastes to be burned in industrial and apartment furnaces. In addition, it made more stringent the requirements for the siting of new hazardous waste disposal facilities and for the continued operation of existing facilities.

Later, RCRA was modified by the Land Disposal Program Flexibility Act of 1996 (P.L. 104-119), which eliminated a mandate that required the EPA to promulgate stringent and costly treatment requirements for certain low-risk wastes that already were being regulated by the Clean Water Act or the Safe Drinking Water Act. Although the EPA had previously indicated that these wastes presented little or no risk, a court decision had required the agency to promulgate far more stringent requirements than were necessary. In fact, the wastes being regulated were not even classified as hazardous (CEQ, 1997).

Comprehensive Environmental Response, Compensation, and Liability Act

The Comprehensive Environmental Response, Compensation, and Liability Act (CERCLA) of 1980 (P.L. 96-510), better known as the "Superfund" law, addressed the problem of existing disposal sites that were producing unacceptable environmental releases. It was initially thought that there were only a few hundred such sites in the United States, but subsequent investigations revealed that the number was in the thousands. Under terms of this act, the EPA is required to identify such sites, prepare a preliminary assessment report for each, and conclude with a site investigation. From the information generated, the sites are then ranked on the basis of their potential for causing human health impacts and/or environmental damage. Those sites with the highest ranking are placed on what is called the national priorities list (NPL) and are subject to mandatory cleanup actions, funded either by the responsible parties or by the allocation of tax-derived Superfund monies.

Certain aspects of CERCLA, particularly its liability provisions, have proved to be extremely controversial. Liability was referred to as "retroactive, strict, and joint-and-several." This meant, among other things, that both past and current property owners were liable for the cleanup of contamination regardless of when it had been created. If the responsible parties could not be found, a single producer who shipped a small quantity of waste to a site might be held liable for the entire cost of cleanup, including remediation of the larger volumes of wastes disposed of by other producers. With passage of the Asset Conservation, Lender Liability, and

Deposit Insurance Protection Act of 1996, these problems were partially rectified. This act provided protection from liability to lenders and fiduciaries involved in such activities so long as they did not participate in the management of the contaminated facility. Protection was subsequently extended to recyclers of paper, plastic, glass, textiles, rubber, metal, and batteries by the Superfund Recycling Equity Act of 1999. This law absolves recyclers from liability unless the person had reason to believe that the material would be burned, the consuming facility was not in compliance with environmental laws, hazardous substances had been added to the material, or the recyclers failed to exercise care in managing the material. The liability exemption, however, is not applicable if the recyclable material contains PCBs in excess of federal standards (CRS, 2001).

Small Business Liability Relief and Brownfields Revitalization Act

The Small Business Liability Relief and Brownfields Revitalizations Act, which was signed into law (P.L. 107-118) in 2002, is directed primarily to the cleanup of less contaminated sites, such as those resulting from spills and other discharges of hazardous substances or from the disposal of small amounts of related materials. It further amended the Superfund Act by alleviating some of the legal liabilities for small businesses under that act and providing financial assistance for the revitalization of these types of sites. It also established the statutory authority for, and provided separate funding to support, the brownfields program (Isler and Lee, 2002). Experience has demonstrated that this program has stimulated the cleanup and restoration of many of the estimated half a million or more abandoned, idled, or underused industrial and commercial facilities in the United States that have low levels of contamination. In many cases, these sites are located in areas that are economically depressed. The overall goal is to revitalize these properties and restore them to usefulness (Chapter 9).

Emergency Planning and Community Right-to-Know Act

The Emergency Planning and Community Right-to-Know Act (EPCRA) was signed into law as Title III of the Superfund Amendments and Reauthorization Act of 1986 (P.L. 99-499). The impetus for its passage was the accident that occurred in an industrial plant in Bhopal, India, in 1984 (Chapter 19). Subsequent investigations revealed that the people who were exposed had little or no knowledge of the potential danger of the operations that were being conducted within the plant. The EPCRA includes three programs. One, the emergency planning function, provides funds for communities to establish local emergency planning councils to work with industry officials in the development of disaster preparedness programs to cope with accidental releases of toxic chemicals. A second, the toxic substances registry, contains community right-to-know provisions that grant local emergency response personnel and the general public access to information on the chemicals present in local facilities. A third, the Toxics Release Inventory (Chapter 2), requires companies to provide EPA and state officials with an annual accounting of toxic chemicals that are routinely released into the environment (CRS, 2001). Facility operators must also report their on-site waste-management activities. More than 600 different substances are covered by the law. One of the major benefits of the EPCRA is that it has made company officials aware of the impacts of their operations on the environment. In the past, many of them had no idea of the magnitude of their releases (Chapter 2).

Radioactive Wastes

For purposes of classification, radioactive wastes in the United States have been divided into two categories: high level and low level. Each of these has been addressed through separate laws and regulations.

NUCLEAR WASTE POLICY ACT (HIGH-LEVEL WASTE)

Through the Nuclear Waste Policy Act of 1982 and the Nuclear Waste Policy Amendments Act of 1987 (P.L. 97-425, 100-203), Congress established procedures for the design and construction of a geologic repository for the permanent disposal of U.S.-generated high-level radioactive waste and spent nuclear fuel. Although initially several proposed disposal sites were considered, Congress later specified that the studies be limited to the proposed Yucca Mountain site in Nevada. Through the Energy Policy Act of 1992 (P.L. 102-486), Congress subsequently assigned responsibility for establishing radiation protection standards for the repository to the EPA. On the basis of these standards, which were published in 2001, the U.S. Nuclear Regulatory Commission (USNRC) promulgated regulations that will serve as the basis for judging whether the proposed repository is acceptable (Federal Register, 2001).

Responsibility for conducting the necessary studies and gathering and interpreting the data that will serve as a basis for determining the acceptability of the facility has been assigned to the U.S. Department of Energy (DOE). If the preliminary data indicate that Yucca Mountain can isolate high-level waste in a satisfactory manner, the DOE, with approval by the U.S. Senate, will apply to the USNRC for a license to construct the facility. The USNRC will, in turn, review the application and determine whether the proposed site and facility meet the applicable federal regulations; if they do, construction will be authorized. As would be anticipated, many residents in Nevada oppose having the proposed repository located in their state. Nonetheless, in 2002 the U.S. Senate granted approval for the DOE to proceed with the submission of a license application. If all requirements are met, the schedule calls for the repository to begin accepting spent fuel and high-level radioactive waste in 2010.

LOW-LEVEL RADIOACTIVE WASTE POLICY AMENDMENTS ACT

Due to the closure of several low-level radioactive waste disposal facilities that had previously operated, by the early 1980s there were only three states—Nevada, South Carolina, and Washington—that continued to provide such services. In due time, the governors and members of the public in these three states began to question whether they should continue to be used as a "dumping ground" for wastes from the rest of the country. Seeking to alleviate this situation, Congress passed the Low-Level Radioactive Waste Policy Amendments Act (LLRWPAA) of 1985 (P.L. 99-240). Under this law, each state, either individually or as part of a regional compact, was made responsible for the disposal of all low-level radioactive wastes (LLRW) produced through commercial or federally sponsored activities within its borders. To ensure compliance, Congress established a series of milestones, penalties, and incentives to encourage additional states to move forward with the siting, construction, and operation of such facilities.

Even so, little if any progress was made during the years that followed. In fact, the Nevada facility was closed in 1992, and the Washington facility discontinued accepting wastes from outside that state in 2003. In the meantime, only one new facility, established in Utah in 1995, has been opened. Although several other states have appeared to be making progress, obstacles inevitably arise. A major one is the public's opposition to the establishment of any type of new disposal facility. As a result, many waste generators are faced with the necessity of retaining (storing) their LLRW on-site. This situation, coupled with the increasing fees that are being charged for disposal, has led to some perhaps unexpected benefits. One is an accelerated effort to minimize the volume of waste through prevention strategies and technologies as well as through incineration and compaction. As a result, the quantities of LLRW generated in recent years are far below what they were a decade ago. While it is clear that new disposal facilities will be necessary before too long, it is equally clear that developing a solution to the problem will not be easy.

Toxic Substances Control Act

Recognizing the need for a comprehensive system for controlling toxic substances, in 1976 Congress passed the Toxic Substances Control Act (TSCA) (P.L. 94-469). Though it is cumbersome and rarely used, the TSCA is sweeping in scope and has proved to be a powerful tool in the regulatory arsenal. One of its provisions was to provide federal agencies with adequate authority to prevent unreasonable risks of injury to health or the environment. Having said this, however, the act went on to caution that this authority should be exercised in such a manner as "not to impede unduly or create unnecessary economic barriers to technological innovation while fulfilling the primary purpose of this Act" (Findley and Farber, 1992, 2000). This immediately raises the question of the relative weights that should be given to these conflicting goals. One example is how long to delay the introduction of a new, but not yet approved, chemical that may be better and less toxic than one it could replace (Sigman, 2000).

The principal responsibilities for implementing the TSCA were assigned to the EPA, which was directed to perform two primary functions: (1) to develop information on the toxicity of various substances and (2) to regulate them, if warranted. Under terms of the act, any company that plans to manufacture or import a new chemical must submit to the EPA a premanufacturing notice containing information on its identity, use, anticipated production or import volume, and disposal characteristics. Any hazards to which workers may be subjected in its manufacture and handling must also be reported. Although chemicals used exclusively in pesticides, foods, food additives, drugs, and cosmetics are exempted, since these are covered by other federal laws, the magnitude of the remaining task is enormous. Shortly after the act was passed, for example, the EPA identified more than 60,000 chemicals that needed to be evaluated. Since that time, thousands of additional chemicals have been introduced. A further problem is that as of 1984, data on the toxicity of more than three-quarters of these chemicals, all of which were being produced in quantities in excess of 1 million pounds per year, were nonexistent. Facing this challenge, EPA personnel have had to rely to a large degree on the structure of a chemical, which they, in turn, compare to those of known toxicity. Nonetheless, if the EPA has reason to suspect that a chemical may pose a risk, but lacks sufficient data to take action, it can require the manufacturer to develop the necessary data.

A chemical intended for export can be regulated by the EPA only if it presents an unreasonable risk to the health of people or to the environment of the United States. The EPA is responsible, however, for notifying the governments of countries that import the chemicals of any associated U.S. regulatory restrictions. In contrast, no chemical substance, mixture, or article containing such materials may be imported into the United States if it fails to comply with any U.S. rule or is otherwise in violation of the Toxic Substances Control Act. These types of issues have had important implications in recent years in the deliberations on international agreements such as the North American Free Trade Agreement.

Federal Food, Drug, and Cosmetic Act

Enacted in 1938, the Federal Food, Drug, and Cosmetic Act is one of the oldest of the major health-regulation laws. It replaced the Pure Food and Drug Act of 1906 and remains a cornerstone of U.S. food-safety policies. The act covers foods for humans and animals, human and veterinary drugs, medical devices, and cosmetics. The original act included two prohibitions relative to foods containing hazardous substances, both of which continue to apply today. The first of these forbade the marketing of food containing "any added poisonous or deleterious substance which may render it injurious to health," a provision that the Food and Drug Administration (FDA) has interpreted as barring any foods that present a serious risk. The second forbade the marketing of foods containing naturally occurring toxic substances; however, permission was granted to continue the marketing of foods that, even though known to contain such substances, have traditionally been a part of the American diet. Premarketing approval was not required by either provision.

The Federal Food, Drug, and Cosmetic Act was revised by the passage of the 1958 Food Additives Amendment, which required that the safety of substances classified as "food additives" be demonstrated prior to marketing. The critical standard for approval was confirmation that the substance is "reasonably certain to be safe." The amendment, however, specifically exempted substances that, in the judgment of qualified scientific experts, are "generally recognized as safe" (GRAS). In the main, such substances are those that have been in use for many years without observed adverse health effects. If new evidence raises doubts about the safety of a GRAS ingredient, it becomes a "food additive," whose use requires approval (Merrill, 1986). The 1958 amendment also exempted food ingredients that had been "sanctioned" by the FDA or by the Department of Agriculture prior to 1958. These included some controversial substances such as sodium nitrite, traditionally used in curing certain meat products such as bacon.

Any review of laws regulating food would be incomplete without mention of the Delaney clause, which Congress added to the Federal Food, Drug, and Cosmetic Act in 1958. Recognizing that cancer was a major and widely feared health problem, this clause stated that "no additive shall be deemed safe if it is found to induce cancer when ingested by man or animal" and directed that the FDA not approve any such additives. For more than four decades, this language was interpreted as requiring a "zero-risk" standard for any cancer-causing food additive, including residues from pesticides found in processed foods. A further complicaton was that in the case of raw foods, the law required the FDA to apply a negligible-risk standard. This was done in the belief that if the cancer risk was outweighed by several factors, such as the ability of the pesticide to help in the production of an adequate, wholesome, and economical food supply, the safe use of the pesticide was warranted. The Delaney clause also limited the introduction of lower-risk pesticides that could replace older and potentially more hazardous materials. In addition, it was unduly narrow in its application since it singled out only one endpoint, namely, cancer, and did not cover substances that might cause birth defects, nerve damage, or immune system failures.

The result was a multitude of subsequent debates on how much of a chemical shown to be carcinogenic when administered in large quantities to animals under laboratory conditions should be considered acceptable as an additive to food designated for consumption by humans. As exemplified by the experience with saccharin, these debates extended well beyond those of a technical or scientific nature. In accordance with the Delaney clause, the FDA proposed to ban the use of this chemical, only to have Congress pass the Saccharin Study and Labeling Act of 1977, which permitted it to be added to food provided the food was accompanied by a label warning the consumer of the potential health risk. Examples of other efforts to define safety, as incorporated into various federal regulations, are summarized in Table 14.2.

To rectify these problems, in 1996 Congress passed the Food Quality Protection Act (P.L. 104-170). This law required that all uses of pesticides in both raw and processed foods be safe and defined safety as "a reasonable certainty that no harm will result from aggregate exposure." Harm was defined as including other risks to health as well as cancer. Among the benefits of this new law was that these changes permitted the EPA and other responsible agencies to direct more of their resources to high-priority concerns in public health and environmental protection activities, as opposed to Delaney-related activities. Other mandates of the Food Quality Protection Act included a requirement that manufacturers demonstrate that their products are safe for infants and children, that consumer right-to-know provisions regarding pesticide risks be implemented, and that risks from pesticides that may be endocrine disruptors be evaluated and reduced (CEQ, 1997).

Federal Insecticide, Fungicide, and Rodenticide Act

Closely related to the Toxic Substances Control Act (TSCA) and the Federal Food, Drug, and Cosmetic Act is the Federal Insecticide, Fungicide, and Rodenticide Act (FIFRA), first enacted in 1947, substantially amended in 1972 (P.L. 92-516), and further amended in 1975, 1978, 1980, and 1988. The act is particularly important inasmuch as it addresses substances that are, by design, biologically active, are applied directly to food products, and are deliberately released into the environment (Sigman, 2000).

Under this act, no pesticide may be marketed unless it has been registered by the EPA, that is, the agency has reviewed and evaluated data on the pesticide and has determined that it does not present an unreasonable risk to health or the environment (Merrill, 1986). To obtain such approval, the manufacturer must submit detailed information on the chemical structure and toxicity of the product. The standard that the EPA must use in making its decision is to avoid an "unreasonable adverse effect." Under this standard, the agency can take into consideration the benefits and costs associated with the use of a pesticide. If, for example, a substitute is not available, approval for the use of an existing, but challenged pesticide might be continued. Nonetheless, if the residue of a pesticide that is

Table 14.2 Safety as defined in U.S. federal regulations

Law	Year	Definition of safety
Federal Food, Drug, and Cosmetic Act	1938	Forbids marketing any food containing "any *added* poisonous or deleterious substance that may render it injurious to health"; forbids marketing foods containing *nonadded* toxicants, i.e., natural agricultural commodities, that make them "ordinarily injurious to health"
Delaney Clause	1958	Excludes any additive that has been "found to induce cancer when ingested by man or animal"
Occupational Safety and Health Act	1970	OSHA shall prescribe the standard "which most adequately assures, to the extent feasible, on the basis of the best available evidence, that no employee will suffer material impairment of health or physical capacity"
Consumer Product Safety Commission	1972	The commission is empowered to promulgate safety standards that it finds "reasonably necessary to prevent or reduce an unreasonable risk of injury"
Federal Insecticide, Fungicide, and Rodenticide Act	1972	Pesticide may be marketed only if, "when properly used . . . it will not generally cause unreasonable adverse effects on the environment"
Federal Water Pollution Control Act	1972	Defines "toxic pollutants" as those that individually or in combination "upon exposure, ingestion, inhalation or assimilation into any organism, either directly from the environment or indirectly by ingestion, through food chains, will, on the basis of information available, cause death, disease, behavioral abnormalities, cancer genetic mutations, physiological malfunctions . . . or physical deformations, in such organisms or their offspring"[a]
Safe Drinking Water Act	1974	Requires that public water-supply systems "meet minimum national standards for the protection of public health" and

Table 14.2 (continued)

Law	Year	Definition of safety
		that contaminants "which may have an adverse effect on human health" be regulated
Resource Conservation and Recovery Act	1976	Defines a hazardous waste as one that "may cause, or significantly contribute to an increase in mortality or an increase in serious irreversible, or incapacitating reversible illness; or pose a substantial present or potential hazard to human health or to the environment when improperly treated, stored, transported, or disposed of, or otherwise managed."
Toxic Substances Control Act	1976	Forbids use or disposal of chemicals when there is a "reasonable basis" to conclude that such use on disposal poses an "unreasonable risk of injury to health or the environment"
Comprehensive Environmental Response, Compensation, and Liability Act (Superfund Act)	1980	States that a "hazardous substance" includes any "elements, compounds, mixtures, solutions, and substances which, when released into the environment, may present a substantial danger to the public health or welfare or the environment"
Food Quality Protection Act	1996	Requires that all uses of pesticides in both raw and processed form be safe, with safety being defined as "a reasonable certainty that no harm will result from aggregate exposure"; harm is defined as including cancer as well as other risks to health[b]

a. The stringency of these requirements and the accompanying timetable for their implementation led to problems in implementing the law. This led, in turn, to lawsuits and later revisions of the requirements through passage of the Clean Water Act of 1977, which mandated technology-based, industrywide limits on toxic pollutants while still allowing the U.S. Environmental Protection Agency to prescribe more stringent controls on liquid-waste releases when necessary to protect health (Merrill, 1986).

b. The Food Quality Protection Act permitted the EPA and other responsible agencies to direct their resources to high-priority concerns in public health and environmental protection, as opposed to Delaney-related activities.

deemed to be carcinogenic concentrates in processed foods, the EPA must base its decision on a health-based standard (Hutt and Merrill, 1991).

The EPA engendered considerable controversy in the early 1970s by canceling registration for a number of pesticides that had previously been approved for use on agricultural crops. These included DDT, aldrin, and dieldrin; the cancellations were based primarily on laboratory studies that suggested that these chemicals were carcinogenic in animals. The criticisms that accompanied these decisions led to important changes in the law and in the way in which it was implemented (Merrill, 1986). Further complicating matters, Congress, in a 1972 amendment, required that the EPA retest and reapprove all pesticides already in use. This task, however, was simplified in 1978 through amendments that restricted testing and approval to the so-called active ingredients. Although existing pesticides marketed in the United States contain between 800 and 900 such ingredients, less than about 200 have been approved to date. Nonetheless, these account for about 60 percent of the active ingredients now in use (Sigman, 2000).

Amendments adopted in 1975 required the EPA to submit proposed pesticide cancellations to a scientific panel for review. As part of this process, the EPA established a procedure for public comment on the risks and benefits of specific pesticides prior to initiation of the formal cancellation process (Merrill, 1986).

Occupational Safety and Health Act

The basic law that governs conditions in the workplace is the Occupational Safety and Health Act of 1970. Its goal is "to assure so far as possible every working man and woman in the Nation safe and healthful working conditions and to preserve our human resources." Two of the primary advances provided were the establishment within the Department of Labor of the Occupational Safety and Health Administration (OSHA) and the establishment within the Department of Health and Human Services of the National Institute for Occupational Safety and Health (NIOSH). OSHA was assigned the authority to conduct workplace inspections and to issue citations and impose fines where serious violations are observed; it also had the responsibility of working with the states to develop and implement improved occupational health programs. In response, OSHA established a program for reviewing and approving the occupational safety and health programs within the states. The result has been a shifting of responsibility for primary enforcement from the federal to the state level. As in the control of air and water pollution, OSHA requires that the regulations imposed by the states keep pace with, and be as effective as, those adopted at the federal level.

NIOSH was assigned the responsibilities for conducting research on safety and health problems, providing technical assistance to OSHA, and recommending standards for adoption by OSHA. NIOSH was also instructed to conduct workplace investigations, gather testimony from employers and employees, and require that employers measure and report employee exposure to potentially hazardous substances (Bingham, 1992). It was also empowered to require employers to provide medical examinations and tests to determine the incidence of occupational illnesses among their employees.

Once OSHA has evaluated the recommendations of NIOSH and has developed plans to propose, amend, or delete a standard, these intentions are published in the Federal Register as a "Notice of Proposed Rulemaking." This format, mandated by the Administrative Procedures Act, is designed to provide all interested parties, including employers, employees, and members of the public, an opportunity for input into development of such standards. Although worker safety was the primary focus during the early years, OSHA now directs equivalent attention to the control of occupational illnesses.

While considerable progress has ensued since the Occupational Safety and Health Act became law, there is a limit to what can be accomplished through the minimal enforcement efforts that are conducted under this act (Bingham, 1992). No more than several thousand inspectors at the federal and state levels are responsible for overseeing almost 6 million work sites. Although inspections, followed by citations when violations are observed, will remain an integral part of the program, continuing improvements in worker safety and health will require cooperative efforts by employers and employees. In the long run, improved worker health and safety will be to their mutual benefit.

Atomic Energy Act

The Atomic Energy Act (AEA) of 1954 established the Atomic Energy Commission (AEC). With passage of the Energy Reorganization Act of 1974, the AEC was abolished and replaced by the U.S. Nuclear Regulatory Commission and what, soon thereafter, became the U.S. Department of

Energy (DOE). The USNRC was assigned the regulatory functions of the AEC, including the responsibility for licensing the transfer, manufacture, acquisition, possession, or use of any nuclear facility and for regulating radioactive materials and their by-products. The USNRC is also authorized to enter into agreements whereby a state can regulate nuclear materials within its borders, with certain exceptions such as the construction and operation of commercial nuclear power plants. Currently, more than 60 percent of the states have entered into such agreements (USNRC, 2003).

The USNRC is required by law to ensure that radioactive materials and related facilities are adequately managed to protect public health and the environment. Regulations developed by the USNRC to meet these responsibilities must be in accord with standards developed by the EPA. Another law closely related to the control of radioactive materials is the Uranium Mill Tailings Radiation Control Act, which requires the DOE to designate and assign priorities for the remediation of wastes at former uranium mill tailings processing sites. In so doing, the DOE must seek the concurrence of the USNRC and perform the work in accordance with standards set by the EPA. Additional relevant laws include the previously cited Nuclear Waste Policy Amendments Act and the Low-Level Radioactive Waste Policy Amendments Act (USNRC, 1991).

Other U.S. Environmental Laws

Many other laws have provided significant support to environmental protection. One group of these, passed in the 1970s to encourage energy conservation, includes the Energy Policy and Conservation Act of 1975, which promoted disclosure of efficiency ratings for household appliances; the Energy Conservation Act of 1976, which encouraged energy-conservation measures in new buildings; and the Energy Tax Act of 1978, which provided tax incentives for energy-efficient automobiles and residences. Many states have adopted similar measures (Findley and Farber, 1992, 2000). These laws were followed by the previously cited Energy Policy Act of 1992, which established energy-efficiency standards for government buildings and for offices and private residences, as well as new approaches to improved efficiency in transportation, including the development of alternative-fueled and electric vehicles and increased use of such vehicles by federal agencies.

Another important law is the Endangered Species Act of 1973, which is designed to protect plant and animal resources from the adverse effects of development. Closely related to this act are the Marine Mammal Protection Act which makes it unlawful, except under certain special circumstances, to take, possess, or trade a marine mammal or marine mammal product; the Bald and Golden Eagle Protection Act, which makes it a criminal offense to pursue, wound, capture, kill, or disturb these birds; and the Wild Free-Roaming Horses and Burros Act, which places these animals under the jurisdiction and protection of the Department of the Interior.

Various executive orders are also important, for example, Executive Orders 11988 and 11990, which were issued in 1977. The former outlines federal policies for the management of floodplains and the protection of wetlands; the latter requires that federal agencies identify and reduce potential impacts on wetlands resulting from proposed activities. Also noteworthy are the systematic efforts that have been under way since the 1960s to promote motor-vehicle safety in the United States. One of the initial efforts in this regard was the passage of the Highway Safety Act, which established the National Highway Safety Bureau, later to become the National Highway Traffic Safety Administration.

International Protocols and Treaties

Interestingly, the Charter of the United Nations (UN) contains no specific statements regarding the protection of the environment. Many of the agencies created by the UN, however, exercise considerable responsibilities in this subject area. These include the Food and Agricultural Organization, the World Health Organization, the International Atomic Energy Agency, the World Meteorological Organization, the International Maritime Organization, the International Labour Organization, and the UN Educational, Scientific, and Cultural Organization. Other international organizations, such as the World Bank, and regional development banks also have significant international influence. Through its environmental standards, the World Bank, for example, specifies emission limits for major facilities that it supports financially. Through this process, these limits become de facto international standards. Several regional organizations, such as the Pan American Health Organization, play similar roles (Ferry, 2001).

In addition, multiple protocols, treaties, and agreements have been adopted at the international level. Two examples are the 1992 Rio Declaration on Environment and Development and the agreements reached at the follow-up Conference in Johannesburg in 2002. These include steps to protect wetlands and deserts, reduce air and water pollution, improve the use of energy and technology, and manage toxic chemicals in a safer manner. One of the specific accomplishments of the Rio Declaration was the creation of the UN Commission on Sustainable Development (Ferry, 2001). Earlier examples of international activities to protect the environment include the Limited Test Ban Treaty of 1963 and the subsequent Treaty on the Non-Proliferation of Nuclear Weapons. Another example is the Biological and Toxin Weapons Convention of 1975 that prohibits the development, production, or stockpiling of biological or toxic agents and devices for delivering such agents for other than peaceful purposes. A totally different example of international cooperation is the International Space Station, a jointly sponsored project that, among other activities, has the capabilities to monitor various environmental factors on Earth. These and other examples of international environmentally related agreements are listed in Table 14.3.

One of the more recent treaties, developed in December 2000 under the auspices of the United Nations Environment Programme and involving representatives from 122 nations, seeks to achieve a worldwide phaseout of the use of certain persistent organic chemicals. The chemicals in question, widely referred to as the "dirty dozen," include PCBs, DDT, dioxins, furans, and other pesticides that have been shown to contribute to developmental effects, cancer, and other health effects in humans and animals. Similar agreements have been developed within the European Union (EU). These include the law passed in 1981 that required that all new chemicals undergo premarket testing to assess their potential health and environmental risks. It was later modified to require a more comprehensive review of all new chemicals. One of the goals was that the producers, rather than the national governments, would shoulder the burden of proving that a chemical was safe (McGinn, 2001).

The most successful of the international agreements include those designed to terminate the dumping of wastes into the North Sea, to prevent the disposal of low-level radioactive waste into the oceans (as prescribed by the 1982 Law of the Sea), and to oversee tuna fisheries in the West Central and Southwest Pacific. Also successful were the Vienna Convention of 1985 and the Montreal Protocol of 1987 that were designed to protect the stratospheric ozone layer. Although the United States has yet to endorse it, the Kyoto Protocol of 2000 has nonetheless signified international cooperation on the long-term control of greenhouse gases.

Table 14.3 International environmentally related conferences, agreements, and treaties

Year	Origin of action	Recommendations
1946	International Convention for the Regulation of Whaling	Established International Whaling Commission to provide for the proper conservation of whaling stocks and thus make possible the orderly development of the whaling industry
1968	UN Biosphere Conference	Discussed global environmental problems, including pollution, resource loss, and wetlands destruction
1972	UN Conference on the Human Environment	Issued a series of recommendations for governmental actions on a variety of environmental problems
1973	Convention on International Trade in Endangered Species of Wild Fauna and Flora	Led to restrictions on trade in about 5,000 animal species and 25,000 plant species that are near or threatened with extinction
1973	Convention for the Prevention of Pollution from Ships	Restricted the release of pollutants from oceangoing vessels, including dumping and accidental spills of oil, garbage, plastics, and sewage
1975	UN Conference on Human Settlements	Drafted 65 recommendations on how best to provide shelter, including agreement that adequate shelter was a basic human right
1979	Convention on Long-Range Trans-boundary Air Pollution	Established the framework for combating and regulating the movement of acid rain and other airborne pollutants across national borders: later related protocols addressed emissions of nitrogen oxides, sulfur, heavy metals, persistent organic compounds, and other pollutants
1982	Law of the Sea	Provided a comprehensive framework for ocean use and contained provisions on ocean conservation, pollution prevention, and protection and restoration of species populations
1982	Stockholm +10 Conference	Issued a declaration expressing concern about the present state of the environment and established an independent commission to craft a global agenda for change

Table 14.3 International environmentally related conferences, agreements, and treaties (continued)

Year	Origin of action	Recommendations
1982	Convention on the Conservation of Antarctic Marine Living Resources	Designed to conserve marine life of the Southern Ocean. This does not, however, exclude harvesting that is conducted in a rational manner
1985	Vienna Convention for the Protection of the Ozone Layer	Established mechanisms for international scientific cooperation and committed the parties to take action to protect stratospheric ozone
1987	Montreal Protocol on Substances That Deplete the Ozone Layer	Required industrial countries to phase out production of a number of ozone-depleting chemicals by 1996, developing countries to meet the goal by 2010
1987	Basel Convention	Called for a control on the movement of hazardous wastes across borders and later forbade exports of wastes from developed to developing countries for final disposal
1992	UN Conference on Environment and Development, the "Earth Summit"	Called for improving the quality of life on Earth by using natural resources more efficiently, protecting global commons, managing human settlements in a better manner, and reducing pollutants and chemical waste
1992	Convention on Biological Diversity	Mandated that countries formulate strategies to protect biodiversity and that industrial countries help the developing countries in implementing these strategies
1992	Convention on Climate Change	Called for avoiding human activities that alter the climate and set a nonbinding 2000 goal for the industrial countries to reduce CO_2 concentrations to those of 1990
1993	Consultative Meeting of Contracting Parties to the 1972 London Convention	Prohibited the dumping of all types of radioactive wastes into the sea
1994	Conference on Population and Development	Emphasized the importance of the education of women and ensuring their access to reproductive health care in the control of population growth

Table 14.3 (continued)

Year	Origin of action	Recommendations
1995	Intergovernmental Panel on Climate Change	Concluded that "the balance of evidence suggests that there is a discernible human influence on global climate" a follow-up report in 2001 cited "new and stronger evidence that most of the observed warming of the last 50 years is attributable to human activities"
1997	Kyoto Protocol	Strengthened the 1992 Convention on Climate Change by mandating specific reductions of greenhouse gases
2000	Biosafety Protocol	Recommended a more precautionary approach to trading genetically altered crops and organisms and required exporters to obtain prior consent from the countries to which genetically altered crops are proposed to be sent
2000	Treaty on Persistent Organic Pollutants	Required the phaseout of nine persistent highly toxic pesticides and limited the use of other chemicals, including dioxins, furans, and PCBs
2001	Human Genome Project	Reported that the human gene count, in contrast to the anticipated 100,000, numbered only about 30,000, about the same as that of a weed or a mouse, stimulating concerns about the wisdom of the genetic engineering of plants and animals
2002	Bangkok Statement: A Pledge to Promote the Protection of Children's Environmental Health	Recognized that children are uniquely vulnerable to the effects of chemical, biological, and physical agents, and that every child should have the right to safe, clean, and supportive environments that ensure his/her survival, growth, development, healthy life, and well-being
2002	International Treaty on Plant Genetic Resources	Designed to bring order and regularity to the transfer of crop genetic resources for most major crops; considered particularly significant since the agricultural systems in all countries are highly dependent on nonindigenous crops for food and agriculture.

Table 14.3 International environmentally related conferences, agreements, and treaties (continued)

Year	Origin of action	Recommendations
2002	World Conference on Sustainable Development (Johannesburg, South Africa)	Focused on some of the most critical threats to global security and climate change with emphasis on action, as contrasted to principles and planning; included was a call for the development of partnerships between the United Nations and various worldwide corporations to encourage businesses to "buy into" the basic values of the UN

The General Outlook

In spite of the efforts of Congress to provide laws that cover all facets of the environment, voids remain. One is the lack of requirements for preliminary assessment of the exposures of workers to new chemicals. While the FDA and the EPA can authorize premarket approval of food additives, drugs, and pesticides, no employer is obligated to obtain prior approval of new processes or materials or to conduct tests to assure that changes in operations will not jeopardize worker health. Only if it is discovered that a material already in use threatens the health of workers may OSHA take action (Merrill, 1986). Another void is the lack of comprehensive legislation on indoor pollution, particularly inside the home (Chapter 5). Although Congress has passed laws relating to radon and lead paint (P.L. 102-550), only in recent years has this body begun to show interest in such problems on a broader basis.

As noted earlier, one of the interesting international regulatory trends has been a shift in the burden for documenting the degree of toxicity of new chemical products from the regulatory agencies to the manufacturers. Much the same approach has been adopted in the United States with respect to enforcement of the Clean Air Act. Rather than requiring the regulators to detect noncompliance, the act makes waste generators responsible for certifying that they are in compliance. Another trend has been the transfer of the primary responsibilities for implementing and enforcing many of the newer environmental laws to the individual states. Still another trend is increasing recognition by Congress of the need to support programs designed to minimize the production of waste and to conserve our natural resources. This is exemplified by key features in the RCRA, the Clean Air Act, and the Clean Water Act—and by the emphasis being placed on pollution prevention. In many cases, pollution prevention/waste minimization is a win-win-win situation for government, industry, and the public. Beyond mere protection of the environment, society as a whole benefits through reduced capital and operating costs, decreased liabilities, cleaner and safer working conditions, conservation of energy and material resources, and the opportunity for government and industry to work together cooperatively.

Another significant development is the increasing role of the judiciary in rendering decisions related to environmental pollution. One example was a ruling by the U.S. Supreme Court in 2001 that the Clean Air Act requires the EPA to establish national ambient air-quality standards (NAAQS) solely on the basis of public health considerations without consideration of costs. Simultaneously, however, the Supreme Court acknowledged that it is appropriate for the states and the EPA to continue to consider costs when specifying the procedures for implementing the standards. A second example, previously discussed in Chapter 5, was a closely related ruling in 2002 by the U.S. Court of Appeals for the District of Columbia Circuit. In this case, the court decided that the EPA could continue to establish NAAQS, as it has in the past, but should be careful to use the more prescriptive "subpart 2," in contrast to the less prescriptive "subpart 1," requirements of the Clean Air Act. Such an approach would enable the EPA to consider the associated costs, as appropriate, during the process of developing both the required schedule and the types of controls that must be applied in implementing the standards. A related benefit of the previously cited Supreme Court ruling was the stimulus it provided to the so-called "multipollutant" control strategy. Through this approach, increasing attention is being directed to the simultaneous regulation of emissions of nitrogen oxides (NO_X), sulfur dioxide (SO_2), and mercury from power plants (Mathai, 2003). This strategy served as a basis for the Clear Skies Initiative proposed by the EPA. Even so, implementation of such an approach will not be without controversy (Doniger, 2003).

Although many factors must be considered in evaluating the relative importance of the multitude of laws discussed in this chapter, according to one expert (Portney, 2000), it would be difficult to name a single law that has led to larger achievements in advancing environmental quality, safety, and health than the Clean Air Act. Its importance is shown by the fact that the regulations promulgated under this act affect the "very air we breathe, the rate at which our cities grow (and sometimes shrink), the quality of our visits to national parks and wilderness areas, and the prices we pay for virtually everything we buy." One of the most vivid demonstrations of its unqualified success is that the quality of the air in virtually every metropolitan area within the United States has improved significantly during the four decades since the basic framework of this act was established.

15

STANDARDS

MANY professional organizations and federal and state agencies have developed recommendations, guidelines, standards, and regulations for limiting exposures to a variety of occupational and environmental contaminants and physical agents. Prominent among these are the American Conference of Governmental Industrial Hygienists (ACGIH), which has issued recommendations for limiting exposures to chemical and physical stresses in the workplace (Chapter 4), and the International Commission on Radiological Protection (ICRP) and the National Council on Radiation Protection and Measurements (NCRP), which have developed similar guidelines for protection against ionizing radiation (Chapter 12). Taking these recommendations and those from other organizations, as well as evaluations of a range of other factors, into consideration, federal agencies, such as the Environmental Protection Agency (EPA), the Food and Drug Administration (FDA), the Occupational Safety and Health Administration (OSHA), and the U.S. Nuclear Regulatory Commission (USNRC), promulgate standards and regulations for the control of various contaminants and physical stresses within the workplace, the home, and the ambient environment.

Types of Standards

One basic goal in establishing standards is to protect human health. These standards are generally referred to as *primary* standards. Another, equally important goal is to protect the environment. These standards are generally referred to as *secondary* standards. In the case of air pollution, secondary standards are designed to protect agricultural crops and property, such as buildings and statues; in the case of water pollution, they are designed to ensure that rivers and streams are acceptable for fishing and swimming; in the case of drinking water, they are designed to ensure acceptability of the aesthetic qualities of the product, such as temperature, color, taste, and odor. Interestingly, the secondary standards for many contaminants in air and water are more stringent than the primary standards.

Current Approaches

One of the first requirements in the development of primary standards is to identify or define the exposed member who is to be protected. Is it an adult, a child, an infant, or a fetus? Aware of this need, the ICRP (1996) has published age-dependent guidance for use in protecting members of the public. In this approach, the population is divided into six different age groups, and a different dose per unit intake coefficient is specified for each. Since the processes for absorbing materials into the body vary, the values for the coefficients for materials that are inhaled are different than for those that are ingested. In the case of inhaled particles, there are different intake limits depending on their size (Chapter 5). In addition, where the health effects of a contaminant depend on the short-term concentration (as is the case for ozone), hourly rather than quarterly or annual limits have been set. These and other types of refinements illustrate the increasing degrees of sophistication that are being incorporated into the development of primary standards.

Another major consideration is to designate, in a generic sense, the person who is to be protected. Early on, EPA regulations required protection of the "maximally exposed individual." Those promulgated by the USNRC required that no "individual member" of the public receive dose rates in excess of the limits. Soon it was recognized that the identification of such an individual would place an arduous burden on organizations that were seeking to document compliance with the applicable regulations. Recognizing this fact, the EPA modified its regulations to require protection of the "reasonably maximally exposed individual" (EPA, 2001), and the USNRC adopted the policy that compliance could be based on the dose rate to an *average* member of the "critical group." This was defined as the group of people who, because of their location or living habits, would be most highly exposed. Subsequently, the EPA stated that its use of the phrase "reasonably maximally exposed individual" was intended to imply the same type of approach. This approach has the advantage of ensuring not only that members of the public do not receive unacceptable doses but also that decisions on the acceptability of a given practice are not prejudiced by a small number of individuals with unusual living habits.

Once the *basic* standards have been developed, *derived* or *tertiary* guides must be established for determining through monitoring programs (Chapter 16) whether the basic standards are being met. These guides may include limits either for intakes of individual contaminants by the exposed population or for the concentrations of individual contaminants within various environmental media (air, water, food). This same approach is applied in developing derived guides for the protection of workers. Derived guides can also be developed on the basis of allowable releases of specific contaminants into the environment via the airborne or liquid pathway (Chapters 5 and 8).

LIMITATIONS

The underlying assumptions and scientific bases for occupational and environmental standards are subject to a host of limitations. The following are some of the more prominent:

1. The most common endpoints used in the establishment of standards are different types of cancer. Since cancers produced by one agent are virtually indistinguishable from those induced by another, the induction can be inferred only on statistical grounds, that is, from an analysis of the dose-dependent increase in their frequency in the exposed population.
2. In only a very few instances have dose-response data been developed with a view to setting standards. Most such data are byproducts of descriptive and analytic studies designed to test specific scientific hypotheses. Even where such data are available, they contain a range of uncertainties—not only those commonly encountered in the study of any biological system, but also those involved in extrapolating data from animals to humans.
3. With few exceptions, the dose to the affected tissues is not known well enough to define the dose-incidence relationship except in a general way. Analyses of dose-incidence relationships for chemical carcinogens are also complicated by the fact that the dose of a chemical at its biological site of action depends on a number of

metabolic and pharmacokinetic factors that can vary with route of exposure, age, sex, genetic constitution, physiological state, action of other chemicals, and other variables.

4. Few standards take into account differences in the weight, size, diet, and lifestyle of various population groups (for example, Japanese in comparison to Americans). One possible approach for solving this problem would be to express the limit for a specific contaminant in terms of intake per unit of body weight, one example being the data presented in Table 2.1, Chapter 2.

5. Despite a consensus that standards for the general public should be much more stringent than those for workers, the agencies responsible for occupational standards and those responsible for environmental standards seldom coordinate with one another. Moreover, standards for limiting concentrations of airborne contaminants inside dwellings and office buildings are essentially nonexistent (Chapter 5).

6. Some standards do not apply to all sources of a given contaminant or physical factor. For example, guidelines for acceptable radiation doses to workers and the public do not include exposures from the natural background and medical applications.

7. Standards for workers and the public are commonly set for individual contaminants in specific environmental media—air, water, food, and soil—even though it is the total intake of the contaminant that is critical. In addition, few standards are set with consideration of the effects of exposures to a combination of occupational and environmental contaminants and stresses.

8. In some instances, such as exposures to electric and magnetic fields, definitive data or sound epidemiological evidence on which to base standards have been lacking. Yet public pressure and the fact that exposures are occurring make standards necessary even when their basis is suspect.

9. The risks associated with the exposures permitted by many standards have not been quantified. Unless they are, it is not possible to compare the relative stringency of, or protection afforded by, standards developed for different environmental contaminants or stresses.

10. Except in the case of air and water, few limits have been derived for protection of the natural environment, property, or aesthetic features. Similar considerations need to be applied to various animals and organisms living within the environment. Recognizing this need, the ICRP is developing a framework for assessing the impacts of ionizing radiation on nonhuman species (ICRP, 2003). As noted in Chapter 17, the EPA has initiated a similar program for assessing the impacts of toxic chemicals on ecosystems.

A MAJOR LIMITATION

As noted previously (Chapter 2), it is widely recognized that immediately observable biological effects of high-level acute doses of almost any agent have a threshold, that is, only if the dose is sufficiently high will immediate effects be observed. The primary question is whether there is a similar threshold for the latent effects that sometimes appear following low-level doses received over a long period of time. The situation is complicated by the fact that while there may be no observable effect in the case of certain chemicals taken into the body at low dose rates, a range of nonobservable precursors to these effects may be taking place. As also noted, the published literature contains data from an increasing number of studies that show that the relationship between effects and exposure/dose at low levels for a host of toxic agents, including some that are carcinogenic, follows a J-shaped curve, not a straight line. The implication of such observations is that at very low doses, the affected organism or animal may actually be receiving some benefit. Until this matter is resolved, this subject will continue to be controversial for a host of environmental stresses.

OTHER CHALLENGES, LIMITATIONS, AND PROBLEMS

As an outgrowth of these limitations and the lack of a uniform system for establishing environmental standards, individual federal agencies use different methods for estimating dose and risk, especially in terms of the economic values of serious health effects and associated social and ethical concerns. In addition, they apply different approaches on the degree of conservatism to be incorporated into such calculations, different depictions of how human exposure may occur, and different assumptions about the period over which protection is to be provided. Federal agencies also incorporate into their standards different concepts of how they are to be interpreted and applied.

Two examples are the "top-down" regulatory strategy adopted by some agencies and the "bottom-up" strategy adopted by others. The former involves setting an upper bound or limit and then reducing the limit (on the basis of site-specific considerations) to a reasonably achievable lower level. The USNRC and the U.S. Department of Energy (DOE) have consistently favored this approach in the development and/or implementation of radiation protection standards. Conversely, the bottom-up strategy has been used to control a variety of other environmental exposures. It involves initially setting a lower, relatively stringent dose or risk goal with the understanding that the goal is to be considered a desirable target, not a limit. If the goal is not achievable on the basis of technical feasibility, cost, and other factors, the regulatory agency may decide to accept a less stringent level. This strategy is reflected in certain EPA regulations, such as those pertaining to toxic air pollutants and to the cleanup of contaminated soil and groundwater. With two such opposite strategies, standards that ultimately result in comparable risk control may, on the surface, appear to be quite different.

Another challenge is the increasing frequency with which federal agencies incorporate considerations of ethical and social factors into the establishment of dose and risk limits. As noted previously (Chapter 13), such an approach may require consideration not only of the economic costs of serious health effects but also of less quantifiable factors, such as the equitable sharing of costs and benefits, perceived public aversion to the given contaminant at any exposure level, and the costs and benefits that could accrue to those outside the at-risk population. A further complication is the lack of interagency consensus on the amount of risk that is acceptable to the public and/or the environment. These types of unresolved issues raise serious questions about the precision, credibility, and overall effectiveness of federal standards and guidelines. Clearly, interagency guidance is needed to derive a structured approach to the process of incorporating cost and benefit considerations into various protective strategies.

Radiation Protection Standards

To provide insights that may be useful in developing occupational and environmental standards, the current approaches relative to protection against ionizing radiation are reviewed in the sections that follow. As will be noted, a framework has been established for coordinating radiation standards throughout the world, and a risk-based approach is being employed to establish dose-rate limits.

SCIENTIFIC BASIS

Shortly after the discovery of x rays in 1895, radiation injuries were reported. Recognizing the need for protection, physicists recommended limits on the allowable doses from x-ray generators. Their initial concern was to avoid direct physical symptoms. As early as 1902, however, scientists suggested that radiation exposures might also have latent effects, such as the development of cancer. This hypothesis was confirmed for external sources during the next two decades and for internally deposited radionuclides during the late 1920s, when bone cancers were reported among workers who applied radioactive luminous paints containing radium to the dials of clocks and watches (Rowland, 1994).

Initial recommendations for the control of exposures were developed on an informal basis. This changed in 1928 with the establishment of the International X-Ray and Radium Protection Committee (known today as the International Commission on Radiological Protection, ICRP) and the U.S. Advisory Committee on X-Ray and Radium Protection (known today as the National Council on Radiation Protection and Measurements, NCRP). One of the major benefits of the international committee and its successor is the forum it has provided for radiation protection experts from throughout the world to meet regularly, discuss the latest information on the biological effects of ionizing radiation, and formally propose appropriate radiation protection standards. Since there are close ties between the ICRP and the NCRP, the recommendations of the two groups tend to be in close conformity. Similar relationships exist among many other countries.

At about the time the ICRP and NCRP were formed, a report on experiments with *Drosophila* flies aroused concern about possible hereditary effects of radiation exposures in humans (Muller, 1927). This consideration shaped radiation protection guidelines and standards from about 1930 to 1960. While the initial focus was on dose-rate limits for occupational exposures, as public concern increased about worldwide exposures to fallout from atmospheric nuclear weapons tests, attention shifted to dose-rate limits for the public. This led to the establishment of the Federal Radiation Council in 1959 to develop U.S. policy on human radiation exposures and to establish limits for the public (FRC, 1960).

Within a decade after the end of World War II, epidemiological studies revealed an increase in leukemia among the survivors of the nuclear bombings of Japan; at the same time, however, these studies failed to demonstrate the anticipated hereditary effects. As a result, somatic effects, primarily leukemia, became the basis for radiation protection standards. In 1972 the Committee on the Biological Effects of Ionizing Radiation (BEIR) reported that solid tumors (cancers of the lung, breast, bone, and thyroid), not leukemia, were the dominant effects of human exposures to ionizing radiation (BEIR, 1972). Since that time, solid tumors have remained the primary basis for the development of radiation standards (BEIR, 1980, 1990).

In 1977 the ICRP broke new ground by proposing a mathematical system that permitted radiation protection standards to be based on what was considered an acceptable level of risk, namely, the probability of death due to the development of latent cancer (ICRP, 1977). As will be noted in the discussion presented later in this chapter, this approach was later modified to expand the risks by taking into account other factors such as the years of life lost (ICRP, 1991). The evolution of these radiation protection standards is summarized in Table 15.1.

Table 15.1 Evolution of the basis for dose limits for ionizing radiation, 1900–2005

Approximate period	Protection criteria
1900–1930	Avoidance of immediate physical symptoms
1930–1950	Avoidance of longer-term biological symptoms, plus concern for genetic effects
1950–1960	Concern for genetic effects (and leukemia)
1960–1970	Concern for somatic effects (primarily leukemia)
1970–	Concern for somatic effects (primarily solid tumors)
1980–	Application of a risk-based approach to radiation protection standards, taking into account latent cancer mortality
1990–	Expansion of risk approach to include additional effects, such as loss in life expectancy and effects of morbidity
2005–	Simplification of standards using dose rates from natural background as a basis for comparison

Table 15.2 ICRP recommendations for occupational whole-body equivalent dose limits, 1934–2005

	Dose limit		
Year	Per day	Per week	Per year
1934	0.2 roentgen		72 roentgens
1950		0.3 roentgen	
1958		0.1 rem	5 (N − 18) rem[a]
1965			5 rem (maximum)
1977			50 mSv;[b] based on acceptable risk[c]
1990			50 mSv; maximum of 100 mSv in any 5 years
2005			20 mSv per year; based on dose being no more than 10 times natural background[d]

a. N is age of worker receiving exposure.
b. 50mSv is equal to 5 rem.
c. Average dose to all workers recommended not to exceed 10% of limit for individual workers.
d. Exclusive of airborne radon and its decay products.

The ICRP recommendations for occupational whole-body external dose-rate limits from 1934 to 2005 are summarized in Table 15.2. Many of these recommendations, and those of the NCRP, have been incorporated into regulations (USNRC, 1991). As depicted in Figure 15.1, dose-rate limits for workers have undergone systematic reductions during the past past century. The continuing nature of this trend is exemplified by indications that the 1991 ICRP-recommended short-term occupational dose-rate limit, 50 mSv per year, will soon be reduced to 20 mSv per year. Except in the very early years, these changes were due not to observed health effects but to the availability of improved control technologies and a better understanding of the associated risks.

PHILOSOPHICAL APPROACH

Based on its experience, the ICRP has concluded that the procedures available for the control of ionizing radiation exposures, if properly applied,

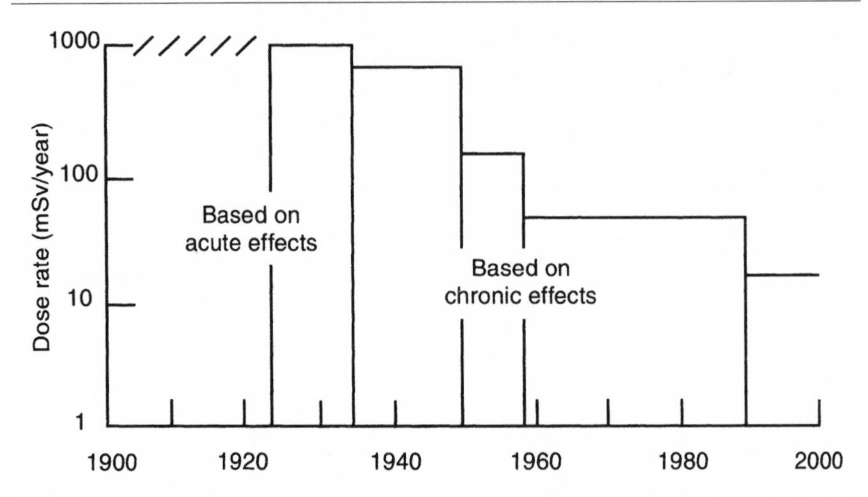

Figure 15.1 Changes in the limits for occupational exposures to ionizing radiation, 1900–2000. The limits since 1925 have been based on recommendations of the International Commission on Radiological Protection and its predecessors.

are sufficient to ensure that the impacts of the accompanying doses remain a minor component of the spectrum of risks to which people are exposed. Of necessity, however, the ICRP recognizes that the basic framework for protection involves judgments pertaining to both social and scientific issues. Therefore, the primary goal is to provide an appropriate standard of protection without unduly limiting the beneficial practices that give rise to radiation exposures. In many cases, this involves maintaining a balance between the benefits to the individual and those to society. When the benefits and detriments do not have the same distributions throughout the population, there is bound to be some inequity. As a result, balancing one versus the other is not a simple process (ICRP, 1991). From a scientific point of view, the goal in establishing radiation protection standards is to prevent the occurrence of deterministic (acute) effects by maintaining doses below the relevant thresholds and to ensure that all reasonable steps are taken to maintain the estimated potential induction of stochastic (latent) effects at an acceptable level. This requires that each application of radiation be justified, that is, it have a positive net benefit; that it be optimized, that is, all exposures must be kept as low as reasonably achievable, economic and social factors being taken into account; and that doses to individuals not exceed the established limits (ICRP, 1991).

CURRENT LIMITS AND THE CONCEPT OF EFFECTIVE DOSE

Dose-rate limits for radiation workers, as recommended by the ICRP, are summarized in column 2 of Table 15.3. To permit some degree of flexibility, limits for workers as well as members of the public (column 3) are provided for short-term (one-year) as well as for longer-term (five-year) exposures. As noted, the limits are expressed in units of the *effective dose.* In reality, this is a dose unit that must be calculated; it cannot be measured. Nonetheless, it is a concept that provides a mechanism for converting a dose to a portion of the body into one to the whole body that would carry with it an equivalent risk. This is necessary because the dose limits for radiation workers and members of the public apply to, and must be expressed in units of, the total dose from both external and internal sources. Without first converting the dose to portions of the body into the equivalent dose to the whole body, one would be, in essence, adding "apples and oranges." Prior conversion of the doses to portions of the body into equivalent whole body doses solves this problem. As will be illustrated in the examples that follow, the effective dose that results from such conversions is far less than the corresponding dose to a single organ or tissue.

For a variety of reasons, limits for members of the public are lower than those for radiation workers (NCRP, 1971). This takes into account the fact that the public includes pregnant women, infants and children, the sick, and the elderly, each of whom may represent a group at increased risk; for example, their metabolism and breathing rates may be different from those of an adult radiation worker. Also, members of the public may be

Table 15.3 Occupational and population dose limits

		General population	
	Radiation workers	Individuals	Total population
Annual limit	50 mSv	5 mSv	
Cumulative limit	20 mSv per year averaged over any 5 years	1 mSv per year, averaged over any 5 consecutive years	<< 1 mSv per year
Relative magnitude	100%	5%	<<5%

exposed 24 hours a day, 7 days a week, for their entire lifetime, whereas workers are generally exposed only during their adult lifetime and presumably only during the time they are on the job. Members of the public also have no choice about their exposure to most environmental sources, and they may receive no direct benefit from that exposure. In addition, they are not subject to the selection, supervision, and monitoring afforded radiation workers. As noted in column 4 of Table 15.3, the average annual dose rate to the population as a whole is expected to be well below the limits for individuals.

RISK OF CANCER FATALITY VERSUS TOTAL DETRIMENT

Early on, the *effective dose* was estimated on the basis of the risk of a latent cancer fatality only (ICRP, 1977). Recognizing that this did not account for the full range of impacts, the ICRP subsequently modified the system to base the effective dose on what was called the *total detriment,* that is, the risks due to mortality and morbidity, potential hereditary effects, years of life lost, and associated health-care and psychological impacts (ICRP, 1991). The basic goal in both cases was to develop what are called *tissue-weighting factors,* which provide a mechanism for calculating effective doses.

In addition to the fact that doses from ingested and inhaled radionuclides commonly affect only portions of the body, as contrasted to the whole body, they differ from external doses in other ways. Whereas external doses cease as soon as the radiation source is removed or the person leaves the area, doses due to internally deposited radionuclides continue until they either decay and/or are biologically eliminated. To account for this fact, the ICRP developed the concept of *committed dose.* In the case of workers, this includes the estimated dose that would be received (accumulated) over the next 50 years following intake; for members of the public, the committed dose includes the estimated dose that would be accumulated from the time of intake to age 70 (ICRP, 1996). This takes into account the fact that members of the public may be exposed from the beginning of life, whereas most workers are not exposed until they are in the range of 20 years of age. Since it is the *committed dose* that is recorded as having been received, it is important to recognize that this approach results in estimated doses that are higher than those that will actually occur. This is especially the case for radionuclides that have long biological and radioactive half-lives, the reason being that workers and adult members of the public who take in such radionuclides in later life will not live long enough to receive the additional 50 years of exposure.

CONCEPT OF COLLECTIVE DOSE

To explain the methodology for calculating tissue-weighting factors and, in turn, effective doses, it will first be necessary to recall the earlier discussion (Table 12.3, Chapter 12) of the concept of *collective dose,* a term that was developed to express the integrated dose to a given population group. As such, it is routinely applied in interpreting the outcomes of epidemiological and biological studies. Nonetheless, certain precautions need to be observed. This dose unit should be applied only if the relationship between the dose and its resulting biological effects can be assumed to be reasonably linear, and the doses are sufficiently high to be statistically significant. In accord with these limitations, the concept is not applicable if the doses to individuals in the target population group are either very high or very low. For example, if 1,000 people each received 10 Sv, the collective dose would be 10,000 person-Sv, yet such a dose received over a short period would be fatal to all members of the exposed group (Table 12.4, Chapter 12). At the other extreme, if 1 billion people each received 10 μSv, the collective dose would likewise be 10,000 person-Sv; however, it would be next to impossible to demonstrate any excess ill effects among a group of this size who received such a small dose.

APPLYING THESE CONCEPTS: FATAL CANCERS ONLY

As an initial step, the number of excess deaths (the fatal cancer probability coefficient) will be estimated for a population group that has received a known dose to an individual organ or tissue. Four examples are described—lung cancer, leukemia, breast cancer, and thyroid cancer. In all cases, and in accord with the limitations expressed earlier, the estimates will be based on data derived from epidemiological studies involving people who received significant doses of ionizing radiation.

Lung cancer. Studies of underground uranium miners and survivors of the nuclear bombings in Japan indicate that if 20,000 people have each received 0.5 Sv (a collective dose of 10,000 person-Sv) to their lungs, about 90 of them will subsequently develop lung cancer. Medical experience has shown that about 95 percent of the lung cancers induced by ionizing radiation are fatal. This is referred to as the cancer lethality fraction (Table 15.4, column 2). On this basis, the risk of death from cancer due to radiation exposures to the lungs would be about

$$\left(\frac{90 \text{ excess cancers}}{10{,}000 \text{ person-Sv}}\right) (0.95 \text{ fatality rate}) = 85 \times 10^{-4}/\text{Sv}.$$

Table 15.4 Fatal cancer and total detriment probability coefficients, tissue-weighting factors, and annual occupational dose limits for individual tissues and organs

Tissue or organ	Cancer lethality fraction	Fatal cancer probability coefficient (10^{-4}/Sv)	Total detriment probability coefficient (10^{-4}/Sv)	Tissue-weighting factor(w_T)[a]	Annual equivalent dose limit (mSv)
Gonads	—	—	133		
Ovary	0.70	10	15	0.20[b]	100
Bone marrow	0.99	50	104	0.12	150
Colon	0.55	85	103	0.12	150
Lung	0.95	85	80	0.12	150
Stomach	0.90	110	100	0.12	150
Bladder	0.50	30	29	0.05	400
Breast	0.50	20	36	0.05	400
Liver	0.95	15	16	0.05	400
Esophagus	0.95	30	24	0.05	400
Thyroid	0.10	8	15	0.05	400
Bone surface	0.70	5	7	0.01	500[c]
Skin	0.002	2	4	0.01	500[c]
Remainder[d]	0.80	50	59	0.05	400
Total		500	725	1.00	

a. To avoid implications of accuracy beyond what the biological data will justify, the tissue-weighting factors for the various tissues and organs have been assigned one of four values: 0.01, 0.05, 0.12, or 0.20.

b. Total for the gonads (including cancer in the ovaries).

c. Based on deterministic effects.

d. The equivalent dose for the remaining body organs is the estimated mean equivalent dose over the whole body excluding the specified tissues and organs.

This number has been designated by the ICRP as the fatal cancer probability coefficient for lung cancer (Table 15.4, column 3).

Bone marrow (leukemia). Studies of the survivors of the World War II nuclear bombings in Japan indicate that if 50,000 people have each received a dose of 0.2 Sv (a collective dose of 10,000 person-Sv) to their bone marrow, after a latency period about 50 excess cases of leukemia will develop in this group. Assuming that 99 percent of these cases are fatal (column 2), the risk of death (fatal cancer probability coefficient) from leukemia due to exposures of the bone marrow would be about

$$\left(\frac{50 \text{ excess cancers}}{10{,}000 \text{ person-Sv}}\right) (0.99 \text{ fatality rate}) = 50 \times 10^{-4}/\text{Sv}.$$

Breast cancer. Epidemiological data indicate that radiation exposures of the breast produce in women an excess of about 80 breast cancers per 10,000 person-Sv. Assuming that breast cancer is fatal 50 percent of the time and that the exposed population consists of 50 percent men (who seldom develop breast cancer) and 50 percent women, the fatal cancer probability coefficient due to breast cancers developing in such a population would be

$$\left(\frac{80 \text{ excess cancers}}{10{,}000 \text{ person-Sv}}\right) (0.5 \text{ fatality rate}) (0.5 \text{ of population}) = 20 \times 10^{-4}/\text{Sv}.$$

Although this example pertains solely to breast cancer, women also experience higher risks from radiation exposure due to cancer of the ovaries. While one might be tempted to conclude that this difference is balanced by the fact that men are subject to an excess risk of testicular cancer, the associated risk is so low that the ICRP includes such cancers in the category of "remainder organs" (Table 15.4). Much the same is true for prostate cancer. In fact, epidemiological studies to date have not shown any increase in cancer in this organ due to radiation exposure. Overall, the risk of fatal cancer in women is up to 20 percent higher than that for men (NRC, 1990; ICRP, 1991). Taking this and other considerations into account, the NCRP has recommended that the accumulated dose limits for female astronauts be approximately 60 percent of those for males (Table 15.5). Pregnant female radiation workers have the added responsibility of ensuring that the fetus is not excessively exposed. For these and other reasons, it might have been appropriate for the ICRP to have placed an asterisk by the breast cancer risk coefficient to acknowledge the assumptions used in its calculation.

Thyroid cancer. Radiation exposures of the thyroid produce an excess of about 80 thyroid cancers per 10,000 person-Sv. Assuming a fatality rate of 10 percent, the fatal cancer probability coefficient would be

Table 15.5 Ten-year career dose limits for astronauts in low-Earth orbit

Age at exposure (years)	Effective dose limit (Sv)[a]	
	Female	Male
25	0.4	0.7
35	0.6	1.0
45	0.9	1.5
55	1.7	3.0

a. These limits are based on an excess lifetime risk of cancer mortality of 3 percent.

$$\left(\frac{80 \text{ excess cancers}}{10{,}000 \text{ person-Sv}}\right)(0.10 \text{ fatality rate}) = 8 \times 10^{-4}/\text{Sv}.$$

Similar calculations can be made to estimate the excess deaths that result from exposures of other body organs, including deaths due to cancers of the reproductive organs (Table 15.4, column 3). The sum of the fatal cancer probability coefficients for radiation-induced cancers in all body tissues and organs is 500×10^{-4}/Sv, the value currently used in radiation protection analyses.

APPLYING THESE CONCEPTS: TOTAL DETRIMENT

As would be anticipated, the probability coefficients for total detriment will be higher than those for fatal cancer alone. Also, as will be noted later, the relative increase in the associated probability coefficients will be less for a cancer that has a high, as contrasted to a low, cancer lethality fraction. Similarly, the increase will be less for a cancer that results in fewer, as contrasted to more, years of productive life lost. For purposes of illustration, two examples are presented: lung cancer, which has a high lethality fraction, and thyroid cancer, which has a low lethality fraction. In these two cases, the years of life lost due to a cancer death (Table 15.6) are approximately the same. As will be noted, the years of life are normalized by assigning a value of 1.00 to 15.0 years, the estimated value for deaths due to cancers of the bone surface, liver, skin, and thyroid.

Lung cancer. Given that the percentage of lung cancers considered to be lethal—that is, the cancer lethality fraction—is 95 percent (Table 15.4, column 2), 5 percent are assumed not to be lethal. In the modification of the fatal cancer probability coefficient for lung cancer (calculated earlier) using the methodology developed by the ICRP (1985), the probability coefficient for total detriment becomes the product of the fatal cancer probability coefficient times 1.00 plus the percentage of lung cancers that are not fatal (0.05), multiplied by a factor of 0.90 to account for the years of life lost (Table 15.6, column 3):

$$(85 \times 10^{-4}/\text{Sv})(1.00 + 0.05)(0.90) = 80 \times 10^{-4}/\text{Sv}.$$

Here the adjustment in the fatal cancer probability coefficient is relatively small.

Thyroid cancer. As noted earlier, 90 percent of the thyroid cancers induced by radiation are considered not to be lethal. In the modification of

Table 15.6 Years of life lost and adjustment factors for various cancers induced by ionizing radiation

Type of cancer	Years of life lost	Adjustment factor
Bladder	9.8	0.65
Bone marrow	30.9	2.06
Bone surface	15.0	1.00
Breast	18.2	1.21
Colon	12.5	0.83
Esophagus	11.5	0.77
Gonads	20.0	1.33
Liver	15.0	1.00
Lung	13.5	0.90
Ovary	16.8	1.12
Skin	15.0	1.00
Stomach	12.4	0.83
Thyroid	15.0	1.00
Remainder	13.7	0.91

the fatal cancer probability coefficient for thyroid cancer, the probability coefficient for total detriment becomes the product of the fatal cancer probability coefficient times 1.00 plus the percentage of thyroid cancers that are not fatal (0.90), multiplied by a factor of 1.00 to account for the years of life lost (Table 15.6, column 3):

$$(8 \times 10^{-4}/\text{Sv})(1.00 + 0.90)(1.00) = 15 \times 10^{-4}/\text{Sv}$$

In this case, the adjustment in the fatal cancer probability coefficient to reflect the total detriment is relatively large.

Probability coefficients for total detriment for other organs within the body can be calculated in the same way. The results of these calculations are recorded in Table 15.4, column 4.

CALCULATION OF TISSUE-WEIGHTING FACTORS

As noted in column 4 of Table 15.4, the sum of the total detriment probability coefficients associated with all types of cancers in all organs of the body, plus an allowance for potential hereditary effects, is about $725 \times 10^{-4}/\text{Sv}$. By using this value, it is possible to calculate tissue-weighting factors, which in turn can be used to estimate not only the whole-body dose that carries with it the same risk as a higher dose to a single organ within the body, but also the dose-rate limits for individual body organs.

As calculated, the total detriment probability coefficient for lung cancer is $80 \times 10^{-4}/\text{Sv}$. Therefore, cancer of the lungs represents 80/725, or about 12 percent, of the sum of the values of the total detriment probability coefficients for each of the various body organs. In other words, if the entire body were exposed to ionizing radiation, the risk and consequences of developing lung cancer would be about 12 percent of the total risk arising through the development of cancers in all organs and the potential probability of hereditary effects. This value, 0.12, is called the tissue-weighting factor (w_T) for the lung.

Similarly, the total detriment coefficient for thyroid cancer is $15 \times 10^{-4}/\text{Sv}$. Therefore, thyroid cancer represents 15/725, or about 2 percent, of the total detriment. Because this value and the prior estimate for the lung imply more accuracy than is justified by the biological data, the ICRP has recommended that tissue-weighting factors for the various organs be assigned one of four values: 0.01, 0.05, 0.12, and 0.20. In this approach, the value assigned to the thyroid is 0.05. Tissue-weighting factors for all the major organs of the body are listed in Table 15.4, column 5.

CALCULATION OF DOSE LIMITS FOR INDIVIDUAL ORGANS

By applying the tissue-weighting factors, it is possible to estimate the effective dose (whole-body equivalent dose) that corresponds to the higher doses received by individual body organs. This is accomplished by multiplying the dose to each individual organ by its tissue-weighting factor. Applying this process in reverse, it is possible to set dose-rate limits for individual organs that are comparable (in terms of risk) to the limit for the whole body. For the lungs, the procedure is to divide the longer-term ICRP dose-rate limit (20 mSv per year) by the relevant tissue-weighting factor (0.12). On this basis, the dose-rate limit for a situation in which the lungs, and only the lungs, were exposed, would be

$$\frac{20 \text{ mSv}}{0.012} = \sim 150 \text{ mSv per year.}$$

A similar annual limit would apply to the bone marrow, colon, and stomach, again on the assumption that each one of these is the only portion of the body that is being exposed at a given time.

Likewise, the dose rate for the thyroid that is comparable to 20 mSv per year to the whole body would be

$$\frac{20 \text{ mSv}}{0.05} = 400 \text{ mSv per year.}$$

A similar limit would apply to the bladder, breast, liver, esophagus, and "remainder" organs.

Although this same approach could be used in estimating a dose-rate limit for the skin and bone surface, the number would be so high as not to be acceptable, that is, the associated dose rate would produce deterministic (acute) effects in that organ. For this reason, the ICRP recommends that the lifetime dose to any body organ for a radiation worker be limited to no more than 20 Sv. If a 40-year working lifetime is assumed, this would amount to 500 mSv per year (Table 15.4, column 6).

New Philosophy and Approach

During the mid 1990s, the ICRP realized that its existing radiation protection recommendations were not only more complicated than necessary, but were also not "user friendly." To correct the situation, it began exploring a new approach in the development of such standards. In so doing, it made a special effort to conduct the process in the open and adopted as a basic philosophy that protection of the individual member of society should be the primary goal because if this goal were accomplished, society as a whole would also be protected. Additionally, it decided to express all limits not only in units of dose rates but also in terms of how these compare to those from natural background radiation (Chapter 12). This information is summarized in Tables 15.7 and 15.8 (Clarke, 2001a, 2001b). For purposes of comparison, the average worldwide dose rate from natural background, exclusive of the contribution

Table 15.7 Individual dose scale (classification system for radiation dose limits proposed by ICRP for 2005)

Classification	Annual effective dose (mSv)	Multiples of dose from natural background[a]
Serious	30–300	30–300X
High	3–30	3–30X
Moderate	0.3–3	0.3–3X
Low	0.03–0.3	0.03–0.3X
Trivial	<0.03	<0.03X

a. Exclusive of exposures from airborne radon and its decay products.

Table 15.8 Protective action levels (proposed by ICRP for 2005)

Category	Multiples of dose from natural backgrounda[a]	Annual effective dose limit (mSv)
Radiation workers	~10X	<20
Intervention to protect the public in an emergency	~10X	<20
Releases into the environment	~0.1X	<0.3
Exemption from regulation	~0.01X	<0.01

a. Exclusive of exposures from airborne radon and its decay products.

from airborne radon and its decay products, is about 1 mSv per year (Chapter 12, Table 12.7).

The General Outlook

This review of U.S. environmental standards clearly demonstrates the need to harmonize the methodologies for developing, and the procedures for applying, occupational and environmental standards. This is true in terms of methods for calculating doses to the public and estimating the associated risks, as well as for determining what levels of risk are acceptable. One of the immediate benefits of a harmonized system would be an enhancement in the cost-effectiveness of environmental controls. Indeed, if all standards had a common risk basis, the current practice of trade-offs (Chapter 13) among various contaminants might be expanded to include trade-offs among releases of the same contaminant to different components of the environment. Ultimately, it might even become possible to effect trade-offs among different types of contaminants and/or stresses—for example, reducing releases of chemical or radiological contaminants to compensate for biological and/or physical stresses. Because of the wide range of associated implications, however, trade-offs of occupational and environmental stresses in this manner would require considerable review, evaluation, and deliberation.

Although it is essential that such harmonization be accomplished among various federal, state, and local agencies within the United States, the effort should not end there. It needs to be expanded so that ultimately all countries of the world use the same approaches and methodologies in establishing environmental standards. Only with such uniformity can one be sure that evaluations of the risks and applications of control of environmental contaminants are conducted on a comparable basis from country to country. One immediate advantage internationally would be that the exportation of hazardous industries from one country to another to escape regulatory controls would no longer be a viable choice. Before such a goal can be achieved, however, there is a need to ensure that the standards within individual countries, most especially the United States, have been harmonized. As noted in Figure 15.2, there are orders-of-magnitude differences in the lifetime risks associated with the existing U.S. limits for ionizing radiation, depending on the source.

The continuing public misconception that compliance with environmental standards will assure risk-free conditions also needs to be ad-

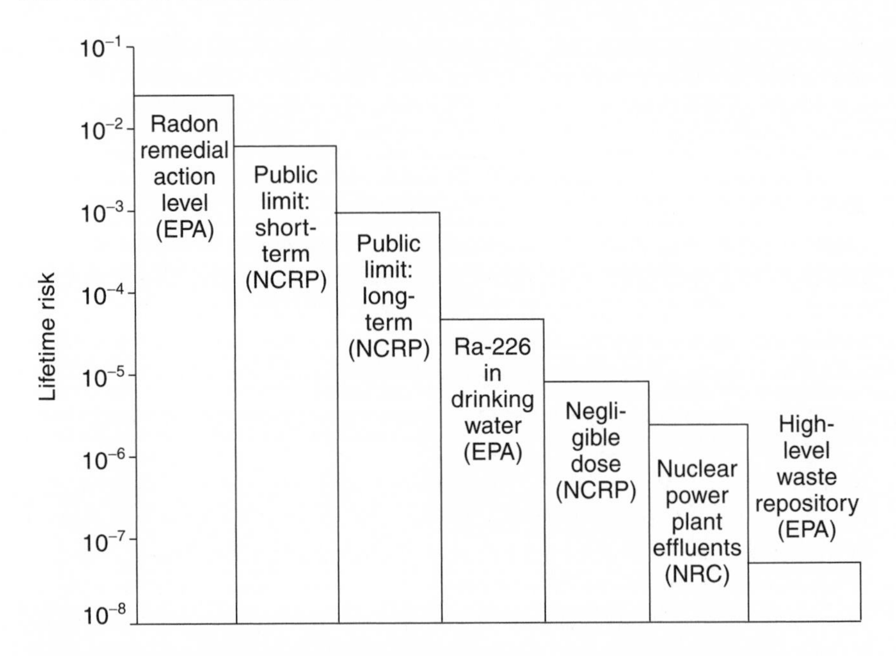

Figure 15.2 Comparison of U.S. standards (lifetime risk) for exposures to various sources of ionizing radiation

dressed. Although many standards are expressed as single, numerical limits, organizations such as the ACGIH stress that in the case of occupational exposures to airborne toxic chemicals, these limits do not represent "fine lines between safe and dangerous concentrations," nor do they represent "a relative index of toxicity" (Chapter 4). Similar qualifying statements accompany the limits recommended by the ICRP (1991) and the NCRP (1993).

One approach for overcoming some of these problems is to provide members of the public, particularly the stakeholders, an opportunity to be involved in the standards-setting process so that they will understand both the goals being sought and the means by which they are to be achieved. Experience has repeatedly demonstrated that although individuals routinely accept risks, such as driving a car, on a voluntary basis, they balk at accepting risks that have been imposed on them without their consent and especially without their knowledge. The approaches that the ICRP is taking (noted earlier) by developing its recommendations in a fully open manner can well serve as a model for other groups to follow. Also to be applauded are the efforts this organization has devoted to ensuring that the resulting standards are expressed in a manner that will provide perspective, as well as be readily understood (Clarke, 2000).

Another area that is in need of increased attention is the development of standards for specific age groups, particularly children. Again, through its development of age-dependent dose coefficients for the intakes via ingestion and inhalation of more than 600 radionuclides, the ICRP has been a leader in this area. Through its efforts, individual dose coefficients are now available for infants, 1-, 5-, 10-, and 15-year-olds and for adults (ICRP, 1996). Where available, age-specific biokinetic data were included in the calculations through which the estimates were developed. Consideration should be given by other organizations to the development of similar information for a range of toxic chemicals, as well as physical and biological stresses.

16

MONITORING

ENVIRONMENTAL monitoring programs were initially conducted on a local basis and had two basic objectives: (1) to estimate exposures to people resulting from certain physical stresses (such as noise and radiation) and from toxic materials that are being, or have been, released and are subsequently being ingested or inhaled and (2) to determine whether the resulting exposures complied with the limits prescribed by regulations. Such programs were either "source" related or "person" related. *Source-related* monitoring programs were designed to determine the exposure or dose rates to a specific population group resulting from a defined source or practice. *Person-related* programs were designed to determine the total exposure from all sources to a specific population group. The latter were particularly useful in instances where several sources contributed to the exposures.

Although programs of these types continue to be important, it is increasingly recognized that assessing risks solely to human health and focusing on problems only on a local scale are inadequate. The purposes and goals of environmental monitoring programs today have expanded far beyond these earlier objectives. Significantly, it is now accepted that some of these programs should have an *environment-related* component, and that conditions should be examined on a regional, local, and global basis. That is, programs should be designed to assess the impact of various contaminants on selected segments of the environment, including ecosystems, and to evaluate factors that may have wide-scale, long-range effects. The types and purposes of current environmental monitoring programs are summarized in Table 16.1.

Table 16.1 Types and purposes of environmental monitoring programs

Type of Program	Purpose
Based on nature of the stress	
Physical stress	To assess the impact of environmental stresses such as noise and external radiation, where the evaluation is based primarily on exposure measurements made in the field, not on samples collected and returned to the laboratory for analysis
Chemical stress	To assess exposures resulting from the ingestion and inhalation of chemical and radioactive contaminants
Based on geographic (spatial) coverage	
Local	To evaluate the impact of a single facility on the neighboring area
Regional	To evaluate the combined impact of emissions from several facilities on a large area
Global	To determine worldwide impacts and trends, such as acidic deposition, depletion of the ozone layer, and potential for global warming
Based on temporal considerations	
Preoperational	To determine potential contamination levels in the environment prior to operation of a new industrial facility; to train staff; to confirm operation of laboratory and field equipment
Operational	To provide data on releases; to confirm adequacy of pollution controls
Postoperational	To assure proper site cleanup and restoration
Based on monitoring objectives	
Source related	To determine population exposures from a single source
Person related	To determine total exposure to people from all sources
Environment related	To determine impacts of several sources on features of the environment such as plants, trees, buildings, statues, soil, water, and ecosystems

Table 16.1 Types and purposes of environmental monitoring programs (continued)

Type of Program	Purpose
Research related	To determine transfer of specific pollutants from one environmental medium to another and to assess their chemical and biological transformation as they move within the environment; to determine ecological indicators of pollution; to confirm that the critical population group has been correctly identified and that models being applied are accurate representations of the environment being monitored
Based on administrative and legal requirements	
Compliance related	To determine compliance with applicable regulations
Public information	To provide data and information for purposes of public relations

Local environmental monitoring programs for industrial facilities are generally conducted by plant personnel or environmental service contractors, whereas regional programs are the responsibility of state and local environmental health and regulatory authorities. The planning and coordination of national programs are handled by federal agencies. Close coordination between the facility operator and the local agencies is necessary if all objectives of the monitoring program are to be met. A well-planned program will usually involve some overlap in the activities of the several monitoring groups, including exchanges of samples and cross-checking of data.

Monitoring Physical Stresses and Toxic Materials

Because of differences in the nature of their exposures, the monitoring of physical stresses and of toxic materials requires significantly different approaches. In the former, monitoring may simply involve identifying their sources and measuring their magnitude. Expanding the scope to include measurements of the distribution of their energies can provide data for estimating the accompanying dose as a function of tissue depth and specific body organ. For some stresses, such as electric and magnetic fields, where there is a lack of information on which attributes of the sources may cause harm, there will be questions as to what types of measurements should be made. Even where the source and effects of a potential physical stress have clearly been identified, challenges will remain.

Measurements to determine exposures from physical stresses, such as noise and ionizing and nonionizing radiation, must commonly be made on a real-time basis. Generally, monitoring instruments are placed near the people being exposed or in concentric rings at various distances from the source. It is important to recognize, however, that the presence of people and monitoring equipment may alter the environment in such a way as to make accurate measurements (say, of electric or magnetic fields) difficult. In addition, the position and location of the people being exposed (for example, whether they are standing on the ground or near a tree or sitting inside an automobile) can nullify the usefulness of the resulting data.

In the case of airborne or waterborne releases of toxic materials, the first step is commonly to measure discharges at the points of release. Additional steps include assessing the movement or transport of specific contaminants within given environmental media (air, water, soil), their transfer from one medium to another, and changes in their physical, chemical and biological characteristics as they interact with various components within the environment (Figure 16.1). Information on such changes, is necessary for estimating their deposition within the body and uptake by various organs. This information, in turn, can be used to estimate the accompanying doses to people. Because many contaminants can cause exposures by several avenues, most environmental monitoring specialists try to identify and trace the movement and behavior of a range of potential contaminants through several environmental pathways so as to identify those contaminants and pathways that are most important.

From the standpoint of health effects, the physical and chemical properties of airborne particles and gases are noteworthy since they are the primary factors that determine their behavior. In the case of particles, their size is very important since it will significantly affect where they deposit in the respiratory tract (Chapter 5), and this, combined with their chemical composition, will determine their movement within the body and potential effects on health. It is also vital to know what other chemicals are associated with a given contaminant, since certain combinations are synergistic. For example, sulfur dioxide, an ubiquitous acidic gas that is

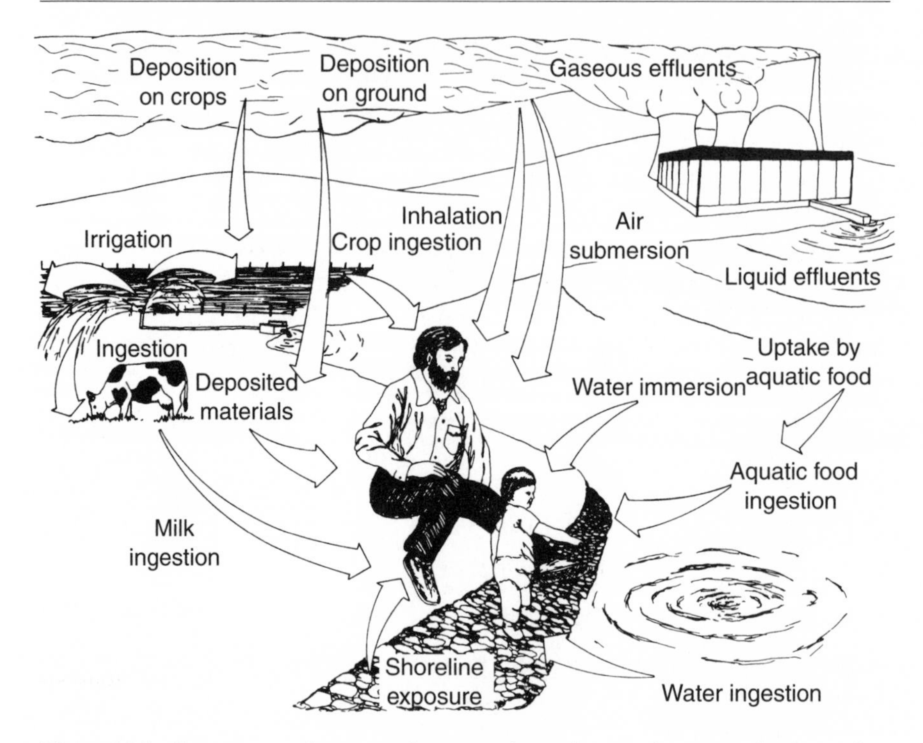

Figure 16.1 Exposure pathways to humans through aquatic, atmospheric, and terrestrial pathways

highly soluble and is ordinarily taken up entirely in the throat and upper airways (where its effects on health are minimal), acutely impairs the functioning of the lungs when it is carried to the alveoli as an acid condensed on the surfaces of small airborne particles. In a similar manner, as noted earlier (Chapter 5), oxides of nitrogen and volatile organic compounds can interact to form ozone which has entirely different toxic effects.

On the basis of these discussions, it can be seen that exposures to physical stresses occur at the source and that quantification of the degree of stress is dependent on measurements made at that location on a real-time basis. The only people subject to the stress will be those located in the vicinity of the source. In contrast, the evaluation of toxic material stresses often involves collecting samples in the field and transporting them to a laboratory for analyses. It is not until these steps have been completed that their potential impacts can be evaluated. Because of the nature of the multiple pathways through which toxic chemicals can interact with people, the accompanying exposures may occur in a variety of settings and at a variety of sites. In this case, however, human exposures will occur only when and if the contaminants interact with people.

Measuring Waterborne and Airborne Exposures

As noted earlier, one of the first steps in assessing potential exposures is to measure the concentrations of individual contaminants in samples of typical releases from the polluting facility. In most cases, air and water serve as the principal pathways for direct exposures (through inhalation and the consumption of drinking water) and as a vehicle for the transport of contaminants from the point of release to other environmental media (such as milk and food). Measurements of the airborne and waterborne contaminants that are leaving a plant can also provide advance information on pending problems in other environmental media. Since critical contaminants can be missed if only the obvious and easily measured effluents are monitored, or if monitoring ceases during key periods such as shutdowns for repairs and maintenance, sample collection and analysis should be conducted during all phases of plant operations.

ASSESSING WATERBORNE RELEASES

A range of samples can be collected to assess the impact of waterborne releases. These include the following:

Grab samples. Collected on a one-time basis, these represent at best a snapshot of the characteristics of the waste. Unless its composition is relatively uniform with time, such samples will not provide useful information.

Composite samples. A blending of a series of smaller samples, these represent only a combination of the characteristics of the waste at the times of collection. Nonetheless, they are better than a single grab sample.

Timed-cycle samples. Collected in equal volumes at regular intervals, these represent the characteristics of the waste only if the flow rate is constant and the characteristics of the waste are relatively uniform.

Continuous flow-proportional samples. Collected on a continuous basis in proportion to the volume of flow, these should provide a sample that is representative of the waste.

Indicator samples. Consisting of various living organisms and plants, these can provide useful data, particularly in identifying contaminants whose concentrations in lakes and streams are below the limits of analytical sensitivity. Information on the history of releases can often be obtained through similar analyses of bottom deposits. Quantifying the amounts of the identified contaminants that are being released, however, can be difficult.

One of the primary advantages of a composite sample is that it minimizes the expense of analyzing liquid-waste streams that have a relatively uniform composition. For waste streams with a wide range of characteristics, oftentimes the collection and analysis of discrete samples is required to determine the temporal nature and concentration of various contaminants. As a general rule, the representativeness of the analytical data improves with sample-collection frequency. In highly variable conditions, samples should be collected as frequently as every five minutes to every hour. The prime consideration is the variability of the composition of the waste at the point of sampling.

The quantity of sample collected depends on the number and nature of the parameters being tested. The quantity should be sufficient to permit all desired analyses, with allowance for possible errors, spillage, and sample splitting for purposes of quality control. If, after collection, the sample is placed in a bottle for transport to the laboratory for analysis, care must be taken to assure that ionic species or small particles suspended in the waste do not attach themselves to the walls of the container. This problem can be avoided by an appropriate choice of bottle, adjusting the pH, or adding stabilizing chemicals to the sample prior to placing it in the container. Similar steps, including refrigeration, will assure that the sample is properly protected against deterioration due to either chemical or biological processes. Such preservation is particularly important when there is a lag between collection and analysis.

ASSESSING AIRBORNE RELEASES

Assessment of the impact of airborne releases generally requires the collection of samples from a variety of sites. As with liquid wastes, an initial step is to sample the various release points at the facility being monitored.

To estimate the upper bound of potential human health effects, the samplers should be located where airborne concentrations and ground deposition of contaminants are likely to be at a maximum. Selection of the sites should be based on the best available meteorological information, coupled with data on local land use. Sites selected for monitoring the impacts on various ecosystems will require a similar approach, but the exposed entity in this case is the environment, not people.

A frequently used sampling system employs a filter or electrostatic precipitator to collect airborne particles and an appropriate set of adsorbers to collect gaseous and volatile contaminants. Cascade impactors or other mechanical separation devices can be utilized to determine the size distribution of airborne particles. The choice of sampler depends on the desired sample volume, sampling rate, power requirements, servicing, and calibration. The minimum amount of air to be sampled is dictated by the sensitivity of the analytical procedure; the amount is often a balance between sensitivity and economy of time. As in any monitoring program, care must be taken to assure that the samples are representative.

At the same time, rapid advances in technology have led to a variety of new techniques for monitoring contaminants in real time on a localized basis. Examples include portable gas chromatographs that are now available for monitoring airborne contaminants in the field. One of the advantages of this technology is that it permits the analyst to identify and quantify the concentrations of organic compounds, such as benzene, toluene, and xylenes, on an individual basis. Because of their accuracy, instruments that incorporate gas chromatography are especially useful during remediation operations to ensure that airborne concentrations are within acceptable limits. Since the analysis is rapid, more samples can be processed, and the boundaries of contamination can be better defined. In some cases, as many as nine instruments can be linked by radio to a central command post at a site. Should the concentration of volatile organic compounds exceed a preset action level, this information will be automatically reported so that appropriate action can be taken (Ebersold and Barker, 2003).

Evaluations and measurements of air pollution on a localized basis can also be made through the use of optical remote-sensing technology. One example is the use of reflected infrared waves to measure the rate at which an atmospheric plume absorbs energy. By applying techniques similar to those used in medical CT scans (Chapter 12), this technology enables scientists to measure the concentrations of hazardous air emissions in multiple single-plane (sliced) images within a plume. These data are then analyzed by using computer models to determine the concentrations of contaminants both vertically and horizontally in order to map the dimensions of the plume and the distribution of pollutant concentrations within it. This technique has proved to be especially effective in analyzing fugitive emissions that previously could not be well characterized. It has proved similarly useful in monitoring area source emissions that are either too numerous to measure individually (for example, those from motor vehicles) or are released from a single source over a large area (for example, large municipal landfills). Other applications include refineries, where this technique permits expensive and troublesome regulation of fugitive releases to be replaced by a "cap" for the entire facility that, in turn, can be monitored continuously (EPA, 2001). The advantages and disadvantages of various sampling methods for the principal types of environmental contaminants and receptor media are summarized in Table 16.2

MONITORING THE INDOOR ENVIRONMENT

While the primary attention in this chapter is on monitoring the ambient (outdoor) environment, there is also a need to monitor the indoor environment, particularly for various airborne contaminants such as those generated by molds. As previously noted (Chapter 5), molds can grow on any moist substance and their source is frequently hidden from view. Because of the contrasting nature of other airborne contaminants that may be present, confirmation of the presence of molds generally requires special sampling techniques. One common method is to collect them using impaction devices that deposit them onto media, such as agar, for subsequent culturing. Other aerosols, such as toxic chemicals, can be collected by passing samples of the air through absorbing liquids. Inhalable particles can be collected on filters.

Whether one is monitoring the indoor or outdoor environment, it is often difficult to interpret the observations in terms of doses to those exposed. Few people, for example, spend significant amounts of times at specific locations either indoors or outdoors. For this reason, increasing efforts have been devoted to the development of methods to evaluate the concentrations of contaminants in the air actually being breathed by people. One of the more significant advances has been the development of personal samplers that can be worn by individual members of the public and are designed to evaluate the quantity of various contaminants being inhaled (Chapter 5). Such samplers are similar, in terms of the data they provide, to the personal monitoring devices available for assessing

Table 16.2 Advantages and disadvantages of various environmental sampling methods

Type of sample	Advantages	Disadvantages
	Atmospheric environment	
Direct measurement		
Real-time field measurements of physical stresses such as noise and radiation	Monitors can be put in place to assess time-integrated exposures	Monitor often disturbs field being monitored; some monitoring equipment (e.g., for assessing electric and magnetic fields) is expensive and complex
Airborne particulates		
Respirable fraction via air sampling	Direct-dose vector; provides data on potential effects on lungs	Omits larger particles that may be significant when deposited in nose, mouth, and throat
Total particulates via air sampling	Provides data for assessing doses to lungs as well as possible effects on skin and intake through ingestion	Not all measured contaminants are respirable
Collection of settled particulates	Represents an integrated sample over known time and geographic area	Weathering may alter results; only large particles are collected by sedimentation
Gases		
Integrated (concentrated) sample	Concentration of samples permits detection of lower concentrations in-air	Samples must usually be analyzed in laboratory; chemical reactions may change nature of collected compounds
Direct measurement	Provides data on real-time basis	Lower limit of detection may not be adequate
	Terrestrial environment	
Milk	Direct-dose vector, especially for children; data easily interpreted	Milk samples are not always available

Table 16.2 Advantages and disadvantages of various environmental sampling methods (continued)

Type of sample	Advantages	Disadvantages
Foodstuffs	Direct-dose vector; data easily interpreted	Samples are not always available from areas of interest; weathering and processing may affect samples
Wildlife	Direct-dose vector	High mobility; not always available; data difficult to interpret
Vegetation	Samples readily available; multiple modes available for accumulating contaminants (by direct deposition and leaf and root uptake)	Data are difficult to interpret; weathering can cause loss of contaminants; not available in all seasons
Soil sampling	Good integrator of deposition over time	High analytical cost; data difficult to interpret in terms of population exposure and dose
	Aquatic environment	
Surface water (nondrinking)	Readily available; indicates possibility of contamination by aquatic plants and animals	Not directly dose related; data difficult to interpret
Groundwater (nondrinking)	Indicator of unsatisfactory waste-management practices	Not always available; data difficult to interpret because of possibility of multiple remote sources
Drinking water	Direct-dose vector; consumed by all population groups	Contaminant concentrations are frequently very low
Aquatic plants	Sensitivity	Data difficult to interpret; not available in all seasons
Sediment	Sensitivity; good integrator of past contamination	Data difficult to interpret because of possibility of multiple remote sources

Table 16.2 (continued)

Type of sample	Advantages	Disadvantages
Fish and shellfish	Direct-dose vector; sensitive indicator of contamination	Frequently unavailable; high mobility
Waterfowl	Direct-dose vector	Frequently unavailable; high mobility; data difficult to interpret

doses from external radiation sources. Personal samplers have been developed to collect airborne gases and particulates or combinations of the two. Important characteristics of such samplers are that they must have minimal power requirements and be relatively quiet and lightweight.

Designing an Environmental Monitoring Program

One of the first steps in designing an environmental monitoring program is to define its objective (what samples are to be collected and where and when) and how the data are to be analyzed (Table 16.3). The program must be planned not only so that the relevant questions are asked at the right time, but also so that the data necessary to answer these questions are obtained. As a result of the increasing sophistication of our understanding of the environment and accompanying technological developments, it is mandatory that these questions be addressed on a continuing basis. In fact, the answers to them will never become final. Other attributes of a successful environmental monitoring program are that (1) it is sufficiently inexpensive to survive unexpected reductions in supporting funds; (2) it is simple and verifiable so that it is not significantly affected by changes in personnel; and (3) it includes measurements that are highly sensitive to changes within the environment.

Most environmental monitoring programs have at least five stages: (1) gathering background data, (2) identifying and evaluating the various pathways of exposure, (3) collecting and analyzing samples, (4) establishing temporal relationships, and (5) confirming the validity of the results. The major steps and/or factors to consider in the design and management of such a program are shown in Figure 16.2. Although the

Table 16.3 Questions to be answered in implementing an environmental monitoring program

Program stage or component	Question
Purpose	What is the goal or objective of the program?
Method	How can the goal or objective be achieved?
Analysis	How are the data to be handled and evaluated?
Interpretation	What might the data mean?
Fulfillment	When and how will attainment of the goal or objective be determined?

information presented is for a nuclear facility, the basic guidance applies to essentially any type of industrial operation. While the figure lists the steps as taking place sequentially, in reality, decisions relative to each of the steps must be made on an iterative basis. Essential in every case is the identification of the potentially critical contaminants that might be released, their pathways through the environment, and the avenues and mechanisms through which they may cause population exposures. Equally essential are the collection of samples that are representative and the application of standard methods in their analyses and quality control (discussed later).

BACKGROUND DATA

Before monitoring begins, background information is needed on other facilities in the area, the distribution and activities of the potentially exposed population, patterns of local land and water use, and the local meteorology and hydrology. These data permit identification of potentially vulnerable groups, important contaminants, and likely environmental pathways whose media can be sampled.

People responsible for the background analysis must take into account the type of installation, the nature and quantities of toxic materials being used, their potential for release, the likely physical and chemical forms of the releases, other sources of the same contaminants in the area, and the nature of the receiving environment. This last item includes natural features (climate, topography, geology, hydrology), artificial features (reservoirs, harbors, dams, lakes), land use (residential, industrial, recreational,

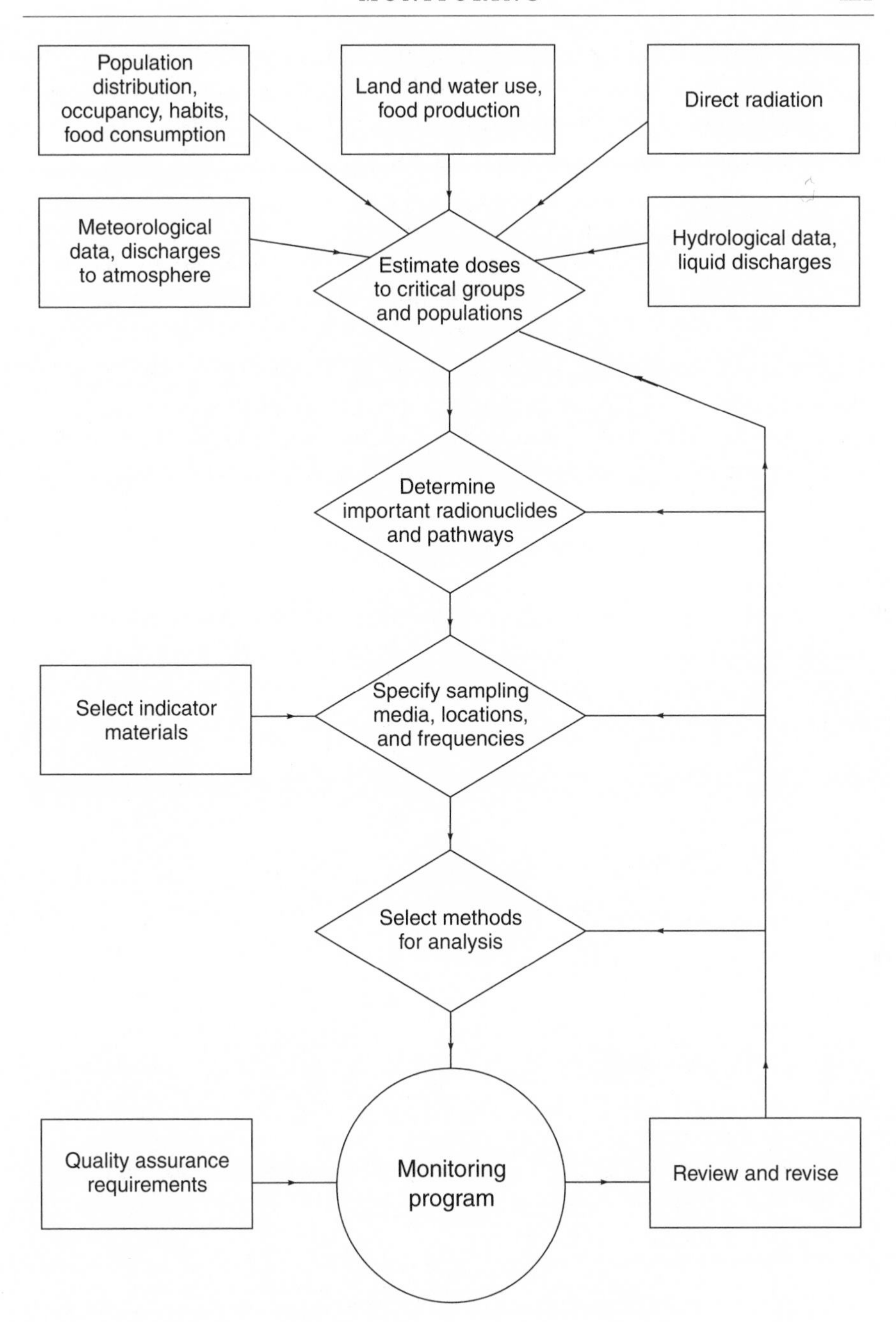

Figure 16.2 Example of the design of an environmental monitoring program and associated dose estimations for a major nuclear facility

dairying, farming of leaf or root crops), and sources of local water supplies (surface or groundwater). Results from a monitoring program conducted before a facility begins operation can be used to confirm these analyses and establish baseline information for subsequent interpretation.

EVALUATION OF PATHWAYS OF EXPOSURE

Contaminants released from an industrial facility may end up in many sections of the environment, and their quantity and composition will vary with time and the nature and extent of facility operation. As a result, discharged materials can reach the public through many pathways. For example, a secondary lead smelter has the potential to release elemental lead and associated compounds into the atmosphere, whereupon they may become an inhalation hazard (Figure 16.3). The same facility can also

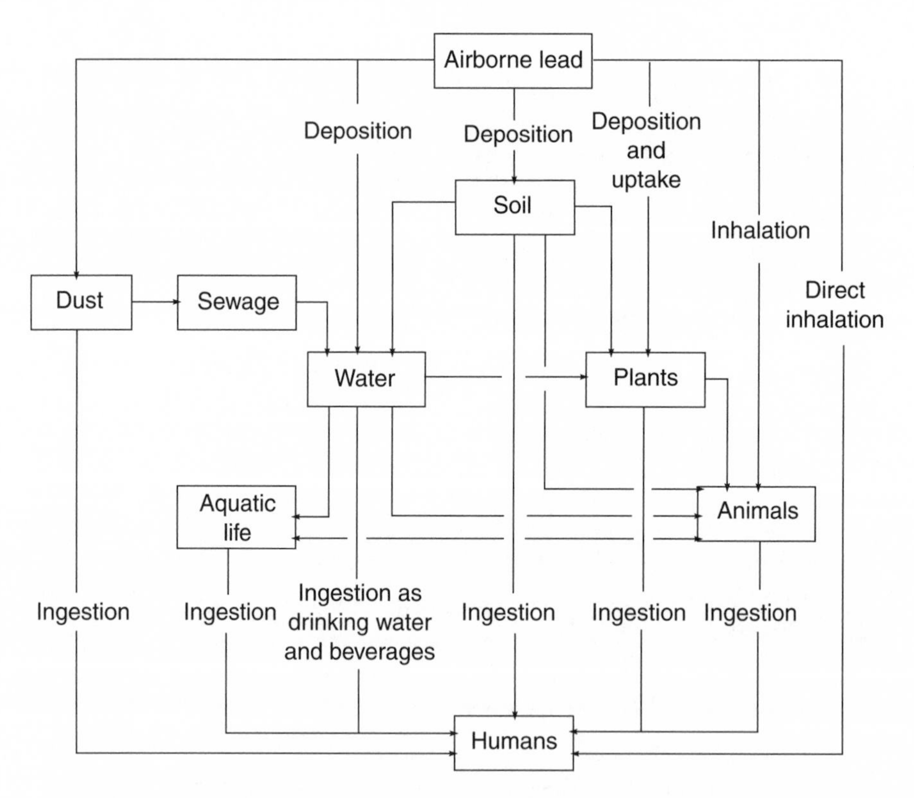

Figure 16.3 Possible pathways to humans of lead released into the atmosphere

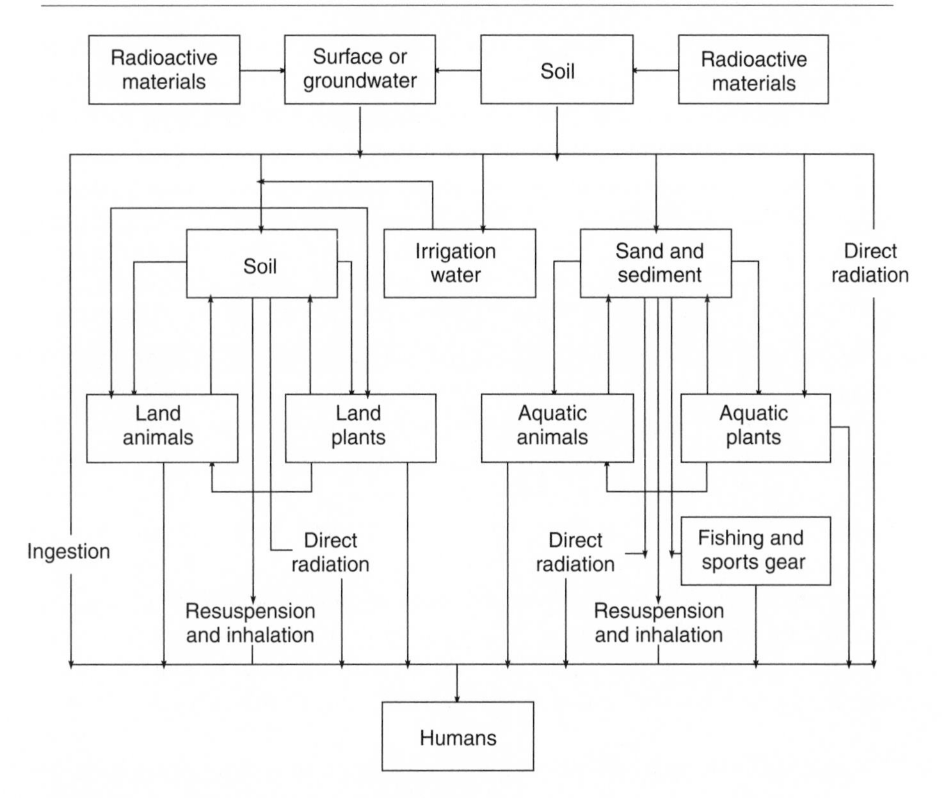

Figure 16.4 Possible pathways to humans of radioactive materials released to ground or surface water (including oceans)

release these contaminants to the soil, either directly or through the air, whereupon they may contaminate groundwater and subsequently be taken up by fish and agricultural products. In a similar manner, the milk from cows and the beef from cattle that graze on pastures adjacent to lead smelters can be expected to have a higher-than-normal lead content. Children who play on contaminated earth near such smelters have shown elevated lead concentrations in their blood. Arsenic emitted by copper smelters follows identical pathways of contamination and human exposure.

Contaminants discharged to the liquid pathway can become a similar source of human exposure. Figure 16.4 outlines the principal pathways to humans for radioactive materials released in this form. Although the figure does not depict this particular aspect, it is important to recognize that surface water can become groundwater through percolation, and that groundwater may become surface water through, for example, groundwater seeps and the use of well water for irrigation.

Tracing the movement of all contaminants through all potential pathways would be physically and economically impossible. Fortunately, in most cases, for example, a nuclear installation, the primary contributors to population dose consist of no more than half a dozen radionuclides that move through no more than three or four pathways. Once these are identified, along with the habits of the people who live or work in the vicinity, it should be possible to identify a "reasonably maximally exposed" group of individuals (the "critical group") whose activities and location would make them likely to receive the largest exposures (Chapter 15). That is not to say, however, that unsuspected pathways may not be important. For years, operators of a major nuclear facility in the United Kingdom disposed of low-level liquid radioactive wastes into the Irish Sea, since theoretical evaluations had shown that this would be acceptable. Later, they discovered to their dismay that certain population groups were consuming larger quantities of radioactive material than had been anticipated. The source was seaweed that they made into flour to make bread.

Another United Kingdom study revealed that pigeons were serving as a pathway for the off-site transport of radionuclides from a nuclear fuel-reprocessing site. Although similar problems had been observed with seagulls, there was little interaction between these birds and the local populace. In the case of the pigeons, however, large numbers were attracted to a garden in a nearby town where local residents regularly fed them. The droppings from the birds led to a local contamination problem both within the garden and on the roofs of nearby buildings. A further problem was that droppings from the birds, as well as grass clippings from the garden, were used to make compost that, in turn, was applied to a small vegetable plot nearby. Resolution of the problem included removing and replacing the topsoil in the garden, establishing measures to prevent access by the pigeons to contaminated areas, and reducing the pigeon flock through culling (Wilkins, 1999). In a similar manner, studies at the Hanford Reservation in the state of Washington revealed another unusual pathway, radioactive materials in desert plants, such as the tumbleweed. Although the groundwater (the source of the contamination) is deep beneath the ground surface, the roots of the tumbleweed readily extend down to that depth. When these plants die, they break off from their roots and can be carried by the wind across the land. Burrowing animals and rabbits can also become contaminated, as well as predatory birds that feed upon such animals.

SAMPLE COLLECTION AND ANALYSIS

In almost any environmental monitoring program, trade-offs must be made to obtain adequate coverage of critical contaminant-pathway combinations at satisfactory analytical sensitivities and costs. Samples collected directly from effluent streams that are discharged into the air and water usually contain the largest number of contaminants at the highest concentrations. Analyses of these samples can provide information on the specific contaminants that are being released as well as the amounts anticipated to be present in the neighboring environment. Once sampling and analyses shift to the environment, it is better, under essentially all conditions, to collect and carefully analyze a small number of well-chosen samples taken at key locations to provide a reliable index of environmental conditions than to process larger numbers of poorly selected, nonrepresentative samples.

At the same time, there may be a need to measure certain other contaminants because of the history of operations at the site or because of specific concerns of the local population. For example, the collection and analyses of possible contaminants in oranges in a major citrus-producing area may be necessary regardless of the concentrations anticipated; the same is true with respect to contaminant levels in cranberries if they are a major source of income to local farmers. Other examples are analysis for radioactive iodine in milk-producing areas near commercial nuclear power plants and for plutonium in the agricultural land near nuclear facilities operated by the U.S. Department of Energy. With the increased interest in assuring that the environment is being adequately protected, it may also be wise to collect and analyze samples that will indicate the range of exposures to plants and animals within various ecosystems.

Because they are faster and less expensive than analyses for specific contaminants, gross measurements of the concentrations of "total suspended particulates" in the atmosphere are sometimes used as a surrogate for, or indicator of, trends in the concentrations of particles in the $PM_{2.5}$ and/or PM_{10} size range. This is acceptable only so long as the relationships between the surrogate and the specific particle size group remain reasonably constant. To ensure that this is the case, the relationship should be verified on an intermittent basis, and especially when the nature of industrial operations or traffic patterns in an area change. Another example of this type of practice is the use of *Escherichia coli* (coliform organisms) as indicators of the possible presence of feces and accompanying disease organisms in drinking water (Chapter 7). Here again, it is important to recognize the limitations associated with such measurements.

Another factor to be considered is the wide range of impacts resulting from continuing reductions in the permissible limits for various contaminants in essentially every type of environmental media. One of these is the need for more sensitive and more accurate analytical capabilities. Although one way to solve this problem is to collect larger samples, this is not always necessary since technological developments have, in many cases, enabled the more sensitive measurements to be made on existing-size samples. At the same time, however, the ability to detect almost any amount of a contaminant, regardless of how small it is, will raise fears in certain segments of the public. Until agreement is reached on what levels of exposure to people are sufficiently small to be considered acceptable, this situation will continue.

USE OF STANDARD PROCEDURES

For many years, various professional environmentally related societies have been active in the development of standard procedures for monitoring the environment and analyzing related samples. Three of the foremost of these are the American Public Health Association, the American Water Works Association, and the Water Environment Federation, which for more than 50 years have published *Standard Methods for the Examination of Water and Wastewater.* This book provides step-by-step procedures for more than 350 testing methods, including detailed procedures for sampling and analyzing a full range of biological and chemical contaminants. A further value of this report is that the techniques presented have been approved by the EPA for generating the types of data required in documenting compliance with federal regulations (Clesceri, Greenberg, and Eton, 1998).

Related types of guidance are needed for many other aspects of environmental monitoring, including documentation of the adequacy of the cleanup of Superfund sites, those being converted into brownfields (Chapter 9), decommissioned commercial nuclear power plants, and facilities that are being remediated by the U.S. Department of Energy. To meet the latter two needs, the EPA and the USNRC, supported by the Departments of Energy and Defense, jointly developed a *Multi-Agency Ra-*

diation Survey and Site Investigation Manual (EPA and USNRC, 1997; Abelquist, 2003). This comprehensive report provides detailed guidance on the basic methods for (1) translating regulatory criteria, expressed in dose rate or risk limits, into derived guides, expressed, for example, in terms of contaminant concentrations in the soil; (2) acquiring scientifically sound and defensible data, using proper field and laboratory equipment and techniques, on the contaminants present at the site; and (3) documenting that the resulting data confirm that the sites meet the regulatory criteria (EPA and USNRC, 1997). As in the case of *Standard Methods*, organizations that apply the techniques described in the cited manual can do so with the knowledge that they are following practices that will be acceptable to regulatory agencies. In 2001 an expanded group of federal agencies and state agencies prepared a *Multi-Agency Radiological Laboratory Analytical Protocols Manual* (EPA et al., 2001).

TEMPORAL RELATIONSHIPS

In the case of nuclear facilities, each type of release generates a characteristic pattern between the time of its occurrence and the time of the subsequent exposures (Figure 16.5). Similar relationships exist with respect to certain aspects of the exposures to toxic chemicals. In the case of radioactive materials, releases to the atmosphere will subject people almost immediately to external exposures (direct radiation) from the cloud as well as exposures to the lungs as a result of the inhalation. Exposures resulting from the deposition of radioactive material from the cloud onto the soil will take longer, and uptake by agricultural crops and pasture grass longer still. Exposures to the bone and other organs such as the thyroid will be delayed, pending uptake and transfer of the material from the lungs. Likewise, exposures to specific organs (other than the stomach and gastrointestinal tract resulting from the ingestion of radioactive material in milk or food) will be delayed until the material is taken up by the blood and deposited in specific body organs. Acute effects from environmental exposures will appear within hours to weeks (depending on the dose); delayed effects (such as latent cancers) from lower-level exposures will not appear for some years.

QUALITY-ASSURANCE REQUIREMENTS

To be effective, an environmental monitoring program must be supported by a sound quality-assurance program. This must include (1) acceptance testing or qualification of laboratory and field sampling and analytic de-

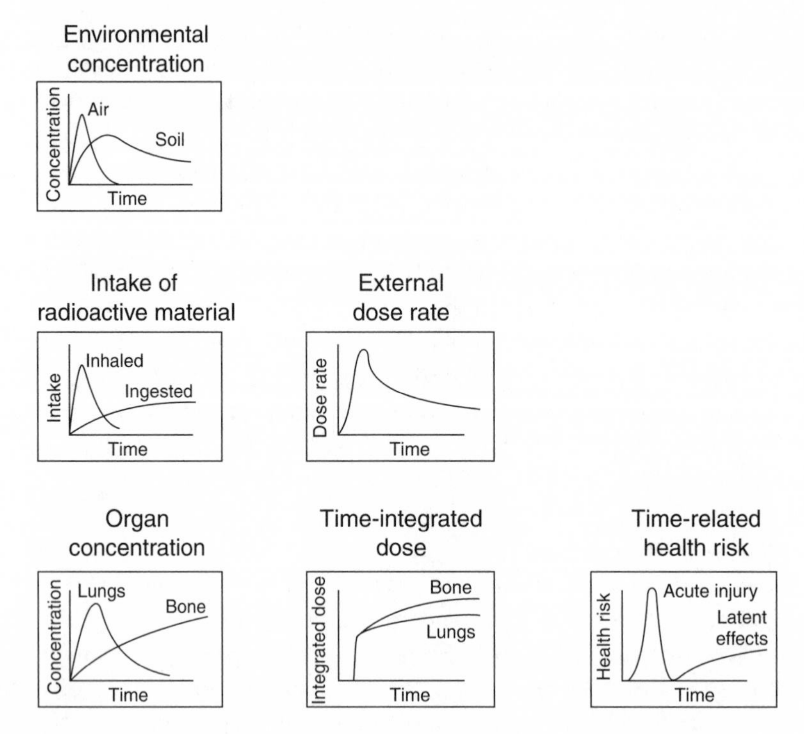

Figure 16.5 Temporal relationships of various types of contamination and accompanying human exposures resulting from environmental releases of radioactive materials

vices; (2) routine calibration of all field-associated sampling equipment and flow-measuring instrumentation; (3) a laboratory cross-check program; (4) replicate sampling on a systematic basis; (5) procedural audits; and (6) documentation of laboratory and field procedures and quality-assurance records.

To facilitate meeting these requirements, the EPA, the NIOSH, and the National Institute for Standards and Technology make available a variety of standard and cross-check samples and provide guidance for the establishment and operation of such programs. The USNRC, for example, requires all laboratories that perform analyses of environmental samples from commercial nuclear power plants to participate in the EPA program or its equivalent. Similar procedures are mandatory in support of data collected for demonstrating compliance with environmental regulations.

Meeting Special Needs

In certain cases, monitoring programs can be designed or modified to meet special needs. Several examples are described here.

CONFIRMING SOURCES OF CHEMICAL CONTAMINANTS

Careful analysis of specific contaminants can provide data to confirm their source. Analyses for vanadium, a characteristic component of fuel oils, can be used, for example, to determine whether airborne contaminants in the environment resulted from the combustion of oil or coal. Another example is the use of the ratios of ^{134}Cs to ^{137}Cs in samples to determine whether a radioactive release resulted from discharges from a nuclear power plant or the detonation of a nuclear device. The energy source in both cases is nuclear fission, one of the products of which is radioactive ^{137}Cs. Also produced is ^{133}Cs, a stable isotope of cesium. If the ^{133}Cs is produced through detonation of a nuclear device, it remains in that form. If the ^{133}Cs is produced within a nuclear reactor, it is subjected to intense neutron bombardment, and a portion of it is converted into radioactive ^{134}Cs. Unless the analysis of an environmental sample reveals the presence of ^{134}Cs, it is not possible to attribute the contamination to releases from a nuclear power plant.

CONFIRMING SOURCES OF BACTERIOLOGICAL CONTAMINANTS

As with chemical compounds, it is often a challenge to determine the source or sources of bacteriological contaminants, for example, in lakes and streams. While *E. coli,* as noted earlier is indicative of the possible presence of disease organisms, this leaves unanswered the question whether the source is humans, farm animals, or wildlife. One possible approach that has been proposed would be to identify the specific bacterial contaminant and seek to match it with a similar organism in a previously created library of bacteria from known sources in the area. Another approach would be to identify the specific contaminant based on the fact that strains of bacteria in people, farm animals, and wildlife respond differently to antibiotics (Malakoff, 2002).

IDENTIFYING SOURCES OF UNANTICIPATED RELEASES

When disruptions in facilities that process valuable metals, such as silver, gold, or plutonium, lead to the release of unusually large amounts of the product in the liquid-waste stream, it is important to know exactly when and where the release occurred. In cases where the processing system is automated, workers may either not be aware that a system failure has occurred or be reluctant to acknowledge that something went wrong. One approach for confirming the source of such a release is to install individual monitors on the effluent streams from each of the buildings within a facility. If the monitors are designed to collect samples 24 hours a day, with an aliquot of the effluent stream from each hour being collected in a separate container, when such an incident occurs, it will be possible to determine where and when the release occurred. The ability to provide such information will also demonstrate to plant officials the benefits of a well-designed and supported monitoring program.

Computer and Screening Models

Computers offer an enormous capacity for collecting, organizing, and storing information that can assist in understanding the environment and the accompanying impacts of human activities. In fact, computer models have for some years been used to estimate the transport and accompanying concentrations of various pollutants in the atmosphere near emission sources. Since information on the meteorology, terrain, and other factors for the given site have already been incorporated into the models, the only additional input data required are the nature and quantities of the contaminants being released. So long as ambient air concentrations are periodically measured to validate the estimates that are being generated, this approach has become accepted practice (Weinhold, 2002).

As the field has matured, programmers have incorporated a wide array of features and capabilities into such models, including the ability to analyze dynamic transfers of contaminants within and between the main biotic and abiotic components of freshwater ecosystems (Beaugelin-Seiller et al., 2002). Unfortunately, this has led to models that are not applicable to many everyday situations. Recognizing this fact, the National Council on Radiation Protection and Measurements (NCRP) has developed a series of what are called "screening techniques." The primary goal is to make available simple methods for evaluating, on a broad basis, whether releases from a given facility comply with the applicable regulations. Although these technique were originally designed to apply only to analyses of releases from point sources, they can readily be modified for analyses of releases from area (nonpoint) sources (NCRP, 1996).

The application of screening techniques has been enhanced by the development of separate models for radionuclide discharges into the atmosphere, surface water, or the ground. Because detailed information on the important transport mechanisms, exposure pathways, and dosimetry parameters have been incorporated into the models, users are able to complete the associated calculations in only a few steps. Another feature is that the values of the input parameters used in developing the models are conservative. Therefore, if the generated dose/exposure estimates show that compliance with the applicable regulations is being achieved, additional calculations using more sophisticated techniques are not necessary. Because of their general applicability, many of these models are being modified for assessments of nonradioactive contaminants (NCRP, 1996).

National and Global Monitoring Systems

The preceding discussion has been directed to programs designed to evaluate the impacts of environmental contaminants on a local or regional basis. Because of the recognition that many contaminants have widespread effects, major efforts are under way to develop systems capable of monitoring releases that have impacts on national or global scales. Several of the more prominent of such programs are discussed here.

NATIONAL MONITORING PROGRAMS

One of the primary networks for monitoring the atmosphere in the United States is the National Atmospheric Deposition Program (NADP). The primary goals of the NADP, which is being developed by the EPA, are to characterize geographic patterns and temporal trends in atmospheric chemical deposition and to apply the resulting data in support of research on the productivity of managed and natural environmental systems, including surface- and ground-water chemical interactions and pollutant source-receptor relationships. Ultimately, such data should provide a foundation for assessing other environmental impacts, such as visibility and materials degradation, and effects on the health of humans as well as domestic animals, wildlife, and fish. The NADP currently has more than 300 monitoring stations located throughout the continental United States, Alaska, Hawaii, Puerto Rico, the Virgin Islands, and parts of Canada. Support for the program is enhanced by the widespread backing of the scientific community for both the data being generated and the accuracy of the program in tracking air-quality trends in wet deposition, including acid rain (Lambert and Bowersox, 2002).

The NADP is made even more effective by the fact that multiple other federal agencies are cooperatively supporting the program, major components of which include the following atmosphere-related networks:

National Trends Network, through which measurements are made on a daily basis of the amounts of precipitation, with analyses of weekly samples for acid rain species, nutrients, base cations, sodium, and chloride

Atmospheric Integrated Research Monitoring Network, through which samples are collected and analyzed to support research and atmospheric modeling of the effects of sources of air pollutants and meteorological conditions on precipitation chemistry

Mercury Deposition Network, which is designed, as the name implies, to provide data on the total mercury content in all weekly precipitation samples, with analyses for methyl mercury in a subset of samples

In addition, there is a fourth national air-pollution network operated by the EPA that is not part of the NADP. This is the Clean Air Status and Trends Network, which is designed to measure gaseous and particulate matter. When it was originally established in 1987, its primary purpose was to measure ground-level ozone concentrations and dry material that contributes to acid rain. In a cooperative venture with the U.S. Geological Survey, this network was used during 1990–1991 to examine the presence of commonly used herbicides, such as atrazine and alachlor, in precipitation in 26 states ranging from the upper Midwest to the East Coast of the United States. Interestingly, herbicides were detected at essentially every site in the study area, primarily during the late spring and summer (Lambert and Bowersox, 2002).

The EPA also supports the Regional Environmental Monitoring and Assessment Program (REMAP), which is designed to determine the extent (numbers, miles, acres) and geographic distribution of each ecosystem class of interest; assess the proportions of each such class that are in good or acceptable condition; evaluate what proportions are degrading or improving, in what regions, and at what rate; and appraise the likely causes and identify methods for improvement (EPA, 1993). As is the case with the NADP, this program depends on ground-based remote-sensing and fast-response instruments to gather the required data. Some monitoring specialists have described EMAP's goal as the determination of the health of an ecosystem in much the same way as a doctor determines the health of a patient (Table 16.4). Proponents believe that such a system, properly applied, would provide information that would enable environmental scientists to anticipate the point at which ecosystems might begin to break down. Other scientists oppose the analogy because ecosystems, unlike organisms, are not consistently structured, do not behave in a predictable manner, and do not have mechanisms such as the neural and hormonal systems of organisms to maintain homeostasis (Griffith and Hunsaker, 1994).

As progress is achieved, other national needs and possible applications of such programs are being identified. One is a system for monitoring the use of water on a national basis. While it is widely recognized that there are major water shortages in many areas of this country, it is simultaneously acknowledged that the data necessary to quantify the extent of these shortages are not available. Today, for example, government officials predominantly rely on stream gauges, river runoff monitors, and similar tools to assess the amount of water available locally. In most cases, no one measures or maintains accurate records on how much water is being withdrawn from wells, rivers, and aquifers. In fact, consumption is often estimated simply on the basis of how many hectares of agricultural crops are being irrigated, and national estimates are simply a compilation of the

Table 16.4 Comparison of ecological health research and human health diagnosis

Ecological research issue	Analogous human health area
Early warning of ecosystem transformation (e.g., localized fish kill in a river)	Early warning of disease, such as PSA (prostate-specific antigen) test for prostate cancer
Exotic plant/animal/virus invasion or outbreak of native indigenous pathogens	Epidemiological studies of disease outbreaks within a population group
Presence of "sensitive zones" in ecosystems	Study of certain body organs that are crucial to the functioning and well-being of the whole
Possible development of ecosystem immunity to particular classes or combinations of stress	Immune antibody responses to foreign antigens

data the individual states report, the quality of which varies considerably from state to state. A further problem is that the infrastructure that supports such activities is loosely organized and often in chaos. If the supply of water is to be effectively used, much more accurate data are needed (Brown, 2002).

Also under way on a national basis, under the auspices of the EPA, is what is called the National Human Exposure Assessment Survey (NHEXAS). As contrasted to a typical monitoring program, it is a multiple-component effort that includes (1) the distribution of questionnaires to provide baseline information on the lifestyles, activities, and sociodemographics of population groups; (2) the collection of soil, house dust, indoor air, tap water, and diet samples; (3) the analysis of these samples for some 30 compounds, including airborne particulates in specific size ranges; and (4) the collection of samples of blood, urine, and hair as biological indicators of human uptake of individual contaminants (Newman, 1995). Component 4 is closely aligned with comparable efforts under way within the Agency for Toxic Substances and Disease Registry (Chapter 2). Although NHEXAS is labor intensive, the data it generates are essential for making longer-range and fuller-scale assessments of the impact of environmental pollutants on an individual basis.

SATELLITE MONITORING SYSTEMS

While satellite monitoring systems are finding increasing applications at the national level, they are almost mandatory for monitoring key environmental factors on a global basis. Such systems, for example, are being used to gather data on the rate at which the world's humid rain-forest cover is disappearing. Such data are essential for estimating the carbon fluxes in the global budget, particularly in terms of quantifying the capacity of the terrestrial sink for absorbing carbon dioxide emissions. Data generated through such systems show that almost 6 million hectares of such forests were lost each year from 1990 through 1997. More than 2 million additional hectares were annually visibly degraded. The need to improve efforts to conserve such resources is clear (Achard et al., 2002). Satellite sensing was also used on a national level to determine which of hundreds of confined animal-feeding operations in eastern North Carolina were in areas flooded as a result of rains that accompanied Hurricane Floyd in September 1999. This enabled researchers to assess the potential for dispersion of animal wastes and the possible pollution of ground- and surface-water supplies. It also offered an opportunity to compare such assessments to those made on the basis of ground-based observations (Wing, Freedman, and Band, 2002).

The possible applications of satellite systems, however, do not end here. Insurance companies, financial institutions, real-estate companies, and other stakeholders often require that environmental assessments be made of sites that are being considered for transfer from one owner to another. Oftentimes, such assessments are time consuming and are complicated by the presence of contaminants on adjoining properties. One possible method for resolving some of the problems is the use of remote sensors, such as digital airborne imaging spectrometers. In addition to the example cited earlier, such sensors have been used for years by forestry personnel to analyze the effects of reforestation, forest-fire monitoring, and the identification of invasive species. Because of the detail and accuracy of the data such sensors provide, they would appear to be ideally suited to assist in resolving some of the problems associated with site assessments. Through the provision of high-resolution hyperspectral imaging data of the terrain that is being surveyed, for example, such sensors could provide information on contaminated areas on the site under consideration, as well as on adjoining sites (Howard, Pacific, and Pacific, 2002).

Stakeholder Involvement

Although the technical aspects of environmental monitoring programs conducted under the auspices of federal agencies are important, there is increasing recognition of the need to provide state and local agencies and, most important, stakeholder groups with opportunities for active involvement in such activities. The success of such interactions is demonstrated by the increasing frequency with which states and municipalities in the United States are being funded by federal agencies and facility operators to conduct independent assessments of the impacts of government and commercial facilities. Examples range from federal agencies that provide support to enable state agencies to conduct independent monitoring and assessment of federal facilities to facility operators who provide monitoring equipment to neighboring schools so that students can conduct independent assessments of their airborne liquid releases. For maximum effectiveness, every effort should be made to complete such arrangements sufficiently early so that the involved groups can initiate their independent monitoring programs prior to start-up of activities within such facilities. This is essential, as discussed earlier, if baseline data on conditions in the area under preoperational conditions are to be available.

To ensure that the data that are being reported are comparable, it is important that all such groups involved apply uniform methods both for collecting samples and conducting the relevant analyses. It is also essential that the stakeholder groups, the governmental monitoring organizations, and the facility operators analyze triplicate samples on a regular basis both for purposes of quality control and to document that the data that are being generated are comparable. It is also important that the several groups agree on the methods that are being used to assess the potential impacts of the releases on public health and the environment. Since the values of the input parameters into the models used for assessing such impacts change as new information and epidemiological data become available, such information will need to be updated on a regular basis. If these precautions are observed, the data generated by all such groups can be combined to yield a comprehensive evaluation of the impact of the given facility on the neighboring public and the environment.

One outstanding example of the benefits of such cooperative relationships is the New Mexico Environmental Evaluation Group, which conducts independent technical evaluations of operations at the Waste Isolation Pilot Plant Project, the federal repository where long-lived transuranic wastes are being disposed of (Chapter 9) (Gray and Ballard, 2000). Another example is the Pilgrim Nuclear Power Plant, located in Plymouth, Massachusetts, where the operating utility has provided the State Radiation Control Program with on-line monitors that supply real-time information on the status of the plant. Such information includes dose rates in the neighboring environment, information on local meteorology, and key factors related to operating conditions within the plant. Monitoring equipment has also been provided by the utility to high schools in the region so that students can measure radiation levels on a daily basis. These activities enhance the scientific educational opportunities for the students, increase public understanding of the impacts of the plant, and serve as an effective public relations tool (Hallisey, 2001).

The General Outlook

One of the major changes in environmental monitoring systems during the 1990s has been to adopt an ecosystem and more global approach. This has led to basic changes in the identity of the environmental factors or trends that need to be assessed and thus to changes in the types of data that will be required. Although the need for fundamental information on the concentrations of various contaminants in the air, water, and soil will remain, this still leaves open the question as to what contaminants should be measured, where, and how frequently. It will be even more difficult to determine the data required to assess the condition of equally important but less visible indicators, such as the ecological processes and conditions that yield food, fiber, and building materials and those that provide "services," such as water purification and recreation. Specific factors that might be monitored to meet these needs include land cover and productivity, species diversity, and key ecological processes, such as those exemplified by the condition of wetlands, marshes, coastal zones, coral reefs, and marine estuaries. A further requirement is a better definition of the indicators needed as input parameters for the various environmental models used to assess the condition of the environment, the meaning of the data that are generated, how best to gather the data, and the effects of emerging technologies on the measurements. The enormity of these challenges is illustrated by the monitoring requirements associated with the assessment of global warming. Until the required models are further developed, and more of the key factors that influence temperature trends are identified, scientists will remain in a quandary as to exactly what types of data are required (NRC, 2000).

In the meantime, enormous amounts of data are being generated through existing environmental monitoring programs, and applications of these data continue to expand. One example is the previously discussed increasing use of information obtained by Earth-orbiting satellites. In accordance with a 1995 agreement, officials in Russia and the United States are exchanging data obtained through the operation of spy satellites during the cold war. Scientists believe that long-term records could provide valuable information on the effects of clouds on heating and cooling of the Earth, as well as insights into the effects of airborne contaminants on cloud formation and on the development of better methods for modeling these effects. Other information provided by satellites may lead to methods for issuing warnings about the occurrence of volcanic eruptions, earthquakes, and other types of natural disasters.

With the increase in magnitude and comprehensiveness of the databases that are being established, far more efforts will need to be directed to their integration and the development of more systematic methods for their compilation and storage. Otherwise, some of the key potential benefits of these data, especially in terms of assessing the long-term effects of various environmental contaminants, may not be achieved. Computer models also need to be refined not only to predict human exposures but also to determine the interrelationships among environmental factors and, as noted earlier, to evaluate their potential impacts on ecosystems.

17

RISK ASSESSMENT

In a personal sense, risk can be defined as the probability that an individual will suffer injury, disease, or death under a specific set of circumstances. In the realm of environmental health, this definition must be expanded to include possible effects on other animals and plants, as well as on the environment itself. Knowing that a certain risk exists, however, is not enough. People want to have some idea of how probable it is that they or their environment will suffer and, if they do, what the effects will be. Determination of the answers to these questions involves the science of risk assessment.

Risk assessment ranges from evaluation of the potential effects of toxic and/or radioactive chemical releases known to be occurring to evaluation of the potential effects of releases due to events whose probability of occurrence is uncertain. In the latter case, the risk is a combination of the likelihood that the event will occur and the likely consequences if it does. In essence, the process of risk assessment requires addressing three basic questions (Kaplan and Garrick, 1981):

- What can go wrong?
- How likely is it?
- If it does happen, what are the consequences?

As might be anticipated, there is a spectrum of endpoints that can be considered in evaluating whether a risk is acceptable (Table 17.1). Once the endpoint has been defined, and the associated risk has been assessed, it can be expressed in qualitative terms (such as "high," "low," or "trivial")

Table 17.1 Spectrum of adverse consequences and factors to consider in risk assessments

1. Shortening of life (mortality)
    - Cancer versus other causes
2. Illness or injury leading to disability
    - Acute versus chronic
    - Permanent versus temporary disability
    - Serious versus minor disability
3. Illness or injury with temporary disability followed by recovery
    - Chronic versus acute
    - Serious versus minor disability
4. Physical discomfort without disability
5. Psychological disorder with behavior consequences
    - Post-traumatic stress disorder
    - Anxiety reaction
    - Stress reaction
    - Chronic frustration and anger
6. Emotional discomfort

Note: Each category should be weighted by the number of people involved.

or in quantitative terms, ranging in value from 0 (certainty that harm will not occur) to 1 (certainty that harm will occur). At the same time, it must be recognized that a given assessment provides only a snapshot in time of the estimated risk of a given toxic agent and is constrained by our current understanding of the relevant issues and problems. To be truly instructive and constructive, risk assessment should always be conducted on an iterative basis, being updated as new knowledge and information become available.

Once a risk has been quantified, the next step is to decide whether the estimated risk is sufficiently high to represent a public health concern and, if so, to determine the appropriate means for control. Such control, which falls under the rubric of what is called *risk management,* may involve measures to prevent the occurrence of an event as well as appropriate remedial actions to protect the public and/or the environment in case the event occurs. Because of the nature of these processes, each step in the risk-management process is accompanied by a multitude of uncertainties. The same is true for risk assessment.

Applications

As described in Chapter 2, two basic concepts are applied in assessing risks from toxic agents. These are exemplified by the two graphs shown in Figure 17.1. The graph on the left represents the linear nonthreshold dose-response curve that is generally assumed to apply to carcinogenic agents: any dose, regardless of how small it is, carries an associated risk. Through risk assessment, scientists seek to quantify the risk associated with a given dose and to use that information to establish an acceptable level of exposure. The graph on the right represents the threshold type of response that is assumed to apply to many noncarcinogenic agents. Although a person is assumed to be "safe" as long as the dose is below the threshold, there is still need for a quantitative estimate of the risk associated with any dose above the threshold. It is in this range that scientists apply risk-assessment procedures. This will be discussed in more detail later.

Among the earliest applications of the techniques of risk assessment were those of the National Aeronautics and Space Administration in assessing the safety of travel in space and those of the U.S. Nuclear Regulatory Commission in assessing the safety of nuclear power plants

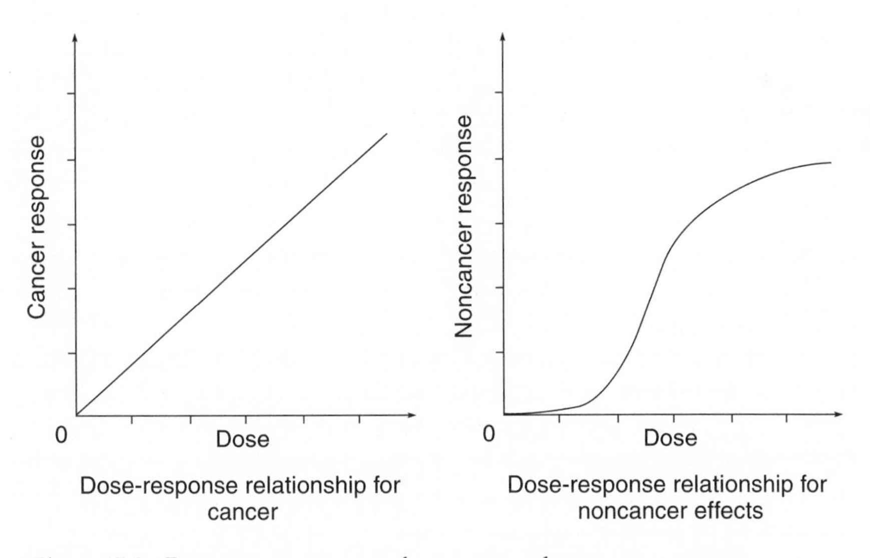

Figure 17.1 Dose-response curves for cancer and non-cancer agents

(USNRC, 1975). Although these applications were designed to provide quantitative estimates of the risk to the public in case of an accident, the same basic techniques have been applied in subsequent years to assess the risks associated with a wide array of industrial operations and products.

As a result of the issuance of Executive Order 12866 (Federal Register, 1993), as well as subsequent laws passed by Congress, such applications will undoubtedly increase. This executive order requires that all federal agencies that are developing new regulations compare the risks each such action is intended to address to other risks within that agency's jurisdiction and provide cost-benefit analyses of the impacts of the proposed actions. Among the stimuli for its issuance were the severe budgetary constraints under which most regulatory agencies and the regulated industries operate. Risk assessment offers such organizations one of the best techniques for comparative evaluation of the risks associated with the use of various toxic agents and for deciding how to address these risks in the most efficient manner.

In spite of these apparent benefits, applications of risk assessment are not without difficulties and controversies. The public, in particular, finds the concept of risk difficult to understand. For many people, the concept of a "safe" dose is far more acceptable. Individuals also fail to realize that the risks associated with the hazards of everyday life vary widely and that the risks most feared are not necessarily those that are most important. As noted earlier (Chapter 15), people generally voice far more concern over risks that they are asked to accept on an involuntary basis and that derive from human-made sources. They have far less concern about risks that they accept voluntarily and that derive from natural sources or have been commonly accepted for years. An excellent example of the latter is the risk of lung cancer due to the presence of naturally occurring radon in the home (Chapter 5). When college students were asked to indicate their perception of the risks from pesticides and nuclear power plants, their rankings were much higher than those of a group of experts. When they were asked to express their perception of the risks of motor-vehicle accidents and coal-fueled electricity-generating stations, their rankings were well below those of the experts (Wilson and Crouch, 2001). Although individual members of the public may be concerned about contaminants (such as chloroform) in their drinking water or toxic substances (such as aflatoxin) in their food, the data in Table 17.2 show that many other risks in their daily lives should be of far more concern. For example, on a quan-

Table 17.2 Estimates of comparative risks to individuals

Hazard	Probabilistic risk estimate	Degree of uncertainty
Death before age 85, all causes	0.70	Low
Death from cancer, lifetime risk	0.25	Low
Death from cigarette smoking, one pack per day for 40 years	0.13	Medium
Death from homicide (black male), lifetime risk	0.05	Low
Death from occupational exposure to benzene, 10 ppm concentration, for 30 years	0.05	Medium–High
Death from motor vehicle-crash, lifetime risk	0.02	Low
Death of police officer on active duty, 30 years of service	0.007	Low
Death from eating 4 tablespoons of peanut butter per day, lifetime risk (aflatoxin)	<0.0005	Medium
Death from drinking water with EPA limit of chloroform, lifetime risk	<0.0005	High
Death from inhaling formaldehyde in urban air, 5 ppb concentration, lifetime risk	<0.000008	High

titative basis, the risk of death from motor-vehicle accidents is more than 40 times as high, the risk of death due to violence (as exemplified by homicide) is more than 100 times as high, and the risk from cigarette smoking is more than 250 times as high as the risks associated with the cited contaminants in drinking water and food.

Although initial applications of risk assessments were primarily directed to the effects on people, experience has demonstrated that these techniques are equally applicable to assessments of ecological and environmental impacts. In line with this thinking, the Environmental Protection Agency (EPA) is (as will be discussed later) directing specific attention to the development of ecological risk-assessment guidelines. The intent is that regulatory agencies will then be able to make better policy decisions on proposed developmental activities in geographic areas that might be particularly sensitive. The types of problems that need to be addressed range from regional issues, such as the possible impact of the drainage of wetlands on the local ecology, to global issues, such as the impacts on the environment of chemicals that can destroy the ozone layer and/or lead to worldwide warming.

In evaluating the effects of toxic agents on people, researchers must determine whether the goal is to assess the impacts on individuals or on population groups. Although some may assume that it would be easier to estimate the risks to an individual, this is not necessarily the case. Most measurements of air pollution, for example, are designed to determine the average concentrations of specific contaminants in the ambient environment (Chapter 16). Consequently, the resulting data are primarily limited to assessing the exposures of members of the public who spend large amounts of time outdoors. An equally important factor is that the accuracy and/or applicability of estimates of the risks either to individuals or to population groups depend on the nature of the toxic agent, the extent and availability of exposure measurements, the range and duration of the exposures, and other factors such as the physical characteristics and lifestyles of those exposed. The necessity of making assumptions concerning each of these factors adds to the uncertainties in the associated risk estimates.

Whereas the risk to an individual, as noted in Table 17.2, is generally expressed in terms of the likelihood of suffering a given detrimental effect, the risk to a population group is generally expressed in terms of the estimated number of excess deaths that will occur due to exposure of the group to a given agent or stress. The goal of this type of assessment is to express the risk in terms of the impact on society as a whole. Examples of the impacts of various risks, in this case expressed in terms of the estimated numbers of resulting annual deaths in the U.S. population, are shown in Table 17.3. This table, like the previous one, includes an estimate of the degree of uncertainty for each number. As noted later, it is important that this type of information be included; in fact, a "complete" risk characterization will include not only the uncertainties of the given estimates, but also the estimates that would have resulted had alternative assumptions and methods been utilized (Ropeik and Gray, 2002).

Table 17.3 Estimates of comparative risks to populations

Risk	Annual total deaths in U.S.[a]	Degree of uncertainty
Cigarettes and other uses of tobacco	450,000	Low
Alcohol abuse	150,000	Low
Highway travel	45,000	Low
Homicide	20,000	Low
Airline travel	1,000	Low
Outdoor air (particles, $PM_{2.5}$)	60,000	Medium
Indoor air (radon)	15,000	High
Pesticide residues in foods	3,000	High

a. Values are approximate.

Qualitative Risk Assessment

Although some regulatory agencies seek to develop *quantitative* risk assessments, the large number of facilities that have the potential for toxic chemical releases makes universal application of such assessments impossible. For this reason, the common approach is to apply as an initial step some type of *qualitative* or *semiquantitative* assessment. Possibilities include (1) qualitative characterizations where health risks are identified but not quantified; (2) qualitative risk estimations where the chemicals present are ranked or classified by broad categories of risk; and (3) semiquantitative approaches where effect levels (for example, "no observable effect") are used in combination with uncertainty factors to establish "safe" exposure levels.

Perhaps the best example of a qualitative approach is the public health assessment methodology that has been developed by the Agency for Toxic Substances and Disease Registry (ATSDR), primarily for evaluating the potential health hazards of various substances present at toxic waste (Superfund) sites (Chapter 9). Such an assessment includes a review and evaluation of data and information about hazardous substances at a site and a characterization of the nature and extent of the associated risk to human health. Frequently, the assessment includes recommendations on actions needed to prevent or mitigate any potential health effects and any additional studies that may be needed (ATSDR, 2003). Although fundamentally qualitative in nature, such assessments are based on an evaluation of a large amount of scientific data.

These data are supplemented in the case of the higher-risk Superfund sites, by in-depth reviews and evaluations of the frequency of occurrence, toxicity, and potential for human exposure for each of the substances present. On the basis of this information, ATSDR officials maintain a priority list of the 10 most hazardous substances known to exist at Superfund sites (Table 17.4a). The characteristics of these substances are, in turn, used to identify the priority human health effects that should receive specific attention in the evaluation of the potential impacts of the associated toxic chemical releases (Table 17.4b). On the basis of the resulting public health assessment, each Superfund site on the priorities list is placed in one of five categories in terms of its overall significance to public health and the associated requirements for follow-up action: (1) urgent public health hazard, (2) public health hazard, (3) indeterminate public health hazard, (4) no apparent public health hazard, and (5) no public health hazard (Johnson, 1992).

Table 17.4a Top ten substances on the priority list

Rank	Substance
1	Arsenic
2	Lead
3	Mercury
4	Vinyl chloride
5	Polychlorinated biphenyls
6	Benzene
7	Cadmium
8	Benzo(a)pyrene
9	Polyaromatic hydrocarbons
10	Benzo(b)fluoranthene

Table 17.4b Priority health conditions to be considered

Birth defects and reproductive disorders
Cancers
Immune function disorders
Kidney dysfunction
Liver dysfunction
Lung and respiratory diseases
Neurotoxic disorders

Quantitative Risk Assessment

In contrast, the EPA concentrates on quantitative risk assessments. Its goal is to characterize in numerical terms the potential adverse health effects of human exposures to toxic agents. As noted in Figure 17.2, such assessments involve four primary steps and serve as one of the principal elements of risk assessment and risk management, of which the former is an essential preparatory step to the latter. Fundamental to these two steps is the conduct of the laboratory and field research to provide the necessary input data. Each of the primary risk-assessment steps is described here.

HAZARD IDENTIFICATION

Hazard identification is a qualitative determination of whether human exposure to a specific agent has the potential for adverse health effects. This generally requires information on its identity and the outcomes of related mutagenesis and cell-transformation studies, animal research, and human epidemiological studies. Information on the physical and chemical properties of the agent is also required for assessing the degree to which it can become airborne and be inhaled and absorbed into the body and for evaluating its solubility in water and availability for transport through the food chain.

DOSE-RESPONSE ASSESSMENT

Dose-response assessment is a quantitative estimate of the hazard potency (power to produce adverse effects) inherent in receiving a dose from a

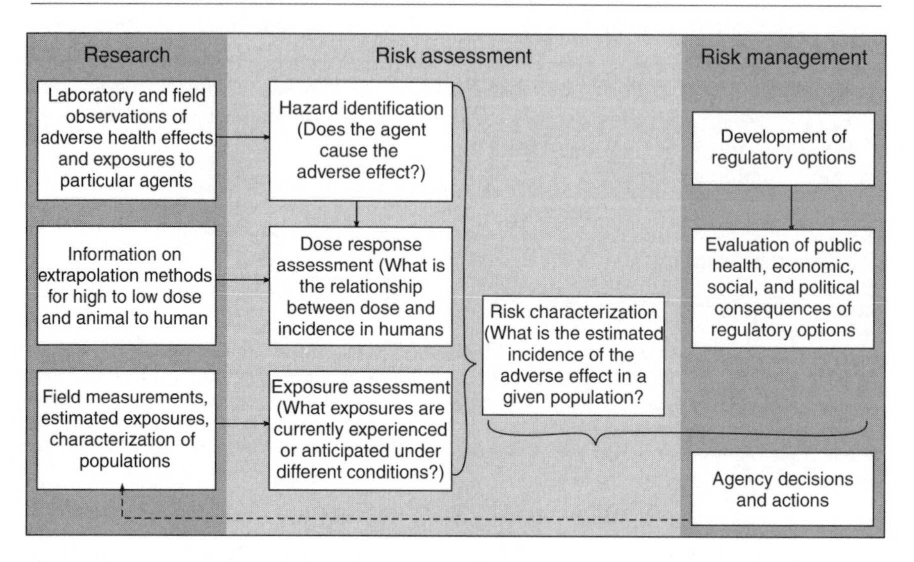

Figure 17.2 Elements of risk assessment and risk management

specific toxic agent. If available, dose-response estimates based on human data are preferred. In their absence, information from studies of other animal species that respond like humans may be used. As previously discussed (Chapter 2), however, the use of animal data introduces multiple uncertainties into the accompanying risk estimates.

EXPOSURE ASSESSMENT

Exposure assessment is an estimate of the extent of exposure to the agent and the accompanying dose to people and the environment. The assessment may be directed to normal releases from a facility, or it may, in the case of accident assessments, involve estimating both the probability of the event and the magnitude of the accompanying releases. Factors considered in performing such assessments include the following:

1. The chemical and physical characteristics of the agent, for reasons similar to those identified in the hazard-identification step. Key parameters include partition coefficients, retardation factors, bioaccumulation factors, and degradation rates. To the extent possible, the values assigned to such parameters should be specific for the system and site being analyzed.

2. Identification and characterization of the person to be protected. This is especially important in cases where it is necessary to protect special groups, such as children and pregnant women, who may be more highly susceptible to a given agent.
3. Recognition of the difference in the exposure measured and the dose that will actually be received by the exposed individuals. Although several people may be exposed to the same agent, the accompanying dose depends on a number of factors. In the case of airborne materials, the dose depends on the age and breathing rate of the person exposed and on whether he/she breathes through the mouth or nose.

RISK CHARACTERIZATION

Risk characterization involves estimating the dose and accompanying adverse risk to people who have been exposed to a specific agent. This process requires integrating the results of the previous processes to produce quantitative estimates of the associated health and environmental risks. Because risk estimates have significant limitations, the EPA requires that in addition to the estimate itself (usually expressed as a number), the risk characterization contain a discussion of the "weight of the evidence" for human carcinogenicity (for instance, the EPA carcinogen classification); a summary of the various sources of uncertainty in the estimate, including those that arise from hazard identification, dose-response evaluation, and exposure assessment; and a report on the range of risks, using the EPA-based risk estimate as the upper limit and zero as the lower limit. Because of the magnitude of the accompanying uncertainties, it is important that those who perform these exercises not be overly conservative in their assumptions. Otherwise, the associated risk estimates may be far in excess of what will be experienced in the real world. For this reason, many public health officials urge that such assessments be designed so that the outcomes are as realistic as possible.

Ecological Risk Assessments

As some of the major initial challenges in risk assessments were resolved, it became apparent that there was a need to address the impacts of pollution on a broader basis, taking into consideration, for example, environmental impacts such as habitat alteration, the loss of biodiversity, and, most important, the overall impacts of the combination of such activities on ecosystems. As a result, the use and application of such assessments have during the 1990s been slowly but steadily expanded. This is illustrated by recognition of the adverse impacts of acid deposition on lakes and forests, the damaging effects of ozone on agricultural crops, and the need to evaluate the long-term potential impacts of global warming on a host of environmental systems. Today ecological risk assessment is playing an ever-increasing role as an important consideration in the regulatory and policy decision-making process (EPA, 2002).

To support this effort, the EPA has developed a range of guidance documents (EPA, 2002). A foundation for these activities is *Future Risk: Research Strategies for the 1990's,* a report prepared by the EPA Science Advisory Board (EPA, 1990). In response to the guidance provided, the EPA expanded its approach in this field by focusing on the full range of resources at risk, coupled with consideration of the assessment of the impacts of multiple environmental stressors taking place at multiple scales and with multiple endpoints. This led to the publication of the *Wildlife Exposure Factors Handbook,* which provides data, references, and guidance for assessments of the impacts of toxic chemicals on wildlife species (EPA, 1993), and *Guidelines for Ecological Risk Assessment,* which is designed to improve the quality and consistency of such assessments among the EPA's program offices and regions (EPA, 1998).

To provide an organizational component for coordinating and directing such activities, the EPA subsequently created the National Center for Environmental Assessment. One of the immediate realizations of the new center's staff was the lack of information on how ecosystems respond to multiple stressors. Until this void could be filled, progress could not be made on the development of techniques for estimating the accompanying consequences. While it was recognized that short-lived stressors may produce transient and frequently immeasurable effects, longer-lived stressors, in contrast, may for a time appear to be without impacts, only to have seemingly minor effects later emerge in the form of damaging endpoints. At this stage, the center's staff is seeking to identify those resources that are most vulnerable and to assign priorities to the research needed to obtain necessary new data (EPA, 2002).

Accident Situations

The previous discussions have been directed primarily to assessments of risks associated with routine releases from a facility or source. For major facilities that contain large amounts of toxic agents, it may be equally important to assess the risks associated with accidental releases. Evaluations of the risks of such releases include not only an estimate of the magnitude of the associated consequences but also, as noted in the introduction to this chapter, the likelihood that such an accident can occur. The primary differences in risk assessments for accidents are described here.

ACCIDENT SCENARIO DEVELOPMENT AND SCREENING

The objective of accident scenario development and screening is to postulate physically possible sequences of events and processes that could lead to a major disruption in a facility and the release and transport of toxic materials into the neighboring environment. Since industrial facilities are located in areas widely diverse in population distribution and land use, and since seismic and tectonic events frequently serve as initiators of disruptive events, the risk of an accident is generally site specific and facility specific. Human error can be a major source of disruptive events, so the magnitude of the risk may depend on the training and skill of the facility operators. For large industrial facilities, the assessment involves identifying the events and processes that could initiate a release and combining these events and processes into physically reasonable scenarios that can be analyzed and evaluated.

CONSEQUENCE ASSESSMENT

For each scenario that is selected, a suite of models must be identified and/or developed to estimate the associated consequences. Typically, the first suite is designed to examine various mechanisms through which the toxic agent might be released. Once the release is assumed to have occurred, a follow-up analysis describes its transport from the source into the environment. One of the ancillary benefits of such exercises is the insight gained in identifying not only voids in the data needed to assess the accident potential of a facility, but also the research required to better understand how the facility will behave in a variety of nonnormal situations. In cases where such voids cannot be filled through laboratory or field experiments, risk assessors may be forced to depend on solicitations from experts to provide the missing information. Depending on the nature of the voids, such solicitations may prove to be sources of added uncertainty.

A related vital step is sensitivity analysis. This permits a quantitative evaluation of changes in the output of the risk-assessment model as a function of differences in the assumptions that are made in establishing the values assigned to the various input parameters. In essence, such an analysis provides a means for identifying those parameters that are most significant in terms of the final product of the analysis—the risk estimate. This, in turn, helps the assessors identify those parameters where added effort in quantifying them will be most cost effective in improving the accuracy of the assessment.

Continuing Challenges

Beyond those previously discussed, multitudes of additional challenges face the analysts who are involved in the conduct of risk assessments. These include evaluations of mixtures of toxic agents, for example, combinations of hazardous chemicals and radioactive materials (NCRP, 2002), considerations of special groups, particularly children (Schmidt, 1999), and the need to harmonize risk classifications of toxic chemicals. The last of these challenges, which has widespread international implications, is discussed in Chapter 2.

Risk Management

The distinction between risk assessment and risk management is obvious in the matter of regulatory decision-making. Risk management (Figure 17.2) is the process of integrating the risk-assessment results with other information (engineering data, socioeconomic and political concerns), weighing the alternatives and the potential negative impacts of the proposed management approaches, and selecting the most appropriate action for reducing or eliminating the risk. In fact, these factors often play a more significant role in risk-management decisions than the nature and magnitude of the risks themselves.

A successful program in risk assessment and risk management requires that certain aspects of the two processes be kept separated (Wiener and Graham, 1995). Otherwise, it is difficult not to confuse the scientific conclusions about the nature of a risk with the social, political, and economic concerns over how the risk should be managed. This does not mean, however, that there should be no communication between the two groups. Once an assessment has been made, those responsible for managing the risk must be provided with both the risk estimate and the context under which it was developed. The primary areas of uncertainty must be defined, along with the degree to which they may influence the accuracy of the risk estimate (NRC, 1994). The techniques of risk management can then be used as aids in setting priorities for action and in analyzing alternative control strategies.

ACCEPTABLE LEVELS OF RISK

One of the primary considerations in risk management is the level at which controls should be applied. Such a level is required, for example, in determining when cleanup operations at a contaminated site can be terminated, or whether a product is sufficiently safe for consumption or use by the public. Such a level has also been proposed in deciding the probability at which hypothesized future events, such as seismic events, is so "unlikely" that they need not be considered in evaluating the risks of a given facility. If scientific evidence were the only basis for establishment of these levels, such decisions would be relatively straightforward. As will be noted later, public input and the need for flexibility have played a major role, with the result that acceptable levels are generally established on a case-by-case basis.

One example of the application of risk levels is in the determination of the degree of cleanup required for Superfund toxic waste sites. In its initial efforts, the EPA ruled that if an existing site did not impose a lifetime excess cancer risk in excess of 1 in 1 million, no cleanup was required. In some situations, the same limit served as the goal for the degree of cleanup for sites that did not initially comply with this level. More recently, the EPA has defined the acceptable excess cancer risk for the sites that are being evaluated under the Superfund program as a range from 1 in 10,000 to 1 in 1 million. This approach was adopted to provide risk managers the flexibility to consider site-specific factors such as the number of people being exposed and the feasibility and cost-effectiveness of cleanup. In other cases, it has adopted the following limits: state water-quality standards, 1 in 100,000 to 1 in 10 million; and air-quality standards, 1 in 1 million for as many people as possible, and 1 in 10,000 for the maximally exposed individual (Graham, 1993).

With respect to seismic events, the USNRC has stipulated in its siting requirements for commercial nuclear power plants that a "capable" fault is one in which "movement at or near the ground surface" has occurred at least once within the past 35,000 years, or such movements have been of a recurring nature within the past 500,000 years" (USNRC, 1962). In a similar manner, the EPA ruled in establishing its standards for the proposed high-level radioactive waste repository at Yucca Mountain that performance assessments need not include consideration of very unlikely events, that is, those that are estimated to have "less than one chance in 10,000 of occurring within 10,000 years" after disposal of the waste (EPA, 2001).

APPLYING THE REGULATORY LIMITS

Although the establishment of regulatory limits is important, that is only an initial step in the process of risk management. Equally important is the development and implementation of the system through which the limits are applied. In fact, experience has shown that the techniques of application can influence not only the amount of protection provided to the public, but also the manner (favorably or unfavorably) in which the public views the limits. These aspects are vividly illustrated by previously discussed differences in the regulatory and risk-management policies of the EPA and the USNRC, namely, the "bottom-up" versus the "top-down" approach (Chapter 15).

Both of these management policies have appealing features. One of the most important in the bottom-up EPA approach is the incorporation of community-based right-to-know programs, which mandate that the identification of the problem, the associated risks, and the options available for reducing the risks be communicated to, and shared with, the public in plain, easy-to-understand language. This also means that throughout the decision-making process, the public must not only be informed but must also be involved. The top-down approach used by the USNRC also has many good features, including the fact that the dose limits are risk-based, that the data supporting them were derived from human studies, and that they are applied using a safety culture where simple compliance with the standard is not considered sufficient. In concert with this approach, the USNRC requires (Chapter 12) not only that licensees meet the regulatory requirements, but that once they have done so, they demonstrate that all readily available means have been adopted to ensure that any releases to the environment and/or doses to their workers and members of the public are as low as reasonably achievable (ALARA) (Gage, 2001).

There are, nonetheless, situations where confirmation of compliance with the regulations can be extremely difficult. A prominent example is the steps required with regard to the adequacy of the long-term disposal of high-level radioactive wastes. In the case of the proposed repository at Yucca Mountain, Nevada (Chapter 9), the regulations require that the facility be designed to contain these wastes and that the expected annual dose be below the regulatory limit for a period of at least ten thousand years (USNRC, 2001). Under the standard approach, documentation of the adequacy of the facility to comply would have been essentially impossible. Recognizing this fact, the EPA in establishing the applicable standards adopted a much more realistic approach, stating that the Department of Energy, which is responsible for building the proposed facility, need document only that there is a *reasonable expectation* that the limits will be met. Such documentation requires less than absolute proof because such proof is impossible to attain (EPA, 2001).

THE PRECAUTIONARY PRINCIPLE

One of the principles that has served as a foundation for ensuring the protection of human health and the environment is that there need not be scientific certainty that an agent or activity causes harm before prudent action is taken to limit or avoid exposure to it. Although this principle had been enunciated decades earlier, international attention was drawn to it through Principle 15 of the 1992 Rio Declaration on Environment and Development. Now identified as the "precautionary principle," it was expressed as follows:

> In order to protect the environment, the precautionary approach shall be widely applied by states according to their capabilities. Where there are threats of serious or irreversible damage, lack of full scientific certainty shall not be used as a reason for postponing cost-effective measures to prevent environmental degradation.

Various expressions of this principle have been subsequently incorporated into international trade and environmental agreements, particularly in those of the European Community and the national laws of some of the countries in Europe. In many instances, however, applications of the principle have proved to be controversial because of widespread disagreement on what it calls for risk managers to do and the degree to which its application should supersede a fully developed risk analysis. Supporting groups believe that it means that no environmental releases should be permitted until they can be unconditionally proved safe, even though such a goal is not attainable. Other supporters, seeking to ensure adequate margins of safety, have called for the use of worst-case assumptions in the conduct of the associated risk assessments. In rebuttal, opponents emphasize that any precautionary actions must be cost effective, appropriate, and proportionate to the threat that is to be avoided. They also point out that the precautionary principle in no way was intended to obviate the need to perform an analysis of the risk, and that any precautionary actions should be viewed as provisional until such time as the necessary research is conducted to confirm whether the assumed risks are real (Rhomberg, 2001).

Regardless of these "flaws," there are cases where application of this principle may clearly be warranted. Prominent among these are situations that involve what might be called "creeping" environmental change, for example, the possible long-term impacts of greenhouse gases on global climate, and low-grade cumulative environmental problems such as acid precipitation, air and water pollution, soil erosion, tropical deforestation, and habitat destruction. Another example is the ongoing debate on genetically engineered foods. In each of these cases, there is no clear threshold that distinguishes negligible from serious harm. There is also a risk that without action, the problem will languish without adequate attention until it is too late to prevent serious damage (Graham, 1999). From the perspective of many environmentalists, application of the precautionary principle can serve both as an effective tool in developing an integrated approach to environmental protection and as a means for identifying and enunciating disagreements on the management of associated potential risk (Rhomberg, 2001). Nonetheless, experience has shown that application of the principle can lead to unanticipated and unfavorable results. A classic example was the addition of methyl tertiary-butyl ether (MTBE) to gasoline with the intention of reducing air pollution. Subsequent observations of MTBE contamination in groundwater supplies (Chapter 5) led to a retraction of this action (Graham, 1999).

Case Studies: Risk Management

As noted earlier, risk management involves the integration of the outputs of risk assessments with a variety of other types of information, coupled with the weighing of a host of contributing factors, to reach a decision on the most appropriate and/or most effective way for managing a specific risk. Because they represent a wide range of the types of considerations and decisions that enter into such evaluations, two examples are summarized here.

CONTROL OF MAD COW DISEASE

In 1986, outbreaks of mad cow disease (bovine spongiform encephalopathy) were detected in cattle in the United Kingdom. Investigations showed that the disease could be spread to healthy cattle if the feed they ate contained rendered waste from the processing of meat (particularly tissues from the brain and spinal cord) from other infected cattle. Therefore, a major component of the control program was to prohibit the consumption of these specific tissues. To accomplish this goal, hundreds of thousands of cattle that either died of the disease or became infected and were killed were disposed of under carefully controlled conditions. The need for control was emphasized by the potential identification of a related fatal human variant of the disease, whose most likely mode of transmission is the same as that for cattle, namely, the human consumption of meat from infected cattle. Through these efforts, the epidemic was subsequently brought to an end. Because of fears that the disease could spread to the United States, the U.S. Department of Agriculture (USDA) asked the staff of the Harvard Center for Risk Analysis to review and evaluate methods for managing the risk.

Basic to the response of the center was the development of a simulation model that took into account the key factors in the transmission of the disease and the control measures that were in place in the United States at that time. The latter included (1) a ban on the importation of certain animals and products from countries where the disease was present and (2) restrictions on certain animal-feed practices in this country. Also factored into the model was an accounting of the degree of anticipated compliance with these restrictions. In addition, at the request of the USDA, the model included capabilities for evaluating the relative importance of various control measures and for identifying additional measures that might be warranted (Gray, Cohen, and Kreindel, 2002).

On the basis of their analyses, the center's staff concluded that under all foreseeable scenarios, the prevalence of mad cow disease, even if it were introduced into the United States, would subsequently decrease over time and would be virtually eliminated within 20 years. At the same time, however, they identified areas where more information was needed. These included data on the disposition of potentially infected animals that might die in this country and on the degree of compliance by meat-processing plants with existing rules that require control of the disposition of the previously cited high-risk tissues from animals that are slaughtered. Factors identified as having the most influence on the likelihood that the disease could spread among animals in the United States included (1) incomplete compliance with the ban on the use of cattle protein in cattle feed and (2) mistakenly providing cattle with feed intended for pigs or poultry (which can legally contain rendered cattle protein). Included among the latter possibilities was that feed and feed products for cattle might have been mislabeled. Factors identified as having the most effect on the potential risk to the human food supply were the consumption of (1) brain and spinal cord tissues from infected cattle and (2) meat from advanced recovery systems (which extract the meat left on the bone after hand butchering). In the latter case, such meat might be contaminated with potentially infected nervous system tissues (Gray, Cohen, and Kreindel, 2002).

Although one case of the variant disease was reported in a human in the United States in 2001, the patient had been born in the United Kingdom and had moved to this country in 1992. Since the time between exposure and onset of illness was estimated to range from 9 to 21 years, medical specialists concluded that it is likely that exposure occurred in the United Kingdom (CDC, 2002). On the basis of the results of the studies of the Harvard Center staff and other information, the USDA announced a series of additional actions to reduce the risk posed by the disease. These included (1) increasing by a factor of more than two the number of cattle in this country to be tested for the disease and (2) identifying additional regulatory actions that might be taken to reduce the risk of exposure and to ensure that U.S. cattle feed remains free of potentially infectious materials.

The simulation model was subsequently used to evaluate the effectiveness of each such proposed action in terms of the reduction in risk. Although the chance that the disease would occur in the United States could not be ruled out, the results of the Harvard Center analyses suggest that the steps that have been taken will ensure that the disease will not spread widely, even if it is introduced. Most important, the study showed that steps that were being taken for controlling the disease made it virtually impossible for an epidemic similar to that in the United Kingdom to occur in this country (Gray, Cohen, and Kreindel, 2002). Their assessment was substantiated in December 2003 when, as noted earlier (Chapter 6), a case of the disease occurred in a cow that had been imported into the United States from Canada. Due to the plans in place, the infected animals were quickly identified and destroyed and the potential of an epidemic was eliminated.

DISPOSAL OF CATHODE-RAY TUBES

One of the common components of older television sets and computers is the cathode-ray tube (CRT). Although such tubes are now being replaced by screens that incorporate liquid crystal technology, disposal of the existing CRTs is a special concern because they (as well as their printed circuit boards and batteries) contain lead to shield the user from ionizing radiation generated by the monitor (Chapter 9). Because the newer screens do not generate ionizing radiation, they do not need to be shielded. To evaluate the risks associated with CRTs, staff members at Resources for the Future (Macauley et al., 2001) constructed a simple model to track what happens once the monitors are destined for disposal and to assess the relative impacts of the available options. A basic assumption in the analysis was that all consumers selected the least costly approach, without explicit account of either social costs or subsequent health effects.

Two groups of users, residential and commercial, were considered in the assessment. In the former case, the monitors were assumed to be disposed of in a municipal landfill, incinerated, or recycled. In the case of large commercial users, it was assumed that the CRTs were handled, as required by federal law, under the regulations pertaining to hazardous waste. Small-quantity commercial users were assumed to use the regular trash-collection service. Also included as an option for all types of users was the selection of storage as a short-term option. Since there are significant differences in the costs of the several disposal options, the analysts divided the consumers into six groups. Four of these covered different types of residential consumers: (1) those living in apartments, (2) those living in houses, (3) those who explicitly must cover the cost of the disposal options selected, and (4) those whose waste fees are hidden among other local taxes and charges. Apartment and house dwellers were separated into two groups since they were assumed to face different costs for storage. The remaining two groups were (5) large commercial consumers, who were, as noted earlier, classified as hazardous waste generators, and (6) small commercial consumers, who were classified as nonhazardous waste generators.

The evaluation showed that the private and social costs associated with disposal depended on where the disposer lived and the disposal options selected. It is important to note in all cases, however, that someone bears the costs. If the consumer selected storage as a disposal option and lived in a home, the added cost was assumed to be zero. If he/she lived in an apartment, the cost was based on rental rates per square foot. If the consumer was assumed to leave the monitor at a drop-off center, account was made of the associated transportation and travel-time costs and of the expenses to the community as a whole in operating the center. With curbside recycling, the costs to the householder may be less, but the community bears the pickup costs. In those cases, where these costs are paid through general governmental revenues, they were defined as community costs. The ultimate goal was to compare these costs, both private and community, with the estimated monetary costs of the health effects that would be avoided if various disposal policies were implemented. The latter costs were computed using a model developed by the EPA (Macauley et al., 2001).

On the basis of practices applied to the nearly 16 million monitors retired in the United States during 1998, the estimated private and community disposal and recycling costs were somewhat less than $1.00 per monitor. The health damages that were avoided through recycling were estimated to range from $0.05 to $0.15 per monitor, depending on the amount of lead each CRT monitor was assumed to contain. Nationwide, the estimated total cost of handling the monitors that year was about $13.5 million and the monetary equivalent of the avoided health damages ranged from $0.8 to $2.7 million. From the standpoint of the environment, banning the incineration and the disposal of monitors in landfills or coupling such bans with various financial incentives to recycle the monitors would have two benefits. They would eliminate the health effects associated with disposal and increase the rate of recycling. At the same time, however, they would dramatically increase the waste-management costs for consumers and the local community. In fact, estimates are that this approach would result in an estimated management cost of $300 million, with the monetary equivalent of the avoided health effects remaining at $0.8 to $2.7 million. If the objective were to avoid health effects entirely, the most economical approach would be to ban incineration. In this case, however, the added cost would be about $50 million to avoid the previously cited health damages (Macauley et al., 2001). In reviewing this study, it is important to note that the values for many of the parameters required as input into the assessment models are estimates at best. For this reason, the quoted monetary values encompass a wide range of uncertainties.

International Risk Policy

As the preceding discussion has shown, there are major differences in the approaches to risk assessment and risk management used by various agencies within the U.S. federal government. Although in some cases these are due to the nature of a toxic agent and the number of people likely to be exposed, there are other major influencing factors. In many cases, for example, Congress prescribes the risk-based elements for regulating a particular facet of the environment within the laws that mandate the action.

One example is the 1 in 10,000 to 1 in 1 million lifetime risk limit specified within the Comprehensive Environmental Response, Compensation, and Liability Act (Beck, 2001). Another influencing factor is that the EPA, in promulgating regulations in accord with various environmental laws, is required to support them with in-depth scientific and risk-based analyses. This further restricts the opportunities for any type of interpretation of the laws.

In contrast, the approaches that are being proposed in the EU member countries for risk assessment and management are largely hazard based. This is particularly true in the control of toxic chemicals. Therefore, the primary risk-management measures are limited to labeling requirements and associated restrictions that, in the main, are based solely on the characteristics of the chemical, for example, whether it has been deemed a carcinogen. Another contrast is that whereas judicial interpretation of environmental statutes in the United States still occurs on a frequent basis as a result either of litigation involving challenges to existing environmental regulations or of demands for new and/or additional regulations, the political system within the EU is based on a civil code. Although the governments in the member countries are responsible for implementing and administering the accompanying directives, they have wide latitude in interpreting and enforcing them (Borchardt, 2000). That is, they can essentially tailor their directives to meet their individual needs. In addition, the judiciary in the EU member countries is relatively weak. As a result, suits by individuals and organizations regarding the implementation of environmental statutes are relatively infrequent. Overall, environmental policy making in the EU member countries may be characterized as less formal, more flexible, and less analytical than that in the United States (Beck, 2001).

Another reason for the contrasts is that there are fundamental social, legislative, and judicial differences between the United States and the EU countries. Environmental programs in the United States are designed to address toxic chemicals in individual media, for example, in meat and vegetables; within specific environments, for example, the workplace; or in specific uses, for example, the application of pesticides. In addition, the regulatory system within the United States involves a dynamic between relatively specific environmental statutes, coupled with a strong judiciary, and a litigious society.

In spite of these potential drawbacks, it is anticipated that risk assessments will be increasingly applied both in the United States and throughout the world. While there are differences among the approaches used in the EU countries, they probably do not exceed those previously described among various federal agencies within the U.S. government. Obviously, a strong effort is needed to resolve these differences both nationally and internationally. Without some type of agreement, chemicals viewed as unacceptable for application in one country might be deemed acceptable in another. In the meantime, the resulting confusion will continue to have a negative impact on international trade.

Risk Communication

Risk-management actions can range from a program of public education to termination of the activity that is the source of the risk. Essential parts of the risk-management process are to communicate the nature and extent of the risk in understandable terms to both experts and laypersons and to share background information fully with all affected groups to make it possible to arrive at an informed estimate of the risk and a wise approach to its management (NRC, 1996).

Even so, risk communication can be a very difficult task. A primary contributor to these difficulties is the manner in which members of the public perceive risk. People are far less fearful of a risk they believe they can control, for example, driving a car; they have less fear of a risk that is natural (radon) that of one that is human made (nuclear power); and they are less fearful of a risk that is old (AIDS) than of one that is new (SARS) (Ropiek and Slovic, 2003). As a result, the perceptions of individuals regarding technological risks, in particular, often bear very little relation to the risks as assessed by scientists. A further problem is the fact that many people have difficulty in (1) dealing with risk estimates that are expressed in terms of numbers with negative exponents and (2) understanding the difference between numbers that are intended to represent upper bounds and those that are intended to represent best estimates of risk, or those that are intended to represent annual versus lifetime risks.

Until recently, scientists and engineers were inclined to believe that the public was simply ignorant of the relevant facts or was irrational and that its views could be disregarded. In reality, many (if not most) of these types of problems are directly attributable to the inadequate manner in which the scientific community communicates this type of information. One possible approach would to be shift to new methods for expressing the nature of risk, for example, expressing various activities in terms of the associated one-in-a-million risk of death (Table 17.5). Another would be to express

Table 17.5 Activities with a one-in-a-million risk of death

Group 1. Living in the United States: time to accumulate a one-in-a-million risk of death

Activity	*Time*
Falls (average over 70 years of age)	15 hours
Firearms	3 days
Falls (average over life)	6 days
Fires	13 days
Drowning	19 days
Floods	2 years
Animal bite or sting	4 years
Tornadoes	5½ years
Lightning	6 years

Group 2. Work-related risks: time to accumulate a one-in-a-million risk of death

Occupation	*Time*
Coal mining (black lung disease)	2 hours
Firefighting	11 hours
Police officer	1¼ days
Coal mining (accidents)	1½ days
Agriculture	1½ days
Transport and public utilities	3 days
Airplane pilot	3 days
Frequent business airline flyer	10 days
Manufacturing	10 days
Government	16 days

Group 3. Personal lifestyle: activities required to represent a one-in-a-million risk of death for adults

Drinking and eating

30 diet sodas with saccharin
30 liters (quarts) per day of water containing EPA limit (10 ppb) of arsenic
4 tablespoons of peanut butter every 10 days (aflatoxin)
35 slices of fresh bread (formaldehyde)
75 pounds of charcoal-broiled steak (benzopyrene and other aromatic hydrocarbons)

Other activities

Smoking 2 cigarettes (cancer and heart disease)
Nonsmoker living in a home with smoker for 2 weeks
Living in Denver versus New York for 40 days (cosmic radiation)
Paddling a canoe for 6 minutes
Traveling 10 miles by bicycle
Traveling 100 miles by car

risks in terms of the "years of life lost." This, for example, would be one means of conveying the fact that a high percentage of those killed in motor-vehicle accidents are in the younger age groups. A similar approach would be to express the relative risks of various activities in terms of the estimated amounts by which they shorten one's life (Table 17.6). Above all, risk assessments should be conducted in the open, and the public should be permitted, indeed, encouraged, to participate. In support of such an approach, the EPA has published what it describes as the seven cardinal rules of risk communication (Table 17.7). Another very useful item is the list of key principles that must be considered in developing risk estimates and conveying the results to the public (Table 17.8).

Table 17.6 Days of life lost through various activities and events

Activity and/or event	Days lost
Natural hazard	
Earthquake	0.2
Hurricane	0.3
Flood	0.4
Tornado	0.8
Lightning	1.1
Sports activity	
Snowmobiling	2
Professional boxing	8
Hang gliding	25
Parachuting	25
Mountain climbing	110
Accident—general	
Electrocution	4.5
Fire	20
Drowning	24
In the home	74
Motor vehicle	207
Accident occupational	
Trade and services	27
Manufacturing	40
Government	60
Mine quarrying	167
Construction	227
Agriculture	320

Table 17.7 The EPA's seven cardinal rules of risk communication

1. Accept and involve the public as a legitimate partner.
2. Plan carefully and evaluate your efforts.
3. Listen to the public's specific concerns.
4. Be honest, frank, and open.
5. Coordinate and collaborate with other credible sources.
6. Meet the needs of the media.
7. Speak clearly and with compassion.

Table 17.8 Key principles of risk assessment

1. Estimates of attributable health risk should make use of the best available science.
2. Since reputable scientists often do not agree about how to assess risk, scientific disputes should be acknowledged.
3. When hard data are lacking, risk assessments should be explicit about any assumptions and should indicate the degree of sensitivity of results to plausible changes in assumptions.
4. Meaningful risk assessments usually develop a central estimate of risk, as well as upper and lower bounds on risk that acknowledge the extent of scientific uncertainty.
5. Public policy decisions about acceptable risk require public participation and application of democratic principles.
6. No quantitative level of risk exists that is universally acceptable or unacceptable; acceptable depends on the circumstances, the people affected, and the decision context.
7. Valid decisions about health risk require consideration of other cherished values such as quality of life, equity, ecological health, personal choice, and economic welfare.
8. Programs to reduce risk should be designed to avoid unintended side effects that may increase risk.
9. When risk reduction is desired, economic incentives and information should be considered in addition to conventional command-and-control regulation.
10. The context in which risk occurs (e.g., voluntary versus involuntary risk) may influence public reaction to risk as much as the magnitude of the risk in question.

Table 17.9 Lessons learned about exposure assessments

1. In seeking to be prudent, risk assessors have placed too much emphasis on the so-called maximally exposed individual.
2. While it is important to characterize the risks to major segments of the population, attention needs also to be directed to evaluations of exposures to special groups who, because of their location and/or living habits, may be exposed to unusually high doses.
3. The repeated use of conservative assumptions should not dictate the results of the assessment.
4. Both risk managers and the public want to understand the statistical confidence in estimates of risk; application of sensitivity analyses can yield important information about and help identify the critical exposure variables.
5. Care should be exercised to ensure proper consideration of data from samples that have no detectable amount of contamination; otherwise the impact of a few samples with detectable contamination can lead to improper conclusions about the actual level of risk.
6. Applications of new models, measured data, and technological developments now enable exposure assessors to estimate with confidence exposures that occurred 40 to 50 years ago.
7. The need to estimate indirect pathways of exposure is now recognized; these include the ability to estimate uptake via inhalation of volatile contaminants released from water while a person is taking a shower.
8. Children are not miniature adults and their exposure patterns are not the same as those of adults.
9. Advances in analytical chemistry have enabled biological monitoring to serve as an excellent tool for validating or confirming the predicted degree of human exposure.
10. In most cases, the most significant risks due to exposure to chemicals occur in the workplace.
11. For most persons, exposures to chemicals and bacteria in the home pose a higher risk than to those in the ambient air or through ingestion of water.

The General Outlook

Ideally, risk assessment can provide structure to the process of setting standards and serve as useful input into public debates on the acceptability of proposed industrial facilities and other types of activities that may have both positive and negative impacts on the public and the environment. At the same time, however, regulators must be careful not to give disproportionate weight to the estimates generated through this process. In particular, they must not let quantitative risk assessments relegate human values and perceptions to a secondary role. Of key importance is that the efforts be conducted in an atmosphere (as just described) that increases public perception of the legitimacy of the regulatory process.

For optimal benefit from the risk-management perspective, it would appear to be appropriate to maximize the number of toxic agents regulated rather than the stringency with which they are controlled on an individual basis. It would also be wise to concentrate on controlling the principal uses of such agents. A pesticide, for example, may have a dozen applications, but only two of these may account for 98 percent of the amounts actually used. In the face of limited resources, regulatory efforts should be concentrated on these major applications. Concurrently, it is important to examine the total system to which a given risk-management procedure is being applied. Unless care is exercised and all interacting factors are considered, risk assessments directed at single issues, followed by ill-conceived management strategies, can create problems worse than those they were designed to correct. A prime example is the previously cited addition of MTBE to gasoline. The single-issue approach can also create public myopia by excluding the totality of alternatives and consequences needed for an informed public choice.

In spite of these caveats, our understanding of many components of risk assessment and risk management has been considerably enhanced in recent years. This is especially true with respect to our ability to analyze, quantify, and interpret exposures. A summary of the lessons that have been learned during the past several decades in this component of the risk-assessment process is presented in Table 17.9.

18

ENERGY

MEETING energy needs and protecting the environment are inseparable goals. Strip mining of coal, for example, can degrade the environment; the drilling, acquisition, and transportation of oil can lead to spills that contaminate vast areas of land, water, or both; and the use of gasoline in cars leads to air pollution and smog. Internationally, the production and consumption of oil can lead to conflicts among nations and even to wars. In a similar manner, the generation of electricity, whether through the combustion of fossil fuels, the fissioning of nuclear fuels, or the harnessing of waterpower, leads to air pollution, problems of waste disposal, and/or other effects on the environment. Longer-range impacts that arise through the burning of fossil fuels include acidic deposition and global warming.

After years of the consumption of energy resources as if they were unlimited, policy makers throughout the world recognize that continued health and safety and environmental protection will be possible only if these resources, particularly the supplies of nonrenewable fossil fuels, are carefully managed, conserved, and protected. The urgency of coping with energy consumption in the United States is illustrated by the fact that this country, with only about 6 percent of the world's population, consumes about 25 percent of the world's energy and about 40 percent of the world's gasoline production (Zauderer, 2002). In fact, domestic sources of oil have been depleted to the extent that the United States imports almost 60 percent of the oil it consumes, the sources of which are shown in Figure 18.1. Overall, this country produces less than three-quarters of the energy it consumes, 85 percent of which is produced by the burning of coal (Truly,

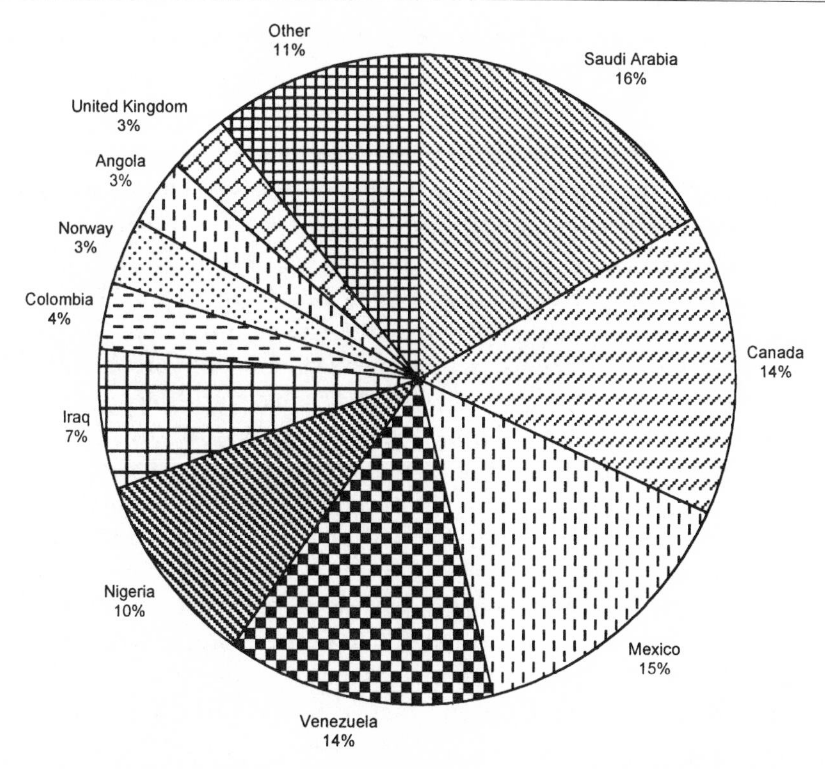

Figure 18.1 Non-domestic sources of oil consumed in the United States, 2002

2002). Other major consumers are the seven largest countries of the Organization for Economic Cooperation and Development (OECD), which consume more than 40 percent of the world's production of fossil fuels.

Energy Uses and Conservation

As many environmental organizations have emphasized, large amounts of energy are wasted. A major contributing factor in the case of the United States is that the prices charged for various types of fuels have traditionally been far below what they should be if even a modest amount of the environmental and public health impacts of their uses were taken into consideration. For these reasons, there have been major efforts in this country during the last several decades of the twentieth century and first decade of the twenty first century to encourage conservation. Steps that have been taken within each of the major consuming sectors are discussed in what follows.

INDUSTRIAL SECTOR

Relying on a mix of fuels, the industrial sector accounts for 38 percent of end-use energy consumption in this country. Of this, 70 percent is used to provide heat and power for manufacturing. In all, this leads to the consumption of 25 percent of the nation's petroleum. Although progress is being made in reducing energy requirements in essentially all types of industries, accomplishments in the chemical industry are particularly noteworthy. In fact, experience shows that for this particular segment, energy productivity is just as important as labor productivity. One of the major reasons is that the chemical industry consumes 23 percent more electricity than primary metals, the second-largest such user. Although electricity use within this industry increased by 20 percent between 1986 and 2000, production rose by more than 50 percent. This led to a 21 percent reduction in the consumption per unit of product, almost 60 percent better than the industrial average. Although there were multiple ways in which these reductions were achieved, one of the most important was the increasing use of cogeneration plants, that is, plants in which the waste heat from one operation is used as input into another stage in the process. One of the outgrowths of this experience is that increasing numbers of industrial leaders are recognizing that in reducing their consumption of energy, they are saving money and reducing the release of contaminants into the environment.

COMMERCIAL SECTOR

About 19 percent of the total energy in the United States is consumed in the commercial sector. A major portion of this is for lighting, which represents, for office buildings, more than 25 percent of electricity usage. To conserve such use, it is important to install energy-efficient lights, make better use of daylight, and incorporate features such as those that turn off the lights when the occupants leave a room. Compact fluorescent units consume only 25 percent as much electricity as incandescent bulbs and can last 12 times as long. Although their initial price is higher, the total cost of the lighting provided over the long term is far less. To provide additional conservation, Congress in 2002 approved support for a 10-year, $500 million research effort to revolutionize the methods available for providing artificial light. A stimulus to this effort was the development of new light-emitting diodes and organic thin films that could provide the basis for the development of solid-state lights. Estimates are that these systems, once in widespread use, could reduce global electricity consumption by 10 percent or more (Malakoff, 2002).

Another major consumer of energy in the commercial sector is the use of electricity to operate electronic equipment such as personal computers (PCs), printers, copiers, and facsimile machines. If the additional electricity used in ventilation and air conditioning is included, the combined load represents some 10–12 percent of the total electricity demand of a typical office. To reduce such usage, computers, their printers, and other desktop devices should be turned off at the end of every workday. Fortunately, the monitors on most computers now automatically shift into a standby mode after a period of inactivity. Not only does this save energy, but it also reduces thermal stresses, improving the reliability of the electronic components within such units. Other steps that can be used to conserve electricity in the operation of computers and their accessories are summarized in Table 18.1.

RESIDENTIAL SECTOR

Data show that one-sixth of the energy in the United States is consumed in the home. Therefore, the home is obviously an important place in which to practice conservation. This can be accomplished through multiple avenues. Energy usage through heating and air conditioning can be significantly reduced by weather-stripping doors and caulking windows. Another step is to set the thermostat at the lowest comfortable setting in winter and the highest comfortable setting in summer. Cooling costs can also be reduced by installing a more efficient air-conditioning unit. Improvements have been made at such a rapid pace that the cost of such units can be reclaimed within as short a time as 5 to 10 years simply on the basis of the reduced costs for electricity. As is the case in the commercial sector, residents should switch to more efficient lighting systems. If every household in the United States replaced four 100-watt incandescent light bulbs with compact fluorescent units, the savings would be equivalent to the output of about 10 1,000-megawatt electricity-generating stations.

Residents should also insist on energy-efficient home appliances. Such products are readily available in the United States. One example is the clothes washer, which, in a typical home, consumes almost 13,000 gallons

Table 18.1 Recommendations on office equipment purchase and use for maximizing energy efficiency

Item	Guidance for purchasing	Guidance for operating
Personal computer	Buy a laptop computer Buy an energy-efficient unit	Turn off at night and weekends Activate power-management features. Turn off when not in use during the day
Computer monitor	Buy an energy-efficient unit Consider an active-matrix color liquid crystal display Buy a monitor only as large as needed Buy only as much screen resolution as needed	Turn off at night and weekends Activate power-management features Turn off when not in use during the day
Computer printer	Consider an inkjet printer Buy an energy-efficient printer Consider sharing a printer Consider a unit with double-sided printing	Turn off at night and weekends Activate power-management features Reuse paper Use electronic mail
Copier	Choose a properly sized unit Consider a copier not based on heat and pressure fusing technology Compare ratings provided by the American Society for Testing and Materials Buy a unit with power-management features Choose a unit offering convenient two-sided copying	Turn off at night and weekends Activate power-management features Use two-sided copying whenever possible Batch copy jobs

of water each year. On average, washers are the highest consumers of water within a home, representing about 22 percent of the total usage. Switching to a more efficient unit will not only save water during the wash cycle but will also more effectively extract the water that remains in the clothes after the final rinse. This will reduce the amount of energy required to dry them (Consumers Union, 2002). The second-highest consumption of water in the home is that for the flushing of toilets. This can be dramatically reduced through the use of low-flush and/or pressure-flush units.

TRANSPORTATION SECTOR

The transportation sector consumes the remaining 26 percent, a major share of which is in the form of fuel for the more than 200 million automobiles now in use in this country. Not only is usage high (accounting for more than 65 percent of the oil consumed in this country), but it is increasing rapidly, having tripled during the past 50 years. Recognizing the implications of this increase and the contributions of automobile emissions to air pollution, the U.S. Congress in 1978 passed legislation that required improved corporate average fuel economy (CAFE), that is, the average miles per gallon that vehicles must achieve. Although this resulted in a 30 percent improvement during the following decade (1978–1987), thereafter the rate of improvement steadily began to decrease; by 2002 the CAFE of the nation's fleet was down by 7 percent from the peak in 1987. A major source of the problem was that the 1978 legislation applied only to passenger automobiles and light trucks; minivans and sport utility vehicles (SUVs) were exempted. Another contributing factor was the installation in motor vehicles of multiple energy-consuming accessories, including, for example, air-conditioning units, radios, and CD players. After lengthy debate, Congress in late 2002 addressed the loophole in the original legislation by requiring an increase in fuel economy for SUVs and minivans. Also included were pickup trucks, which had by that time become one of the largest-selling vehicles (Grant, 2002).

Because more stringent standards will undoubtedly be forthcoming, automobile manufacturers are exploring a variety of possibilities for additional improvements, since most of the advances through weight reductions and engine refinements have already been achieved. One approach being actively pursued is the marketing of what are called hybrid cars that are powered by a combination of a gasoline engine and electric motor. The batteries that power the electric motor are recharged through harnessing the kinetic energy of the car during braking. Some of these cars have outstanding fuel economy ranging above 20 kilometers per liter (50 miles per gallon). In fact, such a vehicle was selected by the editors of one automobile magazine as the 2004 American Car of the Year (Gritzinger, 2004). The same company that manufactures this car has announced that it is intends to market a luxury version that incorporates these same features. Other approaches being adopted include the wider incorporation of navigation systems that enable drivers to determine exactly where they are and to select the best route to reach their destination, saving both fuel and time. Another is the previously described upcoming shift from 12- to 36-volt batteries (Chapter 11), which will enable the engine to stop instead of idling at traffic lights, plus the installation of features such as the "steer-by-wire" approach (Hamilton, 2002). This saves weight by eliminating the mechanical and hydraulic linkages to the front wheels. Also being marketed are automobiles with engines that can be operated on all or a portion of their cylinders. Called "displacement on demand," this would enable an eight-cylinder engine, for example, to be operated on either four, six, or eight cylinders, depending on the power demands at a given time. Since no fuel would be used by the cylinders that are idle, this could yield an 8 to 25 percent increase in fuel economy (Bolig, 2003).

Energy Resources

The fuels available to meet the world's energy needs fall into two broad categories: (1) nonrenewable sources, such as fossil and nuclear fuels, and (2) renewable sources, such as solar and geothermal energy.

FOSSIL FUELS

Fossil fuels are the best examples of nonrenewable sources. The most common forms in use today are coal, oil, and natural gas. Estimates of the worldwide and domestic resources for each of these are as follows:

Coal. Perhaps sufficient to meet needs for two to four centuries on a worldwide basis. At present, the United States has abundant reserves.

Oil (petroleum). Perhaps sufficient to meet needs for three to five decades on a worldwide basis. If worldwide reserves thought to exist but still not discovered are considered, the supply could last more than 150 years (Voss, 2002).

Natural gas. Perhaps sufficient for five or more decades on a worldwide basis. However, new supplies are being discovered, and the total recoverable reserves are not known at this time. In fact, some experts predict that supplies yet to be discovered could prove adequate for as long as 200 years.

In evaluating this information, it is important to recognize that the estimated reserves of these sources are not static quantities. The technology for resource exploration, development, and extraction continuously improves. The proven oil reserves in the United States today, for example, are double those thought to be available 50 years ago. At the same time, new technological developments and discoveries are producing dramatic changes in the amounts of such reserves that can be extracted. With current techniques, as much as two-thirds of the oil present in a well remains in the ground. It is too thick to recover. New techniques, such as the use of bacteria, are being applied on an experimental basis to break up the complex carbon molecules in the remaining oil so that it can be pumped out. Such a process could double the production in up to 40 percent of the existing oil wells (Greene, 2001). Another favorable factor is the discovery that some conventional oil and gas reservoirs that had previously been thought depleted appear to be refilling through seepage from deeper reservoirs. Yet to be determined are the anticipated volumes of such seepage or the proportionate volume of the depleted reservoirs that may ultimately be refilled.

Much the same situation exists with respect to natural gas. Enormous amounts, for example, are believed to be buried deep beneath the ocean floor. Some estimate that such supplies could last tens of thousands of years. Nonetheless, questions remain. Will it be possible, for example, to transport natural gas discovered in faraway areas in a safe and efficient manner to where it is needed? Since it is volatile, natural gas is difficult to handle. If it is shipped as a liquid, it must be maintained at a temperature of −130°C (−200°F), or at tens of atmospheres pressure. The fact is that much of the natural gas in the world in located in such remote areas that it is currently worthless from a commercial point of view (Voss, 2002).

In the case of oil, federal officials have periodically recommended that exploration be conducted within the Arctic National Wildlife Refuge. Although similar recommendations in the past have been rejected as a result of, for example, the *Exxon Valdez* oil spill (Chapter 19), pressures generated by the nation's increasing dependence on foreign oil continue to lead to renewals of such proposals. Proponents indicate that technological improvements, such as multilateral drilling, have significantly reduced the environmental impacts of such activities. They also point out that exploring for oil is now a more exact process, and that the oil can be transported through pipelines located above the ground surface, thereby eliminating the need to disrupt the land surface for repairs. In contrast, opponents caution that the process of extracting crude oil can still mar the landscape and introduce the risk of oil spills, and they emphasize that Alaska's North Slope sustains a rich Arctic ecosystem that provides a habitat for polar bears, musk oxen, and several caribou herds. Other opponents have noted that projections of the total amount of oil that would be produced by the proposed drilling would meet only about 5 percent of our national needs. This debate will undoubtedly continue for years to come.

NUCLEAR FUELS

Nuclear fuels are another example of nonrenewable energy sources. The energy available from existing U.S. sources of uranium, if consumed exclusively in the current generation of nuclear power plants, is estimated to be about equal to that available through the combustion of the existing sources of natural gas or petroleum. With the development of breeder reactors (and the recycling of plutonium), however, nuclear fuels could become a much larger energy source than any of the fossil fuels (Figure 18.2). This is discussed in more detail in the sections that follow. Even so, none of these sources begins to approach the amount of energy that would be made available by the effective harnessing of the fusion process. After having pursued research on this process on an independent country-by-country basis, leaders in these efforts concluded about 10 years ago that the magnitude of the required effort was so large that the best hope for success was to pursue the development of a successful machine on a cooperative basis. This culminated in plans for the International Thermonuclear Experimental Reactor (ITER), a cooperative venture involving scientists from the European Union, China, Japan, Russia, South Korea, and the United States. Projected for completion in 2014, the ITER will be twice the size of any existing fusion experimental device. If successful, it will

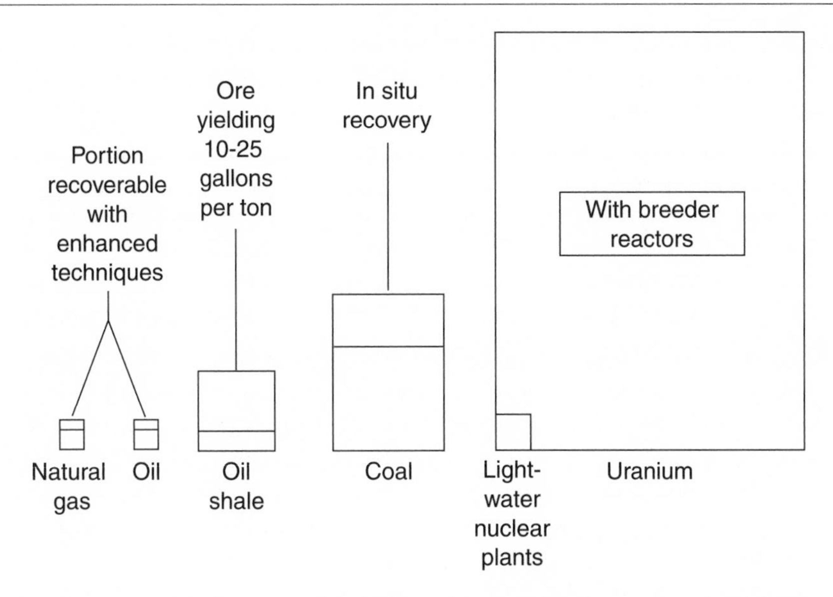

Figure 18.2 Comparison of available energy resources in the United States

pave the way for fusion power stations. Nonetheless, major challenges, both of a technical and political nature, remain (Clery and Normile, 2004).

GEOTHERMAL ENERGY

Although geothermal energy is available in enormous quantities, it is often difficult to tap and is limited to certain geographic areas, most of which are remote from population and industry. The main potential of geothermal energy in the United States appears to lie in localized use as a source of inexpensive heat on a relatively small scale.

SOLAR ENERGY

Solar sources offer tremendous potential for meeting the world's energy needs. Since such sources are renewable, once the technology for a given use has been established, it can be applied more or less indefinitely.

Hydropower. Hydropower is one of the most common sources of solar energy. A prominent example is the enormously successful series of hydroelectric power plants that were constructed by the Tennessee Valley Authority on major waterways in the south central United States during the mid-twentieth century. Still, hydroelectric plants can have severe impacts on environmental and public health. These will be discussed in the section on "Environmental Impacts" that follows.

Tidal Power. For years, it has been known that the crashing of waves upon beaches and the rise and fall of the ocean tides represent vast sources of energy. What was lacking was a system for converting this energy into electricity. Systems have now been developed for accomplishing this goal. In fact, the world's first commercial system for harnessing the energy in waves was installed in November 2000 on the Scottish island of Islay. The plant includes a large, partially submerged, concrete chamber built into the shoreline. The incoming water forces air trapped in a chamber 25 meters (about 80 feet) wide onto the blades of a turbine that is connected to a generator. When the waves recede, the vacuum created in the chamber sucks air back through the turbine in the opposite direction, permitting electricity to be generated both when the waves come in and when they recede. The plant can reliably produce 500 kilowatts of power. Similar projects are in various stages of evaluation in Australia, Great Britain, Japan, and New Zealand and on the west coast of the United States (Staedter, 2002). Other innovations include the use of small helical turbines to generate electricity by harnessing the flow of water in streams and tidal basins without the need for a dam, and the development of a hydrofoil system for harnessing the energy of tidal streams beneath the sea.

Wind power. Considerable progress is being made in harnessing wind power. In fact, it is now the most rapidly expanding energy source in both the United States and the world. A major stimulus for these developments is the dramatic reduction (by a factor of more than 10) during the past 20 years in the costs of producing electricity by this means. As a result, its cost today is about one-quarter of that for electricity generated using natural gas. A new generation of ultrahigh-efficiency plants, soon to be available, holds promise of having double the efficiency of existing plants. Another advantage is that whereas the costs of gas and oil fluctuate, the cost of wind power (once a plant has been built and placed in operation) is steady.

One of the commonly cited disadvantages of this source of energy is that the wind does not always blow. Nonetheless, there are ample areas in the United States, for example, where the wind is relatively dependable. Other disadvantages are that the whirling blades can kill birds and bats (Holden, 2004) and interfere with television reception in nearby homes. Still other disadvantages are the large land areas required, the accompanying noise, and the aesthetics, especially of so-called wind farms. This last complaint is exemplified by the somewhat unexpected opposition voiced by many people, including environmentalists, immediately following a proposal to install a wind farm in the sea between Cape Cod and the islands of Martha's Vineyard and Nantucket off the southern coast of the state of Massachusetts. Among the complaints were that the windmills would interfere with marine fishing and navigation, the towers would endanger marine life and migrating birds, during the day the windmills would be clearly visible, and during the night their warning lights would interfere with the beauty of the area. Air-traffic controllers also expressed concern that the towers, each taller than the Statue of Liberty, would cause airplane accidents (Roosevelt, 2002).

Interestingly, experience has shown that opposition to such projects frequently subsides once they are completed and placed in operation. Prior to the installation of the Tuno Knob wind farm project, which is located offshore within three kilometers (two miles) of coastal vacation homes in Denmark, several thousand residents submitted complaints. After the farm was installed in 1995, the opposition subsided, particularly because the sounds of the turbines proved not to be audible from shore, tourism increased, and property values remained high. As of 2003, Denmark was meeting 15 percent of its energy needs through wind power.

On the basis of these trends, the use of wind power is anticipated to continue to expand, perhaps even more rapidly. Almost 1,700 megawatts (MWe) of new wind electricity-generating capacity were installed in the United States during 2001, increasing the nation's capacity from this source by 40 percent and raising the total generating capacity to almost 4,300 MWe (Blankinship, 2002). A major contributor to this increase was the state of Texas, where more new wind electricity-generating capacity was installed that year than had been previously installed in the entire United States during any previous single year. More than 20 offshore wind farms have been proposed for installation along the U.S. east coast, extending from Massachusetts to Virginia. What would be the world's largest wind farm is in the planning stages for construction in Iowa (Rao, 2003). By 2020, the estimated wind electricity-generating capacity in this country will provide 6 percent of the nation's needs, up from 0.3 percent in 2002. This pales, however, in comparison to what is being accomplished in Europe. During 2002 countries in that area spent more than $5.8 billion on new wind power installations, an increase of 33 percent over 2001. In fact, Germany, Denmark, and Spain now obtain more than 20 percent of their electricity via wind power. Overall, Europe now has more than three-quarters of the world's wind electricity-generating capacity (Sawin, 2003a).

Solar cells. The advantages of solar power are obvious. Every minute of every day the sun transfers to the earth more energy than the entire world uses in a year. In essence, solar energy is a potential source of virtually unlimited, clean, and free electricity. One of the major disadvantages is the relatively high cost of converting sunlight into electricity. Nonetheless, this problem is being solved through the development, for example, of solar panels that are made of ultrathin films or wafers of silicon, which is far less expensive (Fairley, 2002). Another achievement has been to develop "dye-sensitized" solar cells that incorporate water-repelling, light-absorbing dyes, coupled with a novel charge-conducting electrolyte that is able to withstand the relentless heat of the midsummer sun (Service, 2003). Other advances include solar collector systems that, through the incorporation of supplemental battery storage units, can provide electricity during both night and day. Such systems can operate independent of the electricity grid (Fies, 2001).

These and other improvements have led to a major increase in the use of solar photovoltaic collectors, including units that provide 1 to 3 kilowatts of electricity output (kWe) for homes and college buildings; 75 kWe for hotels; and 500 kWe or more for large commercial complexes (Fies, 2001). More than 200,000 homes in the United States now derive some or all of their power from solar cells (Fairley, 2002). Buoyed by this progress, the solar cell industry is anticipated to grow at an annual rate of 40 to 50 percent over the next few years (Sawin, 2003b). Although solar power currently accounts for only about 1 percent of the U.S. electrical output, its contribution could increase dramatically in the very near future (Service, 2003).

The potential applications of solar energy extend far beyond the examples discussed here. In the less developed countries, the sun is used for heating and distilling water and for drying crops. Another example is the use of biomass, which, in the form of firewood, is used as a fuel throughout the world. Another approach is the conversion of biomass in the form of agricultural crops into various types of fuel. These include the production of ethanol from sugarcane and methanol from wood materials and the generation of methane through anaerobic digestion of animal and plant wastes. This gas, for example, is a by-product of the operation of municipal sanitary landfills. The estimated annual production by such facilities in the United States approaches 10 million metric tons (Chapter 9).

To complete the above discussion, there is a need to comment on the efforts underway within the United States to shift to hydrogen as a primary energy source. While the use of hydrogen as a source of energy, for example, in fuel cells, could conceivably reduce the release of harmful contaminants into the environment (Schultz et al., 2003), there is still the challenge of producing and distributing the hydrogen which, depending on how it is accomplished, could be a major source of environmental contamination on its own. This is further discussed later in the final section of this chapter.

Environmental Impacts

One topic that inevitably arises in any discussion of energy use is the environmental impact of conventional electricity-generating power plants. Numerous studies have shown that airborne emissions from coal- or oil-fired plants have deleterious health impacts. The use of nuclear power, the primary alternative, remains controversial. Yet demand for electricity continues to grow far more rapidly than the relative increase in population (Figure 18.3). As in many other countries, the use of electricity in the United States has far outstripped population growth or total energy use. From 1973 to 1988 alone, electricity consumption in this country increased by more than 50 percent. Worldwide, electricity consumption is expected to increase by nearly 70 percent between 2000 and 2020, with most of this increase being in the less developed countries (Sawin, 2003b).

To provide perspective, the environmental and public health impacts of the primary methods of generating electricity, namely, hydroelectric power, geothermal energy, fossil-fueled power plants, and nuclear-powered plants, are assessed in the sections that follow.

HYDROELECTRIC POWER

Harnessing water power on a major scale generally involves the construction of a dam on a river or stream. The potential energy in the water behind the lake thus formed is used to turn one or more turbines and generate electricity. The first of the world's large such units, the Hoover Dam on the Colorado River in the southwestern United States, inaugurated what some call the "age of dams," a period of time that spanned three-quarters of a century. Worldwide, an estimated 45,000 large dams

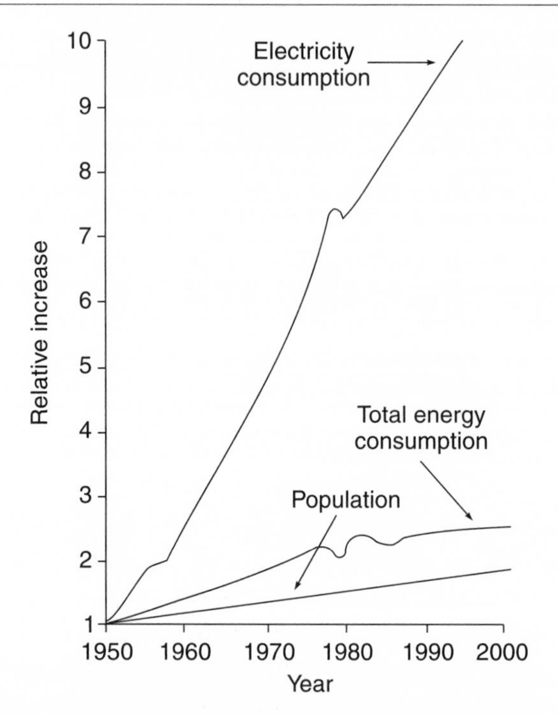

Figure 18.3 Relative increase in U.S. population, total energy use, and electricity consumption, 1950–2000 (data normalized to 1950)

today hold back about 14 percent of all runoff from precipitation. The accompanying hydroelectric systems provide more than half of the electricity needs for some 65 countries. The water made accessible by dams also irrigates the lands on which 40 percent of the world's food is produced.

In addition to the generation of electricity, hydroelectric dams provide reservoirs that permit large numbers of people to inhabit forbiddingly arid regions. They also protect plains from periodic flooding and permit the construction of cities that otherwise could not exist.

Large-scale dams, however, can also alter the environment and people's lives drastically. The classic example is the Aswan High Dam on the Nile River in Egypt. As planned, this project had two objectives: to produce electric power and to irrigate the nearby desert. But many unforeseen results ensued. In its original state, the Nile River served as a mechanism for the downstream transport of tremendous quantities of silt and organic matter. During the annual spring floods, portions of this material were deposited on the banks of the river, where it served as fertilizer for agricultural crops. The material that remained ultimately reached the Mediterranean Sea, where it served as food for large numbers of fish.

After the Aswan High Dam was completed, the river no longer flooded its banks, and the fertility of the farming areas declined. Without the discharge of nutrients into the Mediterranean, the fishing industry was essentially destroyed. Construction of this facility also displaced 90,000 people who had lived in the areas that were flooded. A further problem was that Lake Nasser, created by the dam, raised the water table and brought dissolved salts from the ground up into the desert topsoil, making it less usable for agriculture than anticipated. Control of the flow rate in the river led to increased growths of algae and phytoplankton, which adversely affected water quality. The accompanying quiescence also promoted the growth of the snail population, which dramatically increased the incidence of schistosomiasis among people who lived nearby.

Numerous other large hydroelectric projects have had similar impacts. The Grand Coulee Dam and its sister units on the Columbia River in Washington State have essentially eliminated salmon fishing in that region. Many biologists predict that without major changes, these fish will someday become extinct. Another facility that has the potential of major impacts is the Three Gorges Dam under construction on the Yangtze River in the People's Republic of China. Scheduled for completion some time between 2010 and 2015, it will generate an estimated 18,200 megawatts of electricity, will avoid repetitions of the catastrophic flooding that has occurred in this area in previous years, and will enable much larger ships to reach the cities on the upper end of the reservoir it creates, hopefully leading to a new age of commerce in central China. At the same time, however, it will increase the water level of the river for some 370 miles upstream, affect the habitats of wildlife, cover many acres of rich farmland, and force the relocation of up to 2 million people. Because of these and other impacts and concerns, many people believe that this will be the last such project ever built (Kosowatz, 1999).

The consequences of sudden failure of a dam must also be considered. More than 400 people were killed in the failure of the Malpasset Dam in southern France during a flood in 1959. Studies have shown that some of the larger dams in the United States may in time be subject to similar failures.

GEOTHERMAL ENERGY

The use of geothermal energy has a variety of environmental impacts. The pressure in pressurized hot-water reservoirs is often a result of the weight of the overlying land; withdrawal of water can lead to subsidence. Accidental spills of the withdrawn water, which often has a high mineral content, can lead to salination of the soil and water pollution. Where water is injected to be converted into steam by the use of heat from dry rocks in the vicinity, induced seismicity is a risk. In addition, significant problems may arise from the release of radon and volatile gases such as hydrogen sulfide that accompany the steam. The quantities of radon that are being released from such operations in northern California equal or exceed the airborne releases of other types of radionuclides from nuclear-powered plants of comparable generating capacity.

NUCLEAR AND FOSSIL FUELS

Nuclear- and fossil-fueled power plants are based on the same principles. In both instances, the fuel is used to produce heat, which converts water into steam. The steam then turns a turbine, which is connected to a generator that produces electricity. Cooling water, usually from a nearby river or lake, condenses the steam that leaves the discharge side of the turbine, and the water (condensed steam) is returned to the heat source for reheating. Condensation of the steam that leaves the turbine produces a vacuum so that the incoming steam will have the necessary pressure to turn the turbine. In this process, however, twice as much energy—that is, twice as much heat—is discharged to the environment with the cooling water as is converted into electricity. For this reason, power plants of this type are increasingly being challenged as sources of thermal pollution. To avoid thermal pollution of water sources, many such plants have cooling towers into which the water used for condensing the steam is sprayed so that it can release its excess heat into the atmosphere instead of into a lake or river. These plants also release various gases and particles into the atmosphere, along with smaller quantities of liquid wastes. The environmental and public health impacts of fossil-fueled and nuclear-fueled power plants can be divided into five stages:

1. Fuel acquisition
2. Fuel transportation
3. Power-plant releases
4. Processing and disposal of spent fuel or ashes
5. Power transmission

Fuel acquisition. The acquisition of fossil fuels has considerable environmental impacts. If coal is strip-mined, the air is polluted with dust, and the surface of the ground is defaced unless the land is restored to its original state. Underground mining frequently produces "acid mine drainage"—sulfuric acid and iron salts that drain or seep from the mine during operation and for some years thereafter. When these materials flow into surface streams, they are toxic to most forms of aquatic life. When the coal is obtained from underground sources, the miners experience an array of occupational health problems. On average, the mining of sufficient coal to provide fuel for one 1,000 MWe power station results in two to four accidental deaths and two to eight cases of black lung disease and other respiratory ailments among coal miners each year.

Underground mining of uranium can also have serious effects on health. Of the 6,000 people who have worked in underground uranium mines in the United States, an estimated 15–20 percent, or about 1,000, have already died or will die from lung cancer. Although this problem is substantially reduced when the uranium is strip-mined, regardless of the method used, subsequent processing of the uranium ore produces large quantities of tailings that can be sources of liquid and gaseous (radon) releases to the environment (Chapter 9).

In a like manner, drilling for oil has environmental impacts both on land and offshore. Leaks from offshore drilling operations can contaminate the marine environment, as did the spills near Santa Barbara, California (1969), and in the North Sea (1977) and the Gulf of Mexico (1979). During subsequent refining of the oil, airborne wastes are produced.

Fuel transportation. A standard 1,000 MWe coal-fired power plant requires about 8,000 tons of fuel per day, enough to fill at least 100 railroad cars. Transporting this amount results in an estimated 2–4 deaths and 25–40 injuries each year, primarily as a result of accidents at railroad crossings. For this reason and for economic considerations, the concept of the "mine-mouth power plant" is being implemented for coal-fired plants in several western states: the power plant is constructed near the source of the coal, and although the electricity must subsequently be transmitted over longer distances, this is judged preferable to transporting the coal.

Enormous quantities of oil are also transported, primarily in oceangoing tankers. As recently as November 2002, one such tanker, the *Prestige,* broke up off the western coasts of Spain and Portugal, resulting in the immediate release of several million gallons of fuel oil into the Atlantic Ocean. Another 20 million gallons sank to the bottom with the vessel and continue to leak out slowly and rise to the ocean's surface. The accompanying impacts forced officials to close about 100 kilometers (about 60 miles) of the coastline to fishing. In the United States, the worst such accident in recent years was that involving the *Exxon Valdez,* which in 1989 released 11 million gallons of oil into Prince William Sound in Alaska (Chapter 19).

Ocean shipments of natural gas pose similar problems. A modern tanker can transport about 125,000 cubic meters of liquefied natural gas at a temperature of −160°C. When the tanker reaches its destination, the liquefied gas is allowed to warm and regasify as it is transferred to storage tanks for later distribution. During this process, the volume of the gas increases by more than 600 percent. An accident while a tanker is unloading its cargo could release millions of cubic meters of natural gas (in expanded cloud form) that could blanket a city. If the cloud were subsequently ignited, widespread death and destruction could result.

A 1,000 MWe plant powered by nuclear fuel requires only about 30–50 tons of fuel per year. Because the original uranium fuel is sealed within fuel rods and is not a significant radiation source, its transportation to the power plant does not present any unusual occupational or environmental health problems.

Power-Plant Releases. Fossil-fueled plants release sulfur oxides, nitrogen oxides, carbon monoxide, and some naturally occurring radioactive material originally present in the fuel. In fact, the United Nations Scientific Committee on the Effects of Atomic Radiation estimates that the collective dose from a coal-fired electricity-generating plant is equivalent to 25 percent of the local and regional population dose from a nuclear-powered station of the same generating capacity (Mettler et al., 1990). Fossil-fueled plants also release significant amounts of carbon dioxide into the atmosphere. The annual amount from a 1,000 MWe power plant is estimated to be about 3.7 million tons.

Table 18.2 summarizes the estimated quantities of major airborne pollutants released by coal-fired, oil-fired, and gas-fired 1,000 MWe plants. The estimated quantities of each listed emission, however, can vary significantly, depending on a range of plant conditions. The amount of particulate matter, for example, depends on the nature of the removal systems. In the case of sulfur dioxide (SO_2), the amount released depends not only on this factor but also on the sulfur content of the coal, oil, or natural gas that is being burned. Therefore, the values listed are primarily for purposes of illustration. Even so, the numbers indicate that coal-fired plants discharge about 1.5 times as much SO_2 per unit of electricity generated as do oil-fired plants and more than 6 times that from gas-fired

Table 18.2 Airborne emissions from fossil-fueled electric power plants (assumed capacity 1,000 MWe)

Pollutant	Emissions (thousands of tons per year)[a]		
	Coal	Oil	Natural gas
Sulfur dioxide	43	29	7
Nitrogen oxide	25	11	7
Carbon monoxide	0.7	1.0	1.1
Particulates	120	0.6	<0.1

a. Emissions will vary depending on nature of fuel, plant design, and operating parameters.

plants. Coal-fired plants also release more than twice as much nitrogen oxide (NO) as oil-fired plants and more than three times that released by gas-fired plants; they release several hundred times as much particulate matter as oil-fired plants and 1,000 or more times as much as plants fueled by natural gas. The discharges of carbon monoxide (CO) by the various plants are not significantly different.

On this basis, an electricity generating plant fueled by natural gas has far less impact on the environment than one fueled by either coal or oil. Nonetheless, coal-fired plants (the worst offenders) produce more than half of the nation's electricity supply today. One could logically ask: Why does natural gas not play a more important role? In years gone by, there were at least three reasons. First, the cost of natural gas was relatively high; second, the adequacy of the supply was in doubt; and third, gas-fired plants were relatively inefficient. Several subsequent developments have brought about a change in this situation. One is that the supplies of natural gas are now believed to be far more abundant than originally estimated. Equally important is that the efficiencies of gas-fired plants have been dramatically increased during the last 10 years so that they are economically competitive with other types of plants. Another factor in their favor is that the time required for obtaining the necessary environmental approvals and constructing this type of plant is much shorter than that for other types.

Two basic types of nuclear-powered plants are in use in the United States today: boiling-water reactors (BWRs) and pressurized-water reactors (PWRs) (Figure 18.4). This is significant because the types and quan-

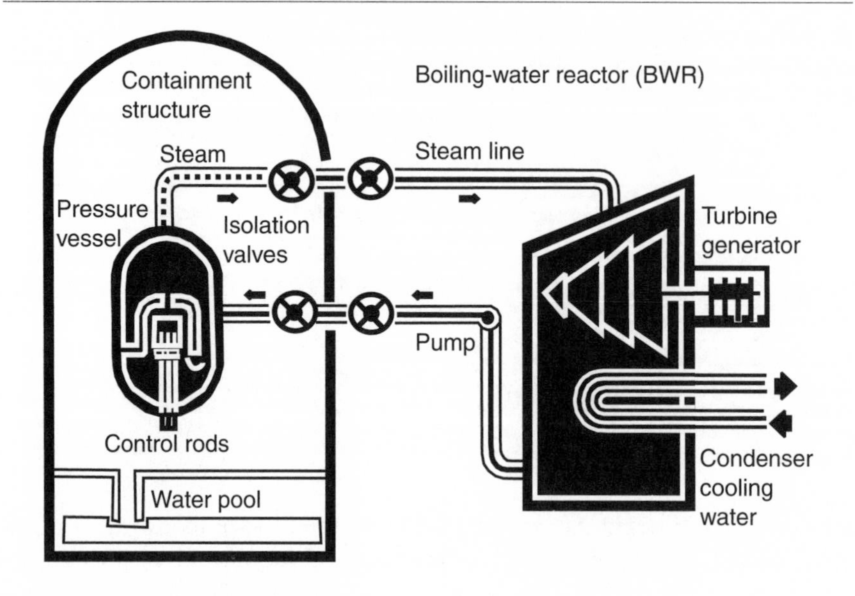

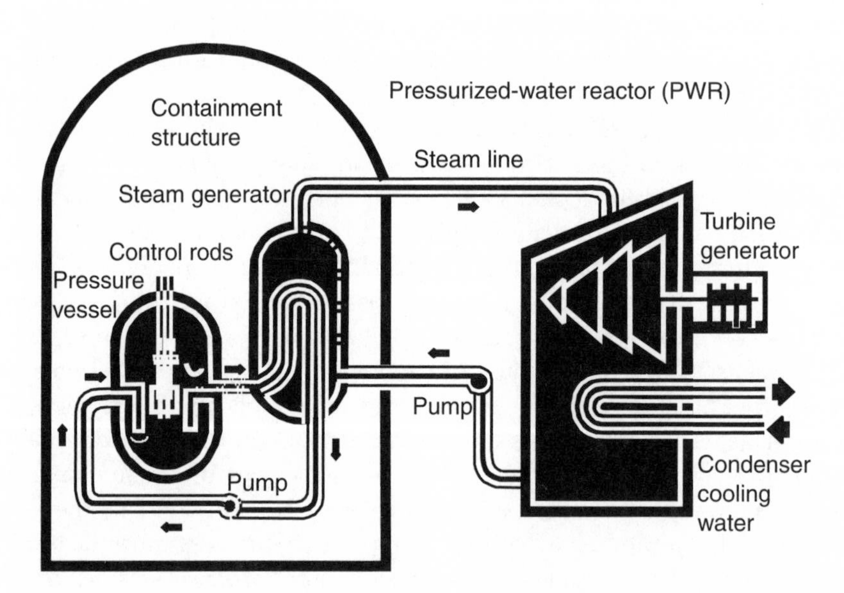

Figure 18.4 Schematic diagrams of boiling water and pressurized water nuclear power plants

tities of their releases differ. In a BWR, the water is heated by the fuel and converted into steam, and this steam turns the turbine. Neutron irradiation of the cooling water as it passes through the reactor core converts stable oxygen into radioactive nitrogen, which is transported with the steam to the turbine. As a result, personnel cannot work in the vicinity of the turbine during plant operation. If leaks occur in the fuel cladding, the water and steam will also contain radioactive fission products.

In a PWR, the water heated by the reactor is kept under sufficient pressure that it is not converted into steam. Through use of an intermediate heat exchanger (steam generator), this water in turn transfers heat to water in a secondary system that is subsequently converted into steam and turns the turbine. Under normal conditions the water in the secondary system will be clean, and any leakage will not release radioactive material. It is also not subjected to neutron irradiation. The tubing in the steam generators, however, sometimes fails, releasing radioactive water and gases from the primary into the secondary system. Although the liquid releases can be readily controlled, some gases are released into the atmosphere.

Table 18.3 summarizes the estimated airborne releases from representative 1,000 MWe PWRs and BWRs that were constructed in 1979 or later. As may be noted, these two types of plants release about equal amounts of carbon 14 (^{14}C), iodine 131 (^{131}I), and tritium (^{3}H); in general, however, PWRs release more krypton 85 (^{85}Kr) and xenon 133 (^{133}Xe).

One approach for gaining perspective on the relative environmental and public health impacts of fossil- and nuclear-fueled electric power plants is

Table 18.3 Annual airborne emissions from nuclear power plants (assumed-capacity 1,000 MWe)

	Pressurized-water reactor		Boiling-water reactor	
Pollutant	Becquerels	Curies	Becquerels	Curies
Krypton-85[a]	7×10^{11}	20	—	—
Xenon 133	7×10^{13}	2,000	4×10^{12}	100
Carbon 14	4×10^{11}	10	4×10^{11}	10
Iodine 131	7×10^{7}	0.002	1×10^{8}	0.003
Tritium	6×10^{12}	150	4×10^{12}	100

a. Releases from boiling-water reactors are negligible.

to compare them on the basis of the quantities of air required to dilute the annual volume of their most critical airborne releases to concentrations that will comply with federal standards for the ambient environment. These data, presented in Table 18.4, are based on the assumptions that all the sulfur is released as sulfur dioxide, that all the nitrogen is released as nitrogen dioxide, and that all the particulate emissions have diameters less than 10 micrometers (PM_{10}; Chapter 5).

As may be noted, the releases that require the largest dilution from a coal-fired plant are the particulates, for which the federal standard is 50 $\mu g/m^3$ (Table 5.1, Chapter 5). Applying the applicable limit for the estimated emissions of SO_2 from plants fueled by oil and natural gas, one can estimate the comparable dilution requirements for the most critical pollutant in each of these two cases. By using a similar approach for ^{133}Xe, the annual volume of air required to dilute the most critical airborne emissions from each type of nuclear-powered electricity-generating station can be estimated. As may be noted, the resulting data show that the volume of air required to dilute the most critical airborne release from a PWR to the acceptable concentration in the ambient environment is 0.005 percent of that needed for the most critical pollutant from a comparable-sized facility fueled by coal, oil, or natural gas. For a BWR, the volume of diluting air would be even less. As was previously noted, more than half of all the electricity in the United States is now being generated from coal

Table 18.4 Annual dilution requirements for a 1,000 MWe power plant

Type of plant	Limiting pollutant	Permissible Concentration	Required dilution[a] Cubic meters	Cubic miles
Coal	PM_{10}	50 $\mu g/m^3$	2×10^{15}	500,000
Oil	SO_2	80 $\mu g/m^3$	3.3×10^{14}	80,000
Natural gas	SO_2	80 $\mu g/m^3$	8×10^{13}	20,000
Nuclear (pressurized-water reactor)	^{133}Xe	2×10^4 Bq/m^3	4×10^9	1

a. Approximate volume of air required to dilute the most critical pollutant from each type of plant to the permissible concentration as prescribed by federal standards for the ambient environment.

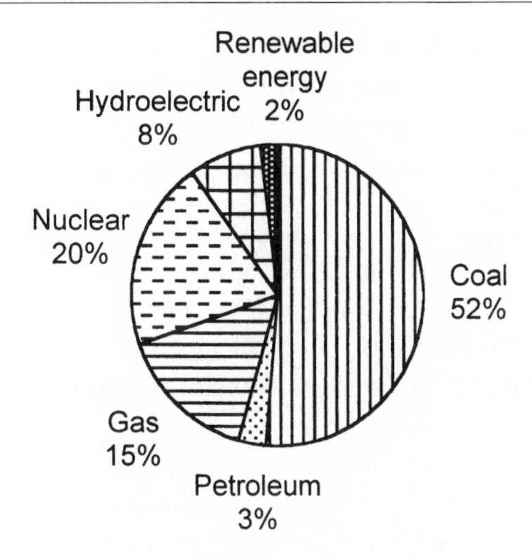

Figure 18.5 U.S. electricity generation by energy source

(Figure 18.5). Although this source of fuel is plentiful and reasonably economical, it nonetheless is a major source of air pollution.

Having conducted such an exercise, one might conclude that a nuclear plant is much safer in terms of airborne releases than any other type. However, these calculations do not include the associated dilution requirements for the airborne gases and dusts that would be released from the tailing piles produced in conjunction with milling the uranium ore. They also do not include any consideration of the airborne releases in the course of enriching the fuel used in these plants or in chemically processing the spent fuel. Although chemical processing is not currently being done in the United States, it is common practice in many other countries and should be taken into account in those cases. Even if these other emissions are considered, on a comparative basis the associated emissions for a nuclear-powered plant are small. At the same time, if such comparisons are to be equitable, one would also need to consider other environmental impacts, such as mining emissions and transportation issues, associated with coal- and oil-fired electricity generating stations.

The outcome of any such comparisons also depends on the applicable environmental and public health standards. Calculations of the volume of air required to dilute a given contaminant to an acceptable level assume the existence of a single universally accepted concentration to which a population group can be exposed. If a higher acceptable concentration is chosen, the calculated volumes of diluting air will be lower. There is also no assurance that the standards for the permissible airborne concentrations of each of the critical pollutants are based on equivalent risks. Still another factor is that natural processes remove pollutants such as SO_2, NO_2, and particulates from the atmosphere. If these processes were taken into account, the quantities of diluting air required for the coal-fired, oil-fired, and gas-fired plants would be less. Similarly, the half-life of ^{133}Xe, the critical airborne release from a nuclear power plant, is only about 5.3 days; it too will quickly dissipate (through radioactive decay) in the ambient environment.

Finally, a comparison of the impacts of the different types of plant depends on whether the potential health impacts of the airborne contaminants are evaluated individually or collectively. Coal-fired plants release both particulates and gases. If the combination has the potential to produce synergistic effects, then they should be evaluated as a totality, not simply on the basis of the most critical individual contaminant. Even so, still other factors need to be considered. The long-term impacts of coal-mine dust kill several thousand miners in the United States every year. In a similar manner, the air pollution from coal-fired plants causes acid deposition, smog, visibility degradation, and global warming, as well as increased cases of asthma, respiratory and cardiovascular disease, and mortality among the general population. In the case of nuclear-powered plants, a major additional consideration is the potential for serious accidents that might release large amounts of radionuclides into the atmosphere. As illustrated by the accident at the Chernobyl nuclear power plant in 1986, such events can have far-reaching impacts.

When all these factors are taken into account (including the atmospheric pollution associated with the processing of spent fuel), it appears that a nuclear-powered plant has far less environmental impact than a plant fueled by oil or natural gas. All three types of facilities appear to be superior to a plant fueled with coal. If adsorption systems are used to delay the release of radioactive gases from spent-fuel chemical-reprocessing facilities or if, as in the United States, the spent fuel is not reprocessed, then nuclear plants remain far superior in terms of airborne releases.

Processing and disposal of spent fuel or ashes. Power plants fueled by natural gas or oil have no spent-fuel disposal problems because these fuels burn cleanly and produce no ash. In coal-fired plants, however, 12–25 percent of the fuel ends up as ash. Thus a 1,000 MWe plant would require

12–25 railroad cars for the daily removal of ash. Where and how this ash is disposed of is also important, since it contains many toxic compounds. As indicated earlier, a nuclear-powered plant produces some 30–50 tons of spent fuel each year. Because it is intensely radioactive, spent fuel poses significant problems of radiation protection and waste disposal. In the United States, such fuel is being stored at the power-plant sites either in water pools or in aboveground dry casks. It will subsequently be transported off-site for disposal in a geologic repository (Chapter 9). The volume of nuclear power-plant wastes is relatively low, but the risks they pose if they are not properly disposed of are relatively high.

Power transmission. The type of power plant has no effect on the efficiency with which electricity is distributed to consumers. Nonetheless, an estimated 12–14 percent, perhaps even 20 percent, of the electricity generated in the United States is lost during transmission. Development of more efficient transmission systems has long been discussed. One promising system would be the use of superconducting cables installed underground. The power losses in such a system would be much less than those currently experienced in high-tension overhead power lines. In addition, an underground system would eliminate much of the visible/aesthetic environmental degradation caused by overhead lines. The United States has more than 325,000 miles of the latter type, 160,000 miles of which operate at high voltage (Truly, 2002). Each mile of transmission line requires up to 100 acres of land as right-of-way.

A National Energy Plan

As the preceding discussion indicates, there is an urgent need to examine energy requirements on a global basis. That longer-range and more thorough planning is needed is illustrated by two observations: (1) an adequate supply of energy is a major determinant of the structure and lifestyle of society; and (2) the manner in which energy needs are met significantly influences economic growth as well as public health and the environment.

An initial step would be to define what a national energy plan must achieve. Among the major considerations is that it provide a coordinated policy for addressing the needs of the three primary energy-consuming sectors—industry, commercial and residential use, and transportation—both from a near-term and long-term perspective. Every effort should also be made to ensure that the plan provides an adequate balance among potentially competing forces such as environmental protection, public health and safety, and economic viability. Other considerations are the increasingly dominant role that the consumption of energy in the form of electricity plays in society and the fact that if the environment is to be protected, hydrogen fuel cells (as discussed earlier) may well prove to be the best source of power. Enhancing the compatibility of these last two observations is that one of the best methods for producing hydrogen is electrolysis, a process that requires electricity (Starr, 2002).

In applying these concepts to the United States, one approach for ensuring that energy is available when and where it is needed would be to establish a network so that electricity and hydrogen fuel can be efficiently transmitted and/or transported to and shared among all parts of the nation. The need for such a network for electricity distribution was clearly demonstrated by the interruption in power that occurred on 14 August 2003 and left more than 50 million people in the north central and eastern portions of the United States without electricity. This, however, will not be an easy task. To reduce environmental impacts and the risks of terrorism, all major parts of the network should probably be located underground. Because of the extended time required to complete such a system, it would be necessary to begin on a regional basis and gradually move to a full nationwide network. One of its principal advantages would be the flexibility it would provide for transmitting electricity back and forth between the coal-based eastern and the hydropower-based western portions of the United States. Because the consumers would be located in four time zones, and electricity demands would be moderated, the generating plants would be able to operate with higher capacity factors and as base-loaded units that is, those that are designed to meet the basic demand as contrasted to those needed to provide additional electricity at times of peak demands. Another major advantage is that the generating plants could be located anywhere as long as they were tied into the system. This would enhance the use of wind, solar, and nuclear power since the plants would not need to be located close to population centers. While such a system would not be applicable in every country, it would hopefully serve as a model for the development of systems that were (Starr, 2002).

The General Outlook

Regardless of the problems and difficulties in meeting the energy needs of the world, progress is being achieved. The increasing recognition by major international energy corporations of the need to understand the concept of sustainable development and incorporate its principles into their long-range plans is especially encouraging. Several worldwide petroleum suppliers, for example, have included alternative energy sources and energy conservation in their plans for resource development. Their reasoning is that in the face of global warming, they must adopt such policies if they are to continue to be profitable (Raven, 2002). In a similar manner, regional electric utility companies are implementing positive environmental measures in their corporate policies. One example is their efforts to encourage customers to adopt more efficient lighting on the grounds that conservation is preferable to the alternative—developing new electricity-generating capacity.

Progress is also being made in other areas, some of which involve innovative concepts and approaches. In the building sector, it has been predicted that electrochemical windows that can be darkened or lightened to control heat from the sun will become commonplace. It is also anticipated that air-circulation systems will include desiccant-based dehumidification systems to reduce cooling loads, and that solar cells will be integrated into the roof and windows to reduce the energy demand in buildings. Eventually, it may be possible to develop buildings that produce the equivalent of the amount of energy they consume (Truly, 2002). Stimulated, in part, by major support from the U.S. government, manufacturers are developing cars that will be powered by fuel cells. Some have envisioned a day when such vehicles could become an integral part of the power-grid operation by functioning as distributed energy resources when they are "parked" near buildings. Although numerous hurdles remain, some people predict that within a decade or two, a fuel cell, comparable in size to a refrigerator, will be available to meet the electricity requirements for a home (Lloyd, 1999). While plans (discussed earlier) for completion of the first step toward a workable fusion reactor are underway, the most optimistic date for bringing this project to fruition is 2014. Based on this fact and the technical difficulties to be surmounted, estimates are that it will require several decades before fusion power plants will be available for routine operation and wide scale application.

Concurrently, it is anticipated that energy will increasingly be consumed in the form of electricity. In fact, because it is so easy to transmit and so convenient to use, electricity has become an essential contributor to the quality of modern life. It is necessary to clean the air, to operate water-purification facilities and sewage-treatment plants, to dispose of old automobiles, and to recycle other types of solid waste. It powers radios, televisions, microwave ovens, computers and office equipment, and labor-saving home appliances. Better lighting reduces accidents on highways and crime in cities. Although conservation can help reduce the overall demand for energy, the need for electricity, and thus for more power plants, will continue. The basic challenge is to educate people to use electricity and other forms of energy more efficiently and more conservatively and to encourage the commercial sector to design, construct, and operate generating stations that function at maximum efficiency with minimal impacts on public health and the environment.

19

DISASTERS

In a broad sense, disasters can be classified as *natural* and *human made.* The former include floods, tornadoes, hurricanes, earthquakes, tsunamis, and volcanic eruptions. The latter include accidents involving industrial and technological facilities. Primary examples are major releases of toxic agents from industrial facilities and oil spills from, or the sinking of, ships at sea. Upon close examination, however, few disasters are either totally natural or human made. The consequences of many disasters, for example, are exacerbated by inadequacies on the part of humans—witness the failure to design a building to resist an earthquake, a ferry to resist heavy seas, or an oil tanker not to leak even though its outer hull has been ruptured. In a similar manner, the frequency and severity of events such as droughts and floods may be influenced by human destruction of forests.

Another approach is to classify disasters on the basis of the nature of their impacts on human health and the environment. In this sense, such events can be classified as (1) those, such as hurricanes, floods, and earthquakes, that are characterized by strong physical forces and whose impacts affect both people and the environment; (2) those, such as oil spills, that involve the release of toxic materials and primarily affect the aquatic and marine environments and associated wildlife; and (3) those, such as the disasters at Bhopal and Chernobyl (discussed later) that, while also involving releases of toxic materials, do so in a manner that enables them to be transported over relatively large distances, through various environmental pathways, and to affect both humans and the environment in a significant manner.

With the globalization of terrorism, however, the nature and potential magnitude of such events are changing rapidly. This was exemplified by the events of 11 September 2001 in the United States, which led to the deaths of about 2,800 people. Although acts of terrorism are also human made and, in some ways, the havoc and destruction wrought by such attacks do not differ from those that result from other types of disasters, there are distinct differences in their nature and characteristics. For these reasons, they will be discussed separately.

As a combination, disasters have a significant impact on people and the environment. Natural disasters alone cause an estimated average of some 150,000 deaths worldwide each year. As would be anticipated, the economic and human costs of such events vary widely. The principal factors that affect these costs include the concentration of the population, the existence of emergency response capabilities, the availability of and accessibility to outside assistance, the efficiency of rescue operations, building design and construction practices, and soil conditions. Nonetheless, in the United States, the overall impacts of such events, in terms of human fatalities, are relatively minor. In fact, an average of only about 350 people in this country die each year due to floods, lightning, tornadoes, hurricanes, earthquakes, and volcanic eruptions combined. Other far less publicized events often have far larger impacts. For example, the drought and heat wave of 1988 caused some 5,000 to 10,000 deaths and losses of nearly $40 billion (Parfit, 1998). A heat wave in France during the summer of 2003 led to the deaths of more than 19,000 people (Luterbacher et al., 2004).

Natural Disasters

For purposes of analyses, natural disasters can be subdivided into two groups: events, such as earthquakes and volcanic eruptions, that are related to geological factors; and those, such as hurricanes and floods, that are related to climatological factors. Of the two, those due to climatological factors, especially hurricanes and floods, occur more frequently and often have larger impacts, especially in terms of the size of the areas affected. Other sources of natural disasters include landslides and avalanches; tsunamis that are generated by earthquakes; wildfires; tornadoes; winter storms and hailstorms; and droughts. The characteristics of each of these sources and some of the more prominent such events that have occurred are discussed in the following sections.

HURRICANES AND FLOODS

Floods can occur after heavy rains, a rapid snow melt, or a hurricane. In some situations, the rains and floods that accompany hurricanes can be more destructive than the winds. This was the case for Hurricane Floyd, which struck North Carolina in 1999. On a worldwide basis, floods account for about 40 percent of all natural disasters. As a combination, hurricanes and floods have the largest impacts of all natural disasters in terms of the numbers of people affected. For the United States, hurricanes and floods are also the highest contributors of all such disasters to fatalities, causing an average of about 165 deaths each year. Next are lightning, with an average of 80 deaths, and tornadoes, with 70 (Parfit, 1998). An added impact of floods is that they often damage and destroy homes, displacing the occupants, and lead to crowded living conditions and compromised personal hygiene. This, in turn, can lead to the spread of infectious diseases, particularly in cases where floods cause drinking-water sources to be contaminated, sewer systems to be disrupted, toxic chemicals to be released from storage tanks, and solid-waste collection and disposal services to be interrupted. Floods also enhance the opportunities for mosquito breeding as well as for snake and dog bites. The magnitudes of the impacts of several recent hurricanes and floods are summarized in Table 19.1.

LANDSLIDES AND AVALANCHES

Landslides and avalanches occur when imbalances of rock, snow, or earth on steep slopes suddenly give way and fall to lower elevations. Historic examples include the some 90 million tons of limestone that roared down Turtle Mountain in Alberta, Canada, in 1903, crushing 70 people. In a similar manner, snow crashed down a slope in Wellington, Washington, in 1910, knocking two stranded trains into a gorge and killing 96 passengers. In 1959, rain-soaked soils slid off hills into Minatitian, Mexico, and buried 800 villagers (Parfit, 1998).

EARTHQUAKES AND TSUNAMIS

In terms of deaths, earthquakes can have major impacts. A classic example is the one that occurred in Turkey in 1999 and killed tens of thousands of people. It also destroyed the homes of another 250,000, and caused billions of dollars in damage (Gore, 2000) (Table 19.1). In contrast, an earthquake in Peru in June 2001 caused a relatively small number of deaths (77), but enormous destruction of property. In Moquegua, in the southern part of the country, for example, half of the dwellings were damaged or de-

Table 19.1 Examples of major natural and man-made disasters, 1998 and later

Event	Location	Date of occurrence	Impact
Natural Disasters			
Earthquakes	Turkey	17 August and 12 November 1999	Tens of thousands killed; 250,000 homeless; billions of dollars damage
	Gujarat, western India	26 January 2001	>30,000 deaths; tens of thousands injured; massive destruction
	Algeria	21 May 2003	1,875 deaths; >7,600 people injured; massive destruction
	Bam, Iran	26 December 2003	28,000 or more deaths; >30,000 injured; thousands homeless
	Imzouren, Morocco	25 February 2004	>560 deaths; ~20,000 people homeless
Hurricanes	Mitch: Central America	October–November 1998	>11,000 deaths; >2 million homeless
	Floyd: Eastern coast of United States	September 1999	63 deaths; ~25,000 families flooded; tens of thousands of hogs, chickens, and turkeys killed
Floods	Yangtze River, People's Republic of China	1998	~4,000 deaths; ~225 million people displaced
	Dominican Republic and Haiti	May 2004	~1,000 deaths; situation exacerbated by deforestation and resulting landslides
Tornadoes	Eastern India	March 1998	145 deaths
	Oklahoma, Kansas, Texas, and Tennessee	May 1999	54 deaths
	Ohio and eight other mid-western states	November 2002	35 deaths; widespread destruction

Table 19.1 (continued)

Event	Location	Date of occurrence	Impact
Volcanic eruption	Goma, Congo	January 2002	40 deaths; >300,000 homeless; ~200,000 people marooned; thousands of buildings destroyed
Human-Made Disasters			
Explosion of military arms	Lagos, Nigeria	27 January 2002	>1,000 killed; exacerbated by ensuing panic
Explosion of train laden with oil and chemicals	Ryongchon, North Korea	22 April 2004	161 killed, half of whom were school children; 1300 injured; tremendous damage and devastation
Ferry sinking	Senegal, Africa	26 September 2002	>950 people drowned
	Inongo, Congo	25 November 2003	182 confirmed deaths; scores missing; due to storms and heavy crowding
Fires	Australia	December 2001–January 2002	>150 homes destroyed; thousands of people evacuated; >1 million acres of forest and farmland razed; heavy loss of wildlife
	Reqa Al-Gharbiya, Egypt	20 February 2002	363 people killed due to fire in train during holiday trip; believed due to stoves used by passengers
	Warwick, Rhode Island	20 February 2003	95 people killed due to fire in nightclub during pyrotechnics display; building was old and crowded

Table 19.1 Examples of major natural and man-made disasters, 1998 and later (continued)

Event	Location	Date of occurrence	Impact
Oil tanker spills	*Erika*; hull split near Brittany	13 December 1999	~4 million gallons of oil released; 100,000 sea birds killed; widespread contamination of French coast
	Jessica, San Cristobal Island, Galapagos Islands	January 2001	160,000 gallons of diesel fuel discharged; widespread contamination
	Prestige; hull split off northwestern coast of Spain near A Coruña	November 2002	1.3 to 2.6 million gallons of fuel oil spilled initially, damaging beaches, and threatening the shellfish, scallops, and mussels industries; oil continued to leak slowly thereafter from sunken vessel
Rupture of gas well	Chongqing, China	23 December 2003	233 killed due to toxic fumes; >9,000 poisoned; 41,000 evacuated
Terrorism	New York, New York, and Washington, DC	11 September 2001	~2800 deaths; World Trade Centers destroyed; Pentagon extensively damaged
	Bali, Indonesia	12 October 2002	>188 deaths and hundreds injured due to explosion of car bomb in nightclub district
	Madrid, Spain	11 March 2004	>200 deaths and ~1,800 injured in bomb attacks on trains

stroyed, and all electric systems failed partially or totally. In larger cities, such as Arequipa and Tacna, the hospitals suffered extensive damage, and about 75 percent of the almost 500 public health facilities in the affected area were damaged. Equally important, 30 percent of the water-supply systems in Tacna, Arequipa, and Moquegua were damaged, and 15 percent were destroyed. Thus even without a high number of deaths, the impacts of earthquakes can be significant. Fortunately, experience is demonstrating that this need not be the case. A good example is the earthquake that occurred in Northridge, California, in 1994. The reasons for this will be discussed later in this chapter.

Another concern is that earthquakes that occur beneath the ocean can create tsunamis that travel thousands of kilometers. Waves from an earthquake in Chile in 1960 killed 61 people in Hawaii and hundreds more on the Pacific Rim (Parfit, 1998).

WILDFIRES

More than 100,000 wildfires occur in the United States and Canada each year. The most common causes are lightning, arson, or carelessness. One of the worst fires, in this case set by an arsonist (and therefore, in a sense, human made), occurred in a canyon above Laguna Beach, California, in 1993. The resulting inferno burned almost 15,000 acres and destroyed 441 houses. In 2000, wildfires in Los Alamos, New Mexico, destroyed about 400 homes and threatened facilities in which toxic chemicals and radioactive materials were being stored and/or handled. In contrast, forest fires set in Indonesia in the fall of 1997 as part of slash-and-burn agricultural practices led to the destruction of more than 8,000 square miles, the deaths of hundreds of people as a result of accidents due to reduced visibility, and illnesses, such as asthma, bronchitis, emphysema, and eye, skin, and cardiovascular diseases, of almost 50 million people (Simons, 1998).

VOLCANOES

Volcanic eruptions are among the most spectacular natural events. Historically, the violent eruption of Mount Krakatau on a small island in Indonesia in 1883 stands as a classic. People throughout the world observed the clouds created by the dust released in that event. Almost equally spectacular, the eruption of Mount Pinatubo in the Philippines in 1991 produced a mushroom-shaped cloud 500 kilometers (310 miles) in diameter, the crown of which reached an altitude of 35 to 40 kilometers (22 to 25 miles) and contained an estimated 20 megatons of sulfur dioxide

(SO_2). The accompanying effects on global climate were evident for several years. These included a 2°C (3.6°F) decrease in surface air temperatures over the Northern Hemisphere continents during the summer of 1992 and up to a 3°C (5.4°F) increase during the winters of 1991–1992 and 1992–1993 (Robock, 2002).

TORNADOES

Tornadoes are a menace throughout the world. They are, in fact, among Earth's most violent natural acts and can have accompanying winds of up to 325 kilometers (200 miles) per hour. On average, about 1,000 touch down in the United States each year, more than in any other country in the world. Most commonly, they occur in a region that stretches from Texas to Nebraska, and cause an annual average of 50 deaths and approximately 1,000 injuries. Experience shows that tornadoes tend to occur in clusters. In 1974, for example, almost 150 tornadoes occurred in less than 48 hours and caused more than 300 deaths in 13 states and Canada. A major storm in South Dakota on 24 June 2003 led to nearly 70 individual tornadoes (Vesilind, 2004). On 1 March 1997, a group of approximately nine tornadoes, originating from two separate thunderstorms, swept a distance of some 160 miles across the state of Arkansas. These caused 26 deaths and more than $100 million in property damage (CDC, 1997a). As is common in such events, essentially all of these tornadoes moved in a northeasterly direction (CDC, 1997b).

WINTER STORMS AND HAILSTORMS

Winter storms and hailstorms occur on a very frequent basis. The buildup of ice from a major winter storm in Canada and the northeastern United States in 1998 caused trees and utility towers to fall and roofs to collapse. A similar event occurred in Arkansas in late 2000. One of the most disruptive impacts of such events is the loss of electricity service. The "Storm of the Century," which occurred in 1993, affected the entire eastern seaboard from Florida to Maine, including eastern Canada. Every major airport on the east coast was closed. In a similar manner, hailstorms can cause major damage. One storm that occurred in Fort Worth, Texas, in 1995 rained hailstones as large as softballs. The resulting impacts included people being pelted with ice and glass being shattered in buildings and cars. Hailstones also kill livestock and destroy agricultural crops. The economic damage from the 1995 event approached $2 billion (Parfit, 1998).

DROUGHT

About 10 percent of the Earth experiences drought in any given year. Although droughts require a long time to develop, they spread farther, last longer, and touch more lives than any other type of natural disaster. Such conditions are common occurrences in the prairie provinces of Canada, the western and central portions of the United States, and northern and central Mexico. The impacts include loss of agricultural crops and livestock and increases in the frequency and severity of wildfires (Parfit, 1998). Droughts can also lead to thousands of human deaths particularly in the less developed countries of the world.

Because their homes are less sturdy and their resources are more limited, people in less developed countries suffer far more devastation from natural disasters than those in developed countries. In fact, more than 95 percent of the deaths from natural disasters during the last 15 years have occurred in the less developed countries of the world. At the same time, more than three-quarters of all deaths, 90 percent of all homelessness, and almost half of all economic losses occurred in Asia (Abramovitz, 2001). On a geographic basis, Asia is most prone to natural disasters; Latin America and Africa are intermediate; North America, Europe, and Australia are least prone. For each major natural disaster in Europe and Australia, 10 occur in Latin America and Africa and 15 in Asia.

Although natural disasters have been a problem for centuries, records show that the number of such events of major consequence, that is, those that result in deaths or losses so high that outside assistance is required, has increased dramatically in recent years. In most cases, this is due to human activities that exacerbate the impacts of natural events. Such activities include the destruction of forests and the construction of buildings along coastlines, along unprotected riverbanks, and on steep hillsides (Abramovitz, 2001). At present, an estimated 60 percent of the world's population lives within 100 kilometers (62 miles) of a coastline. The number of deaths from earthquakes and volcanic eruptions is also increasing. Mexico City, for example, has nearly 20 million people living in a region subject to earthquakes and volcanoes. At the same time, the migration of the U.S. population to California, an active seismic area, continues unabated (Parfit, 1998). The probabilities of occurrence of a range of health effects caused by various types of natural disasters are summarized in Table 19.2.

Table 19.2 Health effects of natural disasters

Health effect	Earthquake	Hurricane, high wind	Volcanic eruption	Flood	Tidal wave, flash flood
Deaths	Many	Few	Varies	Few	Many
Severe injuries (requiring extensive medical care)	Overwhelming	Moderate	Variable	Few	Few
Increased risk of infectious disease	A potential problem in all major disasters; probability increases with overcrowding and deteriorating sanitation				
Food scarcity	Rare (may occur as a result of factors other than food shortages)	Rare	Common	Common	Common
Major population movements	Rare (may occur in heavily damaged urban areas)	Rare	Common	Common	Common

Human-Made Disasters

Human-made disasters also include a diversity of types. Examples of those that can cause immediate and widespread effects are releases of toxic chemicals from an industrial plant or the widespread distribution of radioactive materials as the result of a nuclear power-plant accident. Other types of disasters are caused by oil-tanker spills, the sinking of passenger ships, and fires. The problems of such events are compounded by the fact that emergency response planning is more difficult for them than for natural disasters. Several of the more prominent sources of human-made disasters are discussed here.

INDUSTRIAL-PLANT ACCIDENTS

One of the worst industrial-plant accidents was the previously cited release of toxic fumes that occurred in Bhopal, India, in 1984, which killed between 2,500 and 7,000 people. The associated impacts were exacerbated by several factors: public housing located next to the plant; the fact that the release occurred at night; delays in alerting the public; little public knowledge of the potential toxicity of the emissions; and the fact that the only medical facility in the area was a small community hospital unable to cope with the thousands of affected people. Even today, few countries provide the public with adequate information about the location of chemical-manufacturing plants or the nature and quantity of chemicals that are manufactured at them.

NUCLEAR POWER-PLANT ACCIDENTS

Just as Bhopal hopefully represents the most devastating industrial accident that will ever occur, the accident at the Chernobyl nuclear power plant hopefully represents the worst possible accident of this type. Although immediate deaths due to acute exposures and injuries from this latter accident were fewer than 50, the impacts were enormous in terms of the size of the affected areas, the number of people exposed, and the potential long-term latent biological effects. During the aftermath, an estimated 240,000 workers received an average dose of 100 mSv in cleaning up and stabilizing the reactor, and an additional 116,000 members of the public who were evacuated from contaminated areas received an average dose of 30 mSv. Several million other people who continued to reside in contaminated areas received an average dose of about 10 mSv. Follow-up epidemiological studies have revealed that several thousand children have developed thyroid cancers due to the intake of radioactive iodine (UNSCEAR, 2000).

OIL-TANKER SPILLS

From the standpoint of direct effects on the environment, few disasters are more dramatic than those that involve major oil spills. Such events, unfortunately, continue to occur on a regular basis. One of the classic such events was the oil spill from the *Exxon Valdez* in Prince William Sound, Alaska, in 1989. This spill, which impacted some 500 miles of the coastline of the Gulf of Alaska, had a devastating effect on the ecology of that area. While major efforts were initiated to clean up the oil, later analyses showed that in reality, less than 15 percent was recovered, and what remains is tenacious. While recent studies have shown that many of the wildlife species are recovering, others (such as harbor seals, herring, harlequin ducks, marbled murrelets, and pigeon guillemots) continue to struggle. In some cases, this may be due to the lack of recovery of the marine life that served as food for these particular species (Mitchell, 1999). Follow-up studies have also shown that long-term exposure of fish embryos to weathered oil, even at extremely low concentrations, has had population consequences through indirect effects on growth, deformities, and behavior that have resulted in long-term negative consequences relative to their mortality and reproduction (Peterson et al., 2003).

More recently (November 2002), the *Prestige* broke up off the western coasts of Spain and Portugal, resulting in the immediate release of several million gallons of fuel oil into the Atlantic Ocean. During subsequent recovery attempts, the vessel was towed about 200 kilometers (125 miles) off the coast, where it broke into two pieces and sank to the bottom. Since the ocean there is about 3.6 kilometers (2 miles; 11,800 feet) deep, and since the water at that depth is very cold, government officials concluded that the estimated more than 20 million gallons of fuel oil that remained on board would to some degree solidify and remain in the ship's hull. Unfortunately, the oil continued to rise slowly to the ocean's surface (Bohannon and Bosch, 2003). Recognizing the need for action, the Spanish government approved a plan to decant the oil from the holds of the tanker into giant plastic bags and haul them to the surface. During the interim, divers were able to seal some of the holes in the wreck using remotely operated submersibles. This reduced the leakage rate from 700 to 10 liters per day. As a result of this event, more than 5,000 Galician fishers and aquaculturists were forced to discontinue activities, pending confirmation that the contaminated seafood was safe to eat. The net economic damage was estimated at $1 billion (Bosch, 2003).

ACCIDENTS IN UNDERGROUND MINES

Mine disasters continue to occur both in the United States and elsewhere. One such event was a fire in a mine in Huntington, Utah in December 1984, that killed 27 miners (NSC, 2001). Unfortunately, many much more severe disasters have occurred in other parts of the world. An explosion in an underground mine in the People's Republic of China in July 2001 is reported to have killed 92 workers, and a flood in a mine a month later is reported to have killed 70 workers, with as many as 200 more missing. The total deaths for the year 2000 in coal-mine disasters alone in that country apparently numbered in the thousands.

Terrorism

As noted earlier, there are basic differences in the nature and characteristics of terrorist attacks as contrasted to natural and other human-made disasters. While the targets of natural disasters are random in nature, terrorists seek to identify and target a nation's vulnerabilities by identifying weak spots in its social infrastructures. While natural disasters can be predicted only statistically and not prevented at all, terrorist attacks can, through the effective application of science, intelligence gathering, and improved investigative techniques, hopefully be predicted and prevented or at least mitigated. At the same time, there are similarities. A primary one is that the impacts of terrorist attacks (other than those of a psychological nature) are generally similar to those of natural and other human-made events. Therefore, systems that have been developed for recovering from other types of disasters can be effectively applied in coping with the aftermath of terrorist attacks (Kennedy, 2002).

Terrorist attacks can range from subtle acts, such as the clandestine dispersal of an infectious agent within a populated area, to clearly visible acts, such as the destruction of buildings and the deaths of the occupants. One of the major objectives in all cases is to disrupt the normal operations of the affected population and their government and, where possible, to incite fear and panic. Another objective is to cause the affected groups to divert major resources to developing countermeasures to prevent such acts, or to restoring the facilities and social order if the attacks prove successful.

POTENTIAL TOXIC AGENTS AND DISPERAL MECHANISMS

The primary toxic agents that might be used are chemical, biological, and/or radiological in nature. The selection of the specific agent will probably depend on the ease with which it can be disseminated, the extent to which it can be expected to cause mortality, its potential for inciting fear and panic, and the difficulty in instituting countermeasures.

Chemical Agents. Potential agents in this category range from those that might be used in warfare to those used in industry (Table 19.3). Because the effects of chemical agents that are absorbed

Table 19.3 Examples of rapidly acting potential chemical agents

Category	Examples
Nerve and incapaciting agents	Sarin (isopropyl methylphosphonofluoridate) Soman (pinacolyl methylphosphonofluoridate) Tabun (ethyl N,N-dimethylphosphoramidocyanidate) BZ (3-quinuclidinyl benzilate) GF (cyclohexylmethylphosphonofluoridate) VX (o-ethyl-[S]-[2-diisopropylaminoethyl]-methylphosphonothiolate)
Pulmonary agents	Benzene Chlorine Chloroform Phosgene Trihalomethanes
Blood agents	Hydrogen cyanide Cyanogen chloride
Cutaneous (blister) agents	Lewisite (an aliphatic arsenic compound, 2-chlorovinyldichloroarsine) Nitrogen and sulfur mustards Phosgene oxime
Poisonous and/or corrosive industrial compounds	Cyanide Nitriles Nitric acid Sulfuric acid
Nitro compounds and oxidizers	Ammonium nitrate combined with fuel oil

Table 19.4 Characteristics of three groups of biological agents

Agents	Characteristics
Variola major (smallpox), *Bacillus anthracis* (anthrax), *Yersinia pestis* (plague), *Clostridium botulinum* toxin (botulism), *Francisella tularensis* (tularemia)	Easily disseminated or transmitted, cause high mortality, could cause public panic and social disruption, require enhanced epidemiology, vaccines and drugs, and diagnostic tests
Hantaviruses, tick-borne hemorrhagic fever viruses, tick-borne encephalitis fever, yellow fever, multi-drug-resistant tuberculosis	Available, easy to produce and disseminate, and have potential for high morbidity and mortality with major health impacts
Coxiella burnetti (Q fever), *Brucella* species (brucellosis), *Burkholderia mallei* (glanders), ricin toxin from *Ricinus communis* (castor beans), epsilon toxin of *Clostridium perfringens*, *Staphylococcus* enterotoxin B	Relatively easy to disseminate, cause moderate morbidity and low mortality, require specific diagnostic capacity and disease surveillance

through inhalation, the skin, or mucous membranes will normally be immediate and obvious, attacks involving such agents are likely to be overt in nature. One example of such an attack was the dispersal of sarin gas in a subway in Tokyo on 20 March 1995 (CDC, 2000).

Biological Agents. The characteristics and relative ranking of agents in this category, based on their potential impacts, are presented in Table 19.4 (CDC, 2000). One of their unique characteristics is that some such agents can be carried and transmitted by rodents and insects. Due to the delay between the time of exposure and the onset of illness, the dissemination of such agents will likely be covert in nature. If, after dispersing such an agent, the responsible parties made known their actions, this would be extremely effective in creating fear and panic among those who were either exposed or were thought to be exposed.

Radiological Agents. One of the most discussed of these is the so-called dirty bomb, through which radionuclides might be dispersed using conventional explosives (Ring, 2004). Another possibility is the detonation of a nuclear device itself. The advantage to the terrorists in the first example is that they could select a specifically hazardous radionuclide to disperse (Table 19.5). Obviously, the overall impacts on the victims of a dirty bomb, however, would be far less than for those exposed to a nuclear detonation.

OTHER DISPERSAL MECHANISMS

Drinking water and food supplies are two other specific dispersal pathways. The key points of vulnerability in drinking-water systems include the open reservoirs that hold the raw water prior to treatment, the treatment plant itself, and the elevated tanks in which the treated water is stored prior to distribution (Chapter 7). Fortunately, in most cases the quantities of toxic agents required to inflict acute effects are large and would be difficult to disperse effectively. To increase security, many plant operators have installed equipment to provide continuous surveillance of reservoirs, as well as physical barriers to deter access.

Biological agents that could potentially be used to contaminate food supplies are included among those listed in Table 19.4. Of particular importance are agents such as *Clostridium botulinum* toxin (botulism) and *Staphylococcus* enterotoxin B (Chin, 2000). Interestingly, viruses such as Ebola are not likely candidates since they kill their human victims rapidly and thus provide little opportunity for effective transfer of the disease from person to person. To enhance the capabilities of physicians to respond to such events, the American Medical Association has developed and distributed a CD-ROM that provides state-of-the-art medical and clinical information for dealing with both biological and chemical attacks (AMA, 2002). In a similar manner, the National Institute for Occupational Safety and Health has issued a guidance document on filtration and air-cleaning systems for protecting the inhabitants of buildings within which airborne toxic agents have been dispersed (NIOSH, 2003).

Another possible dispersal mechanism, as noted earlier, would be the detonation of a "conventional" nuclear weapon. This would also provide a mechanism for major physical destruction. Such a device might be stolen, illicitly purchased from the stockpile in some country, or built by terrorists from enriched uranium and/or plutonium that they had acquired. Of these three options, the first two would probably be the most logical since it would be unlikely that terrorists could manufacture an

Table 19.5 Characteristics of potential radiological agents

Radionuclide	Characteristics	Potential effects
^{60}Co	Reasonably long lived, concentrates in liver, would be difficult for terrorists to handle	Represents both an internal and external hazard, high risk on local basis
^{90}Sr	Long lived, deposits in bone, retained by body	Possible latent leukemia among exposed people, taken up by agricultural plants
^{99}Tc	Extremely long lived, but not retained in the body, mobile within the environment	Contaminated land might not be usable for years
^{131}I	Short lived, would require detonation of nuclear device to produce	Latent thyroid cancers in exposed children, effective in inciting fear and panic
^{137}Cs	Long lived, readily taken up by the body, exposes whole body	Represents both external and internal hazard, can be taken up by agricultural products
^{226}Ra	Naturally occurring, long lived, retained by the body	Contaminated land might not be usable for years, can produce latent bone carcinomas in exposed people
^{238}U	Extremely long lived, commonly referred to as depleted uranium	Effects would primarily be psychological, primarily a chemical rather than radioactive toxin, contaminated land might not be usable for years
^{239}Pu	Extremely long lived, retained by the body, feared by the public	Contaminated land might not be usable for years, could incite fear and panic in exposed population

effective weapon on their own. Even so, depending on how many lives a terrorist organization is willing to sacrifice, the possibility that it could obtain or construct a nuclear weapon cannot be ignored (Fleming, 2002).

Prominent among specific targets, particularly in terms of the potential dispersal of radioactive materials, are commercial nuclear power plants. Such plants, however, are enclosed in large protective concrete containment domes that are about 1.5 meters (5 feet) thick. To ensure their integrity, the concrete domes contain reinforcing steel bars. Inside each dome is a heavy steel liner, and the fuel for the reactor is, in turn, enclosed within a steel pressure vessel 30 to 35 centimeters (12 to 14 inches) thick (NCRP, 2001). To confirm that the containment domes will perform as anticipated, they have been subjected to a variety of tests under real-world conditions. One involved flying a military fighter aircraft into a test wall that simulated a containment dome. Although the engines of the plane penetrated approximately 5 centimeters (2 inches) into the wall, the test results showed that the reactor would not have been damaged (Chapin et al., 2002). Follow-up analyses showed that even a large passenger plane, fully loaded with jet fuel, would not penetrate the containment structure. Another possible target would be a major structure such as a hydroelectric dam. Since they are constructed of such large amounts of concrete, their very nature inherently protects them from the effects of most conceivable types of attacks.

Mitigating and Preventing the Impacts of Natural Disasters

One of the characteristics of certain types of natural disasters is that people can be warned of their impending occurrence. Days of advance warning can be provided in cases of hurricanes and floods; minutes of advance warning can be provided in cases of tornadoes. In either case, timely and effective responses to these warnings can significantly reduce losses of both lives and property. As will be noted later, however, much work remains in terms of both increasing the timeliness of such warnings and minimizing the impacts.

HURRICANES

Experience shows that hurricanes have three primary impacts: loss of life, direct property destruction, and associated loss of commerce. Mitigative measures that can alleviate these problems include (1) technological forecasting; (2) reduced evacuation times; and (3) the provision of refuge. In each case, however, the available systems and/or facilities are inadequate. For example, in spite of computer models and satellite-based tracking systems, improved capabilities for providing longer-range forecasts need to be developed. The same is true with respect to road systems, controlled residential and commercial development, better building practices, and safe in-place shelters for people who might otherwise have to leave impacted areas. Ironically, experience shows that the location of shelters should not be publicized in advance; otherwise people will delay evacuating, knowing that such shelters exist. Other measures that can be used to mitigate the effects of hurricanes include restricting initial development and redevelopment in high-risk areas, enforcing hurricane-resistant building codes, and educating the public on the successful implementation of steps to reduce the loss of life.

FLOODS

During the late summer of 1999, major areas within eastern North Carolina and Virginia, as well as New Jersey, were flooded as an aftermath of Hurricane Floyd. Although prompt response by public health and disaster-control agencies prevented the usual historical causes of death (drownings, lack of accessible medical care), an estimated 63 people died and an estimated $3 to $6 billion were lost through the deaths of tens of thousands of chickens, turkeys, and hogs. That the loss of human lives was as low as it was is a tribute to the effective emergency response of public health and relief agencies and the cooperative efforts of many groups, including the military services. Fortunately, in this case the available meteorological and communication equipment permitted the National Weather Service to deliver timely information to response agencies. This allowed time for evacuation and reduced the risk of entrapment by flash floods.

EARTHQUAKES

The primary reasons for the much higher number of deaths in the previously described earthquake in Turkey were the density of the population, the magnitude of the event, and inadequate building codes (or the failure of contractors to observe them). Other contributing factors were insufficient rescue and debris-removal equipment and either an inadequate number of local medical facilities or the fact that they were damaged by the earthquake and no longer in operation. Although the basic reasons for the low death toll in Peru in 2001 are not clear, the relatively low level of damage caused by the previously cited earthquake in California in 1994 was primarily the result of strict building codes, sound construction techniques, well-organized and well-rehearsed emergency response capabilities, ample communication facilities, and the increasing reliability of forecasting techniques.

TORNADOES

In the previously cited tornadoes that occurred in the United States on 1 March 1997, more than half of the deaths occurred among persons who were in mobile homes. Because other forms of housing are not economically available to many people, increasing numbers of people are living in such homes (Parfit, 1998). For this reason, it is imperative that arrangements be made to enable such people to have access to underground shelters. For people who are living in permanent types of homes, comparable protection can be provided if they go to the basement, if one is available, or seek shelter in hallways and interior rooms (CDC, 1997a).

Mitigating and Preventing the Impacts of Human-Made Disasters (Including Terrorist Events)

As is the case with natural disasters, it is important to take steps to minimize the impacts of human-made disasters. In most cases, these steps can be applied with equal effectiveness to acts of terrorism.

INDUSTRIAL-PLANT ACCIDENTS

As noted earlier, the accident at the plant in Bhopal showed that such events could be devastating. Stimulated by this event and a subsequent much smaller release from a similar plant in West Virginia, the U.S. Congress passed the Emergency Planning and Community Right-to-Know Act (EPCRA) as part of the Superfund Amendments and Reauthorization Act of 1986. Affected manufacturers and plant operators must evaluate the impacts of accidental toxic chemical releases, using worst-case scenarios, and develop plans for preventing such disasters (Neville, 2001). Adding support, the U.S. chemical industry has initiated programs both to reduce the types of accidents that can be caused by chemical releases and to limit their consequences. This has led to reductions in the amounts of both hazardous materials present within various plant systems at any given time and those produced as intermediate products. Efforts are also being directed to improvements in the transportation of hazardous materials.

Data show that of the more than 25,000 hazardous material incidents reported annually in the United States, almost a third are transportation related (ATSDR, 1994).

NUCLEAR POWER-PLANT ACCIDENTS

In much the same way as the Three Mile Island accident provided the stimulus for improvements in the safety of nuclear power plants in the United States, the Chernobyl accident led to similar actions throughout the world. These have included the development of new nuclear power-plant designs that contain features to ensure improved reliability and maintainability of the control of the reactor, increased resistance of the plant structure to seismic events, and the incorporation of three separate and independent systems for cooling the reactor core in case of an accident. Other advances include better inspections and oversight, with each change being closely tied to its importance in terms of risk reduction. These actions have been supplemented by vastly improved training programs for nuclear power-plant operating personnel and programs established by the nuclear industry itself to provide independent reviews of the safety of such operations. Although the activities of the U.S. chemical industry (cited earlier) are akin to these, the programs in the nuclear power industry are far more active and widespread. Within the United States, these activities are conducted under the leadership of the Institute of Nuclear Power Operations (INPO); globally, they are conducted under the leadership of the World Association of Nuclear Operators (WANO). Through these programs, periodic rigorous on-site safety evaluations are conducted at every nuclear plant in the world. The goal is not only to ensure that every plant meets national and international regulations, but also that the responsible utilities are promoting the attainment of excellence in their operations.

OIL-TANKER SPILLS

Assessments and experience have demonstrated that the amount of oil spilled in an accident involving a tanker depends on (1) its capacity and design; and (2) the amount and type of structural damage it suffers. One of the most important factors is whether it has two hulls or one, the former being far superior in avoiding the loss of oil even in relatively large accidents, especially those involving collisions and groundings (NRC, 2001). Recognizing this fact and responding to public outcries for action following the *Exxon Valdez* oil spill, the U.S. Congress passed the Oil Pollution Act of 1990. This required, among other things, that all oil tankers traveling in U.S. waters be equipped with double hulls. Because most countries either conduct or desire to conduct trade with the United State, this requirement quickly became the de facto standard tanker design worldwide. An added stimulus in achieving this goal is a United Nations treaty that bans single-hull vessels. The phase-in period for this treaty, however, extends to 2015. As of 2002, more than half of the estimated 10,000 tankers operating in the world continued to be the old-style, single-hull variety. In the meantime, European Union officials have announced their intention to impose stricter inspections on tankers that dock in their ports.

ACCIDENTS IN UNDERGROUND MINES

Improvements are being made in the safety of mining operations conducted underground, and the probability of events that result in worker deaths is being reduced. A more significant factor, however, in reducing deaths and injuries in such operations is the shift from underground to surface mining. While in 1950 more than 75 percent of the coal mined in the United States was obtained through underground mining, today less than 38 percent is being obtained through this approach. During the same time period, the total production of coal in the United States doubled (CEQ, 1998). This is another instance where one of the remedies to a problem may come about in an unexpected way—in this case, the recognition of vast supplies of coal near the ground surface and the fact that in many instances, the coal could be recovered in a simpler and less expensive manner. Nonetheless, it must be recognized that surface, or strip, mining can have devastating impacts on the environment. In most cases, however, regulations require that the ground surface be restored in terms of soil, trees, and vegetation.

TERRORIST EVENTS

In spite of the nature of terrorism, there are multiple actions that governments can take to enhance security and thereby reduce the probability of terrorist events. One is to apply more fully the capabilities of existing technologies. Such steps include deploying better systems for tracking and ensuring the security of nuclear and other materials, such as chlorine gas, that could be used as weapons; increasing the production of medicines for treating victims of attacks that involve biological weapons; and improving communications among emergency personnel who will be called upon to respond to such events (NRC, 2002). Even so, the challenges of coping with the threat of global terrorism will continue to be enormous and complex.

Emergency Plans

Regardless of the type of disaster, a well-designed and well-executed emergency plan is essential for the rapid mobilization necessary to respond in an effective manner to the immediate health-care needs of the people affected and to restore disrupted services. The plan should be clear, concise, and complete. It should also be dynamic, flexible, and subject to frequent evaluation and update. It should designate precisely who does what and when, and everyone involved should be thoroughly familiar with its contents. Its top priority should be to provide an immediate response to the event by locating and providing emergency medical services to the victims, controlling fires, removing downed power lines, and controlling leaks of natural gas. On a longer-range basis, the goals should be to provide health care and shelter for victims and, as noted later, to restore important services such as a safe water supply and basic sanitation. Next in importance are arrangements to provide a safe food supply.

TYPES, CHARACTERISTICS, AND PHASES

In general, there are two types of emergency plans. One is national or regional in scope and defines the responsibilities and mobilization procedures of personnel in key public and environmental health departments and emergency preparedness agencies. Planning at this level frequently includes coordination of civil defense and military services. The second, which is local in scope, is much more definitive and includes detailed listings of the personnel involved, their individual responsibilities, and the range of countermeasures available for implementation. Properly coordinated, the two plans can provide a cadre of well-trained personnel to cope with natural disasters and human-made events of almost any magnitude.

Experience demonstrates that disaster plans can generally be divided into four phases: (1) the years before the event (the pre-event phase); (2) the warning or alerting phase, just prior to events that can be predicted; (3) the response phase, immediately following an event; and (4) the recovery (rehabilitation) phase. Although many aspects of the required responses are similar, there are sufficient differences to require somewhat different approaches in responding.

PRE-EVENT PHASE

The objectives during the pre-event phase are to anticipate that accidents and disasters will occur and to plan for responding to them. Specific steps should be taken to identify all available organizational resources; inventory the types and locations of available supplies and equipment, including hardware and medical supplies; identify private-sector contractors and distributors who can provide otherwise scarce specialized personnel and equipment; review essential community and industrial facilities to identify those that may be vulnerable to a disaster; and define the responsibilities of each agency or group and establish lines of communication. In support of these activities, an emergency operations center should be designated and properly equipped.

WARNING OR ALERTING PHASE

In the case of hurricanes, tornadoes, and floods, where advance warning will be available, there will be an opportunity to alert emergency planning personnel and to have them move, where appropriate, to the emergency operations center. Hopefully, through applications of science and technology, it will be possible to provide similar warnings for certain types of terrorist-related events, particularly those of major magnitude. Prominent among activities during this phase should be the provision of timely and accurate information to the media and the public on what to anticipate. This information should include specific details on preparations that should be made.

RESPONSE PHASE

Usually fire, emergency medical, and police personnel will be the first official response personnel at the site of a major disaster. One of their first objectives will be to provide security to the affected area to assure the safety of both the victims and workers. Another objective will be to provide guidance to the laypeople already present who have volunteered their services. The most experienced senior person should take charge, immediately surveying the area and carefully assessing the scene, the number of victims, and their injuries. This person should relay information to the emergency operations center and make recommendations for action. Officials at the center must then determine what additional support is needed.

RECOVERY PHASE

During the recovery phase, substantial numbers of injured people may need follow-up care. All survivors will require food, water, shelter, clothing, and sanitation facilities. Floods, in particular, promote unsanitary conditions through the buildup of debris and blockage of sewer systems. This is the reason, as discussed later, that it is essential that the restoration of water-supply and sewage systems be given priority.

Emergency Plan Development

As the preceding discussion has demonstrated, coping with the impacts of any type of disaster requires a careful, well-designed emergency preparedness plan. An essential part of its development in the case of industrial facilities is an assessment of the potential internal, external, and natural phenomena that can cause accidental releases of toxic materials. Therefore, the institution of protective measures for industrial facilities requires full consideration of the potential impacts of natural events, such as hurricanes, floods, and earthquakes, on such facilities. One law that has proved to be a valuable asset in such planning is the previously cited Emergency Planning and Community Right-to-Know Act. Under this act, operators of industrial facilities in which certain chemicals are used are required to report the amounts and locations to local emergency planning committees, who, in turn, can use this information in developing emergency preparedness plans.

The first steps in the development of such plans are to consider the full spectrum of accidents that can occur; estimate their individual likelihood of occurrence; analyze the potential consequences; assess the systems that have been incorporated into the facility to prevent such releases; identify significant structures, systems, and components that are present to mitigate the consequences; and identify a selected subset of accidents and related scenarios that will need to be formally considered. Applications of these principles are discussed here.

PRE-EVENT PHASE

The initial conduct of what is called hazard identification and evaluation is largely a qualitative exercise. It is designed to provide a comprehensive evaluation of natural phenomena and other external events, as well as internal events, that can initiate one or more failures within the facility that can affect the public, the workers, and the environment. Among the more important items is an inventory of all hazardous materials and energy sources within the facility. Once this is in hand, the next steps are to evaluate the consequences of an unmitigated release from the facility and to use this information to classify the facility and/or its operations into one that has the potential for significant (1) off-site consequences, (2) on-site consequences, or (3) consequences that have impacts on a localized basis only. Since this ranking, as the description implies, is based on the assumption that none of the safety systems mitigate the release, it represents what might be called a bounding calculation. This is one of the requirements of the Clean Air Act Amendments of 1990. Detailed evaluations and follow-up are required only for those facilities that fall into the first or second category. Those responsible for such evaluations should consider all possible modes of facility operation, including start-up, shutdown, and abnormal testing or maintenance configurations.

Estimates of the accompanying potential doses to the public can be made on the basis of "realistic" or "conservative" assumptions, or they may involve the bounding calculations just described. Due to the underlying assumptions, any such estimates will be accompanied by large uncertainties. For this reason, field surveys should be conducted as soon as possible following any unanticipated releases to confirm not only that the theoretical estimates of exposures/doses are as anticipated, but also that adequate responses have been initiated.

COUNTERMEASURES

The application of countermeasures is critical in mitigating the impacts of a disaster. To be acceptable, a countermeasure must first be *effective* (that is, it must substantially reduce population exposures below those that would otherwise have occurred); second, it must be *safe* (that is, it should introduce no health risks with potentials worse than those presented by the releases); third, it must be *practical* (that it, it must be capable of being administered at a reasonable cost and without creating legal problems), and last, it must be *defined* (with no jurisdictional confusion about responsibility and authority for applying the measure). In addition, it must be recognized that almost any countermeasure will carry with it health risks and social and economic disruption, depending on when and where it is applied. On the basis of these criteria, the following countermeasures have proved most useful.

Evacuation. Evacuation can be effectively applied in many types of industrial accidents, especially those that result in releases of some type of toxic material into the atmosphere. The feasibility of evacuation, however, depends on a wide range of factors. These include the magnitude and likely duration of the release; the weather and time of day; the time interval between the accident and the order to evacuate; the availability of transportation and suitable shelter for those evacuated; the potential for vehicle accidents and personal injuries; and the potential for increased uptake of releases as a result of the exertions involved. Evacuation is particularly effective in cases involving natural events, such as hurricanes and flooding (Figure 19.1). As might be expected, the protective value increases with the rapidity with which evacuation is executed, with the distance between the airborne release and the evacuated population, and the length of time the airborne cloud remains in the area.

Sheltering and respiratory protection. The simplest and least disruptive of all proposed countermeasures is instructing people to remain indoors. This can be effective in responding both to airborne releases of certain toxic materials and to weather-related emergencies. The value of sheltering can be enhanced by encouraging the potentially exposed population to use common household materials for respiratory protection. In the Bhopal accident, for example, those exposed could have protected themselves simply by placing a wet cloth (e.g., a handkerchief, washcloth, or

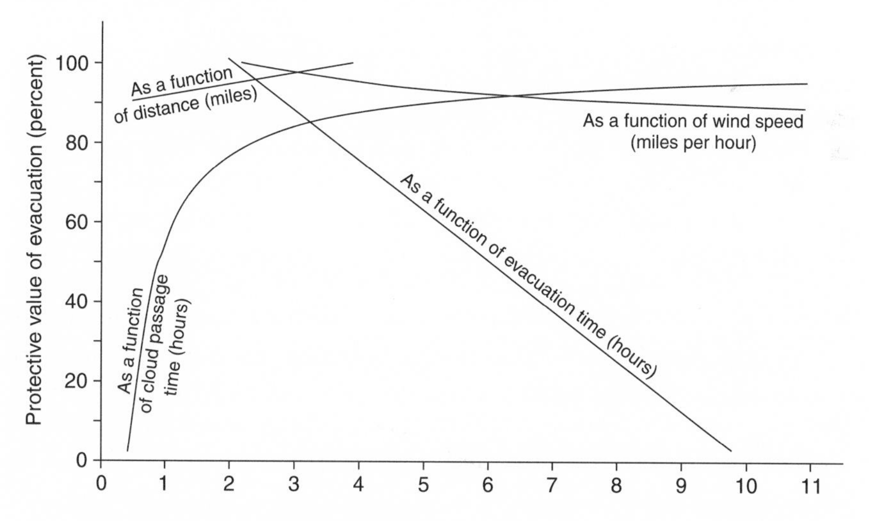

Figure 19.1 Protective value of evacuation as a function of time required for evacuation, wind speed, distance, and cloud passage

bath towel) over their face to protect their eyes and to prevent inhalation of methyl isocyanate into their lungs. Unfortunately, this information was not made available. Sheltering and respiratory protection can be especially effective for people who are close to an accident site and would need to move through the airborne cloud during evacuation, and for those for whom data concerning the magnitude, duration, and direction of an airborne release are unavailable or unreliable.

Protective prophylaxis. If a toxic chemical that has been released is identifiable and a known antidote is available, it is possible that it can be administered to counteract or negate the effects of the exposures. One prominent example is stable iodine, which will reduce the uptake of radioactive iodine by the thyroid, depending on when it is administered (Figure 19.2).

Other countermeasures. The major agricultural crops of immediate concern after an accidental airborne release will be those that grow above-

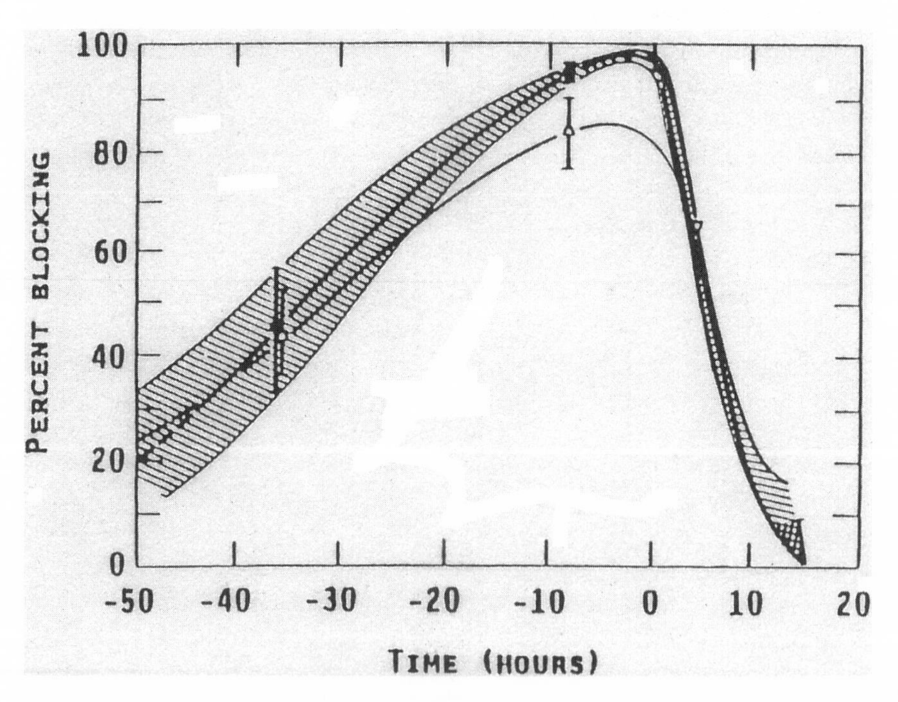

Figure 19.2 Percentage of thyroid blocking afforded by the administration of 100 milligrams of stable iodine as a function of time before and after an assumed intake of 3.7 x 10 becquerals of ^{131}I

ground, such as lettuce and cabbage, since they may have been directly contaminated. If they are harvested immediately, and the outer membranes are removed, this may provide a simple, low-risk method of averting the intake, for example, of airborne particulates. For crops such as tomatoes or squash that can be washed, this may provide similar protection. Similarly, it may be possible to make fruit, such as oranges and bananas, acceptable for human consumption simply by removing the outer peel. Crops that grow underground, such as carrots, beets, and potatoes, should pose no problem until enough time has passed for the contaminants to gain access to them via root uptake through the soil. The use of previously packaged foods is another simple method for avoiding the intake of contaminants.

On farmland that has suffered extensive contamination, the soil must be treated so that the contaminants are reduced to acceptable levels. Idling, or nonuse of the soil, for a specific period is one strategy for averting exposure for those contaminants that naturally degrade or dissipate with time. Another is deep plowing to move the contaminants below the root uptake level of agricultural crops. In some circumstances, it may be possible to grow alternative crops, such as cotton and flax, that do not contribute to the human diet. Highly contaminated soil, however, may have to be removed and disposed of. The water in covered wells will probably not be contaminated. Since it may require time for contaminants deposited on the surface of the soil to travel to groundwater sources, these should be monitored to ensure that they continue to be acceptable for consumption. One special need is the establishment of permissible levels of contamination in (1) foodstuffs that are acceptable for consumption and (2) soils that are acceptable for agricultural use and/or human habitation.

At the same time, however, it is important to note that airborne toxic materials can be transported many miles from the accident site and can lead to the contamination of agricultural products, including milk, over large geographic areas. Since these foods are widely distributed, those produced in the more highly contaminated areas may be consumed by people some distance away. A further concern is the fact that the movement of contaminants through the food pathway can be extremely rapid. Radioactive iodine, for example, can appear in milk within hours of being deposited on pasture grass (Figure 19.3). As a result, protective action needs to be taken promptly. In the case of milk, one of the most effective countermeasures is to shift cows from outdoor feed to stored (noncontaminated) feed.

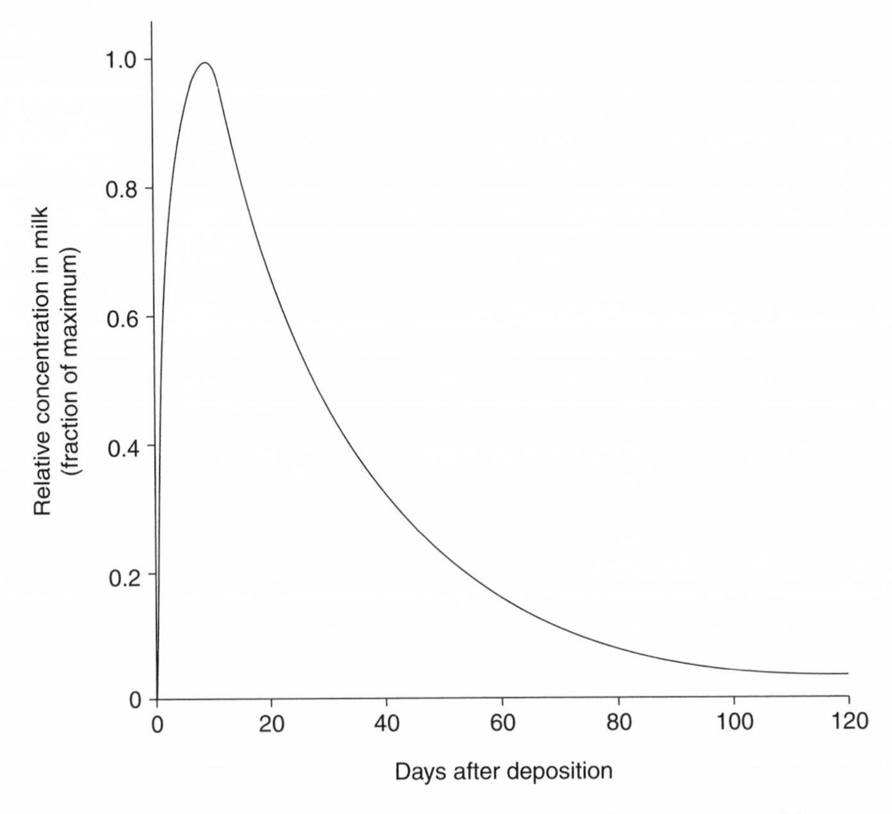

Figure 19.3 Concentration of radionuclides in milk as a function of time following a single deposition on pasture grass

Response Coordination and Follow-Up

Recognizing that all types of disasters, including acts of terrorism, share common requirements, emergency planners at all levels are now combining such activities within one organization. One of the primary benefits of consolidation is that preparedness and response capabilities are better organized, staffed, and focused. In terms of details, one of the most important aspects of local response organizations is to ensure that attempts to mitigate the impacts of a disaster do not add to its negative consequences. Unfortunately, this is often not the case, especially when governments and organizations outside the affected area attempt to assist. Physicians and nurses frequently are sent into the region in numbers that far exceed the need. In actuality, for most disasters it is unlikely that medical personnel from outside the affected country will be required. Another frequent problem is an influx of volunteers who neither speak the local language nor have disaster relief experience. Unless these types of situations are avoided or mitigated, problems during the days, weeks, and months after the initial event can often lead to what some have called the *secondary disaster.* A summary of some of the positive actions that should be taken in response to a disaster and some of the negative actions that should be avoided is presented in Table 19.6.

Table 19.6 Responses to disasters: positive and negative actions

Positive actions	Avoiding negative actions
Recognize that every disaster is unique. Its effects on health depend on the degree of development in the affected country.	Do not overreact to media reports for urgently needed international assistance. Take care to ensure that donations meet real needs.
It is incumbent on the part of the affected people to inform donors about what is needed. This is just as important as specifying what is not needed.	Recognize that the quality and appropriateness of the assistance are more important than its amount, its monetary value, or how rapidly it arrives.
Emergency assistance should be designed to complement, not duplicate, what is being done by those in the affected country.	Do not promote shipments of medical or paramedical personnel, medical equipment, or field hospitals.
Cash donations are preferable. These will enable the affected groups to purchase at the local level what they need. This will save time and storage and transportation costs.	Do not promote the shipment of used items, such as clothes, shoes, and food. Never donate medicines that have expired or are about to expire.
Recognize that the affected populations will continue to need assistance during long-term rehabilitation and reconstruction.	If a product is not acceptable in the donor country, it is also not acceptable as a donation.
Because of its importance, a system for effective communication with the affected groups needs to be established prior to the time when a disaster strikes.	Donors should not compete with each other.

In many cases, one of the major concerns on the part of the impacted populace, especially following hurricanes and floods, is that the event itself will be followed by outbreaks and even epidemics of infectious disease. In reality, experience shows that such outbreaks are rare. In order for such a risk to exist, the disease of concern must have existed in the population prior to the disaster. Some of the myths that inevitably come to the surface following essentially every disaster, and the opposing realities, are summarized in Table 19.7 (PAHO, 2002). Far more important

Table 19.7 Myths and realities of natural disasters

Myths	Realities
Any kind of international assistance is needed and it is needed now!	A hasty response that is not based on an impartial evaluation only contributes to chaos. It is better to wait until the real needs have been determined.
Foreign medical volunteers with any kind of medical background are needed.	The local population is almost always able to cover immediate lifesaving needs. Only medical personnel with skills not available in the affected country may be needed.
Disasters are random killers of people.	Disasters strike hardest on the most vulnerable groups: the poor, women, children, and the elderly.
Epidemics and plagues are inevitable after every disaster.	Epidemics do not occur spontaneously after a disaster, and dead bodies will not lead to catastrophic outbreaks of exotic diseases. The key to preventing disease is to improve sanitary conditions and educate people.
The affected population is too shocked and helpless to take responsibility for its own survival.	On the contrary, many people find new strength during an emergency, as evidenced by the thousands of volunteers who have spontaneously united during the immediate aftermath.
Conditions will be back to normal within a few weeks.	The effects of a disaster last for a long time. Much of the financial and material resources of the affected countries is depleted during the immediate postimpact phase.

is the need to restore potable water supplies and sanitary services. To strengthen the resistance of such facilities to disasters, the Pan American Health Association (PAHO) recommends what is described as a *vulnerability analysis,* the basic objective of which is to identify measures that can be taken prior to a disaster to strengthen such systems.

The protection of the health and safety of workers who are involved in rescue and recovery operations must also be considered. Those who respond to volcanic eruptions and wildfires are frequently subjected to respiratory hazards and ocular problems. Those who are involved in responding to industrial accidents and terrorism events frequently suffer facial injuries and exposures to carbon monoxide, musculoskeletal hazards, thermal stresses, and death from electrical hazards. This was vividly demonstrated by the terrorist attack on the World Trade Centers in New York City in 2001. At the time of the event, primary attention was directed, as might have been anticipated, to rescue efforts. Subsequent evaluations, however, showed that, during the fires that ensued, the towers spewed toxic gases and particulates into the air similar to in some respects similar to, but in quantities many times, those that come from a chemical factory. These pollutants included pulverized and vaporized concrete, wallboard, ceiling tiles, computers, electrical equipment, and office furniture. Fortunately, nearby residents escaped serious exposures because most of the toxic substances were transported far above the city by the heat of the fires. This was not the case, however, for the rescue workers (Service, 2003). Another aspect to be considered is that some rescue- and recovery-operation personnel also suffer stress-related illnesses, while still others may suffer severe forms of distress, including anxiety and depression. The same is true for the members of the public who are victims of disasters. All too often, these types of problems are overlooked.

The General Outlook

Although more work is needed, the previous discussions show that there have been significant advances in recent years in the methods for addressing disasters. For natural events, these include methods for forecasting their occurrence. For human-made events, these include applications of engineering technology to enable industrial plants to respond in a manner that mitigates a potential triggering event, supplemented by the development of designs that limit the releases of harmful substances in case such an event occurs. Other efforts have led to the development of improved communication systems that enable information to be made available more rapidly to those responsible for making decisions after an event has occurred, and of models into which real-time monitoring data can be fed after an accident for updating predictions of the magnitude of the consequences and the geographic areas that will be affected. Another advance has been the development of methods for factoring into accident prevention and mitigation the possible contributions from malevolent acts such as terrorism.

In the meantime, experience has demonstrated the importance of informing the public on what to expect and measures to take to avoid harm. This includes providing advance information about the possible effects of, and countermeasures for, exposures to chemicals, as well as timely warnings so that people can seek shelter or evacuate (Table 19.8). In addition, emergency planning officials are increasingly involving the public in the development of disaster response plans. Not only does such an approach ensure their cooperation and support, but it also leads to the development of plans that are superior to those that would otherwise be produced. One reason for this is that in most cases, the first people to respond in a disaster

Table 19.8 Sources of information on disasters

World Health Organization

- Virtual Health Library for Disasters (2001 Edition), available on CD-ROM and on the Internet at http://www.helid.desastres.net/

Pan American Health Organization

- *Natural Disaster Mitigation in Drinking Water and Sewerage Systems: Guidelines for Vulnerability Analysis* (1998)
- *Humanitarian Assistance in Disaster Situations: A Quick Guide for Effective Donations* (2002)
- *Self-Study Course on Prevention, Preparedness, and Response to Chemical Accidents* (2002)
- General information: *Disasters: Preparedness and Mitigation in the Americas,* issued monthly through the Washington, DC, office

American Council on Science and Health

- *A Citizen's Guide to Terrorism and Preparedness and Response: Chemical, Biological, Radiological, and Nuclear* (2003). 1995 Broadway, 2nd Floor, New York, NY 10023–5860.

will be those in the affected community. Another is that while the development of building codes, early-warning systems, and approaches to training and community preparedness require international support, they are best organized and implemented at the regional and national levels.

With regard to prevention, analyses show that most industrial accidents occur through a combination of several factors, the most important of which are hidden or apparently minor design errors in a piece of equipment, poor maintenance, taking shortcuts to maintain schedule, and bad communications. In fact, disasters at the human-machine interface rarely have one cause. Sometimes a cause is a technical "blind spot," such as the absence of coolant-level indicators in the Three Mile Island nuclear power plant; sometimes it is the perceived need to complete a mission quickly, as in the case of the launch of the space shuttle *Challenger.* This realization indicates that the universal features of a sound safety culture must be supported at the corporate, managerial, and individual levels. As confirmation of this attitude, senior management must generate and enthusiastically support a distinct definition of the lines of responsibility and authority for safety; work practices must be adequately defined and controlled; the staff must be appropriately qualified and trained; and a system must be in place that includes rewards and sanctions that promote good safety practices. A further support for such developments is the increasing recognition that preparedness for and mitigation of the impacts of disasters represent a sound economic investment. In fact, data show that every dollar spent on disaster preparedness saves seven dollars in economic losses. If the social and ecological costs that are averted are also taken into consideration, the estimated rate of return would be even higher (Abramovitz, 2001).

20

A GLOBAL VIEW

MANY of our environmental problems—air and water pollution, solid waste, and food contamination—are consequences of large-scale cultural patterns. Some are the net result of millions of people making individual decisions; others are triggered by a small number of people with key decision-making powers in industry, government, and academia. Although many of the problems are local, ozone depletion, acidic deposition, ecosystems impacts, global warming, and the resulting climatic and environmental changes have global implications. Solutions will require cutting across national jurisdictions and a shift in focus from protection and restoration to planning and prevention.

Our global problems reflect three major trends. First, as the result of a fourfold increase in the world's population and an eighteenfold increase in the values of goods produced since 1900 (Sachs, 2004), the quantity of pollutants being generated has significantly increased. Second, there has been a shift from the use of natural products to the production and use of synthetic chemicals. For example, a billion pounds of synthetic pesticides are used every year in the United States. Many have proved to be highly toxic, and some persist and accumulate in biological systems and in the atmosphere. Third, expanded technological capabilities and in some cases the export of hazardous technologies have led to a situation in which the less developed countries are now as polluted as the developed nations.

Even though there is no consensus on how to solve these increasingly difficult problems, progress is being made on both the national and international levels. Several examples are discussed in the sections that follow.

ation

Closely relatec which the world's population is increasing is the proble ation (Chapter 1). In fact, this is the most powerful and visible anthropogenic force on Earth, with the portion of the world's population that lives in cities projected to reach 60 percent within the next 20 years. Ideally, cities exist as places where technology, population, culture, economics, and natural systems intersect and interact. Nonetheless, the space they occupy and the resources required to fulfill their needs absorb, transform, and/or consume either directly or indirectly ever-larger amounts of forests and arable land. Although cities are essential instruments of social advancement, wealth creation, globalization, creativity, psychic energy, and birthrate reduction, many are dysfunctional (Bugliarello, 2001).

Recognizing these facts, many experts believe that the concept of a city must be rethought in terms of efficiency, manageability, and quality of life, including the emotional aspects. If these challenges are to be successfully addressed, it will be necessary to learn how to organize and engineer the city as an organic whole along its three dimensions: biological, social, and machine. The challenges that this poses are larger in scale and complexity and involve more disciplines than any previously encountered. As such, they need to be addressed both nationally and globally.

Ozone Depletion

Ozone depletion is an instructive problem to highlight because it has been effectively addressed through cooperative efforts at the international level. The problem was created through the release into the air of chlorofluorocarbons (CFCs), which for some 80 years had been used as refrigerants in household appliances and air conditioners, as industrial solvents, and as propellants for aerosol sprays. Once released, CFCs mix with other compounds and rise slowly into the stratosphere. Ultraviolet (UV) radiation from the sun subsequently destroys the CFC molecules and releases highly reactive chlorine atoms. These react with ozone and convert it into normal oxygen. A single CFC molecule can destroy tens of thousands of molecules of ozone. Although ozone is considered a pollutant when it is near the ground (Chapter 5), in the stratosphere it shields the earth's surface from UV radiation. Destruction of this shield increases the amount of UV radiation that reaches the Earth. The resulting effects include increased skin cancers and cataracts, lower crop yields, and damage to materials such as vinyl plastics.

Recognizing the problem, the international community took action that led to the Montreal Protocol on Substances That Deplete the Ozone Layer, an international treaty developed in 1987 that committed 47 nations to reducing their production of CFCs by 50 percent by 1998. Subsequently, the developed nations agreed to eliminate all CFC production by 2000, a date that was later advanced to 1996. Studies conducted in 2003 show that the hole in the ozone layer that stimulated the initial action to correct the problem is being reduced. Due to their more significant technical and economic challenges, less developed nations were given an additional 10 years to accomplish this goal. To assist them in this regard, a fund was established to help them make the transition to replacement chemicals. As frequently is the case, the Environmental Protection Agency (EPA) later estimated that the $45 billion cost to rid the United States of CFC emissions was more than offset by the $32 trillion in crop damage, skin disease, and ecological problems that were averted.

Acidic Deposition

As is well known, airborne contaminants have a wide range of effects, and their movement does not respect national boundaries (Chapter 5). A further problem is that subsequent to their release into the atmosphere, pollutants such as nitrogen oxides can be converted into nitric acid, and sulfur oxides can be converted into sulfuric acid. Subsequent deposition of these acids, either dry or as nitric or sulfuric acid in rain or snow, has been shown to impose an unprecedented burden on forests, streams, and lakes throughout the world.

The principal measures for controlling acidic deposition and its effects are (1) to reduce the discharges of nitrogen and sulfur oxides into the atmosphere and (2) to treat sensitive ecosystems to make them less susceptible to damage. Recognizing that sulfur dioxide (SO_2) accounts for about two-thirds of the acidic deposition in the northeastern United States and eastern Canada, Congress, in passing the Clean Air Act Amendments of 1990, mandated a 50 percent reduction by the year 2000 in the releases of this contaminant from coal-fired plants in the Midwest. Similar attention was directed to the control of nitrogen oxide releases from automobiles in the southwestern United States, including cities such as Los Angeles. Although no remedial actions have been developed for treating terrestrial ecosystems damaged by acidic deposition, it is common practice to add lime to lakes to neutralize the acids that have been added to them. While emissions of SO_2 were reduced by 31 percent from 1993–2002, emissions of nitrogen oxides were reduced by only 12 percent (Chapter 5). Obviously, more work needs to be done.

Biodiversity

Biologists have identified and assigned formal names to more than 1.5 million species of plants, animals, and microorganisms; the total for all species on Earth is estimated to range up to 100 million (Wilson, 2002). In fact, some form of life occupies essentially every available niche on Earth. These forms include photosynthetic bacteria, microscopic invertebrates, fungi, and mites that inhabit the cold and dry environment of Antarctica; specialized microbes that flourish in volcanic hydrothermal vents on the ocean floor with water temperatures near the boiling point; and marine organisms that survive at depths with pressures a thousand times higher than at the surface. Still other species prosper at altitudes equivalent to that of Mount Everest (Myers, 2002).

Even so, multitudes of organisms, while vast in numbers, are extremely fragile. As such, they are poorly equipped to withstand the relentless assault of humanity on their habitats. The net result is that the current rate of extinction of organisms, plants, and animals is estimated to be 100 to 1,000 times as high as it was before the coming of humanity. Concurrently, the birthrate of new species has declined (Wilson, 2000). As the negative effects of other factors, such as climate change (discussed later) take place, the situation could well become far worse. A further problem is that once a species is destroyed, it cannot be restored.

BENEFITS OF BIODIVERSITY

Even Charles Darwin in his *Origin of Species* noted the benefits of biodiversity in his statement that "It has been experimentally proved that if a plot of ground is sown with one species of grass, and a similar plot be sown with several distinct genera of grasses, a greater number of plants and a greater weight of dry herbage can thus be raised" (Hector and Hooper, 2002). That is to say, the yield of the biodiverse plot is superior. Experience shows that this is but one of multitudes of such benefits. These include blood thinners, based on the venom of the deadly Asian saw-scaled viper and the European leech; digitalis, derived from foxglove and used to boost the pumping of the heart; quinine, obtained from the cinchona tree and used to treat malaria; and a blood-clotting agent from the horseshoe crab that is used to detect potentially fatal bacteria in vaccines, drugs, and medical devices. One of the most prominent examples is Paclitaxel, which is used to treat ovarian and breast cancers. It was discovered in the bark of the Pacific yew tree. Once it was found, chemists acknowledged that it was too complex a chemical structure to have been discovered by them.

PROBLEMS OF NONNATIVE SPECIES

While biodiversity, if left alone, provides many benefits, the introduction by humans of nonnative, foreign, or alien species into an area can have negative effects equivalent to those caused by the loss of diversity. Although in some cases such introductions were inadvertent, in other cases they were intentional, with the goal of controlling an existing problem. Examples include so-called Jackson's chameleons, native to East Africa, which were released in Oahu, Hawaii, in the 1970s. These invaders subsequently spread throughout the islands and have jeopardized many of the species that are unique to the area and had evolved during the archipelago's eons of isolation. They and other alien invaders are costing the state millions of dollars in damage to forests, crops, and buildings (Leslie, 2002). In a similar manner, introduction of the yellow crazy ant, *Anoplolepis gracilipes,* into Christmas Island (located in the Indian Ocean) about 70 years ago created a major problem when their population suddenly exploded in the late 1980s, producing supercolonies that infested one-fifth of the island. Within a two year period, the trophic dynamics of an entire ecosystem was altered (Hurtley, 2003). Another example is an invasive strain of the tropical alga *Caulerpa taxifolia,* which was introduced, perhaps accidentally through the discharge of aquarium waste, into a San Diego County lagoon in 2000. Once it is present, *C. taxifolia* has runners that grow several centimeters per day, forming a dense green carpet that excludes all other plants. Herbivores that might otherwise check its growth are deterred by the toxins it produces. Attempts to eradicate the strain in the Mediterranean and southeast Australia have failed (Withgott, 2002).

The problems, however, do not end here. For example, more than 100 different species of aquatic plants and animals are being raised in aquaculture farms in the United States. Unfortunately, such activities have led to the introduction of nonnative aquatic plants, fish, invertebrates, parasites, and pathogens into many areas of this country. If this trend continues, aquaculture may well become the leading vector of aquatic invasive species worldwide, and many such introductions may lead to unpredictable and irreversible ecological impacts.

PRESERVING BIODIVERSITY

Noting that a major share of the world's biodiversity is concentrated in a relatively small number of coral reefs, forests, savannas, and other habitats, environmentalists are undertaking vigorous efforts to preserve these areas. Prominent examples are those in Madagascar, the Philippines, and the Mediterranean-climate coast of California. Less well known are specific areas in Ecuador, India, and South Africa. In all, some 25 such areas, which occupy only about 1.4 percent of the Earth's land surface, are believed to be the exclusive homes of 44 percent of the plant species and 35 percent of the birds, mammals, reptiles, and amphibians that exist (Wilson, 2002).

One of the largest such preservation efforts is a cooperative venture of Mexico and seven countries in Central America, called the Mesoamerican Biological Corridor. The goal is to link existing protected areas and provide additional ones so that scores of corridors will be established that will ultimately enable animals to have safe passage from Chiapas, Mexico, in the north to Darien Gap, Panama, in the south. Similar efforts are under way throughout the world (Kaiser, 2001). Even so, such projects do not guarantee that endangered species will receive the attention they require. In fact, scientists continue to report that elsewhere in the world the loss and degradation of natural wildlife habitats are continuing at essentially an unabated rate. One possible approach to halt this trend is to provide financial incentives, ranging from tax breaks to future development rights, to owners who choose to protect such habitats. Another approach is to develop and initiate programs to help people recognize the marked economic benefits that habitats generate. Many analyses show that the economic benefits provided through the preservation of habitats are, at a minimum, 100 times as valuable as those that will be gained through continuing to destroy them for short-term gains (Balmford et al., 2002).

Ecosystems

According to the Council on Environmental Quality (CEQ, 1997), there are four major types of ecosystems (Table 20.1). To make the list complete,

Table 20.1 Major types of ecosystems

Type	Examples
Terrestrial	Forests; agricultural lands
Aquatic	Freshwater rivers
Coastal and marine	Coral reefs, major bay and ocean fishing areas
Riparian	Floodplains and wetlands

one might add humans since essentially every person, at one time or other, serves as a semipermanent host for organisms, such as mites, lice, bedbugs, and fungi, or an intermediate host for other organisms, such as mosquitoes.

As scientists have explored ecosystems in more depth, they have increasingly recognized that they are "extraordinarily complex and dynamic, poorly understood, and prone to unforeseeable behavior that may alter their functionality" (Prugh and Assadourian, 2003). It is therefore not surprising to learn that none of the major ecosystems has been immune to the impacts of humans, and that these impacts are, in almost all cases, due primarily to a lack of recognition and understanding of the interplay with, and interdependence of, each of the several subunits within such systems. Excessive harvesting of one species of fish from the ocean, for example, has ramifications far beyond the depletion of that species. Each species within such an ecosystem is linked to many others as a predator, as a scavenger, or as a source of food or shelter. If the species that is removed is peripheral to the system, it may be possible for the ecosystem to continue to function. If, however, it is a major or keystone player and is overly impacted, the ecosystem will be forced to establish a new equilibrium that inevitably will not be as functional as the original (Hayden, 2001). Some of the specific ways in which humans have impacted each of the four major ecosystems are discussed here.

TERRESTRIAL ECOSYSTEMS

Forests are a major constituent of terrestrial ecosystems. Among their better-known benefits is that they provide both ecological services and economic goods, ranging from soil and watershed protection and timber and firewood to wildlife habitat and recreation. They also moderate climate, capture and store precipitation, serve as sinks for carbon dioxide, and are home to two-thirds of all species. In fact, as much as half of the carbon in the world's biomass may be stored in forests; any loss of this resource thus reduces the Earth's capacity to absorb carbon dioxide from the atmosphere. The cutting and hauling of trees from a forest also removes essential nutrients and topsoil, the water and nutrient cycles are destabilized, and the soil itself is left without protection from flooding and erosion.

In spite of these attributes, forests are being rapidly destroyed by human activities. From 1960 to 1990, 10 percent of the forests in the United States were destroyed. The impacts on tropical rain forests have been even more devastating. Worldwide, these types of forests are being destroyed at a rate of 40–50 million acres (60,000–75,000 square miles) per year. Fortunately, steps are being taken to correct this situation. For example, in its long-range plans for developing the Amazonian region, the Brazilian government hopes to expand its system of national forests by some 50 million hectares, an area equivalent to the size of Spain. Ancillary benefits are that this initiative will complement ongoing programs for protecting other land areas and will enhance conservation of the biodiversity in that region of the world (Verissimo, Cochrane, and Souza, 2002).

Agricultural lands represent a second major type of terrestrial ecosystem. Poor farming techniques, such as deep plowing, followed by wind and water erosion, lead to the runoff of billions of tons of topsoil in the United States every year. Once removed, topsoil requires hundreds to thousands of years for regeneration. Equally important, the topsoil that enters water bodies causes turbidity, silting, and deterioration of aquatic habitats and reduces the storage capacity in lakes and reservoirs. Fertilizers, pesticides, and salts present in eroded sediment reduce water quality. Soil particles blown by wind cause dust storms that physically damage crops and buildings. Windblown particles also contribute to air pollution, which exacerbates respiratory ailments and impairs vision. Urban and suburban sprawl also contribute to these problems. Farmlands with acres of the world's richest topsoil are being covered by asphalt parking lots, housing developments, and shopping malls.

Multiple techniques can be used to correct soil erosion. Among the most successful are various forms of conservation tillage, such as no-till farming, in which the residue from a previous crop is left on the field, and new seeds and/or plants are placed in the ground without prior plowing. Other techniques include contour plowing, maintenance of vegetative buffer strips between fields and along waterways, planting trees or grass cover on highly erodible soils, and maintaining a vegetative cover on idle land.

AQUATIC ECOSYSTEMS

Although the destruction of aquatic ecosystems continues to be a problem, the consequences are being increasingly recognized. This is illustrated by a detailed assessment of the impacts due to the construction during the past half century of a series of dams for hydropower and flood control on the Missouri and Columbia Rivers. While the discussion that follows will emphasize the negative impacts of these installations, it is important to keep in mind that the hydropower units that were constructed have served as major sources of low-priced electricity for people who live in these areas. In addition, the reservoirs created by the dams have enhanced recreational opportunities for millions of people and have enabled similar numbers to inhabit otherwise forbiddingly arid regions. Another benefit has been a major increase in food production through the water made available for irrigation (Reisner, 2000).

In the case of the Missouri River, the benefits just described were achieved through the construction of a system of multiple levees and six dams (Kearney, 2002). Experience, however, has revealed that these undertakings have had many negative impacts. The most prominent is the loss of habitats of dozens of native species of fish, birds, and other wildlife. The primary reasons for these losses are that the lack of quiescent periods of water flow has kept fish hatchlings from drifting to the shore to develop. Another change is that the sediment that formerly moved with the water now settles out in the reservoirs created by the dams, thus eliminating the creation of shoreline habitats and shallow shoals for fish and other aquatic life. In a similar manner, the removal of plants and trees along the river to make room for farms and developments has destroyed the habitats of other types of wildlife.

Actions that are being considered for restoring the river include opening one or more of the dams. While this would overcome some of the problems just noted, there would obviously be many negative impacts. For example, the upriver reservoirs created by the dams would be lost, thus removing major fishing and recreational areas that are utilized by about 1 million visitors annually. Also, the homes of many people who live on what were formerly the downstream flood plains would now be subject to flooding. For these and other reasons, any such action would be highly controversial. Therefore, it is doubtful that any of the proposed restoration actions will be implemented within the immediate future (Thigpen, 2002).

In the case of the Columbia River, the major negative impacts have been on the fish industry. Prior to installation of the hydropower dams there, some 10 to 15 million adult salmon annually swam from the ocean to the upper-river watershed, which provided about 40 percent of their ancestral spawning habitat. Although salmon that spawn on tributaries of the Columbia can successfully make it up the fish ladders around the smaller dams, sooner or later they encounter a dam that is too imposing to pass. Additional problems are logging, livestock grazing, and the diversion of water for irrigation, all of which contribute to the loss of spawning habitat. Even though efforts have been made to restore the fisheries, many prominent biologists predict that the salmon fishery will some day become extinct (Reisner, 2000).

COASTAL AND MARINE ECOSYSTEMS

As noted earlier (Chapter 19), people throughout the world are migrating to coastline areas. This shift is most dramatically illustrated by the giant coastal cities in Africa and Asia. These and cities in other parts of the world frequently lack provisions for adequately treating their domestic and industrial liquid wastes (Tibbetts, 2002). As a result, the coastal zones of the world are being subjected to increased nutrient loading, toxic contamination, and habitat alteration. The resulting impacts are illustrated by the Chesapeake Bay, where the clearing of the land and the establishment of plantations along its shores have led to increased loss of soil and organic matter via runoff. Early on, the oyster population was sufficient to filter and cleanse the water in the bay in an estimated time of less than a week, with the result that the pollution was kept in balance. With the passage of time, however, the harvesting of oysters became a major occupation, ultimately reducing the population to the point where the estimated time for the oysters to cleanse the water had increased to more than a year. The increase in suspended matter soon thereafter made the water uninhabitable for the rockfish and blue crabs, which, in turn, were replaced by less desirable sea nettle jellyfish and toxic algae (Hayden, 2001).

A similar episode has occurred in the shallow bays of southern Florida and the Caribbean. When settlers first arrived in that region, green sea turtles were abundant. By the late 1600s and early 1700s, their population was almost depleted. Concurrently, turtle grass, which had served as the primary food for the turtles, began to flourish. As the plants died and decayed, the oxygen content of the water was reduced to such an extent that fish and shrimp could no longer live in it. Ironically, by the 1980s the turtle grass was itself destroyed by molds that grew in the decaying organic matter. Only by restoring the turtle population and the grass they eat will it be possible to revitalize what was once a major ecosystem (Hayden, 2001).

Similar problems have been caused in marine fishing areas, such as Georges Bank off the coast of New England and Canada. One of the major culprits in this case is thought to be the use of trawls and scallop dredges that compact and kill bottom dwellers, such as sponges, deep-sea corals, bryozoans, and other sedentary animals. Without the protection against predators provided by these types of sea life, baby fish are unable to survive. By the 1990s, the populations of adult cod, hake, and other groundfish had dramatically declined (Hayden, 2001). Recognizing that all would suffer unless these problems were solved, fishermen, environmentalists, scientists, and the courts agreed to cooperatively seek a solution. The answer came through the application of scientific principles. The basic approach was to evaluate the situation in terms of population dynamics. This led to establishment of a schedule for opening and closing the region's scallop grounds to permit the marine life to recover. Within three years, the mass of scallops had increased sufficiently to support a stable and productive scallop industry. Buoyed by this success, these groups subsequently applied the same approach to the harvesting of some dozen bottom-dwelling species of fish, such as flounder, cod, and haddock (Greene, 2002). Although setting limits on the amounts of fish that can be caught is essential to maintaining an ecosystem, this is not the total answer. As stated repeatedly in this book, their restoration requires taking a systems approach, that is, it requires an understanding of the total ecosystem (Hayden, 2001). That is not meant to imply, however, that solving these types of problems will be simple.

Other problems of a marine nature can originate through activities of a military and/or industrial nature. One specific type of event that occurred several times during the early years of the twenty-first century was the beaching and deaths of whales following underwater tests of a sonar system that was being developed by the U.S. Navy. Autopsies revealed hemorrhages consistent with acoustic trauma in and around the ears of those that died. A federal judge, in response to legal objections filed by the National Resources Defense Council (NRDC), subsequently called for a halt to the tests and mandated that the U.S. Navy negotiate with the NRDC and other conservation groups over when and where such systems can be tested and used during peacetime (Malakoff, 2003). Other sources of potential stresses on marine life include offshore activities associated with seismic exploration for oil, and drilling rigs and the operation of supertankers, icebreakers, cruise ships, and even tugboats and ferries (Carpenter, 2002).

RIPARIAN ECOSYSTEMS (WETLANDS)

Riparian ecosystems are commonly described as those that are influenced by the intermittent interaction of freshwater systems with low-lying land, or that are at the interface of freshwater and saltwater ecosystems. Wetlands, which include tidal marshes, swamp forests, peat bogs, prairie potholes, and wet meadows, are among the best examples. Biologically, they are the most productive ecosystems in the world, serving as nurseries and feeding grounds for a range of commercial fish species, as nesting and feeding grounds for waterfowl and migratory birds, and as habitats for many other forms of animal life (otters, turtles, frogs, snakes, and insects). Wetlands also trap nutrients and sediments; purify water by removing coliform bacteria, heavy metals, and toxic chemicals; provide flood protection by slowing and storing water; and anchor shorelines and provide erosion protection (CEQ, 1997). Without the removal of nutrients, the resulting excessive growth of phytoplankton or algae can destroy coral reefs and other coastal environments that serve as habitats for fish, seabirds, and other animals. Later, when the algae die and sink to the bottom, their decay can deplete the concentrations of oxygen in bottom waters and cause environmental hypoxia (Stegeman and Solow, 2002).

In spite of their benefits, by the late 1980s less than half of the wetlands that had originally existed in the contiguous United States remained. Fortunately, major efforts are under way to protect such resources. In accordance with the Clean Water Act, a permit must be obtained from the Army Corps of Engineers to alter, fill, or otherwise change the characteristics of such resources. As part of this process, an applicant must first avoid, or at least minimize, any ensuing damage. If damage cannot be avoided, the Army Corps of Engineers requires the permit holder, or a third party paid by the permit holder, to restore or replace the impacted wetlands (Kearney, 2001). This can be accomplished by the creation of new wetlands, the enhancement of existing wetlands, the guaranteed preservation of existing wetlands, or payment of a fee to compensate for the loss.

One of the unfortunate examples of the negative impacts of human activities on wetlands is the Florida Everglades, a river of grass that once covered almost 12,000 square kilometers (5,000 square miles) of that state. With the increasing demand for land, water within this area was gradually diverted to urban and agricultural use. As part of this process, a major channel was dug to facilitate the flow of the water into the ocean. As a result, bird populations have decreased, and 68 species, such as the manatee and the panther, that lived there are now endangered. Recognizing what was being done, the Army Corps of Engineers has undertaken an effort to undo the earlier changes and restore the ecosystem. Hopefully, this will repeat the highly successful efforts of the cooperative multinational group that restored the vitality of the wetlands of the Romanian Danube delta (Karen F. Schmidt, 2001).

In the past, groups that have sought to develop plans for protecting ecosystems have discovered that information for evaluating the success of their efforts was often inadequate and/or not relevant. In the late 1990s, about 150 ecologists and policy experts from universities, environmental groups, industry, and government undertook an effort to fill this void. The results of their initial efforts revealed that the situation was much worse than imagined (Powell, 2002). While these experts recognize that their report is far from complete, they are hopeful that it represents a first step both in providing the required information and indicating what additional types of measurements are needed. Unfortunately, techniques for conducting about half of the required measurements are not available (Shouse, 2002).

For purposes of their analyses, the experts divided the nation's ecosystems into six broad categories (Table 20.2). As may be noted, these differ to some extent from those presented previously (Table 20.1). For each category, 10 key characteristics (Table 20.3) were identified, each of which is thought to be an important measure of the health of a specific ecosystem. The results were not favorable. At least one measurable contaminant was present in virtually all freshwater streams, in stream sediments, and in the flesh of freshwater fish. In addition, for more than one-quarter of the groundwater in the nation, the concentration of at least one contaminant exceeded the applicable human health standard (Powell, 2002). The assembled data also revealed that exotic and invasive species have already altered the American environmental landscape. Every river system, for example, has at least one invasive species, and 60 percent have between 1

Table 20.2 Alternate categorization of ecosystems

1. Coasts and oceans
2. Forests
3. Farmlands
4. Fresh waters
5. Grasslands and shrublands
6. Urban and suburban areas

Table 20.3 Key characteristics of ecosystems

1. Ecosystem extent, that is, whether it is growing or shrinking
2. Fragmentation
3. Presence or absence of key chemicals needed for life
4. Contaminants
5. Physical conditions, including factors such as erosion or depth to groundwater
6. Plants and animals
7. Biological communities
8. Plant growth and productivity
9. Production of food and fiber and use of water
10. Recreational use and other services provided by the ecosystem

and 10 invasive species living in them. On the positive side, the study showed that 85 percent of U.S. streams meet human health standards, that agricultural production has doubled since the 1950s, and that the amount of land threatened by erosion has been reduced by one-third since 1985 (Shouse, 2002). While the authors of the report recognize that policy making about the environment will always be contentious, they believe that debates on how best to manage our nation's resources should not be sidetracked through needless disputes about the facts (Powell, 2002).

Global Warming

One of the most contentious and debated problems that face the world today is global warming. The reasons are that the subject is complex, its potential impacts will affect the entire world, and there are difficult questions that need to be answered (Table 20.4). One of the major leaders in addressing this subject is the United Nations, as exemplified by the discussions held and decisions made in Japan in 1997. The outcome of this effort, which followed the 1992 Earth Summit on Sustainable Development, was the Kyoto Protocol (Table 14.3, Chapter 14), a key part of which was the establishment of a plan for stabilizing the atmospheric concentrations of the so-called greenhouse gases. This action was based on scientific evidence that chemical compounds such as carbon dioxide (CO_2), methane, and CFCs, when present in the Earth's atmosphere, are transparent to incoming shortwave electromagnetic radiation from the sun. Once the radiation scatters off the surface of the Earth, however, its energy is reduced such that it is now readily absorbed by these compounds. In essence, the heat from the incoming radiation is trapped in a manner similar to that of a greenhouse. For this reason, any increase in the concentrations of these gases in the atmosphere will cause the temperature near the surface of the earth to increase. Although the degree to which greenhouse gases are the cause of global warming is not known, it is clear that the Earth is warming. This has been demonstrated in multiple ways.

- During the past half century, the dates on which 385 species of British plants have flowered has advanced by an average of 4.5 days (Fitter and Fitter, 2002). In a similar manner, the leaves of most deciduous plant species in various Mediterranean ecosystems are un-

Table 20.4 Questions to be addressed relative to global warming

What are its causes and what is their relative importance?
What can be done to reduce and/or alleviate its impacts, and which of the various methods will be the most effective?
Considering various levels of control, to what degree will the impacts be reduced?
What are the possibilities that controlling one aspect of global warming will exacerbate other types of effects?

folding an average of 16 days earlier and falling an average of 13 days later than they did 50 years ago. In western Canada, *Populus tremuloides* is now blooming 26 days earlier than a century ago (Penuelas and Filella, 2001).

- During the past four decades, there have been measurable changes in the species composition of certain marine plankton in the North Atlantic. These include a northward extension of more than 10 degrees latitude of warm-water species, accompanied by a decrease in the number of cold-water species (Beaugrand et al., 2002). There have also been significant advances in the dates for first breeding in a number of bird species. Migratory birds that winter south of the Sahel are now arriving later in Europe (Penuelas and Filella, 2001).

When these observations are combined with the physical data that indicate an increase both in average global temperatures and the rates at which glaciers are receding or the ice on rivers is breaking up, it is clear that global warming is a reality. One of the most important observations with regard to the biological implications is that not all species respond to global warming at the same rates. The result could well be that the blooming of a particular flower will no longer coincide with the presence of the butterfly that previously pollinated it. If there is no nectar for the insect, it could well be that at some time in the future there will be no seeds to grow more plants (Kennedy, 2002). Even though other observations show that both plants and insects can adapt to global warming, and fossil evidence indicates that many plants have survived numerous episodes of climate change in the past, such changes were one to two orders of magnitude slower than those predicted for the future.

POTENTIAL FUTURE IMPACTS

Some of the observations to date of the evidence of global warming have just been described. Other potential short-term future impacts include the following:

- Coral reefs throughout the world are bleaching, a process that can be caused by as little as a 1°C increase in temperature. If this occurs over several consecutive warm seasons, there is a risk of major modification and even permanent loss of some species, even if account is taken of their ability to adapt. In fact, some scientists predict that even if atmospheric concentrations of CO_2 were stabilized at 450 parts per million by 2100, which is consistent with the Kyoto Protocol, full protection of the reefs would probably not be feasible (O'Neill and Oppenheimer, 2002).

- Glaciers in Alaska are melting at a much more rapid rate than previously estimated. In fact, the amount of water that is being released into the Bering Sea and northern Pacific Ocean is estimated at 96 ± 35 km^3 per year. This could produce an annual rise in the sea level of 0.27 ± 0.10 mm (Arendt et al., 2002). In addition to the devastating flooding of low-lying small island nations, such a rise could lead to the incursion of saltwater into coastal aquifers, the extension of the saltwater wedge into estuaries, and the increased probability of damage from storm surges.
- Evidence shows that the ice cover is melting in the Arctic and that ice shelves in Antarctica are crumbling. In similar manner, glaciers in temperate regions are disappearing (Albert, 2004). Losses of the ice cover in the Arctic could reduce the mats of diatoms—microscopic, silica-encased algae—that hang from the bottom of the ice. Under normal conditions, these organisms sink to the sea floor upon death and serve as food for worms and crustaceans, which, in turn, serve as a primary source of food for bottom-feeding whales (Kerr, 2002). This series of coupled events exemplifies the far-reaching extent of the impacts of global warming.
- Data also show that as little as a one- to two-degree increase in temperature can lead to insect-borne disease outbreaks, such as malaria, dengue, and yellow fever, in areas previously not affected, for example, at higher elevations where the insects and microbes were previously absent. In Hawaii, the movement of mosquitoes into higher elevations led to an epidemic of avian malaria that has killed multitudes of native birds that were not immune to the disease. Today there are almost no native birds in Hawaii below 1,370 meters (4,500 feet) (Harvell et al., 2002).

ASSOCIATED UNCERTAINTIES

While there are uncertainties in estimates of the causes and impacts of global warming, the following is known. With regard to human-made contributions, it is clear that the concentrations of greenhouse gases in the atmosphere are increasing due to human activities. It is equally clear that these increases are causing surface air temperatures and subsurface ocean temperatures to rise. Potential secondary effects suggested by computer model simulations and basic physical reasoning include higher rainfall rates in wet areas and increased probabilities of drought in semiarid regions. The situation is complicated, however, by the fact that the accompanying impacts of these changes will be critically dependent on the magnitude of the warming and the rate at which it occurs (NRC, 2001). Ultimately, resolution of the issues on the contributions of human activities to global warming will require the establishment of a global observation system for monitoring the climate. Until this can be done, estimates related to any aspect of this subject will have to be accompanied by a range of caveats (NRC, 2001).

CONTROL OF HUMAN-MADE CONTRIBUTIONS

One of the gases that is thought to play a major role in global warming is CO_2. For this reason, methods are being explored for reducing the emission of this gas into the atmosphere. One approach would be to increase the absorption of CO_2 by phytoplankton. One method that has been proposed for accomplishing this goal is to add iron to areas of the ocean where the densities of these organisms are low. Unfortunately, subsequent computer evaluations indicated that plankton blooms exhibit characteristics of an excitable system. As a result, it is possible that if the perturbations created by the additions of iron exceeded a certain threshold, the net result could be a larger-than-expected, iron-induced bloom with detrimental consequences of its own. As in another suggested approach, namely, increasing the vegetative cover within the terrestrial environment, it is extremely important that prior to conducting any such experiments on a broad scale, the potential impacts of geoengineering interventions of this type on the ecosystem as a totality be thoroughly analyzed (Chin, 2002). Another illustration of the range of factors to be considered is the magnitude of the contribution to global warming of atmospheric aerosols, for example, black carbon or airborne soot. Since this substance absorbs sunlight, it could play a significant role. In fact, reductions in the emissions of black carbon aerosols may be an important step in reducing the rate of global warming.

Sustainable Development

The concept of sustainable development plays a major role in the future of the Earth. On the assumption that humanity and nature are separate entities, yet closely related, the usual definition of this term is "development that consumes the dividends of nature without impairing its capital" (Starr, 1997). If this goal is to be achieved, however, two facts must be recognized: first, that the natural resources of the Earth are finite; second, that if, as noted earlier, the world's population continues to increase in an unrestrained manner, these resources will inevitably be depleted. The need to address this second factor cannot be overemphasized. A thousand years ago, the global population was roughly stable at about 300 million people, 1 person for every 20 present on Earth today. By 1600, the population began to double roughly every 200 years; by 1900, this note had decreased to 100 years. Unless future population-control efforts are dramatically more successful than in the past, the current population could double before the year 2050. The resulting demands for food, clothing, shelter, and manufactured goods will ultimately place intolerable pressures on the world's resources of energy, air, water, land, and biota.

The urgency in addressing these and related types of issues was one of the primary reasons for convening the World Conference on Sustainable Development in Johannesburg, South Africa, in 2002. One of the principal objectives of the conference was to emphasize action and accountability, as contrasted to principles and planning (Charles W. Schmidt, 2002). One innovative decision was to encourage the development of partnerships between the United Nations (UN) and various worldwide corporations. Through this approach, UN officials hope to encourage businesses to "buy into" the basic values of that organization, including the need to promote human rights and economic growth in an environmentally friendly way. That this is proving successful is shown by the fact that industrial organizations worldwide are increasingly recognizing that it is not adequate simply to develop methods for treating waste after it has been produced. There is a need to examine the industrial processes themselves and modify them to cope with the wastes before they are generated (Chapter 9).

In seeking to define what is needed to achieve sustainability, engineers and ecologists have developed a variety of terms to describe this new discipline. These include "industrial ecology," "ecological engineering," and "Earth systems engineering." Regardless of the name, the primary goals are to gain a better understanding of the challenges posed by complex, nonlinear systems of global importance, most notably, environmental systems, and to create and apply the tools that will respond effectively to these challenges. Some of the approaches that are being proposed are technical; some involve policy. One of the major objectives in all cases is to understand the fundamental character of interactions between nature and society (Karn and Bauer, 2001). Examples of the successful application of these approaches and principles are the programs described earlier to restore the Everglades and the Romanian Danube delta; to enable marine life in the Georges Bank off the coast of New England to recover; and to protect and preserve existing habitats that host a major share of the world's biodiversity. Because these examples all relate to the correction of problems that have been permitted to develop, it is obvious that significantly more efforts need to be directed to protection, preservation, and prevention.

The General Outlook

Although a variety of topics were discussed in the preceding sections, the challenges in the field of environmental and public health extend far beyond these examples. The magnitude of these challenges is exemplified by the topics identified by the National Research Council as deserving immediate attention from the standpoint of research (Table 20.5). While certain aspects of the items on this list can be addressed by the scientific community, essentially all of them will, during some phase or time, require cooperative efforts on the part of the leaders of the global community of nations. At the other extreme, certain challenges and problems will always remain that must, of necessity, be addressed at a local or regional level. Examples of several of these issues are discussed here.

SOLVING PROBLEMS INTERNATIONALLY

One of the best examples of how a global environmental problem can be successfully controlled followed the realization during the late 1950s and early 1960s that radioactive materials that were released during atmospheric testing of nuclear weapons were being transported throughout the world. In fact, tests of the air, water, and soil in all countries in the Northern Hemisphere showed the presence of these materials. Cooperative efforts on an international basis led to the ban on the testing of such devices in the atmosphere. Later the ban was expanded to include underground tests.

Under the leadership of the United Nations, several other environmental problems have been similarly addressed in recent years. These include the successful efforts to halt the depletion of the ozone layer; the treaty, concluded in 2000, that initiated a worldwide phaseout of the use of certain persistent organic chemicals; and the Kyoto Protocol, which is

Table 20.5 Research needs in environmental health

Subject area	Description of need
Deserving immediate attention	
Biodiversity and ecosystem functioning	Improved understanding of the factors, including human activities, that affect biodiversity and how biodiversity relates to the overall functioning of an ecosystem
Hydrologic forecasting	Capability to help predict changes in freshwater resources and changes in the environment caused by floods, droughts, sedimentation, and contamination.
Infectious diseases and the environment	Better understanding of how pathogens, parasites, and disease-carrying species, as well as the humans and other species they infect, are affected by changes in the environment, with the goal of preventing outbreaks of infectious diseases in plants, animals, and humans
Land-use dynamics	Methods for applying recent advances in data collection and analysis to document and understand the causes and consequences of changes in land cover and use
Also important	
Biogeochemical cycles	Understanding of how changes in the balance of carbon, oxygen, hydrogen, sulfur, and phosphorus in soil, water, and air affect the functioning of ecosystems, atmospheric chemistry, and human health.
Climate variability	More complete comprehension of how the Earth's climate varies over a wide range of time scales, from extreme storms that develop quickly to changes in weather patterns that occur over several decades
Institutions and resource use	More information about how the condition of natural resources is shaped by markets, governments, international treaties, laws, and informal rules that govern environmentally significant human activities
Reinventing the use of materials	Additional data on the forces that drive human use of reusable metals such as copper and zinc, hazardous metals such as mercury and lead, reusable plastics and alloys, and ecologically dangerous compounds such as CFCs and pesticides

designed to combat global warming. Each of these was discussed earlier. The first two efforts were eminently successful; the third has not been, at least to date. In all three cases, the outcomes can best be explained in terms of the game theory of public policy. As often proves to be the case, the players were the developed and the less developed countries. With respect to the ozone layer, the developed nations were willing to take a lead role in reducing CFCs because the overall benefits outweighed the costs of eliminating CFC production. The less developed nations also benefited through being allocated extra time to reduce their CFC emissions and being granted funds to help them pay for the transition. No country was granted big concessions, but neither did any country suffer a significant loss. The result was, and continues to be, that a significant problem was addressed and its long-term consequences were averted.

Efforts to phase out the use of certain persistent organic chemicals appear to be following a similar road to success. This was largely due to almost universal recognition that elimination of their production would significantly reduce potential health effects in billions of people. The immediate endorsement of the action by the United States and the availability of less persistent substitutes at comparable costs added impetus to the effort.

The Kyoto Protocol negotiations represented an entirely different situation. In this case, the developed countries viewed themselves as the bigger losers. While they could bear the brunt of the costs of reducing greenhouse-gas emissions, compliance would require significant changes in the lifestyles of their inhabitants. In the case of the United States, this would necessitate implementing measures to dramatically reduce the consumption of gasoline. This might well prove to be politically unacceptable for the nation's leaders. The negative reaction was exacerbated when the less developed countries requested that fines be imposed on the countries with the largest emissions. This immediately created resentment on the part of the developed countries, especially those in which national leaders viewed the required changes as an indictment of their domestic environmental policies. On the other hand, if the less developed countries are forced to contribute to emission reductions, they have a readily available reason not to cooperate, namely, that such action would have an extremely negative impact on their economic progress. For these and other reasons, negotiations to implement the protocol are likely to remain at an impasse (Gruebel, 2002).

SOLVING PROBLEMS NATIONALLY AND LOCALLY

As noted earlier, many environmental problems are regional or local in nature. Their effective prevention and control will require input from all levels—industry, governmental agencies, and society.

Responsibilities of industry. Of primary importance is that industrialists and industrial engineers learn to incorporate sound environmental thinking into the initial selection and design of manufacturing processes and products. Pollution controls must be designed into industrial equipment, not added on later. Multinational corporations and financial institutions also have an obligation to set a new moral tone for the world. They must commit themselves to a sustainable future and be prepared to sacrifice a portion of their profits to do so. Close examination will demonstrate that such an approach is in their best interests, as well as essential to their survival.

Responsibilities of governmental agencies. Governmental agencies likewise need to assume increased responsibility for protecting the environment and improving its two-way relationship with people. As enumerated by the EPA's Science Advisory Board (Johnson, 1995), the principal needs are as follows:

1. To develop programs that will provide continuing evaluation of key environmental areas, such as ecosystem sustainability, non-cancer human health effects, nontraditional environmental stressors, and the health of the oceans;
2. To emphasize the avoidance of future environmental problems as much as the control of those that exist;
3. To stimulate coordinated efforts among federal, state, and local agencies and the private sector to develop the capability to anticipate and respond to environmental change; an integral part of this effort is the establishment of an early-warning system to identify emerging environmental risks;
4. To recognize that global environmental quality is a matter of strategic national interest and to adopt policies that link security, foreign relations, environmental quality, and economic growth.

If, as is recommended in item 3, various governmental agencies, as well as the private sector, are to have the capabilities for anticipating and responding to environmental changes, it will be necessary for them to expand databases in both the scientific and policy sectors. One step in this direction is the previously cited UN program related to biodiversity.

Responsibilities of society. Concurrently, society as a whole needs to develop a forward-looking attitude in dealing with environmental problems. Societal behavior will change only if enough people become aware of environmental problems and act, both as individuals and through their elected governmental representatives. The American Medical Association (AMA, 1989), in its policy statement on *Stewardship of the Environment*, suggests that the United States play a leading role in effecting such change:

> The U.S. and the world at large appear to be facing environmental threats of unprecedented proportions, and scientists, environmental activists, health professionals, politicians and world leaders are beginning to realize the need for changes in societal behavior (i.e., human behavior as well as the conduct of business and industry) as a means of forestalling these potential threats. Societal changes must be initiated worldwide if they are to have any significant effect overall. However, their implementation will need a model, most suitably a national model. . . . The U.S. could well become a model for environmental stewardship if a grassroots movement were to develop to encourage and endorse a protective and nurturing philosophy towards the environment at both the personal and societal levels.

Accomplishing these objectives will require action on several fronts, most notably in the field of education. If such efforts are to be successful, the inculcation of environmental values must begin with young people at the primary- and secondary-school levels. Even then, such a program can be effective only if it has the support of the entire community—schools, churches, business and industry, trade associations and professional groups, advertising and news media, and government at all levels.

The increasing efforts being made by religious organizations to examine specific environmental issues and to encourage the peoples of the world to take action to solve them are especially encouraging. One example is the convening by the World Council of Churches and the Parliament of World Religions of several international meetings to discuss the role of socioeconomic structures and personal lifestyles in global climate change. Other such activities include the conferences on Religions of the World and Ecology, convened from 1996 to 1998; the Religion, Science, and Environment Symposia held in 1994, 1997, 1999, and 2002 to discuss regional water-related environmental issues; and the Millennium World Peace Summit of Religious and Spiritual Leaders held in 2000 to discuss a new ethic for global stewardship (Gardner, 2003).

In spite of what may often seem an insurmountable task, progress is being made. Polls show that a clear majority of people in the United States, for example, are concerned about the environment. Indeed, the latest data show that more than two-thirds of the population, if asked to choose between protection of the environment and economic growth, would select the former. This has proved true even in times of economic downturns. Problems that were listed as of highest concern included pollution of drinking water; contamination of soil and water by toxic wastes; pollution of rivers, lakes, and reservoirs; and air pollution. It is interesting to note that those who were polled were also knowledgeable about essentially all of the global problems discussed earlier, although they were farther down on the list in terms of priorities for concern (Greenberg, 2001). The will is there; the knowledge and technology continue to evolve. The primary need is for leadership to marshal existing support and to set priorities for action.

References • Credits
Index

REFERENCES

1. The Scope

Brink, Susan. 2002. "Phys Ed Redux." *U.S. News and World Report* 132, no. 19 (3 June), 50–52.

Bugliarello, George. 2001. "Rethinking Urbanization." *Bridge* 31, no. 1 (Spring), 5–12.

CDC. 1998. "Youth Risk Behavior Surveillance—United States, 1997." *Morbidity and Mortality Weekly Report* 47, no. SS-3 (14 August), 1–89.

———. 1999. "Ten Great Public Health Achievements—United States, 1900–1999." *Morbidity and Mortality Weekly Report* 48, no. 12 (2 April), 241–243.

———. 2002a. "Annual Smoking-Attributable Mortality, Years of Potential Life Lost, and Economic Costs—United States, 1995–1999." *Morbidity and Mortality Weekly Report* 51, no. 14 (12 April), 300–303.

———. 2002b. "Guidelines for School Programs to Prevent Skin Cancer." *Morbidity and Mortality Weekly Report* 51, no. RR-4 (26 April), 1–21.

Colditz, G. A., M. Samplin-Salgado, C. T. Ryan, H. Dart, H. L. Fisher, A. Tokuda, and B. Rockhill. 2002. "Fulfilling the Potential for Cancer Prevention: Policy Approaches." *Cancer Causes and Control* 13, no. 3 (April), 199–212.

DeKay, Mark, and Micheal O'Brien. 2001. "Gray City, Green City." *Forum for Applied Research and Public Policy* 16, no. 2 (Summer), 19–27.

EPA. 2001. "EPA's Commitment to Environmental Justice." Memorandum from the Administrator, Washington, DC (9 August).

Eskenazi, Brenda, and Philip J. Landrigan. 2002. "Environmental Health Perspectives and Children's Environmental Health." *Environmental Health Perspectives* 110, no. 10 (October), A 559–A 560.

Fields, Scott. 2002. "Urban Issues: If a Tree Falls in the City." *Environmental Health Perspectives* 110, no. 7 (July), A 392.

Fisher, Laurie. 2001. "New Non-Smoking Policies in College Dorms." *Source* 8, no. 1 (October/November), 3. Harvard Center for Cancer Prevention, Boston.

Forastieri, Valentina. 1997. *Children at Work: Health and Safety Risks.* Geneva: International Labour Office.

Friedman, Jeffrey M. 2003. "A War on Obesity, Not the Obese," *Science* 299, no. 5608 (7 February), 856–858.

Gill, Thomas M., Christianna S. Williams, Julie T. Robison, and Mary E. Tinetti. 1999. "A Population-Based Study of Environmental Hazards in the Homes of Older Persons." *American Journal of Public Health* 89, no. 4 (April), 553–556.

Guyer, Ruth Levy. 2001. "Backpack = Back Pain." *American Journal of Public Health* 91, no. 1 (January), 16–19.

Hesketh, Therese, Qu Jian Ding, and Andrew Tomkins. 2001. "Smoking among Youth in China." *American Journal of Public Health* 91, no. 10 (October), 1653–1655.

HHS. 2000. *Healthy People 2010: Understanding and Improving Health.* 2nd ed. (November). Washington, DC: Government Printing Office.

Holden, Constance. 2001. "Nation Continues to Gain." *Science* 291, no. 5503 (19 January), 429.

Kaiser, Jocelyn. 2002. "Texas Surgeon Vows to Take Next Step in Beating Cancer." *Science* 296, no. 5572 (24 May), 1394–1395.

Kelmer, Katrina, and Laura Helmuth. 2003. "Obesity—What Is to Be Done?" *Science* 299, no. 5608 (7 February), 845.

Late, Michelle. Editorial. 2002. "Cigarette Warning Labels a Success in Canada." *Nation's Health* (March), 13.

Martin, George M. 2001. "Frontiers of Aging" (editorial). *Science* 294, no. 5540 (5 October), 13.

Meister, Kathleen, G. L. Ross, K. L. Schneider, and E. M. Whelan, eds. 2003. *Cigarettes: What the Warning Label Doesn't Tell You.* 2nd ed. New York: ACSH.

Morris, R. D., and W. R. Hendee. 1992. "Environmental Stewardship: Exploring the Implications for Health Professionals." Draft report, Medical College of Wisconsin, Milwaukee.

Shaw, Jonathan. 2004. "The Deadliest Sin." *Harvard Magazine* 106, no. 4 (March–April), 36–43, 98–99.

Sheehan, Molly O. 2002. "What Will It Take to Halt Sprawl?" *World-Watch* 15, no. 1 (January/February), 12–23.

Suk, William A. 2002. "Beyond *The Bangkok Statement:* Research Needs to Address Environmental Threats to Children's Health." *Environmental Health Perspectives* 110, no. 6 (June), A 284–A 285.

Targ, Nicholas, and La Ronda Bowen. 2002. "Environmental Justice: The Nexus of Technology, Sociology, and Legal Authority." *EM* (August), 24–29.

USPHS. 2001. *Surgeon General's Call to Action to Prevent and Decrease Overweight and Obesity.* (December). Washington, DC.

Wakefield, Julie. 2002. "Learning the Hard Way." *Environmental Health Perspectives* 110, no. 6 (June), A 298–A 305.

Willett, Walter. 2002. "The Food Pushers." *Science* 297, no. 5579 (12 July), 198–199.

World Commission on Environment and Development. 1987. *Our Common Future,* p. 393. Oxford: Oxford University Press.

2. *Toxicology*

Ames, B. N. 1971. "The Detection of Chemical Mutagens with Enteric Bacteria." In A. Hollander, ed., *Chemical Mutagens: Principles and Methods for Their Detection,* vol. 1, pp. 267–282. New York: Plenum Press.

Ansari, Armin. 2004. Private Communication. Atlanta, GA: CDC (12 April).

ATSDR. 2001. *Guidance Manual for the Assessment of Joint Toxic Action of Chemical Mixtures (Draft for Public Comment).* Atlanta, GA: HHS.

———. 2003. *Toxicological Profile for Malathion.* Agency for Toxic Substances and Disease Registry. Atlanta, GA: HHS.

Butterworth, B. E., D. C. Dorman, K. W. Gaido, S. J. Sumner, J. C. Corton, S. J. Borghoff, and R. B. Conolly. 1999. "Research at CIIT on the Risks to Human Health from Exposure to Chemicals." *CIIT Activities* 19, no. 10 (October), 1–8. Research Triangle Park, NC.

Calabrese, Edward J., Linda A. Baldwin, and Charles D. Holland. 1999. "Hormesis: A Highly Generalizable and Reproducible Phenomenon with Important Implications for Risk Assessment." *Risk Analysis* 19, no. 2, 261–281.

Casarett, Louis J., Curtis D. Klaassen, and John Doull, eds. 2001. *Casarett and Doull's Toxicology: The Basic Science of Poisons.* 6th ed. New York: McGraw-Hill Medical Publishing Division.

CDC. 2001. *Why Poison Ourselves? A Precautionary Approach to Synthetic Chemicals—National Report on Human Exposure to Environmental Chemicals.* Atlanta, GA: HHS.

———. 2003. *Second National Report on Human Exposure to Environmental Chemicals.* Atlanta, GA: HHS.

Doull, John. 1992. "Toxicology and Exposure Limits." *Applied Occupational and Environmental Hygiene* 7, 583–585.

Doull, J., and M. C. Bruce. 1986. "Origin and Scope of Toxicology." In C. D. Klaassen, Mary O. Amdur, and J. Doull, eds., *Casarett and Doull's Toxicology: The Basic Science of Poisons,* 3rd ed., pp. 3–10. New York: Macmillan Publishing Company.

EPA. 1986. "Guidelines for the Health Risk Assessment of Chemical Mixtures." *Federal Register* 51, 34014–34025. Washington, DC.

Gochfeld, Michael. 1998. "Principles of Toxicology." In Robert B. Wallace, ed., *Maxcy-Rosenau-Last Public Health and Preventive Medicine,* pp. 415–427. Stamford, CT: Appleton & Lange.

Greenlee, William F. 2002. "Message from the President." In *2001 Annual Report,* pp. 7–8. Research Triangle Park, NC: Chemical Industry Institute of Toxicology.

Henry, Carol J., and James S. Bus. 2000. "Long-Range Research Initiative of the American Chemistry Council." *CIIT Activities* 20, no. 7 (July), 1–5.

Kaiser, Jocelyn. 2003. "A Healthful Dab of Radiation." *Science* 302, no. 5644 (17 October), 378.

Kamrin, Michael. 2003. *Traces of Environmental Chemicals in the Human Body: Are They a Risk to Health?* New York: American Council on Science and Health (May), revised edition.

Klaassen, C. D. 1986. "Principles of Toxicology." In C. D. Klaassen, Mary O. Amdur, and J. Doull, eds., *Casarett and Doull's Toxicology: The Basic Science of Poisons,* 3rd ed., pp. 11–32. New York: Macmillan Publishing Company.

Lippmann, Morton. 1992. "Introduction and Background." In Morton Lippmann, ed., *Environmental Toxicants: Human Exposures and Their Health Effects,* pp. 1–29. New York: Van Nostrand Reinhold.

Loewenberg, Samuel. 2003. "Europe Whittles Down Plans for Massive Chemical Testing Program." *Science* 302, no. 5647 (7 November), 969.

Loomis, T. A. 1968. *Essentials of Toxicology.* 3rd ed. Philadelphia: Lea & Febiger.

Lovett, Richard A. 2000. "Toxicologists Brace for Genomics Revolution." *Science* 289, no. 5479 (28 July), 536.

Lu, Frank C. 1991. *Basic Toxicology: Fundamentals, Target Organs, and Risk Assessment.* 2nd ed. New York: Hemisphere Publishing Corporation.

Moriarty, F. 1988. *Ecotoxicology: The Study of Pollutants in Ecosystems.* 2nd ed. New York: Academic Press.

NRC. 1983. *Drinking Water and Health.* Vol. 5. Board on Toxicology and Environmental Health Hazards. Washington, DC: National Academy Press.

———. 1986. *Drinking Water and Health.* Vol. 6. Board on Toxicology and Environmental Health Hazards. Washington, DC: National Academy Press.

———. 1995. *Radiation Dose Reconstruction for Epidemiologic Uses.* Board on Radiation Effects Research. Washington, DC: National Academy Press.

Rhomberg, Lorenz. 1996. "Are Chemicals in the Environment Disrupting Hormonal Control of Growth and Development?" *Risk in Perspective* 4, no. 3 (April) 1–6. Center for Risk Analysis, Harvard School of Public Health.

Schettler, Sunessa, and Mara Seeley. 2002. "Examining Multiple Chemical Sensitivity." *Trends in Risk and Remediation* (Winter), 4. Cambridge, MA: Gradient Corporation.

Schmidt, Charles W. 2002. "Assessing Assays." *Environmental Health Perspectives* 110, no. 5 (May), A 248–A 251.

Seeley, Mara. 2001. "Carcinogen Classification." *Trends in Risk and Remediation* (Spring), 3, 5. Cambridge, MA: Gradient Corporation.

Service, Robert F. 2002. "More on Drug Pollution." *Science* 296, no. 5567 (19 April), 463.

Smith, R. P. 1992. *A Primer of Environmental Toxicology.* Philadelphia: Lea & Febiger.

Society of Toxicology. 1988. Definition of Toxicology. Report of the TOX's 1990s Commission. Reston, VA.

Stone, R. 1993. "FCCSET Develops Neurotoxicology Primer." *Science* 261, no. 5124 (20 August), 975.

3. Epidemiology

Brain, J. D., R. Kavet, D. L. McCormick, C. Poole, L. B. Silverman, T. J. Smith, P. A. Valberg, R. A. Van Etten, and J. C. Weaver. 2003. "Childhood Leukemia: Electric and Magnetic Fields as Possible Risk Factors." *Environmental Health Perspectives* 111, no. 7 (June), 962–970.

Cantor, K. P., A. Blair, G. Everett, R. Gibson, L. F. Burmeister, L. M. Brown, L.

Schuman, and F. R. Dick. 1992. "Pesticides and Other Agricultural Risk Factors for Non-Hodgkin's Lymphoma among Men in Iowa and Minnesota." *Cancer Research* 52, no. 9 (1 May), 2447–2455.

Curriero, F. C., J. A. Patz, J. B. Rose, and S. Lele. 2001. "The Association between Extreme Precipitation and Waterborne Disease Outbreaks in the United States, 1948–1994." *American Journal of Public Health* 91, no. 8 (August), 1194–1199.

Dockery, D. W., C. A. Pope III, X. Xu, J. D. Spengler, J. H. Ware, M. E. Fay, B. G. Ferris Jr., and F. E. Speizer. 1993. "An Association between Air Pollution and Mortality in Six U.S. Cities." *New England Journal of Medicine* 329, 1753–1759.

Doll, Richard, and A. Bradford Hill. 1950. "Smoking and Carcinoma of the Lung—Preliminary Report." *British Medical Journal* 2 (30 September), 739–748.

Doll, Richard, and Richard Peto. 1976. "Mortality in Relation to Smoking: 20 Years' Observations on Male British Doctors." *British Medical Journal* 2 (25 December), 1525–1536.

English, D. 1992. "Geographical Epidemiology and Ecological Studies." In P. Elliott, J. Guzick, D. English, and R. Stern, eds., *Geographical and Environmental Epidemiology: Methods for Small-Area Studies,* pp. 3–13. New York: World Health Organization Regional Office for Europe, Oxford University Press.

Goldsmith, John R. 1986. *Environmental Epidemiology: Epidemiological Investigation of Community Environmental Health Problems.* Boca Raton, FL: CRC Press.

Hande, M. Prakash, Tamara V. Azizova, Charles R. Geard, Ludmilla E. Burak, Catherine R. Mitchell, Valentin F. Khokhryakov, Evgeny K. Vasilenko, and David J. Brenner. 2003. "Past Exposure to Densely Ionizing Radiation Leaves a Unique Permanent Signature in the Genome." *American Journal of Human Genetics* 72, 1162–1170.

Hill, A. B. 1965. "The Environment and Disease: Association or Causation?" *Proceedings of the Royal Society of Medicine* 58, 259–300.

Ishibe, Naoko, and Karl T. Kelsey. 1997. "Genetic Susceptibility to Environmental and Occupational Cancers." *Cancer Causes and Control* 9, 504–513.

McMichael, Anthony J. 1994. "Invited Commentary—'Molecular Epidemiology': New Pathway, or New Traveling Companion?" *American Journal of Epidemiology* 140, no. 1 (1 July), 1–11.

———. 2001. "Global Environmental Change as a 'Risk Factor': Can Epidemiology Cope?" (editorial). *American Journal of Public Health* 91, no. 8 (August), 1172–1174.

Monson, Richard. 1990. *Occupational Epidemiology.* 2nd ed. Boca Raton, FL: CRC Press.

Muirhead, Colin. 2001. "Radiation Risks in Kyoto." *Radiological Protection Bulletin,* no. 231 (September), 34–35. NRPB, Chilton, Didcot, United Kingdom.

NCRP. 2001. *Evaluation of the Linear-Nonthreshold Dose-Response Model for Ionizing Radiation.* Report no. 136. Bethesda, MD. NCRP.

NRC. 1991. *Environmental Epidemiology: Public Health and Hazardous Wastes.* Washington, DC: National Academy Press.

Perera, Frederica P. 2000. "Molecular Epidemiology: On the Path to Prevention?" *Journal of the National Cancer Institute* 92, no. 8 (19 April), 602–612.

Pope, C. A., III, R. T. Burnett, M. J. Thun, E. E. Calle, D. Krewski, K. Ito, and

G. D. Thurston. 2002. "Lung Cancer, Cardiopulmonary Mortality, and Long-Term Exposure to Fine Particulate Air Pollution." *JAMA* 287, no. 9 (6 March), 1132–1141.

Pope, C. A., III, M. J. Thun, M. Namboodiri, D. W. Dockery, J. S. Evans, F. E. Speizer, and C. W. Heath. 1995. "Particulate Air Pollution as a Predictor of Mortality in a Prospective Study of U.S. Adults." *American Journal of Respiratory and Critical Care Medicine* 151, 669–674.

Robinson, David. 2002. *Cancer Clusters: Findings vs. Feelings.* New York: ACSH.

Shigematsu, I. 2000. "The 2000 Sievert Lecture—Lessons from Atomic Bomb Survivors in Hiroshima and Nagasaki." *Health Physics* 79, no. 3 (September), 234–241.

Surgeon General. 1989. "Executive Summary, the Surgeon General's 1989 Report on Reducing the Health Consequences of Smoking: 25 Years of Progress." *Morbidity and Mortality Weekly Report* 38, no. S-2 (24 March), 8.

Terracini, B. 1992. "Environmental Epidemiology: A Historical Perspective." In P. Elliott, J. Cuzick, D. English, and R. Stern, eds., *Geographical and Environmental Epidemiology: Methods for Small-Area Studies,* pp. 253–263. New York: World Health Organization Regional Office for Europe, Oxford University Press.

Trichopoulos, Dimitrios. 1994. "Risk of Lung Cancer from Passive Smoking." *Principles and Practice of Oncology: PPO Updates* 8, no. 8 (August), 1–8.

USPHS. 1964. *Smoking and Health: Report.* Surgeon General's Advisory Committee on Smoking and Health, Publication no. 1103. Washington, DC. USPHS.

WHO. 1983. *Guidelines on Studies in Environmental Epidemiology. Environmental Health Criteria,* Report 27. Geneva, Switzerland.

4. The Workplace

ACGIH. 2004. *2004 TLVs and BEIs—Based on the Documentation of the Threshold Limit Values for Chemical Substances and Physical Agents & Biological Exposure Indices.* Cincinnati, OH: ACGIH.

Burgess, William A. 1995. *Recognition of Health Hazards in Industry.* 2nd ed. New York: Wiley-Interscience.

——— 1997. "Impact of Emerging Technologies on Occupational Health." *AIHAJ* 58, no. 7 (July), 469–473.

CDC. 2001a. "Nonfatal Occupational Injuries and Illnesses Treated in Hospital Emergency Departments—United States, 1998." *Morbidity and Mortality Weekly Report* 50, no. 16 (27 April), 313–317.

———. 2001b. "Pesticide-Related Illnesses Associated with the Use of a Plant Growth Regulator—Italy, 2001." *Morbidity and Mortality Weekly Report* 50, no. 39 (5 October), 845–847.

———. 2002. "Workers' Memorial Day—April 28, 2002." *Morbidity and Mortality Weekly Report* 51, no. 16 (26 April), 345.

Daum, Kent. 2004. "New Study Links Computer Vision and Productivity." *Industrial Hygiene News* 27, no. 2 (March), 15.

Fine, Lawrence J. 1996. "The Psychological Work Environment and Heart Disease." Editorial. *American Journal of Public Health* 86, no. 3 (March), 301–303.

Franco, Giuliano. 2001. "Bernardino Ramazzini: The Father of Occupational Medicine." *American Journal of Public Health* 91, no. 9 (September), 1382.

Hamilton, Alice. 1943. *Exploring the Dangerous Trades.* Boston: Little, Brown.

Herbert, Robin, and Philip J. Landrigan. 2000. "Work-Related Death: A Continuing Epidemic." *American Journal of Public Health* 90, no. 4 (April), 541–545.

Herrick, Robert F. 1998. "Industrial Hygiene." In Robert B. Wallace, ed., *Maxcy-Rosenau-Last Public Health and Preventive Medicine,* 14th ed., pp. 661–667. Norwalk, CT: Appleton & Lange.

Kenoyer, J. L., R. D. Stenner, W. B. Andrews, R. I. Scherpelz, and R. D. Aaberg. 2000. *Estimating Worker Risk Levels Using Accident/Incident Data.* Report PNNL-13338. Richland, WA: Pacific Northwest National Laboratories.

Keyserling, W. Monroe. 2000. "Workplace Risk Factors and Occupational Musculoskeletal Disorders, Part 1: A Review of Biomechanical and Psychophysical Research on Risk Factors Associated with Low-Back Pain." *AIHAJ* 61, no. 1 (January/February), 39–50.

Keyserling, W. Monroe, and Thomas J. Armstrong. 1998. "Ergonomics and Work-Related Musculoskeletal Disorders." In Robert B. Wallace, ed., *Maxcy-Rosenau-Last Public Health and Preventive Medicine,* 14th ed., pp. 645–659. Norwalk, CT: Appleton & Lange.

Levine, Steven P. 2001. "An Industry Safety and Health Forgot: Health Care." *Synergist* 12, no. 4 (April), 33–34.

NIOSH. 2003. *NIOSH ALERT: Preventing Deaths, Injuries, and Illnesses of Young Workers.* DHHS (NIOSH) Publication No. 2003-128. Cincinnati, OH: U.S. Department of Health and Human Services, National Institute for Occupational Safety and Health.

NRC. 2001. *Musculoskeletal Disorders and the Workplace: Low Back and Upper Extremities.* Washington, DC: National Academy Press.

NSC. 2001. *Injury Facts, 2001 Edition.* Itasca, IL: NSC.

OSHA. 1998. "OSHA Finalizes Respiratory Protection Standard." *Today!* 6, no. 1 (January), 1, 11.

———. 2002. "OSHA's New Ergo Plan: Guidelines Favored over Rules." *Industrial Safety and Hygiene News* 36, no. 6 (June), 14.

Patty, F. A. 1978. "Industrial Hygiene: Retrospect and Prospect." (Prologue). In G. D. Clayton and F. E. Clayton, eds., *Patty's Industrial Hygiene and Toxicology,* 3rd rev. ed., vol. 1. New York: John Wiley and Sons.

Pratt, Stephanie G. 2003. *Work-Related Roadway Crashes: Challenges and Opportunities for Prevention.* DHHS (NIOSH) Publication No. 2003–119. Cincinnati, OH: U.S. Department of Heal Human Services, National Institute for Occupational Safety and Health.

Schlecht, Paul C., and P. F. O'Connor (eds). 2003. *NIOSH Manual on Analytical Methods (NMAM).* DHHS (NIOSH) Publication No. 2003–154, Cincinnati, OH: Third Supplement (15 March).

Snook, S. H. 1989. "The Control of Low Back Disability: The Role of Management." In K. H. E. Kroemer, J. D. McGlothlin, and T. G. Bobick, eds., *Manual Material Handling: Understanding and Preventing Back Trauma,* pp. 97–101. Akron, OH: American Industrial Hygiene Association.

U.S. Congress. 1913. *An Act to Create a Department of Labor.* Public Law 426, 62nd Congress, Washington, DC.

———. 1990. *Pollution Prevention Act.* Public Law 101-508, 42 USC 13101 et seq. Washington, DC.

Wassell, James T., Lytt I. Gardner, Douglas P. Landsittel, Janet J. Johnston, and Janet M. Johnston. 2000. "A Prospective Study of Back Belts for Prevention of Back Pain and Injury." *JAMA* 284, no. 21 (6 December), 2727–2732.

5. Air in the Home and Community

Bortnick, S. M., B. W. Coutant, and T. Hanley. 2002. "Public Reporting of an Air Quality Index Using Continuous $PM_{2.5}$ Monitoring Data." *EM* (March), 27–33.

CEQ. 1997. *Environmental Quality: The Twenty-fifth Anniversary Report of the Council on Environmental Quality.* Washington, DC: Executive Office of the President.

Dooley, Erin E. 2002. "Biodiesel Bulldozes Ahead." *Environmental Health Perspectives* 110, no. 8 (August), A 453.

Dumyahn, Thomas S., John D. Spengler, Harriet A. Burge, and Michael Muilenburg. 2000. *Comparison of the Environments of Transportation Vehicles: Results of Two Surveys.* Standard Technical Publication 1393. West Conshohocken, PA: ASTM.

EPA. 2000. *National Emission Trends, 1900 to 1998.* Report EPA 454/R-00-002. Research Triangle Park, NC: EPA.

———. 2002. *National Air Quality: 2000—Status and Trends.* Research Triangle Park, NC, EPA.

———. 2003. *National Air Quality and Emissions Trends Report, 2003 Special Studies Edition.* Washington, DC: U.S. Environmental Protection Agency.

Findley, Roger W., and Daniel A. Farber. 2000. *Environmental Law in a Nutshell.* 5th ed. St. Paul, MN: West Group Publishing Company.

French, Hilary F. 1991. "Eastern Europe's Clean Break with the Past." *World-Watch* 4, no. 2 (March–April), 21–27.

Gruenspecht, Howard K., and Robert N. Stavins. 2002. "New Source Review under the Clean Air Act: Ripe for Reform." *Resources,* no. 147 (Spring), 19–23.

Helfand, William H. 2001. "Donora, Pennsylvania: An Environmental Disaster of the 20th Century." *American Journal of Public Health* 91, no. 4 (April), 553.

Jezouit, Debra J., and Joshua B. Frank. 2002. "BART Sent Back to the Drawing Board." *EM* (August), 30–33.

Krupnick, Alan J. 2002. "Does the Clean Air Act Measure Up?" *Resources,* no. 147 (Spring), 2–3.

Larsen, Ralph E. 2002. Personal communication. Environmental Protection Agency, Research Triangle Park, NC (June).

Latko, Mary Ann. 2000. "Guidelines and Resources for Indoor Air Quality Professionals." *EM* (September), 15–17.

Long, Christopher M. 2002. "Overview: Indoor Air Quality." *Trends in Risk and Remediation* (Spring), 1–2. Cambridge, MA: Gradient Corporation.

NRC. 2001. *Evaluating Vehicle Emissions Inspection and Maintenance Programs.* Washington, DC: National Academy Press.

Pope, C. A., III, R. T. Burnett, M. J. Thun, E. E. Calle, D. Krewski, K. Ito, and G. D. Thurston. 2002. "Lung Cancer, Cardiopulmonary Mortality, and Long-Term Exposure to Fine Particulate Air Pollution." *JAMA* 287, no. 9 (6 March), 1132–1141.

Service, Robert F. 2002. "Cleaning Air While Sparing Water." *Science* 296, no. 5567 (19 April), 463.

Spengler, J. D., and K. Sexton. 1983. "Indoor Air Pollution: A Public Health Perspective." *Science* 221, no. 4605 (1 July), 9–17.

Stone, Richard. 2002. "Air Pollution: Counting the Cost of London's Killer Smog." *Science* 298, no. 5599 (13 December), 2106–2107.

Weinhold, Bob. 2002a. "Air Pollution: U.S. Air Only Fair." *Environmental Health Perspectives* 110, no. 8 (August), A 452.

———. 2002b. "Fuel for the Long Haul?" *Environmental Health Perspectives* 110, no. 8 (August), A 458–A 464.

Wilkening, Kenneth E., Leonard A. Barrie, and Marilyn Engle. 2000. "Trans-Pacific Air Pollution." *Science* 290, no. 5489 (6 October), 65, 67.

6. Food

Ackerman, Jennifer. 2002. "Food: How Safe? How Altered?" *National Geographic* 201, no. 5 (May), 2–51.

Bourdelais, A. J., C. R. Tomas, J. Naar, J. Kubanek, and D. G. Baden. 2002. "New Fish-Killing Alga in Coastal Delaware Produces Neurotoxins." *Environmental Health Perspectives* 110, no. 5 (May), 465–470.

Bright, Chris. 1999. "Super-bugs Arrive." *World-Watch* 12, no. 2 (March/April), 9–11.

CDC. 1998. "Ciguatera Fish Poisoning—Texas, 1997." *Morbidity and Mortality Weekly Report* 47, no. 33 (28 August), 692–694.

———. 2000. "Giardiasis Surveillance—United States, 1992–1997." *Morbidity and Mortality Weekly Report* 49, no. SS-7 (11 August), 1–13.

———. 2001a. "Diagnosis and Management of Foodborne Illnesses—A Primer for Physicians." *Morbidity and Mortality Weekly Report* 50, no. RR-2 (26 January), 1–69.

———. 2001b. "Botulism Outbreak Associated with Eating Fermented Food—Alaska, 2001." *Morbidity and Mortality Weekly Report* 50, no. 32 (17 August), 680–682.

———. 2002a. "Outbreak of *Campylobacter jejuni* Infections Associated with Drinking Unpasteurized Milk Procured through a Cow-Leasing Program—Wisconsin, 2001." *Morbidity and Mortality Weekly Report* 51, no. 25 (28 June), 548–549.

———. 2002b. "Multistate Outbreak of *Escherichia coli* O157:H7 Infections Associated with Eating Ground Beef—United States, June–July, 2002." *Morbidity and Mortality Weekly Report* 51, no. 29 (26 July), 637–639.

———. 2002c. "Norwalk-Like Virus—Associated Gastroenteritis in a Large, High-Density Encampment—Virginia, July 2002." *Morbidity and Mortality Weekly Report* 51, no. 30 (2 August), 661–663.

Chin, James, ed. 2000. *Control of Communicable Diseases Manual,* 17th ed. Washington, DC: APHA.

Codex Alimentarius Commission. 2003. "Global GM Food Standards." *Environmental Health Perspectives* 111, no. 14 (November), A 755.

Consumers Union. 2003. "Of Birds and Bacteria." *Consumer Reports* 68, no. 1 (January), 24–28.

Eubanks, Mary. 2002. "Allergies a la Carte—Is There a Problem with Genetically Modified Foods?" *Environmental Health Perspectives* 110, no. 3 (March), A 130–A 131.

Falkow, Stanley, and Donald Kennedy. 2001. "Antibiotics, Animals, and People—Again!" (editorial). *Science* 291, no. 5503 (19 January), 397.

FDA. 1995. *Food Code.* Springfield, VA: National Technical Information Service.

Ferber, Dan. 2002. "Livestock Feed Ban Preserves Drugs' Powers." *Science* 295, no. 5552 (4 January), 27–28.

Fraser, Claire M. 2004. "An Uncertain Call to Arms." *Science* 304, no. 5669 (16 April), 359.

Goklany, Indur M. 2001. "The Future of Food." *Forum for Applied Research and Public Policy* 16, no. 2 (Summer), 59–65.

Gray, G., J. Cohen, and S. Kreindel. 2002. "Evaluating the Risk of Bovine Spongiform Encephalopathy in the United States." *Risk in Perspective* 10, no. 2 (March), 1–6. Center for Risk Analysis, Harvard School of Public Health.

Huang, J., S. Rozelle, C. Pray, and Q. Wang. 2002. "Plant Biotechnology in China." *Science* 295, no. 5555 (25 January), 674–677.

IFT. 2002. "IFT Expert Report on Emerging Microbiological Food Safety Issues: Implications for Control in the 21st Century." (20 February). Atlanta, GA: IFT.

Kaiser, J., C. Holden, and P. Bagla. 2002. "India OKs GM Cotton." *Science* 295, no. 5564 (29 March), 2345.

Kennedy, Donald, 2003. "Agriculture and the Developing World." Editorial. *Science* 302, no. 5644 (17 October), 357.

Lambert, Craig. 2004. "The Way We Eat Now." *Harvard Magazine* 106, no. 5 (May–June), 50–57, 98–99.

Levine, Samantha. 2004. "Who'll Stop the Mercury Rain?" *U.S. News & World Report,* 136, no. 11 (April 5), 70–71.

Loaharanu, Paisan. 2003. *Irradiated Foods* (Fifth Edition). American Council on Science and Health, New York, NY (May).

Mader, Paul, Andreas Fließbach, David Dubois, Lucie Gunst, Padruot Fried, and Urss Niggli. 2002. "Soil Fertility and Biodiversity in Organic Farming." *Science* 296, no. 5573 (31 May), 1694–1697.

Marcus, Mary B. 2001. "Organic Foods Offer Peace of Mind—At a Price." *U.S. News and World Report* 130, no. 2 (15 January), 48–50.

Marshall, Douglas L., and James S. Dickson. 1998. "Ensuring Food Safety." In Robert B. Wallace, ed., *Maxcy-Rosenau-Last Public Health and Preventive Medicine,* 14th ed., pp. 723–736. Norwalk, CT: Appleton & Lange.

Metcalfe, Dean D. 2003. "Introduction: What Are the Issues in Addressing the Allergenic Potential of Genetically Modified Foods?" *Environmental Health Perspectives* 111, no. 8 (June), 1110–1113.

Pickrell, John. 2003. "U.K. Government Panel Gives GM Crops Cautious Support." *Science* 301, no. 5632 (25 July), 447–448.

Red Tide Research Group. 2002. "The Current of Red Tide Research." *Environmental Health Perspectives* 110, no. 3 (March), A 132–A 133.

Satcher, David. 1999. "Surgeon General's Column." *Commissioned Corps Bulletin* 13, no. 4 (April), 1–2. Washington, DC: HHS.

Sheehan, Emily. 2003. "Food Products Labels to Bear Information on Trans Fats." *The Nation's Health* (September), 1 and 16. Washington, DC: American Public Health Association.

Stokstad, Erik. 2002. "Organic Farms Reap Many Benefits." *Science* 296, no. 5573 (31 May), 1589.

Taylor, Michael R., and Sandra A. Hoffmann. 2001. "Redesigning Food Safety—Using Risk Analysis to Build a Better Food Safety System." *Resources,* no. 144 (Summer), 13–16.

Tick, Jody. 2004. "The Food Safety Research Consortium Rolls Out New Risk-Ranking Model." *Resources,* 152 (Fall/Winter), 4. Washington, DC: Resources for the Future.

Weiss, Giselle. 2002. "Acrylamide in Food: Uncharted Territory." *Science* 297, no. 5578 (5 July), 27.

Willett, Walter. 2001. *Eat, Drink, and Be Healthy.* New York: Simon & Schuster.

7. Drinking Water

Alley, William M., Richard W. Healy, James W. LaBaugh, and Thomas E. Reilly. 2002. "Flow and Storage in Groundwater Systems." *Science* 296, no. 5575 (14 June), 1985–1990.

Canby, Thomas Y. 1980. "Water—Our Most Precious Resource." *National Geographic* 158, no. 2 (August), 144–179.

CDC. 2000a. "Surveillance of Waterborne-Disease Outbreaks—United States, 1997–1998." *Morbidity and Mortality Weekly Report* 49, no. SS-4 (26 May), 1–35.

———. 2000b. "Giardiasis Surveillance—United States, 1992–1997." *Morbidity and Mortality Weekly Report* 49, no. SS-7 (11 August), 1–13.

———. 2001a. "Prevalence of Parasites in Fecal Material from Chlorinated Swimming Pools—United States, 1999." *Morbidity and Mortality Weekly Report* 50, no. 20 (25 May), 410–412.

CDC. 2001b. "Norwalk-Like Viruses-Public Health Consequences and Outbreak Management." *Morbidity and Mortality Weekly Report* 50, no. RR-9 (1 June), 1–17.

———. 2001c. "Responding to Fecal Accidents in Disinfected Swimming Venues." *Morbidity and Mortality Weekly Report* 50, no. 20 (25 May), 416.

———. 2002a. "Populations Receiving Optimally Fluoridated Public Drinking Water—United States, 2000." *Morbidity and Mortality Weekly Report* 51, no. 7 (22 February), 144–147.

———. 2002b. "Progress toward Global Dracunculiasis Eradication, June, 2002." *Morbidity and Mortality Weekly Report* 51, no. 36 (13 September), 810–811.

CEQ. 1998. *Environmental Quality: The World Wide Web: The 1997 Report of the*

Council on Environmental Quality. Washington, DC: Executive Office of the President.

Chin, James, ed. 2000. *Control of Communicable Diseases Manual.* 17th ed. Washington, DC: APHA.

Clesceri, Lenore S., Arnold E. Greenberg, and Andred D. Eton, eds. 1998. *Standard Methods for the Examination of Water and Waste Water.* 20th ed. Waldorf, MD: APHA, AWWA, and WEF.

Consumers Union. 2003. "Clear Choices for Clean Drinking Water." *Consumer Reports* 68, no. 1 (January), 32–35.

EPA. 2004. "RNA-Based Research for Safer Drinking Water." NRMRL News (March). Cincinnati, OH: National Risk Management Research Laboratory, EPA.

Fleming, Hu. 2002. "Out of the Dark: Ultraviolet Disinfection Lights Up Municipal Drinking Water Treatment Systems." *Environmental Protection* 13, no. 3 (March), 46–53.

Frontinus, Sextus Julius. A.D. 97. *The Water Supply of the City of Rome.* Clemens Herschel, trans., 1973. Boston: NEWWA.

Johnson, Nels, Carmen Revenga, and Jaime Echeverria. 2001. "Managing Water for People and Nature." *Science* 292, no. 5519 (11 May), 1071–1072.

Johnson, Warren T. 1999. "The Future of Filtration Technology—Membranes." In *World of Water 2000: The Past, Present, and Future,* pp. 142, 144–146. Supplement to *PennWell Magazine.* Tulsa, OK: WaterWorld and Water & Wastewater International.

Kim, Yong. 2003. "Water Security: The First Line of Defense." "Water and Wastewater Products." Supplement to *Environmental Protection* (May/June), 30–33.

Laughlin, James, ed. 2001. "Alum Replacement Gains Popularity at Municipal Plants." *Water World* (November), 22–23.

Lee, Sherline H., Deborah A. Levy, Gunther F. Craun, Michael J. Beach, and Rebecca L. Calderon. 2002. "Surveillance for Waterborne-Disease Outbreaks—United States, 1999–2000." *Morbidity and Mortality Weekly Report* 51, no. SS-8 (22 November), 1–47.

Leland, D. E., and M. Damewood III. 1990. "Slow Sand Filtration in Small Systems in Oregon." *AWWA Journal* 82, no. 6 (June), 50–59.

Montaigne, Fen. 2002. "Water Pressure." *National Geographic* 202, no. 3 (September), 2–33.

Montgomery, James M. 1985. *Water Treatment Principles and Design.* New York: John Wiley and Sons.

Noah, Marilyn. 2002. "Graywater Use Still a Gray Area." *Journal of Environmental Health* 64, no. 10 (June), 22–25.

Nordstrom, D. Kirk. 2002. "Worldwide Occurrences of Arsenic in Ground Water." *Science* 296, no. 5576 (21 June), 2143, 2145.

Oates, Wallace E. 2002. "The Arsenic Rule: A Case for Decentralized Standard Setting?" *Resources,* no. 147 (Spring), 16–18.

Parfit, Michael. 1993. "Water—The Power, Promise, and Turmoil of North America's Fresh Water—Map Supplement: Water." *National Geographic* special edition, 184, no. 5A (November).

Sawin, Janet L. 2003. "Water Scarcity Could Overwhelm the Next Generation." *World-Watch* 16, no. 4 (July/August), 8.

Schroeder, H. A. 1974. "Role of Trace Elements in Cardiovascular Diseases." *Medical Clinics of North America* 58, no. 2, 381–396.

Simon, Paul. 2001. "Thirsty World." *Environmental Protection* 12, no. 11 (November), 12–13.

Symons, James M. 1992. *"Plain Talk about Drinking Water": Answers to 101 Important Questions about the Water You Drink.* Denver: AWWA.

Tramposch, Walter, and Charles Fluharty. 2003. "Let Your UV Light Shine." *Water and Wastewater Products,* Supplement to *Environmental Protection* 14, no. 7 (September/October), 26, 28, 30.

8. Liquid Waste

Anderson, W. C. 2002. "The Lessons of History." *Environmental Engineer* 38, no. 1 (January), 6.

ATSDR. 2000. *Toxicological Profile for Polychlorinated Biphenyls (Update).* Atlanta, GA: HHS.

CEQ. 1998. *Environmental Quality: The World Wide Web: The 1997 Report of the Council on Environmental Quality.* Washington, DC: Executive Office of the President.

Claudio, Luz. 2002. "The Hudson: A River Runs through an Environmental Controversy." *Environmental Health Perspectives* 110, no. 4 (April), A 184–A 187.

Dix, Stephen P. 2001. "Onsite Wastewater Treatment: A Technological and Management Revolution: Part 1." *Water Engineering & Management* 148, no. 9 (September), 24–26, 28.

Edwards, Peter. 1992. *Reuse of Human Wastes in Aquaculture: A Technical Review.* Water and Sanitation Report no. 2, UNDP–World Bank Water and Sanitation Program. Washington, DC: World Bank.

EPA. 2003. "Clean Watersheds Needs Survey Reveals $180 Billion in Needs." *WaterWorld* 19, no. 11 (November), 1, 27–28.

Francisco, Donald. 2001. Personal communication. Department of Environmental Science and Engineering, University of North Carolina, Chapel Hill (30 October).

Furukawa, David H. 1999. "The Key to Future Water Supplies: Desalination." In *World of Water 2000: The Past, Present, and Future,* pp. 148–150. Supplement to *PennWell Magazine.* Tulsa, OK: WaterWorld and Water & Wastewater International.

Gloyna, Ernest. 1971. *Waste Stabilization Ponds.* Monograph Series no. 60. Geneva: WHO.

Gray, Albert C. 1999. "Wastewater—Watershed Management Is Key to Next Millennium." In *World of Water 2000: The Past, Present, and Future,* pp. 114–118. Supplement to *PennWell Magazine.* Tulsa, OK: WaterWorld and Water & Wastewater International.

Guy, Brenda, and Mike Catanzaro. 2002. "Proper Operation, Maintenance, and

Servicing of Aerobic Wastewater Treatment Systems." *Journal of Environmental Health* 64, no. 8 (April), 23–24.
Hetrick, Scott. 2001. *Bio-kinetic Wastewater Treatment Systems.* Norwalk, OH: Norweco.
Holden, Constance. 2002. "Random Samples: Dead Zone Grows." *Science* 297, no. 5584 (16 August), 1119.
Logan, Terry J. 1999. "Biosolids—Challenges and Options in the Next Millennium." In *World of Water 2000: The Past, Present, and Future,* pp. 130, 132–134. Supplement to *PennWell Magazine.* Tulsa, OK: WaterWorld and Water & Wastewater International.
Lovley, Derek R. 2001. "Anaerobes to the Rescue." *Science* 293, no. 5534 (24 August), 1444–1446.
NRC. 1993. *Managing Wastewater in Coastal Urban Areas.* Committee on Wastewater Management for Coastal Urban Areas. Washington, DC: National Academy Press.
Pitois, S., M. H. Jackson, and B. J. B. Wood. 2001. "Sources of Eutrophication Problems Associated with Toxic Algae: An Overview." *Journal of Environmental Health* 64, no. 5 (December), 25–32.
Ruiz, Gregory M., Tonya R. Rawlings, Fred C. Dobbs, Lisa A. Drake, Timothy Mullady, Anwarul Huq, and Rita R. Colwell. 2000. "Global Spread of Microorganisms by Ships" (letter to editor). *Nature* 408, no. 6808 (2 November), 49.
Sakamoto, Gail. 2000. "UV Disinfection of Reclaimed Wastewater: The North American Experience." *Environmental Protection* 11, no. 10 (October), 20–25.
Satchell, Michael. 1996. "Hog Factories: Cheap Meat, Costly Problems." *U.S. News and World Report* vol. 120, no. 3 (22 January), pp. 58–59.
Sims, Danny, and Dennis E. Bentley. 2001. "Relieving Wastewater Treatment Capacity Constraints through Improved Processing of Aerobically Digested Solids." *EM* (November), 32–36.
Sun, Baolin, Benjamin M. Griffin, Hector L. Ayala-del-Rio, Syed A. Hashsham, and James M. Tiedje. 2002. "Microbial Dehalorespiration with 1,1,1-Trichloroethane." *Science* 298, no. 5595 (1 November), 1023–1025.
Weinhold, Bob. 2002. "Water Pollution: Up a Chemical Creek." *Environmental Health Perspectives* 110, no. 7 (July), A 390.
Wolfe, Pamela. 1999. "History of Wastewater." In *World of Water 2000: The Past, Present, and Future,* pp. 24–36. Supplement to *PennWell Magazine.* Tulsa, OK: WaterWorld and Water & Wastewater International.

9. Solid Waste

Black, Harvey. 2002. "The Hottest Thing in Remediation." *Environmental Health Perspectives* 110, no. 3 (March), A 146–A 148.
CEQ. 1997. *Environmental Quality: The Twenty-Fifth Anniversary Report of the Council on Environmental Quality.* Washington, DC: Executive Office of the President.
———. 1998. *Environmental Quality: The World Wide Web: The 1997 Report of the Council on Environmental Quality.* Washington, DC: Executive Office of the President.

Chen, William. 2002. "Greyfields: Revitalizing Communities through Mixed-Use Redevelopment." *EM* (October), 29–30.

DOE. 1999. *Implementation Guide for Use with DOE M 435.1.* Publication DOE G 435.1. Washington, DC: DOE.

EPA. 1986a. *RCRA Orientation Manual.* Report EPA/530-SW-86-001. Washington, DC: EPA.

———. 1986b. *Solving the Hazardous Waste Problem: EPA's RCRA Program.* Report EPA/530-SW-86-037. Washington, DC: EPA.

———. 1993a. *Safer Disposal for Solid Waste: The Federal Regulations for Landfills.* Report EPA/530-SW-91-092. Washington, DC: EPA.

———. 1993b. "Guidance to Hazardous Waste Generators on the Elements of a Waste Minimization Program." *Federal Register* 58, no. 102 (28 May), 31114–31120. Washington, DC: EPA.

Fox, Robert D. 1996. "Physical/Chemical Treatment of Organically Contaminated Soils and Sediments." *EM* (May), 28–34.

Golaine, Andrea. 1991. "Superfund: Money Squandered in the Name of Public Health." *Priorities* (Fall), 30–31. New York: ACSH.

Hayden, Thomas. 2002. "Science and Technology—Trashing the Oceans." *U.S. News and World Report* 133, no. 17 (4 November), 58–60.

Hughes, Joseph B. 1996. "Biological Treatment of Hazardous Waste." In *Frontiers in Engineering,* pp. 37–39. Washington, DC: National Academy Press.

Isler, Margaret, and Martin R. Lee. 2002. "Environmental Protection Issues in the 107th Congress (Updated July 3, 2002)." Congressional Research Service. Washington, DC: Library of Congress.

Jurdi, Mey. 2002a. "Transboundary Movement of Hazardous Wastes into Lebanon, Part 1: The Silent Trade." *Journal of Environmental Health* 64, no. 1 (January–February), 9–14.

Jurdi, Mey. 2002b. "Transboundary Movement of Hazardous Wastes into Lebanon, Part 2: Environmental Impacts and the Need for Remedial Actions." *Journal of Environmental Health* 64, no. 1 (January–February), 15–19.

Lavelle, Marianne. 2002. "Arsenic and Barbeque." *U.S. News and World Report* 133, no. 10 (16 September), 58–59.

Link-Wills, Kimberly. 2002. "Revolutionary Technology." *Georgia Tech Alumni Magazine* 79, no. 1 (Summer), 45–47.

NCRP. 2002. *Risk-Based Classification of Radioactive and Hazardous Chemical Wastes.* Report no. 139. Bethesda, MD: NCRP.

———. 2003. *Management Techniques for Laboratories and Other Small Institutional Generators to Minimize Off-Site Disposal of Low-Level Radioactive Waste.* Report no. 143. Bethesda, MD: NCRP.

O'Connell, Kim A. 2003. "Poison Planks." *Waste Age* 34, no. 10 (October), 40–44.

Padgett, C. B. 2001. "Oregon Survey Trumpets State's Efforts." *Waste Age* 32, no. 12 (December), 6–7.

Portney, P. R., and R. N. Stavins. 2000. "Introduction." In P. R. Portney and R. N. Stavins, eds., *Public Policies for Environmental Protection,* 2nd ed., pp. 1–10. Washington, DC: RFF.

Saxe, Jennifer K. 2002. "Land Disposal of Non-hazardous Materials." *Trends in Risk and Remediation* (Fall), 1–2. Cambridge, MA: Gradient Corporation.

Shea, Cynthia. 1988. "Plastic Waste Proliferates." *World-Watch* 1, no. 2 (March–April), 7–8.

Sperber, JoAnn. 2002. "Sweden, Finland Pursue Deep Geologic Repository." *Nuclear Energy Insight* (August), 2. Washington, DC: NEI.

Stuckey, H. T., and P. F. Hudak. 2002. "Waste Investment." *Environmental Protection* 13, no. 3 (March), 60–64, 71.

Tom, Patricia Anne. 2001. "Good Wood Gone Bad." *Waste Age* 32, no. 8 (August), 36–51.

U.S. Congress. 1990. *Pollution Prevention Act.* Public Law 101-508, 42 USC 13101 et seq. Washington, DC.

———. 1992. *Federal Facility Compliance Act.* Public Law 102-386, 42 USC 6901 et seq. Washington, DC.

Verbit, S. R. 2001. "New Law May Unlock Potential of Brownfields." *Environmental Protection* 12, no. 8 (August), 31–33.

Wolpin, Bill. 2002. "A Moveable Beast" (editorial). *Waste Age* 33, no. 3 (March), 4.

Zacha, Nancy J. 2003. "Low-Level Waste Disposal in the United States—Status Update." *Radwaste Solutions* 10, no. 4 (July/August), 18–20.

10. Rodents and Insects

ATSDR. 2002. *Toxicological Profile for DDT/DDD/DDE (Update).* Atlanta, GA: HHS.

Canby, Thomas Y. 1977. "The Rat—Lapdog of the Devil." *National Geographic* 152, no. 1 (July), 60–87.

Carson, Rachel. 1962. *Silent Spring.* Boston: Houghton Mifflin.

Carter, Jimmy. 2002. "On the Road with President Carter: Targeting River Blindness." *Update* (Spring), 5. Atlanta, GA: Carter Center.

CDC. 2000. "Human Rabies—California, Georgia, Minnesota, New York, and Wisconsin, 2000." *Morbidity and Mortality Weekly Report* 49, no. 49 (15 December), 1111–1115.

———. 2002. "Lyme Disease—United States, 2000." *Morbidity and Mortality Weekly Report* 51, no. 2 (18 January), 29–31.

Conniff, Richard. 1977. "The Malevolent Mosquito." *Reader's Digest* 111, no. 664 (August), 153–157.

Denholm, I., G. J. Devine, and M. S. Williamson. 2002. "Evolutionary Genetics: Insecticide Resistance on the Move." *Science* 297, no. 5590 (27 September), 2222–2223.

Enserink, Martin. 2000. "Malaysian Researchers Trace Nipah Virus Outbreak to Bats." *Science* 289, no. 5479 (28 July), 518–519.

Hayashi, Alden M. 1999. "Attack of the Fire Ants." *Scientific American* 280, no. 2 (February), 26, 28.

Huepel, Patricia Smith. 2002. "State Targets Fire Ants." *Sun Journal,* New Bern, NC (6 June), A1, A3.

Kaiser, Joyce. 2000. "Global Warming, Insects Take the Stage at Snowbird." *Science* 289, no. 5487 (22 September), 2031–2032.

Lecrubier, Aude. 2002. "Pest Control: How to Fool a Fly." *Popular Science* 260, no. 1 (January), 42.

Leslie, Mitch. 2004. "Killing Bugs Softly." Editorial. *Science* 303, no. 5663 (5 March), 1445.

Matuschka, Franz-Rainer, Stefan Endepols, Dania Richter, and Andrew Spielman. 1997. "Competence of Urban Rats as Reservoir Hosts for Lyme Disease Spirochetes." *Journal of Medical Entomology* 34, no. 4 (July), 489–493.

McCracken, G. F., and Jay Dickman. 2002. "Bat Patrol." *National Geographic* 201, no. 4 (April), 114–123.

Pennisi, Elizabeth. 1996. "U.S. Beefs Up CDC's Capabilities." *Science* 272, no. 5267 (7 June), 1413.

Pimentel, David. 2000. "Biological Control of Invading Species" (letter to editor). *Science* 289, no. 5481 (11 August), 869.

Richardson, R. H., J. R. Ellison, and W. W. Averhoff. 1982. "Autocidal Control of Screwworms in North America." *Science* 215, no. 4531 (22 January), 361–370.

Satchell, Michael. 2000. "Rocks and Hard Places—DDT: Dangerous Scourge or Last Resort?" *U.S. News and World Report* 129, no. 23 (11 December), 64–65.

Shaw, Jonathan. 2001. "The Landscape Infections." *Harvard Magazine* 104, no. 2 (November/December), 42–47.

Spielman, Andrew. 1995. Private communication (June).

Spielman, Andrew, and Michael D'Antonio. 2001. *Mosquito: A Natural History of Our Most Persistent and Deadly Foe.* New York: Hyperion.

Spielman, Andrew, and Robert B. Kimsey. 1997. "Zoonosis." In *Encyclopedia of Human Biology,* 2nd ed., vol. 8, pp. 803–812. New York: Academic Press.

Stokstad, Erik. 2001a. "First Light on Genetic Roots of Bt Resistance." *Science* 293, no. 5531 (3 August), 778.

———. 2001b. "Parasitic Wasps Invade Hawaiian Ecosystem." *Science* 293, no. 5533 (17 August), 1241.

Wilson, E. O. 2002. "Hot Spots—Preserving Pieces of a Fragile Biosphere." *National Geographic* 201, no. 1 (January), 86–89.

11. Injury Control

Associated Press. 2003. "When It's Hot Outside, Remember Kids Inside." Sun Journal, New Bern, NC (26 August), A1 & A2.

Berg, Katherine, Marilyn Hines, and Susan Allen. 2002. "Wheelchair Users at Home: Few Home Modifications and Many Injurious Falls." *American Journal of Public Health* 92, no. 1 (January), 48.

Black, David W. 2004. "People, Places & Things." Road&Track 55, no. 7 (March), 22.

CDC. 1998. "Deaths Resulting from Residential Fires and Prevalence of Smoke Alarms—United States, 1991–1995." *Morbidity and Mortality Weekly Report* 47, no. 38 (2 October), 803–806.

———. 1999a. "Childhood Work-Related Agricultural Fatalities—Minnesota, 1994–1997." *Morbidity and Mortality Weekly Report* 48, no. 16 (30 April), 332–335.

———. 1999b. "Playground Safety—United States, 1998–1999." *Morbidity and Mortality Weekly Report* 48, no. 16 (30 April), 329–332.

———. 1999c. "Achievements in Public Health, 1900–1999: Motor-Vehicle Safety:

A 20th Century Public Health Achievement." *Morbidity and Mortality Weekly Report* 48, no. 18 (14 May), 369–374.

———. 2000. "Unpowered Scooter-Related Injuries—United States, 1998–2000." *Morbidity and Mortality Weekly Report* 49, no. 49 (15 December), 1108–1110.

———. 2002a. "Buckle Up America Week, May 20–27, 2002." *Morbidity and Mortality Weekly Report* 51, no. 19 (17 May), 416.

———. 2002b. "Injuries and Deaths Among Children Left Unattended in or Around Motor Vehicles—United States, July 2000–June 2001." *Morbidity and Mortality Weekly Report* 51, no. 26 (5 July), 570–572.

De Haven, H. 1942. "Mechanical Analysis of Survival in Falls from Heights of Fifty to One Hundred and Fifty Feet." *War Medicine* 2, 586–596.

Durbin, Dee-Ann. 2004. "Road Deaths Up Slightly Last Year, Government Says." *Sun Journal* (April 29), A7. Associated Press.

Dwortzan, Mark. 2000. "Locked and Loaded against Gun Injuries." *Harvard Public Health Review* (Fall), 8–10.

Gibson, J. J. 1961. "Contribution of Experimental Psychology to the Formulation of the Problem of Safety: A Brief for Basic Research." In Herbert H. Jacobs, ed., *Behavioral Approaches to Accident Research.* New York: Association for the Aid of Crippled Children.

Haddon, William, Jr. 1970. "On the Escape of Tigers: An Ecologic Note." *American Journal of Public Health* 60, no. 12 (December), 2229–2234.

IIHS 1994. "Best and Worst—1988–92 Passenger Vehicles with Lowest and Highest Drive Death Rates during 1989–93." *IIHS Status Report* 29, no. 11 (8 October), 2.

———. 1999. *The Year's Work, 1999.* Arlington, VA: IIHS.

Jacobsen, Heather A., Matthew W. Kreuter, Douglas Luke, and Charlene A. Caburnay, 2001. "Seat Belt Use in Top-Grossing Movies vs Actual US Rates, 1978–1998." American Journal of Public Health 91, no. 9 (September), 1395–1396.

Law, Alex. 2002. "Stop—In the Name of the Law!" *AutoWeek* 52, no. 45 (4 November), 8.

Lissy, K., J. Cohen, M. Park, and J. D. Graham. 2000. "Cellular Phones and Driving: Weighing the Risks and Benefits." *Risk in Perspective* 8, no. 6 (July), 1–6. Center for Risk Analysis, Harvard School of Public Health.

Mandelblit, Bruce D. 2001. "Fighting Back—How to Prevent Workplace Violence." *Industrial Safety and Hygiene News* 35, no. 9 (September), 70–71.

McCosh, Dan. 2001. "A Jolt for Your Car." *Popular Science* 259, no. 1 (July), 66–68.

Miller, Matthew. 2002. "Mortal Allies: Guns and Suicide." *Harvard Public Health Review* (Summer), 60. Harvard School of Public Health.

Miller, Matthew, Deborah Azrael, and David Hemenway. 2002. "Firearm Availability and Unintentional Firearm Deaths, Suicide, and Homicide among 5–14 Year Olds." *Journal of Trauma* 52, no. 2 (February), 267–275.

Mullet, Andreas. 2004. "Florida's Motorcycle Helmet Law Repeal and Fatality Rates." *American Journal of Public Health* 94, no. 4 (April), 556–558.

NIOSH. 2001. *Building Safer Highway Work Zones: Measures to Prevent Worker Injuries from Vehicles and Equipment.* DHHS/NIOSH Publication no. 2001-128. Cincinnati, OH: National Institute for Occupational Safety and Health.

———. 2003. *Work-Related Roadway Crashes.* DHHS (NIOSH) Publication 2003–119. Cincinnati, OH: National Institute for Occupational Safety and Health.
NSC. 1990. *Accident Facts, 1990 Edition.* Itasca, IL: NSC.
———. 2001. *Injury Facts, 2001 Edition.* Itasca, IL: NSC.
Phillips, William G., ed. 2001. "Automotive Technology—New Car Option: A Crystal Ball." *Popular Science* 259, no. 6 (December), 37.
Rendell, Julian. 2004. "Euro-Pains: Pedestrian Crash Rules Likely to Impact Sports Car and SUV Design," *AutoWeek* 54, no. 7 (February 16), 6.
Waller, Julian A. 1994. "Reflections on a Half Century of Injury Control." *American Journal of Public Health* 84, no. 4 (April), 664–670.

12. Electromagnetic Radiation

ACGIH. 2004. *2004 TLVs and BEIs—Based on the Documentation of the Threshold Limit Values for Chemical Substances and Physical Agents and Biological Exposure Indices.* Cincinnati, OH: ACGIH.
Agnew, J., K. Grainger, I. Clark, and C. Driscoll. 1998. "Protection from UVR by Clothing." *Radiological Protection Bulletin,* no. 200 (April), 14–17. Chilton, Didcot, Oxon, United Kingdom: NRPB.
BEIR [Committee on the Biological Effects of Ionizing Radiation]. 1980. *The Effects on Populations of Exposure to Low Levels of Ionizing Radiation.* Report no. 3. Washington, DC; National Academy Press.
Bouville, A., S. L. Simon, C. W. Miller, H. L. Beck, L. R. Anspaugh, and B. G. Bennett. 2002. "Estimates of Doses from Global Fallout." *Health Physics* 82, no. 5 (May), 690–705.
Ferruci, Joseph T. 2002. "Clinical Application of Adult Computed Tomography Scanning." In *Proceedings of Symposium on Computed Tomography: Patient Dose* (November), pp. 5–6. Bethesda, MD: NCRP.
ICNIRP. 1998. "Guidelines for Limiting Exposure to Time-Varying Electric, Magnetic, and Electromagnetic Fields (Up to 300 GHz)." *Health Physics* 74, no. 4 (April), 494–522.
ICRP. 1991. *1990 Recommendations of the International Commission on Radiological Protection.* Publication 60. Annals of the ICRP 21, nos. 1–3. New York: Pergamon Press.
———. 1994. *Dose Coefficients for Intakes of Radionuclides by Workers.* Publication 68. Annals of the ICRP 24, no. 4. New York: Pergamon Press.
———. 1996. *Age-Dependent Doses to Members of the Public from Intake of Radionuclides: Part 5—Compilation of Ingestion and Inhalation Dose Coefficients.* Publication 72. Annals of the ICRP 26, no. 1. New York: Pergamon Press.
———. 1999. *Protection of the Public in Situations of Prolonged Radiation Exposure.* Publication 82. Annals of the ICRP 29, nos. 1–2. New York: Pergamon Press.
———. 2000. *Managing Patient Dose in Computed Tomography.* Publication 87. Annals of the ICRP 30, no. 4. New York: Pergamon Press.
INPO. 2003. "Achieving 2005 Goals—WANO Performance Indicators Show Steady U.S. Industry Improvements." *Nuclear Professional* 18, no. 2 (Second Quarter), 8–9. National Academy for Nuclear Training.

Kirschner, Suzanne Kantra. 2002. "FDA, It's Time to Study Cellphone Radiation." *Popular Science* 261, no. 3 (September), 16.

Little, John B. 1993. "Biologic Effects of Low-Level Radiation Exposure." In J. M. Taveras, M. Juan, and J. T. Ferruci, eds., *Radiologic Physics and Pulmonary Radiology*, vol. 1, ch. 13. Philadelpha: J. B. Lippincott.

Long, Michael E. 2001. "Surviving in Space." *National Geographic* 199, no. 1 (January), 6–29.

Malakoff, David, and Dennis Normile. 2004. "U.S. Could Pull Back on Studies of Atom Bomb Survivors." *Science* 304, no. 5667 (2 April), 33.

NCRP. 1987a. *Exposure of the Population in the United States and Canada from Natural Background Radiation.* Report no. 94. Bethesda, MD: NCRP.

———. 1987b. *Radiation Exposure of the U.S. Population from Consumer Products and Miscellaneous Sources.* Report no. 96. Bethesda, MD: NCRP.

———. 1996. *Sources and Magnitude of Occupational and Public Exposures from Nuclear Medicine Procedures.* Report no. 124. Bethesda, MD: NCRP.

———. 2001. *Evaluation of the Linear-Nonthreshold Dose-Response Model for Ionizing Radiation.* Report no. 136. Bethesda, MD: NCRP.

———. 2003a. *Presidential Report on Radiation Protection Advice: Screening of Humans for Security Purposes Using Ionizing Radiation Scanning Systems.* Commentary No. 16. Bethesda, MD: NCRP.

———. 2003b. *Radiation Protection in Dentistry.* Report No. 145. Bethesda, MD.

O'Hagan, John, and Robert Hill. 1998. "Laser Pointers." *Radiological Protection Bulletin*, no. 199 (March), 15–20. Chilton, Didcot, Oxon, United Kingdom: NRPB.

Ropeik, David, and George Gray. 2002. *Risk: A Practical Guide for Deciding What's Really Safe and What's Dangerous in the World around Us.* Boston: Houghton Mifflin.

USNRC. 2001. *Information Digest: 2001 Edition.* Report NUREG-1350, vol. 13. Washington, DC: USNRC.

Tompkins, Elizabeth S. 2004. "World List of Nuclear Power Plants." Nuclear News 47, no. 3 (March), 35–60.

Valberg, Peter A. 2001. "Do Power-Line Electric and Magnetic Fields (EMF) Affect Health?" *Trends in Risk and Remediation* (September), 4–5. Cambridge, MA: Gradient Corporation.

Waters, Martha, T. F. Bloom, and B. Grajewski. 2000. "The NIOSH/FAA Working Women's Health Study: Evaluation of the Cosmic-Radiation Exposures of Flight Attendants." *Health Physics* 79, no. 5 (November), 553–559.

Westerman, Bryan R. 2002. "Manufacturing Influences on Computed Tomography Dose." In *Proceedings of Symposium on Computed Tomography: Patient Dose* (November), 11. Bethesda, MD: NCRP.

13. Environmental Economics

Ayensu, E., D. van R. Claasen, M. Collins, A. Dearing, L. Fresco, M. Gadgil, H. Gitay, G. Glaser, C. Juma, J. Krebs, R. Lenton, J. Lubchenco, J. A. McNeely, H. A. Mooney, P. Pinstrup-Andersen, M. Ramos, P. Raven, W. V. Reid, C. Samper, J. Sarukhan, P. Schei, J. G. Tundisi, R. T. Watson, X. Guanhua, and

A. H. Zakri. 1999. "International Ecosystem Assessment." *Science* 286, no. 5440 (22 October), 68.

Day, Felicia, ed. 2001. "Halting the Worldwide 'Race to Fish'—RFF Researchers Evaluating New Zealand's Tradable Quota System for Fisheries as a Model Approach." *Resources,* no. 124 (Summer), pp. 3–4. Washington, DC: RFF.

Dorfman, Robert, and Nancy S. Dorfman. 1977. "Introduction." In Robert Dorfman and Nancy S. Dorfman, eds., *Economics of the Environment: Selected Readings,* 2nd ed., pp. 1–37. New York: W. W. Norton & Company.

Evans, John, Jonathan Levy, James Hammit, Carlos Santos-Burgoa, and Margarita Castillejos. 2002. "Health Benefits of Air Pollution Control." In Luisa T. Molina and Mario J. Molina., eds., *Air Quality in the Mexico Megacity,* chapter 4, pp. 105–136. Dordrecht: Kluwer Academic Publishers.

Ferraro, P. J., and R. D. Simpson. 2001. "Cost-Effective Conservation: A Review of What Works to Preserve Biodiversity." *Resources,* no. 143 (Spring), 17–20. Washington, DC: RFF.

Frederick, K. D. 2001. "Water Marketing: Obstacles and Opportunities." *Forum for Applied Research and Public Policy* 16, no. 1 (Spring), 54.

Freeman, A. M., III. 1982. "Risk Evaluation in Environmental Regulation." In W. A. Magat, ed., *Reform of Environmental Regulation,* pp. 47–69. Cambridge, MA: Ballinger Publishing Company.

Graham, John. 2001. Personal communication. Harvard School of Public Health, Boston (May).

Hammit, J. K. 2000. "Valuing Lifesaving: Is Contingent Valuation Useful?" *Risk in Perspective* 8, no. 3 (March), 1–6. Center for Risk Analysis, Harvard School of Public Health.

Hardin, Garrett. 1968. "The Tragedy of the Commons." *Science* 162, no. 3859 (13 December) 1243–1248.

Henriquez, B. P. 2004. "Information Technology: The Unsung Hero of Market-Based Environmental Policies." *Resources* 152 (Fall/Winter), 9–12. Washington, DC: Resources for the Future.

Johnson, Nels, Carmen Revenga, and Jaime Echeverria. 2001. "Managing Water for People and Nature." *Science* 292, no. 5519 (11 May), 1071–1072.

Mendelsohn, Robert. 2002. "Environmental Economics and Human Health" (editorial). *Environmental Health Perspectives* 110, no. 3 (March), A 118–A 119.

Mock, Gregory, and Wendy G. Vanassect. 2001. "Ecosystems: Valuing the Invaluable." *National Geographic* 200, no. 5 (November), frontispiece.

Oates, Wallace E. 1999. "Forty Years in an Emerging Field—Economics and Environmental Policy in Retrospect." *Resources,* no. 137 (Fall), 8–11. Washington, DC: RFF.

Pigou, A. C. 1932. *The Economics of Welfare.* London: Macmillan.

Portney, Paul R. 2000a. "Air Pollution Policy." In Paul R. Portney and Robert N. Stavins, eds., *Public Policies for Environmental Protection,* 2nd ed., pp. 77–123. Washington, DC: RFF.

———. 2000b. "EPA and the Evolution of Federal Regulation." In Paul R. Portney and Robert N. Stavins, eds., *Public Policies for Environmental Protection,* 2nd ed., pp. 11–30. Washington, DC: RFF.

Powell, Alvin. 2002. "Stavins Steps Down but Not Out." *Harvard University Gazette* 98, no. 8 (7 November), 9.

Roberts, M. J. 1982. "Some Problems of Implementing Marketable Pollution Rights Schemes: The Case of the Clean Air Act." In W. A. Magat, ed., *Reform of Environmental Regulation,* pp. 93–117. Cambridge, MA: Ballinger Publishing Company.

Ruff, Larry E. 1977. "The Economic Common Sense of Pollution." In Robert Dorfman and Nancy S. Dorfman, eds., *Economics of the Environment: Selected Readings,* 2nd ed., pp. 41–58. New York: W. W. Norton & Company.

Ruffing, Kenneth G. 1999. "Achieving Sustainability." *Forum for Applied Research and Public Policy* 14, no. 4 (Winter), pp. 20–24.

Sedjo, R. A. 1999. "The Unkindest Cut." *Resources,* no. 136 (Summer), 5–6. Washington, DC: RFF.

Watkins, John P. 2000. "Environmental Economics." In Craig N. Allin, ed., *Encyclopedia of Environmental Issues,* vol. 1, pp. 278–281. Pasadena, CA, and Hackensack, NJ: Salem Press.

Webster, Paul. 2002. "Russia Can Save Kyoto, If It Can Do the Math." *Science* 296, no. 5576 (21 June), 2129–2130.

14. Environmental Law

Associated Press. 2002. "Bush Takes Up Review of Landmark Environmental Law." Washington, DC (30 August).

Bingham, Eula. 1992. "The Occupational Safety and Health Act." In William N. Rom, ed., *Environmental and Occupational Medicine,* 2nd ed., pp. 1325–1331. Boston: Little, Brown & Company.

CEQ. 1997. *Environmental Quality: The Twenty-fifth Anniversary Report of the Council on Environmental Quality.* Washington, DC: Executive Office of the President.

CRS. 2001. "Environmental Laws: Summaries of Statutes Administered by the Environmental Protection Agency." CRS Report to Congress, Order Code RL30798. Washington, DC (4 January).

Doniger, David D. 2003. "Not Exactly Clear Skies." *EM* (April), 20–21.

EPA. 1998. "Drinking Water Contaminant Candidate List." *Federal Register* 63 (2 March), 10274–10287. Washington, DC.

Ferry, Steven. 2001. *Environmental Law: Examples and Explanations.* 2nd ed. New York: Aspen Law & Business.

Findley, R. W., and D. A. Farber. 1992. *Environmental Law in a Nutshell.* 3rd ed. St. Paul, MN: West Publishing Company.

———. 2000. *Environmental Law in a Nutshell.* 5th/ed. St. Paul, MN: West Group Publishing Company.

Freeman, A. Myrick, III. 2000. "Water Pollution Policy." In Paul R. Portney and Robert N. Stavins, eds., *Public Policies for Environmental Protection,* 2nd ed., pp. 169–213. Washington, DC: RFF.

Hutt, Peter, and Richard Merrill. 1991. *Food and Drug Law.* 2nd ed. Westbury, NY: Foundation Press.

Isler, Margaret, and Martin R. Lee. 2002. "Environmental Protection Issues in the

107th Congress (Updated July 3, 2002)." Congressional Research Services. Washington, DC: Library of Congress.
McGinn, Anne Platt. 2001. "EU Considers Tougher Stance on Chemicals." *World-Watch* 14, no. 2 (March/April), 9.
Mathai, C. V. 2003, "Multipollutant Legislation for the Electric Power Generation Industry: An Update." *EM* (April), 18–31.
Merrill, Richard A. 1986. "Regulatory Toxicology." In C. D. Klaassen, Mary O. Amdur, and J. Doull, eds., *Casarett and Doull's Toxicology: The Basic Science of Poisons,* 3rd ed., pp. 917–932. New York: Macmillan Publishing Company.
Portney, Paul R. 2000. "Air Pollution Policy." In Paul R. Portney and Robert N. Stavins, eds., *Public Policies for Environmental Protection,* 2nd ed., pp. 77–123. Washington, DC: RFF.
Sigman, Hilary. 2000. "Hazardous Waste and Toxic Substance Policies." In Paul R. Portney and Robert N. Stavins, eds., *Public Policies for Environmental Protection,* 2nd ed., pp. 215–259. Washington, DC: RFF.
USNRC. 1991. *Nuclear Regulatory Legislation, 101st Congress.* Report NUREG-0980, vol. 2, no. 1. Washington, DC: USNRC.
———. 2001. "Disposal of High-Level Radioactive Wastes in a Proposed Geologic Repository at Yucca Mountain, Nevada" (Title 10, CFR, Part 63). Federal Register 66, no. 213 (9 November), 55371–55816. Washington, DC.
———. 2003. *Information Digest, 2003 Edition.* Report NUREG-1350, vol. 15 (June). Washington, DC: USNRC.

15. Standards

BEIR [Committee on the Biological Effects of Ionizing Radiation]. 1972. *The Effects on Populations of Exposure to Low Levels of Ionizing Radiation.* Report no. 1. Washington, DC: National Academy Press.
———. 1980. *The Effects on Populations of Exposure to Low Levels of Ionizing Radiation.* Report no. 3. Washington, DC: National Academy Press.
———. 1990. *The Effects on Populations of Exposure to Low Levels of Ionizing Radiation.* Report no. 5. Washington, DC: National Academy Press.
Clarke, Roger H. 2000. "Control of Low-Level Radiation Exposure: Time for a Change?" Paper presented at the Annual Meeting, NCRP, Crystal City, VA (5–6 April).
Clarke, Roger H. 2001a "Control of Low-Level Radiation Exposure: What is the Problem and How Can it be Solved?" *Health Physics* 80, no. 4 (April), 391–396.
———. 2001b. "Radiological Protection at the Start of the 21st Century." *Radiological Protection Bulletin* 231 (September), 10–12, Chilton, Didcot, United Kingdom: NRPB.
EPA. 2001. "Public Health and Environmental Radiation Protection Standards for Yucca Mountain, NV: Final Rule." U.S. Environmental Protection Agency, Code of Federal Regulations, 40 CFR, Part 197. Federal Register 66, no. 114 12 June), 32074–32135. Washington, DC.

FRC. 1960. *Background Material for the Development of Radiation Protection Standards.* Report no. 1. Washington, DC: HHS.

ICRP. 1977. *Recommendations of the International Commission on Radiological Protection.* Publication 26. Annals of the ICRP 1, no. 3. New York: Pergamon Press.

———. 1985. *Quantitative Bases for Developing a Unified Index of Harm.* Publication 45. Annals of the ICRP 15, no. 3. New York: Pergamon Press.

———. 1991. *1990 Recommendations of the International Commission on Radiological Protection.* Publication 60. Annals of the ICRP 21, nos. 1–3. New York: Pergamon Press.

———. 1996. *Age-Dependent Doses to Members of the Public from Intake of Radionuclides: Part 5—Compilation of Ingestion and Inhalation Dose Coefficients.* Publication 72. Annals of the ICRP 26, no. 1. New York: Pergamon Press.

———. 2003. *A Framework for Assessing the Impact of Ionising Radiation on Nonhuman Species.* Publication 91. Annals of the ICRP 33, no. 3. New York: Pergamon Press.

Muller, Hermann J. 1927. "Artificial Transmutation of the Gene." *Science* 66, no. 1699 (11 July), 84–87.

NCRP. 1971. *Basic Radiation Protection Criteria.* Report no. 39. Bethesda, MD: NCRP.

———. 1993. *Limitation of Exposure to Ionizing Radiation.* Report no. 116. Bethesda, MD: NCRP.

Rowland, R. E. 1994. *Radium in Humans: A Review of U.S. Studies.* Argonne National Laboratory. Springfield, VA: U.S. Department of Commerce.

USNRC. 1991. "Standards for Protection against Radiation." CFR, Title 10, part 20. Washington, DC.

16. Monitoring

Abelquist, Eric W. 2003. "Applications of MARSSIM" (Editorial). Operational Radiation Safety, Supplement to *Health Physics* 84, no. 6 (June), S97–S99.

Achard, F., H. D. Eva, H.-J. Stibig, P. Mayaux, J. Gallego, T. Richards, and J.-P. Malingreau. 2002. "Determination of Deforestation Rates of the World's Humid Tropical Forests." *Science* 297, no. 5583 (9 August), 999–1002.

Beaugelin-Seiller, K., P. Boyer, J. Garnier-Laplace, and C. Adam. 2002. "CASTEAUR: A Simple Tool to Assess the Transfer of Radionuclides in Waterways." *Health Physics* 83, no. 4 (October), 539–542.

Brown, Kathryn. 2002. "Water Scarcity: Forecasting the Future with Spotty Data." *Science* 297, no. 5583 (9 August), 926–927.

Clesceri, Lenore S., Arnold E. Greenberg, and Andred D. Eton, eds. 1998. *Standard Methods for the Examination of Water and Waste Water.* 20th ed. Waldorf, MD: APHA, AWWA, and WEF.

Ebersold, Peter J., and Nicholas Barker. 2003. "Having a Field Day," *Environmental Protection* 14, no. 3 (April), 45–49.

EPA. 1993. *R-EMAP: Regional Environmental Monitoring and Assessment Program.* Report EPA/625/R-93/012. Washington, DC: EPA.

———. 2001. "Measuring Pollution with Infrared Rays." *NRMRL News* (September), 1. Washington, DC: National Risk Management Research Laboratory.

EPA and USNRC. 1997. *Multi-agency Radiation Survey and Site Investigation Manual (MARSSIM).* Report EPA 402-R-97-016; Report NUREG-1575. Washington, DC: EPA.

EPA, USNRC, DOE, DOD, NIST, USGS, FDA, Kentucky, and California. 2001. *Multi-agency Radiological Laboratory Analytical Protocols Manual (MARLAP).* Report EPA 402-B-01-2001; Report NUREG-1576; Report NIST PB2001-106745. Washington, DC: EPA.

Gray, Donald H., and Sally C. Ballard. 2000. *EEG Operational Surveillance of the WIPP Project during 2000.* Report No. EEG-81. Carlsbad, NM: Environmental Evaluation Group, State of New Mexico.

Griffith, Jerry A., and Carolyn T. Hunsaker. 1994. *Ecosystem Monitoring and Ecological Indicators: An Annotated Bibliography.* Report EPA/620/R-94/021. Washington, DC: EPA.

Hallisey, Robert M. 2001. Personal communication. Radiation Control Program, Commonwealth of Massachusetts, Boston (5 April).

Howard, H. T., K. H. Pacific, and J. A. Pacific. 2002. "The Evolution of Remote Sensing." *Environmental Protection* 13, no. 4 (April), 28, 30–34.

Lambert, Kathy F., and Van Bowersox. 2002. "Environmental Monitoring and National Security: Is There a Connection?" *EM* (August), 17–22.

Malakoff, David. 2002. "Microbiologists on the Trail of Polluting Bacteria." *Science* 295, no. 5564 (29 March), 2352–2353.

NCRP. 1996. *Screening Models for Releases of Radionuclides to Atmosphere, Surface Water, and Ground.* Report no. 123 I. Bethesda, MD: NCRP.

Newman, Alan. 1995. "Major U.S. Human Exposure Assessment Survey Gets under Way." *Environmental Science and Technology* 29, no. 9 (September), 398A–399A.

NRC. 2000. "Ecological Indicators for the Nation." Board on Environmental Studies and Toxicology, and Water Science and Technology Board. Washington, DC: National Academy Press.

Weinhold, Bob. 2002. "Air Pollution: U.S. Air Only Fair." *Environmental Health Perspectives* 110, no. 8 (August), A 452.

Wilkins, Bernard. 1999. "Pigeons—A Novel Form of Airborne Radionuclides." *Radiological Protection Bulletin,* no. 215 (September), 7–11. Chilton, DidCot, Oxfordshire, United Kingdom: NRPB.

Wing, S., S. Freedman, and L. Band. 2002. "The Potential Impact of Flooding on Confined Animal Feeding Operations in Eastern North Carolina." *Environmental Health Perspectives* 110, no. 4 (April), 387–391.

17. Risk Assessment

ATSDR. 2003. *The Agency for Toxic Substances and Disease Registry: FY2002 Profile and Annual Report.* Report ATSDR-PE-RP-2003-0001. Atlanta, GA: CDC.

Beck, Barbara D. 2001. "International Risk Policy." *Trends in Risk and Remediation* (Spring), pp. 1–2. Cambridge, MA: Gradient Corporation.

Borchardt, K. D. 2000. *The ABC (?) of Community Law.* Brussels, Belgium: European Commission.

CDC. 2002. "Probable Variant Creutzfeldt-Jakob Disease in a U.S. Resident—Florida, 2002." *Morbidity and Mortality Weekly Report* 51, no. 41 (18 October), 927–929.

EPA. 1990. *Future Risk: Research Strategies for the 1990s.* Report no. SAB–EC–88–040. Science Advisory Board, The Research Strategies Committee. Washington, DC: EPA.

———. 1993. *Wildlife Exposure Factors Handbook.* Report no. EPA/600/R–93/187, vols. 1–2. Office of Research and Development. Washington, DC: EPA. Available at *http://cfpub.epa.gov/ncea/cfm/wefh.cfm.*

———. 1998. "Guidelines for Ecological Risk Assessment." Federal Register 63, no. 93 (14 May), 26845–26924. Washington, DC: EPA.

———. 2002. *NCEA's Ecological Assessment Program.* Washington, DC: National Center for Environmental Assessment. http://cfpub.epa.gov/ncea/efm/ecologic.efm.

Federal Register. 1993. "The President: Executive Order 12866—Regulatory Planning and Review." *Federal Register* 58, no. 190 (4 October), 51735–51744. Washington, DC.

Gage, Stephen D. 2001. "Harmonizing Controls for Chemicals and Radionuclides." *Health Physics* 80, no. 4 (April), 388–389.

Graham, John D. 1993. "The Legacy of One in a Million." *Risk in Perspective* 1, no. 1 (March), 1–2. Center for Risk Analysis, Harvard School of Public Health.

———. 1999. "Making Sense of the Precautionary Principle." *Risk in Perspective* 7, no. 6 (September), 1–6.

Gray, G., J. Cohen, and S. Kreindel. 2002. "Evaluating the Risk of Bovine Spongiform Encephalopathy in the United States." *Risk in Perspective* 10, no. 2 (March), 1–6. Center for Risk Analysis, Harvard School of Public Health.

Johnson, B. L. 1992. "Principles of Chemical Risk Assessment: The ATSDR Perspective." In *Proceedings of a Conference on Chemical Risk Assessment in the DoD: Science Policy and Practices.* pp. –35. Cincinnati, OH: ACGIH.

Kaplan, Stanley, and B. John Garrick. 1981. "On the Quantitative Definition of Risk." *Risk Analysis* 1, no. 1, 11–27.

Macauley, M., K. Palmer, J.-S. Shih, C. Cline, and H. Holsinger. 2001. "The Environment and the Information Age." *Resources* 145 (Fall), 6–9. Washington, DC: Resources for the Future.

NCRP. 2002. *Risk-Based Classification of Radioactive and Hazardous Chemical Wastes.* Report no. 139. Bethesda, MD: NCRP.

NRC. 1994. *Building Consensus through Risk Assessment and Management of the Department of Energy's Environmental Remediation Program.* Committee to Review Risk Management in the DOE's Environmental Remediation Program. Washington, DC: National Academy Press.

———. 1996. *Understanding Risk: Informing Decisions in a Democratic Society.* Committee on Risk Characterization. Washington, DC: National Academy Press.

Rao, Anand. 2003. "World's Largest Wind Farm Planned for Iowa." *WorldWatch* 16, no. 4 (July/August), 9.

Rhomberg, Lorenz R. 2001. "Parsing the Precautionary Principle." *Trends in Risk and Remediation* (Spring), 4. Cambridge, MA: Gradient Corporation.

Ropeik, David, and George Gray. 2002. *Risk: A Practical Guide for Deciding What's Really Safe and What's Dangerous in the World around You.* Boston: Houghtin Mifflin.

Ropeik, David, and Paul Slovic. 2003. "Risk Communication: A Neglected Tool in Protecting Public Health". *Risk in Perspective* 11, Issue 2 (June), 1–6. Boston, MA: Harvard Center for Risk Analysis, Harvard School of Public Health.

Schmidt, Charles W. 1999. "A Closer Look at Chemical Exposures in Children." *Environmental Science and Technology* 4, no. 2 (1 February), 72A–75A.

USNRC. 1962. "Reactor Site Criteria." CFR, Title 10, part 100, Appendix A. Washington, DC.

———. 1975. *Reactor Safety Study: An Assessment of Accident Risks in U.S. Commercial Nuclear Power Plants.* Report WASH-1400, NUREG-75/014. Washington, DC: USNRC.

———. 2001. "Disposal of High-Level Radioactive Wastes in a Geologic Repository at Yucca Mountain, Nevada." Code of Federal Regulations, Title 10, Part 63, Washington, DC.

Wiener, Jonathan Baert, and John D. Graham. 1995. "Resolving Risk Tradeoffs." In John D. Graham and Jonathan Baert Wiener, eds, *Risk vs. Risk: Tradeoffs in Protecting Health and the Environment,* pp. 226–271. Cambridge, MA: Harvard University Press.

Wilson, Richard, and Edmund A. C. Crouch. 2001. *Risk-Benefit Analysis.* 2nd ed. Cambridge, MA: Harvard Center for Risk Analysis.

18. Energy

Blankinship, Steve. 2002. "Update: AWEA Conference Reflects Record U.S. Wind Capacity Installation." *Power Engineering* 106, no. 7 (July), 20.

Bolig, Andy. 2003. "Precisely Controlled Power." *Corvette Fever* 25, no. 5 (May), 78–79.

Clery, D., and D. Norman. 2004. "Compromise Deal Hinges on a Graceful Runner-Up." *Science* 303, no. 5660 (13 February), 940.

Consumers Union. 2002. "Washing Machines." *Consumer Reports* 67, no. 7 (July), 40–41.

Fairley, Peter. 2002. "Solar on the Cheap." *Technology Review* 105, no. 1 (January/February), 48–53.

Fies, Brian. 2001. "Photovoltaics Rising—Beyond the First Gigawatt." *EPRI Journal* 26, no. 1 (Summer), 13–20.

Grant, Lee. 2002. "The Drive for Better Fuel Economy." *Technology Today* 23, no. 2 (Fall), 16–19. San Antonio, TX: Southwest Research Institute.

Greene, Katie. 2001. "Unleash the Bugs." *Popular Science* 259, no. 5 (November), 34.

Gritzinger, Bob. 2004. "Something Old, Something New." *AutoWeek* 54, issue 2 (January 12), 6.

Hamilton, Anita. 2002. "Driving by Wire: The Car of the Future Looks Something like This." *Time Magazine* 160, no. 21 (18 November), 74–75.

Holden, Constance ed. 2004. "Batting at Windmills." *Science* 304, no. 5668 (9 April), 203.

Kosowatz, John J. 1999. "Mighty Monolith: Scientific American Presents Extreme Engineering." *Scientific American* 10, no. 4 (Winter), 15–23.

Lloyd, Alan C. 1999. "The Power Plant in Your Basement." *Scientific American* 281, no. 1 (July), 80–86.

Malakoff, David. 2002. "Lighting Initiative Flickers to Life." *Science* 296, no. 5574 (7 June), 1782.

Mettler, F. A., W. K. Sinclair, L. Anspaugh, C. Edington, J. H. Harley, R. C. Ricks, P. B. Selby, E. W. Webster, and H. O. Wyckoff. 1990. "The 1986 and 1988 UNSCEAR Reports: Findings and Implications." *Health Physics* 58, no. 3 (March), 241–250.

Raven, Peter H. 2002. "Science, Sustainability, and the Human Prospect." *Science* 297, no. 5583 (9 August), 954–958.

Roosevelt, Margot. 2002. "Not in My Back Bay." *Time Magazine* 160, no. 14 (30 September), 62.

Sawin, Janet. 2003a. "European Wind Energy Production Reaches New Highs." *WorldWatch* 16, no. 3 (May/June), 8.

———. 2003b. "Charting a New Energy Future." In *State of the World, 2003,* pp. 85–109. WorldWatch Institute. New York: W. W. Norton & Company.

Schultz, M. G., T. Diehl, G. P. Brasseur, and W. Zittel. 2003. "Air Pollution and Climate-Forcing Impacts of a Global Hydrogen Economy." *Science* 302, no. 5645 (24 October), 624–627.

Service, Robert F. 2002. "Shrinking Fuel Cells Promise Power in Your Pocket." *Science* 296, no. 5571 (17 May), 1222–1224.

———. 2003. "Solar Cells—Tricks for Beating the Heat Help Panels See the Light." *Science* 300, no. 5623 (23 May), 1219.

Staedter, Tracy. 2002. "Wave Power: How the Rolling Sea Generates Electricity." *Technology Review* 105, no. 1 (January/February), 86–87.

Starr, Chauncey. 2002. "National Energy Planning for the Century: The Continental SuperGrid." *Nuclear News* 45, no. 2 (February), 31–35.

Truly, Richard H. 2002. "New Energy Frontiers." *Bridge* 32, no. 2 (Summer), 5–10.

Voss, David. 2002. "Hitting the Natural-Gas Jack-Pot." *Technology Review* 105, no. 1 (January/February), 68–72.

Zauderer, Bert. 2002. "Global Warming Debate" (letter to editor). *Power Engineering* 106, no. 6 (June), 6.

19. Disasters

Abramovitz, J. N. 2001. "Averting Unnatural Disasters." In *State of the World, 2001,* pp. 123–142. WorldWatch Institute. New York: W. W. Norton & Company.

AMA. 2002. "Bioterrorism Awareness: A Definitive Resource for State-of-the-Art Medical and Clinical Information for Responding to Biological or Chemical Attack." CD-ROM. Chicago: AMA.

ATSDR. 1994. "Emergency Incident Risk Communication: The Cantara Loop Spill." *Hazardous Substances and Public Health* 3, no. 4 (January), 1–4.

Bohannon, John, and Xavier Bosch. 2003. "Spanish Researchers Vent Anger over Handling of Oil Spill." *Science* 299, no. 5606 (24 January), 490.

Bosch, Xavier. 2003. "For Spain, Oil Spill Disaster Is in the Bag." *Science* 302, no. 5650 (28 November, 2003), 1485.

CDC. 1997a. "Tornado-Associated Fatalities—Arkansas, 1997." *Morbidity and Mortality Weekly Report* 46, no. 19 (16 May), 412–416.

———. 1997b. "Tornado Disaster—Texas, May 1997." *Morbidity and Mortality Weekly Report* 46, no. 45 (14 November), 1069–1073.

———. 2000. "Biological and Chemical Terrorism: Strategic Plan for Preparedness and Response: Recommendations of the CDC Strategic Planning Workgroup." *Morbidity and Mortality Weekly Report* 49, no. RR-4 (21 April), 1, 5–6.

CEQ. 1998. *Environmental Quality: The World Wide Web: The 1997 Report of the Council on Environmental Quality.* Washington, DC: Executive Office of the President.

Chapin, Douglas M., Karl P. Cohen, W. Kenneth Davis, Edwin E. Kintner, Leonard J. Koch, John W. Landis, Milton Levenson, I. Harry Mandil, Zack T. Pate, Theodore Rockwell, Alan Schreisheim, John W. Simpson, Alexander Squire, Chauncey Starr, Henry E. Stone, John J. Taylor, Neil E. Todreas, Bertram Wolfe, and Edwin L. Zebroski. 2002. "Nuclear Power Plants and Their Fuel as Terrorist Targets." *Science* 297, no. 5589 (20 September), 1997, 1999.

Chin, James, ed. 2000. *Control of Communicable Diseases Manual.* 17th ed. Washington, DC: APHA.

Fleming, Melissa. 2002. "International Symposium Focuses on Global Nuclear Terrorism." *Health Physics* 82, no. 1 (January), 120–121.

Gore, Rick. 2000. "Earthquake in Turkey." *National Geographic* 198, no. 1 (July), 36–51.

Kennedy, Donald. 2002. "Science, Terrorism, and Natural Disasters" (editorial). *Science* 295, no. 5554 (18 January), 405.

Luterbacher, J., D. Dietrich, E. Xoplaki, M. Grosjean, and H. Wanner. 2004. "European Seasonal and Annual Temperature Variability, Trends, and Extremes Since 1500." *Science* 303, no. 5663 (5 March), 1499–1503.

Mitchell, John. 1999. "In the Wake of the Spill—Ten Years after *Exxon Valdez.*" *National Geographic* 195, no. 3 (March), 96–117.

NCRP. 2001. *Management of Terrorist Events Involving Radioactive Material.* Report no. 138. Bethesda, MD: NCRP.

Neville, Angela. 2001. "Seeing Community Right-to-Know Laws in a New Light" (editorial). *Environmental Protection* 12, no. 12 (December), 6.

NIOSH. 2003. *Guidance for Filtration and Air-Cleaning Systems to Protect Building Environments from Airborne Chemical, Biological, or Radiological Attacks.* DHHS (NIOSH) Publication No. 2003–136. Cincinnati, OH: National Institute for Occupational Safety and Health.

NRC. 2001. *Environmental Performance of Tanker Designs in Collision and Grounding: Method for Comparison.* Special Report 259, Transportation Research Board. Washington, DC: National Academy Press.

———. 2002. *Making the Nation Safer: The Role of Science and Technology in Countering Terrorism.* Washington, DC: National Academy Press.

NSC. 2001. *Injury Facts, 2001 Edition.* Itasca, IL: NSC.

PAHO. 2002. *Humanitarian Assistance in Disaster Situations—A Quick Guide for Effective Donations.* Washington, DC: PAHO.

Parfit, Michael. 1998. "Living with Natural Disasters." *National Geographic* 194, no. 1 (July), 2–39.

Peterson, C. H., S. D. Rice, J. W. Short, D. Esler, J. L. Bodkin, B. E. Ballachey, and D. B. Irons. 2003. "Long-Term Ecosystem Response to the Exxon Valdez Oil Spill." *Science* 302, no. 5653 (19 December), 2082–2086.

Ring, Joseph P. 2004. "Radiation Risks and Dirty Bombs." Operational Radiation Safety, Supplement to *Health Physics.* 86: 2 (February), S42–S47.

Robock, Alan. 2002. "The Climatic Aftermath." *Science* 295, no. 5558 (15 February), 1242–1244.

Service, Robert F. 2003. "Chemical Studies of 9/11 Disaster Tell Complex Tale of 'Bad Stuff'." *Science* 301, no. 5640 (19 September), 1649.

Simons, Lewis M. 1998. "Indonesia's Plague of Fire." *National Geographic* 194, no. 2 (August), 100–119.

Taylor, Gregg M. ed., 2004. "Potassium Iodine Should be Made Available to People." *Nuclear News* 47, no. 1 (January), 63.

UNSCEAR. 2000. *Sources and Effects of Ionizing Radiation: United Nations Scientific Committee on the Effects of Atomic Radiation: UNSCEAR 2000 Report to the General Assembly with Scientific Annexes.* New York: UN.

Vesilind, Priit J. 2004. "Chasing Tornadoes." *National Geographic* 205, no. 4 (April), 2–37.

20. A Global View

Albert, Mary R. 2004. "The International Polar Year" (Editorial). *Science* 303, no. 5663 (5 March), 1437.

AMA. 1989. *Stewardship of the Environment.* Council on Scientific Affairs, Report G, 1–89. Chicago: AMA.

Arendt, A. A., K. A. Echelmeyer, W. D. Harrison, C. S. Lingle, and V. B. Valentine. 2002. "Rapid Wastage of Alaska Glaciers and Their Contribution to Rising Sea Level." *Science* 297, no. 5580 (19 July), 382–386.

Balmford, A., A. Bruner, P. Cooper, R. Costanza, S. Farber, R. E. Green, M. Jenkins, P. Jefferiss, V. Jessamy, J. Madden, K. Munro, N. Myers, S. Naeem, J. Paavola, M. Rayment, S. Rosendo, J. Roughgarden, K. Trumper, and R. K. Turner. 2002. "Economic Reasons for Conserving Wild Nature." *Science* 297, no. 5583 (9 August), 950–953.

Beaugrand, Gregory, Philip C. Reid, Frederic Ibanez, J. Alistair Lindley, and Martin Edwards. 2002. "Reorganization of North Atlantic Marine Copepod Biodiversity and Climate." *Science* 296, no. 5573 (31 May), 1692–1694.

Bugliarello, G. 2001. "Rethinking Urbanization." *Bridge* 31, no. 1 (Spring), 5–12.

Carpenter, Betsy. 2002. "Sound and Fury." *U.S News and World Report* 133, no. 24 (23 December), 50–92.

CEQ. 1997. *Environmental Quality: The Twenty-fifth Anniversary Report of the Council on Environmental Quality.* Washington, DC: Executive Office of the President.

Chin, Gilbert. 2002. "Oceanography—Limits to Growth." Editorial. *Science* 297, no. 5578 (5 July), 15.

Fitter, A. H., and R. S. R. Fitter. 2002. "Rapid Changes in Flowering Time in British Plants." *Science* 296, no. 5573 (31 May), 1689–1691.

Gardner, Gary. 2003. "Engaging Religion in the Quest for a Sustainable World." In *State of the World, 2003,* pp. 152–175. WorldWatch Institute. New York: W. W. Norton & Company.

Greenberg, Michael. 2001. "Earth Day Plus 30 Years: Public Concern and Support for Environmental Health." *American Journal of Public Health* 91, no. 4 (April), 559–562.

Greene, Katie. 2002. "Bigger Populations Needed for Sustainable Harvests." *Science* 296, no. 5571 (17 May), 1229–1230.

Gruebel, Marilyn. 2002. Personal communication. Albuquerque, NM (December).

Harvell, C. Drew, Charles E. Mitchell, Jessica R. Ward, Sonia Altizer, Andrew P. Dobson, Richard S. Ostfeld, and Michael D. Samuel. 2002. "Climate Warming and Disease Risks for Terrestrial and Marine Biota." *Science* 296, no. 5576 (21 June), 2158–2162.

Hayden, Thomas. 2001. "Deep Trouble—Overfishing Has Torn the Sea's Web of Life—Mending It Won't Be Easy." *U.S. News and World Report* 131, no. 9 (10 September), 68–70.

Hector, Andy, and Rowan Hooper. 2002. "Darwin and the First Ecological Experiment." *Science* 295, no. 5555 (25 January), 639–640.

Hurtley, Stella, ed. 2003. "Meltdown in an Island Forest." *Science* 302, no. 5642 (3 October), 21.

Johnson, Jeff. 1995. "EPA Must Look to the Future, Says Science Advisory Board." *Environmental Science and Technology* 29, no. 3 (March), 112A–113A.

Kaiser, Jocelyn. 2001. "Bold Corridor Project Confronts Political Reality." *Science* 293, no. 5538 (21 September), 2196–2197, 2199.

Karn, Barbara P., and D. Bauer. 2001. "Sustainability: Insight from Industrial Ecology" (letter to editor). *Science* 293, no. 5537 (14 September), 1995–1996.

Kearney, Bill. 2001. "A Wetland Gained for a Wetland Lost?" *National Academies INFOCUS* 1, no. 2 (Fall/Winter), 17–18. Washington, DC: NAS.

———. 2002. "Bringing Back the Big Muddy." *National Academies INFOCUS* 2, no. 1 (Spring), 15–16. Washington, DC: NAS.

Kennedy, Donald. 2002. "POTUS and the Fish." Editorial. *Science* 297, no. 5581 (26 July), 477.

Kerr, Richard A. 2002. "A Warmer Arctic Means Change for All." *Science* 297, no. 5586 (30 August), 1490–1492.

Leslie, Mitch. 2002. "Trouble in Paradise," *Science* 295, no. 5558 (15 February), 1199.

Malakoff, David. 2003. "Judge Blocks Navy Sonar Plan." *Science* 301, no. 5638 (5 September), 1305.

Myers, Norman. 2002. "A Convincing Call for Conservation" (Review of *The Future of Life* by Edward O. Wilson [New York: Knopf, 2002]). *Science* 295, no. 5554 (18 January), 447–448.

NRC. 2001. *Climate Change Science: An Analysis of Some Key Questions.* Washington, DC: National Academy Press.

O'Neill, Brian C., and Michael Oppenheimer. 2002. "Dangerous Climate Impacts and the Kyoto Protocol." *Science* 296, no. 5575 (14 June), 1971–1972.

Penuelas, Josep, and Iolanda Filella. 2001. "Responses to a Warming World." *Science* 294, no. 5543 (26 October), 793, 795.

Powell, Alvin. 2002. "Agreeing on What to Argue About: Environmental Report Tries to Make Sense of Flood of Data." *Harvard University Gazette* 98, no. 2 (25 September), 27–28.

Prugh, Thomas, and Erik Assadourian. 2003. "What Is Sustainability, Anyway?" *World-Watch* 16, no. 2 (September/October), 10–21.

Reisner, Marc. 2000. "Unleash the Rivers." *Time Magazine,* Earth Day 2000, Special Edition 155, no. 17 (April–May), 66–71.

Sachs, Jeffrey D. 2004. "Sustainable Development." Editorial. *Science* 304, no. 5671 (30 April), 649.

Schmidt, Charles W. 2002. "The Down-to-Earth Summit—Lessening Our Ecological Footprint." *Environmental Health Perspectives* 110, no. 11 (November), A 682–A 685.

Schmidt, Karen F. 2001. "A True-Blue Vision for the Danube." *Science* 294, no. 5546 (16 November), 1444–1445, 1447.

Shouse, Ben. 2002. "U.S. Environment: Report Takes Stock of Knowns and Unknowns." *Science* 297, no. 5590 (27 September), 2191.

Starr, Chauncey. 1997. "Sustaining the Human Environment: The Next Two Hundred Years." In Jesse H. Ausubel and H. Dale Langford, eds., *Technological Trajectories and the Human Environment,* pp. 185–198. Washington, DC: National Academy Press.

Stegeman, John J., and Andrew R. Solow. 2002. "Environmental Health and the Coastal Zone." *Environmental Health Perspectives* 110, no. 11 (November), A 660–A 661.

Thigpen, David E. 2002. "The Fight over Big Muddy's Flow." *Time Magazine* 160, no. 2 (8 July), 72.

Tibbetts, John. 2002. "Coastal Cities—Living on the Edge." *Environmental Health Perspectives* 110, no. 11 (November), A 674–A 681.

Verissimo, Adalberto, Mark A. Cochrane, and Carlos Souza Jr. 2002. "National Forests in the Amazon." *Science* 297, no. 5586 (30 August), 1478.

Wilson, Edward O. 2000. "Biodiversity—Vanishing before Our Eyes." *Time Magazine,* Earth Day 2000, Special Edition 155, no. 17 (April–May), 29–34.

———. 2002. "Hot Spots—Preserving Pieces of a Fragile Biosphere." *National Geographic* 201, no. 1 (January), 86–89.

Withgott, Jay. 2002. "California Tries to Rub Out the Monster of the Lagoon." *Science* 295, no. 5563 (22 March), 2201–2202.

CREDITS

Tables

Table 1.1 Developed by author.

Table 1.2 Based on Harvard University, *Harvard Report on Cancer Prevention,* (Boston: Center for Cancer Prevention, 1996).

Table 1.3 Based on U.S. Department of Health and Human Services, *Healthy People 2010: Understanding and Improving Health,* 2nd ed. (Washington, DC: Government Printing Office, November 2000).

Table 1.4 Centers for Disease Control and Prevention, "Ten Great Public Health Achievements—United States, 1900–1999," *Morbidity and Mortality Weekly Report* 48, no. 12 (2 April 1999), 241–243.

Table 2.1 C. D. Klaassen, "Principles of Toxicology," in C. D. Klaassen, Mary O. Amdur, and J. Doull, eds., *Casarett and Doull's Toxicology: The Basic Science of Poisons,* 3rd ed. (New York: Macmillan Publishing Company, 1986), table 2-1, p. 12. Adapted with permission of The McGraw-Hill Companies.

Table 2.2 C. D. Klaassen, "Principles of Toxicology," in C. D. Klaassen, Mary O. Amdur, and J. Doull, eds., *Casarett and Doull's Toxicology: The Basic Science of Poisons,* 3rd ed. (New York: Macmillan Publishing Company, 1986), table 2-2, p. 13. Adapted with permission of The McGraw-Hill Companies.

Table 2.3 Gail Charnley, lecture outline on "Principles of Toxicology," presentation in course on "Risk," Center for Continuing Professional Education, Harvard School of Public Health, Boston, 2001.

Table 2.4 Gail Charnley, lecture outline on "Principles of Toxicology," presentation in course on "Risk," Center for Continuing Professional Education, Harvard School of Public Health, Boston, 2001.

Table 3.1 From National Research Council (NRC), *Radiation Dose Reconstruction for Epidemiologic Uses* (Washington, DC: National Academy Press, 1995), table 7-2, p. 73.

Table 3.2 Based on National Research Council, *Environmental Epidemiology:*

Public Health and Hazardous Wastes (Washington, DC: National Academy Press, 1991), table 3-4, p. 120.

Table 4.1 Compiled by author.

Table 4.2 Centers for Disease Control and Prevention, "Prevention of Leading Work-Related Diseases and Injuries," *Morbidity and Mortality Weekly Report,* vol. 32, no. 2 (21 January 1983), table 1, p. 25.

Table 4.3 John Tibbetts, "Under Construction: Building a Safer Industry," *Environmental Health Perspectives* 110, no. 3 (March 2002), A 134–A 141.

Table 4.4 ACGIH, *2004 TLVs and BEIs—Based on the Documentation of the Threshold Limit Values for Chemical Substances and Physical Agents & Biological Exposure Indices* (Cincinnati, OH: American Conference of Governmental Industrial Hygienists, 2004).

Table 4.5 ACGIH, *2004 TLVs and BEIs—Based on the Documentation of the Threshold Limit Values for Chemical Substances and Physical Agents and Biological Exposure Indices* (Cincinnati, OH: American Conference of Governmental Industrial Hygienists, 2004).

Table 4.6 ACGIH, *2004 TLVs and BEIs—Based on the Documentation of the Threshold Limit Values for Chemical Substances and Physical Agents and Biological Exposure Indices* (Cincinnati, OH: American Conference of Governmental Industrial Hygienists, 2004).

Table 5.1 Private communication, Tom Helms, EPA (September 2002).

Table 5.2 EPA. *National Air Quality and Emissions Trends Report, 2003 Special Studies Edition.* Chapter 1, Executive Summary (Washington, DC: U.S. Environmental Protection Agency, 2003), p. 2.

Table 6.1 Centers for Disease Control and Prevention, "Diagnosis and Management of Foodborne Illnesses—A Primer for Physicians," *Morbidity and Mortality Weekly Report* 50, no. RR-2 (26 January 2001), 1–69.

Table 6.2 Based on Abram S. Benenson, ed., *Control of Communicable Diseases Manual,* 16th ed. (Washington, DC: American Public Health Association, 1995), p. 184.

Table 6.3 Based on information from Institute of Food Technologists, "Government Regulation of Food Safety: Interaction of Scientific and Societal Forces," *Food Technology* 46, no. 1 (January 1992), 73–80.

Table 6.4 Based on information from Institute of Food Technologists, "Government Regulation of Food Safety: Interaction of Scientific and Societal Forces," *Food Technology* 46, no. 1 (January 1992).

Table 7.1 Prepared by author.

Table 7.2 Lee Harms, Cindy L. Wallis-Lage, and Don Ratnayaka, "What the Future Holds for Water and Wastewater: Disinfection," in *World of Water 2000: The Past, Present, and Future,* supplement to *PennWell Magazine* (Tulsa, OK: WaterWorld and Water & Wastewater International, 1999), 152–156.

Table 7.3 Compiled by author.

Table 7.4 Sandra Postel, "Increasing Water Efficiency," in Lester R. Brown et al., eds., *State of the World,* Worldwatch Institute Report (New York: W. W. Norton, 1986), table 3.6, p. 55. Copyright 1986 by Worldwatch Institute. Reprinted by permission of W. W. Norton and Company, Inc.

Table 8.1 Modification of table in U.S. Environmental Protection Agency, *The Quality of Our Nation's Waters: A Summary of National Water Quality Inventory: 1998* (Washington, DC: 2000), supplemented by information from National Safety Council, *Injury Facts," 2001 Edition* (Itasca, IL: NSC, 2001), p. 147.

Table 8.2 Based on National Research Council, *Managing Wastewater in Coastal Urban Areas,* Committee on Wastewater Management for Coastal Urban Areas (Washington, DC: National Academy Press, 1993), table ES.1, p. 5.

Table 9.1 U.S. Environmental Protection Agency, *Solving the Hazardous Waste Problem: EPA's RCRA Program,* Report EPA/530-SW-86-037 (Washington, DC: EPA, 1986), p. 8.

Table 9.2 Based on U.S. Nuclear Regulatory Commission, *Regulation of the Disposal of Low-Level Radioactive Waste: A Guide to the Nuclear Regulatory Commission's 10 CFR Part 61* (Washington, DC: Office of Nuclear Material Safety and Safeguards, 1989), fig. 1, p. 2a; and idem, *Information Digest: 1995 Edition,* Report NUREG-1350, vol. 7 (Washington, DC: U.S. Government Printing Office, 1989).

Table 9.3 Compiled by the author.

Table 9.4 U.S. Environmental Protection Agency, *RCRA Orientation Manual,* Report EPA/530-SW-86-001 (Washington, DC: EPA, 1986), pp. II-9, III-4, IV-3.

Table 9.5 U.S. Environmental Protection Agency, *Low-Level Mixed Waste: A RCRA Perspective for NRC Licensees,* Report EPA/530-SW-90-057 (Washington, DC: EPA, 1990), p. 23.

Table 9.6 Based on American Institute of Chemical Engineers, "Garbage! The Story of Waste Management and Recycling" (Washington, DC, 1993), unnumbered table, p. 6; Council on Environmental Quality, *Environmental Quality: The World Wide Web: The 1997 Report of the Council on Environmental Quality* (Washington, DC: Executive Office of the President, 1998), tables 8.1 and 8.2, p. 305.

Table 9.7 Compiled by author.

Table 9.8 Compiled by author.

Table 10.1 Based on Pan American Health Organization, *Emergency Vector Control after Natural Disaster,* Scientific Publication no. 419 (Washington, DC: *PAHO,* 1982), p. 18.

Table 10.2 Based on Peter Wehrwein, "Pharmaco-philanthropy," *Harvard Public Health Review* (Summer 1999), 32–39.

Table 11.1 National Safety Council, 2001. *Injury Facts, 2001 Edition,* 46. (Itasca, IL: NSC, 2001).

Table 11.2 National Safety Council, 2001. *Injury Facts, 2001 Edition,* 122. (Itasca, IL: NSC), table "Sports Participation and Injuries, United States, 1999."

Table 12.1 International Commission on Non-Ionizing Radiation Protection, "General Approach to Protection against Non-ionizing Radiation," *Health Physics* 82, no. 4 (April 2002), 540–550, table 1.

Table 12.2 International Commission on Non-Ionizing Radiation Protection, "General Approach to Protection against Non-ionizing Radiation," *Health Physics,* 82, no. 4 (April 2002), 540–550, table 2.

Table 12.3 Compiled by author.

Table 12.4 Based on information in Institute of Medicine, *An Evaluation of Radiation Exposure Guidance for Military Operations,* Interim Report (Washington, DC:

National Academy Press, 1997), National Council on Radiation Protection and Measurements (NCRP), *Management of Terrorist Events Involving Radioactive Materials,* Report no. 138 (Bethesda, MD: NCRP, 2001). October 2001, Appendix A, pp. 154–158.

Table 12.5 Based on Institute of Medicine, *An Evaluation of Radiation Exposure Guidance for Military Operations,* Interim Report (Washington, DC: National Academy Press, 1997).

Table 12.6 Based primarily on information in Wallace Friedberg, Kyle Copeland, Frances E. Duke, Keran O'Brian III, and Edgar B. Darden Jr., "Radiation Exposure during Air Travel: Guidance Provided by the Federal Aviation Administration for Air Carrier Crews," *Health Physics* 79, no. 5 (November 2000), 591–595.

Table 12.7 Based on ICRP, *Protection of the Public in Situations of Prolonged Radiation Exposure,* Publication 82, Annals of the ICRP 29, nos. 1–2 (New York: Pergamon Press, 1999), table A.1, p. 73.

Table 12.8 Compiled by author, using the following sources: ICRP, "Summary of the Current ICRP Principles for Protection of the Patient in Diagnostic Radiology," in *Radiological Protection in Biomedical Research,* ICRP Publication 62, Annals of the ICRP, vol. 22, no. 3 (New York: Pergamon Press, 1993), pp. i–xxiv (for the dose for mammography); ICRP, *Managing Patient Dose in Computed Tomography,* ICRP Publication 87, Annals of the ICRP 30, no. 4 (New York: Pergamon Press, 2000), table 2, p. 20; ICRP, *Radiation and Your Patient: A Guide for Medical Practitioners,* Supporting Guidance 2, Annals of the ICRP, vol. 31, no. 4 (New York: Pergamon Press, 2001) table 2, p. 20; and Stanley H. Stern, 2002. "Nationwide Evaluation of X-Ray Trends 2000–2001: Survey of Patient Radiation Exposure from Computed Tomographic Examinations in the United States," in *Proceedings of Symposium on "Computed Tomography: Patient Dose"* (Bethesda, MD: National Council on Radiation Protection and Measurements, (November 2002), p. 4.

Table 12.9 Based on A. Bouville, S. L. Simon, C. W. Miller, H. L. Beck, L. R. Anspaugh, and B. G. Bennett, "Estimates of Doses from Global Fallout," *Health Physics* 82, no. 5 (May 2002), table 8, p. 699.

Table 13.1 Compiled by author, based primarily on Robert N. Stavins, "Market-Based Environmental Policies," chapter 3 in P. R. Portney and Robert N. Stavins, eds., *Public Policies for Environmental Protection,* 2nd ed. (Washington, DC: Resources for the Future, 2000), pp. 31–76, table 3.1, p. 37, supplemented by information on the "Clear Skies Initiative" from Peter Zaborowsky, "Getting a Head Start: Facilities Should Start Reducing Hazardous Air Emissions Now to Prepare for the Proposed Multi-pollutant Trading Program," *Environmental Protection* 13, no. 4 (April 2002), 26–27, 53.

Table 13.2 Based on information in Bernard L. Cohen, "Society's Valuation of Life Saving in Radiation Protection and Other Contexts," *Health Physics* 38, no. 1 (January 1980), 33–51, table 1, p. 36 (this reference served as a source of the numbers in the categories of "Motor-vehicle safety" and "Medical care"); Evans, John, Jonathan Levy, James Hammit, Carlos Santos-Burgoa, and Margarita Castillejos, "Health Benefits of Air Pollution Control," in Luisa T. Molina and Mario J. Molina, eds., *Air Quality in the Mexico Megacity,* pp. 105–136. (Dordrecht: Kluwer Academic Publishers, 2003) (this reference served as the source for the estimate of the cost per life saved in complying with air-pollution standards in the category on "Environmental protection"); and USNRC, "Numerical Guides for Design Objectives and Limiting Conditions for Operation to Meet the Criterion 'As Low As Is Reasonably Achievable' for Light-Water-Cooled Nuclear Power Reactor Effluents," Code of Federal Regulations, Title 10, Part 50, Appendix I, U.S. Nuclear Regulatory Commission, Washington, DC (this reference served as the source for the estimate of the cost per life saved in preventing thyroid cancers from nuclear power plants in the category on "Environmental protection").

Table 14.1 Compiled by author.

Table 14a Compiled by author.

Table 14.2 Compiled by author, based on Congressional Research Service, "Environmental Laws: Summaries of Statutes Administered by the Environmental Protection Agency," CRS Report to Congress, Order Code RL30798, Washington, DC (4 January 2001); CEQ, *Environmental Quality: The Twenty-fifth Anniversary Report of the Council on Environmental Quality* (Washington, DC: Executive Office of the President, 1997), pp. 94–97; and Richard A. Merrill, "Regulatory Toxicology." In C. D. Klaassen, Mary O. Amdur, and J. Doull, eds., *Casarett and Doull's Toxicology: The Basic Science of Poisons,* 3rd ed. (New York: Macmillan Publishing Company, 1986), pp. 917–932.

Table 14.3 Compiled by author, using information from ATSDR, "A Pledge to Protect Children—The Bangkok Statement," *Public Health and the Environment* 1, no. 1/2 (Fall 2002), 21, Agency for Toxic Substances and Disease Register, Atlanta, GA; Cary Fowler, "Sharing Agriculture's Genetic Bounty," *Science* 297, no. 5579 (12 July 2002), 157 (2002 International Treaty on Plant Genetic Resources); C. Runyan and M. Norderhaug, "The Path to the Johannesburg Summit," *World-Watch* 15, no. 3 (May/June 2002), 30–35 (2002 World Conference on Sustainable Development); and Charles W. Schmidt, "The Down-to-Earth Summit—Lessening Our Ecological Footprint," *Environmental Health Perspectives* 110, no. 11 (November 2002), A 682–A 685 (2002 World Conference on Sustainable Development).

Table 15.1 Compiled by author.

Table 15.2 Compiled by author. Based on data in ICRP, *Recommendations of the International Commission on Radiological Protection,* Publications 1, 9, 26 (New York: Pergamon Press, 1959, 1966, 1977); and idem, 1990 *Recommendations of the International Commission on Radiological Protection,* Publication 60, Annals of the ICRP, vol. 21, no. 1–3 (New York: Pergamon Press, 1991).

Table 15.3 Compiled by author. Based on data in idem, *1990 Recommendations of the International Commission on Radiological Protection,* Publication 60, Annals of the ICRP, vol. 21, no. 1–3 (New York: Pergamon Press, 1991); and National Council on Radiation Protection and Measurements, *Recommendations on Limits for Exposure to Ionizing Radiation,* Report no. 91 (Bethesda, MD, NCRP, 1987).

Table 15.4 Based on data in International Commission on Radiological Protection (ICRP), *1990 Recommendations of the International Commission on Radiological Protection,* ICRP Publication 60, Annals of the ICRP 21, nos. 1–3 (New York: Pergamon Press, 1991).

Table 15.5 Based on National Council on Radiation Protection and Measurements (NCRP), *Operational Radiation Safety Program for Astronauts in Low-Earth*

Orbit: A Basic Framework, Report no. 142 (Bethesda, MD: (December 2002), table 2.4, p. 21.

Table 15.6 Based on data in ICRP, *1990 Recommendations of the International Commission on Radiological Protection,* ICRP Publication 60, Annals of the ICRP 21, nos. 1–3 (New York: Pergamon Press, 1991), table B-18, p. 134.

Table 15.7 Based on Roger H. Clarke, "Control of Low-Level Radiation Exposure: What Is the Problem and How Can It Be Solved?" *Health Physics* 80, no. 4 (April 2001), 391–396, fig. 1, p. 394.

Table 15.8 Based on Roger Clarke, "Radiological Protection at the Start of the 21st Century," *Radiological Protection Bulletin,* no. 231 (September 2001), 10–12, National Radiological Protection Board, Chilton, Didcot, United Kingdom, unnumbered table, p. 12.

Table 16.1 Prepared by the author.

Table 16.2 Prepared by author. Based on data in J. P. Corley, D. H. Denham, R. E. Jaquish, D. E. Michels, A. R. Olsen, and D. A. Waite, *A Guide for Environmental Radiological Surveillance at ERDA Installations,* Report ERDA–77–24 (Springfield, VA: Department of Energy, National Technical Information Service, 1977); and D. W. Moeller, J. M. Selby, D. A. Waite, and J. P. Corley, "Environmental Surveillance for Nuclear Facilities," *Nuclear Safety* 19, no. 2 (January–February 1978), table 3, p. 73.

Table 16.3 Based on Jerry A. Griffith and Carolyn T. Hunsaker, *Ecosystem Monitoring and Ecological Indicators: An Annotated Bibliography,* Report EPA/620/R-94/021 (Washington, DC: EPA, July 1994), p. 5.

Table 16.4 Based on Jerry A. Griffith and Carolyn T. Hunsaker, *Ecosystem Monitoring and Ecological Indicators: An Annotated Bibliography,* Report EPA/620/R-94/021 (Washington, DC: EPA, July 1994),table 1, p. 8.

Table 17.1 Michael Gochfeld and Joanna Burger, "Environmental and Ecological Risk Assessment," in Robert B. Wallace, ed., *Maxcy-Rosenau-Last Public Health and Preventive Medicine,* pp. 435–441 (Stamford, CT: Appleton & Lange, 1998).

Table 17.2 Based on John D. Graham, "Annual Report, 1990," Center for Risk Analysis, Harvard School of Public Health (Boston, 1991), unnumbered table, p. 3.

Table 17.3 Based on John D. Graham, "Annual Report, 1990," Center for Risk Analysis, Harvard School of Public Health (Boston, 1991), unnumbered table, p. 3.

Table 17.4a ATSDR, *The Agency for Toxic Substances and Disease Registry: FY 2002 Profile and Annual Report,* Report ATSDR-PE-RP-2003-0001 (Atlanta, GA: Agency for Toxic Substances and Disease Registry, 2003), table 1, chapter 2, p. 31.

Table 17.4b ATSDR, *The Agency for Toxic Substances and Disease Registry: FY 2002 Profile and Annual Report,* Report ATSDR-PE-RP-2003-0001 (Atlanta, GA: Agency for Toxic Substances and Disease Registry, 2003), chapter 3, text on p. 43.

Table 17.5 Based on Richard Wilson, and Edmund A. C. Crouch, *Risk-Benefit Analysis,* 2nd ed. (Cambridge, MA: Harvard Center For Risk Analysis, 2001), chapter 7, "Lists of Risks," tables 7-4A and 7-4B, pp. 208, 209.

Table 17.6 Based on Bernard L. Cohen, "Catalog of Risks Extended and Updated," *Health Physics* 61, no. 3 (September 1991), tables 3, 7, 8 (pp. 319, 325, 327).

Table 17.7 U.S. Environmental Protection Agency, "Seven Cardinal Rules of Risk Communication," Report 230-K-92-0001 (Washington, DC, 1992).

Table 17.8 Based on John D. Graham, "Annual Report, 1993," Center for Risk Analysis, Harvard School of Public Health (Boston, 1994), inside front cover.

Table 17.9 Based on Dennis J. Paustenbach, "The Practice of Exposure Assessment: A State-of-the-Art Review," in A. Wallace Hayes, ed., *Principles and Methods of Toxicology,* Fourth Edition (Philadelphia, PA: Taylor and Francis, 2001).

Table 18.1 Based on Morton Blatt, "The Energy-Efficient Office," *EPRI Journal* 19, no. 5 (July/August 1994), 16–23.

Table 18.2 Based on Science Concepts, Inc., "The Impact of Nuclear Energy on Utility Fuel Use and Utility Atmospheric Emissions, 1973–1990" (Chevy Chase, MD, December 1991), table 3, p. 14.

Table 18.3 Based on National Council on Radiation Protection and Measurements (NCRP), *Carbon-14 in the Environment,* Report no. 81 (Bethesda, MD: NCRP, 1985), and J. Tichler, K. Norden, and J. Congemi, *Radioactive Materials Released from Nuclear Power Plants: Annual Report, 1987,* Report NUREG/CR-2907, vol. 8 (Washington, DC: U.S. Nuclear Regulatory Commission, 1989).

Table 18.4 Based on data in J. Tichler, K. Norden, and J. Congemi, *Radioactive Materials Released from Nuclear Power Plants: Annual Report, 1987,* Report NUREG/CR–2907, vol. 8 (Washington, DC: U.S. Nuclear Regulatory Commission, 1989); and Ralph Larsen, Senior Scientist, Personal Communication, Environmental Protection Agency, Research Triangle Park, NC (May 2003).

Table 19.1 Compiled by author.

Table 19.2 Adapted from chart provided by Pan American Health Organization, Washington, DC.

Table 19.3 Based on Centers for Disease Control and Prevention (CDC), "Biological and Chemical Terrorism: Strategic Plan for Preparedness and Response: Recommendations of the CDC Strategic Planning Workshop." Agency for Toxic Substances and Disease Registry, *Morbidity and Mortality Weekly Report* 49, no. RR-4 (21 April 2000), box 5, 7–8.

Table 19.4 Based on Centers for Disease Control and Prevention (CDC), "Biological and Chemical Terrorism: Strategic Plan for Preparedness and Response," Recommendations of the CDC Strategic Planning Workshop, Agency for Toxic Substances and Disease Registry, *Morbidity and Mortality Weekly Report* 49, no. RR-4 (21 April 2000), box 3, 5–6.

Table 19.5 Compiled by author.

Table 19.6 Based on PAHO, *Humanitarian Assistance in Disaster Situations—A Quick Guide for Effective Donations* (Washington, DC: Pan American Health Organization, 2002).

Table 19.7 Based on PAHO, *Humanitarian Assistance in Disaster Situations—A Quick Guide for Effective Donations* (Washington, DC: Pan American Health Organization, 2002.

Table 19.8 Compiled by author.

Table 20.1 Based on CEQ, *Environmental Quality: The Twenty-fifth Anniversary Report of the Council on Environmental Quality* (Washington, DC: Executive Office of the President, 1997), chapter 7, pp. 129–131.

Table 20.2 Based on Alvin Powell, "Agreeing on What to Argue About: Environmental Report Tries to Make Sense of Flood of Data," *Harvard University Gazette* 98, no. 2 (25 September 2000), 27–28.

Table 20.3 Based on Alvin Powell, "Agreeing on What to Argue About: Environmental Report Tries to Make Sense of Flood of Data," *Harvard University Gazette* 98 no. 2 (25 September 2000), 27–28.

Table 20.4 Compiled by author.

Table 20.5 Based on National Research Council (NRC), *Grand Challenges in Environmental Sciences* (Washington, DC: National Academy Press, 2001).

Figures

Figure 1.1 *World Almanac* 2000. (Mahwah, NJ: World Almanac Books, 2001) (pp. 878–879).

Figure 1.2 Prepared by author.

Figure 1.3 Wayne R. Ott, "Total Human Exposure: Basic Concepts, EPA Field Studies, and Future Research," *Journal of the Air and Waste Management Association* 40, no. 7 (July 1990), fig. 1, p. 968.

Figure 1.4 Based on information in Curtis Runyan, "Human Motion, Still Pictures," *World-Watch* 13, no. 5 (September/October 2000), 36–40, unnumbered table, p. 38.

Figure 1.5 Based on R. D. Morris and W. R. Hendee, "Environmental Stewardship: Exploring the Implications for Health Professionals," draft report, Medical College of Wisconsin, Milwaukee, 1992, figs. 1a–1c, pp. 3–5.

Figure 2.1 T. A. Loomis, *Essentials of Toxicology,* 3rd ed. (Philadelphia: Lea & Febiger, 1968), fig. 2.1, p. 15.

Figure 2.2 Based on T. A. Loomis, *Essentials of Toxicology,* 3rd ed. (Philadelphia: Lea & Febiger, 1968), fig. 2.2, p. 16.

Figure 2.3 Edward J. Calabrese, Linda A. Baldwin, and Charles D. Holland, "Hormesis: A Highly Generalizable and Reproducible Phenomenon with Important Implications for Risk Assessment," *Risk Analysis* 19, no. 2 (1999), 261–281, fig. 1.

Figure 3.1 National Research Council, *Environmental Epidemiology: Public Health and Hazardous Wastes* (Washington, DC: National Academy Press, 1991), fig. 3–5, p. 122. Copyright 1990 by the American Chemical Society.

Figure 3.2 National Research Council, *Environmental Epidemiology: Public Health and Hazardous Wastes* (Washington, DC: National Academy Press, 1991), fig. 7.1, p. 221.

Figure 4.1 Based on National Safety Council, *Injury Facts, 2001 Edition* (Itasca, IL: National Safety Council, 2001), unnumbered figure, page 47, with modifications from *Morbidity and Mortality Weekly Report,* Centers for Disease Control and Prevention, 47, no. 15 (24 April 1998).

Figure 4.2 Adapted from Barry S. Levy and David H. Wegman, eds., *Occupational Health: Recognizing and Preventing Work-Related Disease,* 2nd ed. (Boston: Little, Brown, 1988), fig. 22.6, p. 21.

Figure 4.3 Prepared by author based on common knowledge.

Figure 4.4 Adapted from American Conference of Governmental Industrial Hygienists, *Industrial Ventilation: A Manual of Recommended Practice,* 20th ed. (Lansing, MI: Committee on Industrial Ventilation, 1988), fig. VS-202, pp. 5–21.

Figure 4.5 Based on Karen Springen, "The Dangerous Desk," *Newsweek* 137, no. 13 (26 March 2001), 66–67, unnumbered figure, p. 67.

Figure 5.1 Richard Wilson and Edmund A. C. Crouch, *Risk-Benefit Analysis,* 2nd ed. (Cambridge, MA: Harvard Center for Risk Analysis, 2001), fig. 2–2, p. 31. Cited as being taken from Beaver, 1953.

Figure 5.2 Adapted from National Council on Radiation Protection and Measurements, *Deposition, Retention, and Dosimetry of Inhaled Radioactive Substances,* Report no. 125 (Bethesda, MD: NCRP, 1997), fig. 9–1, p. 155.

Figure 6.1 Adapted from F. L. Bryan, *Foodborne Diseases and Their Control* (Atlanta, GA: Centers for Disease Control and Prevention, U.S. Department of Health and Human Services, 1980), fig. 4–1, p. 2.

Figure 6.2 Walter Willett, *Eat, Drink, and Be Healthy* (New York: Simon & Schuster, 2001), fig. 2, p. 17. Reprinted with the permission of Simon & Schuster Adult Publishing Group from *Eat, Drink, and be Healthy: The Harvard Medical School Guide to Healthy Eating* by Walter C. Willett. Copyright © 2001 by President and Fellows of Harvard College.

Figure 7.1 Richard Wilson and Edmund A. C. Crouch, *Risk-Benefit Analysis,* 2nd ed. (Cambridge, MA: Harvard Center for Risk Analysis, 2001), fig. 4.6, p. 119. These authors cite the figure as being taken from G. A. Condran and R. A. Cheney, "Mortality Trends in Philadelphia: Age- and Cause-Specific Death Rates, 1870–1930," *Demography* 19, no. 1 (1982), 97–123.

Figure 7.2 CEQ, *Environmental Quality 23rd Annual Report, Council on Environmental Quality,* (Washington, DC: Government Printing Office, 1993), unnumbered figure, p. 226.

Figure 7.3 Roger M. Waller, *Ground Water and the Rural Home Owner* (Washington, DC: U.S. Geological Survey, 1989), unnumbered figure, p. 13.

Figure 7.4 Drawn by author, based on Sandra Postel, *The Last Oasis: Facing Water Scarcity,* World-Watch Environmental Alert Series (Washington, DC: World-Watch Institute, 1997).

Figure 7.5 Developed by author.

Figure 7.6 American Association of Vocational Instructional Materials, *Planning for an Individual Water System* (Athens, GA: AAVIM, 1973), fig. 48, p. 58.

Figure 8.1 Developed by author.

Figure 8.2 Developed by author.

Figure 8.3 CEQ, *23rd Annual Report, Council on Environmental Quality* (Washington, DC: Government Printing Office, 1993), unnumbered figure, p. 5.

Figure 8.4 Developed by author.

Figure 8.5 Developed by author.

Figure 8.6 Developed by author.

Figure 9.1 Washington State Department of Ecology, *Solid Waste Landfill Design Manual* (Olympia: Washington State Department of Ecology, 1987), fig. 4.26.

Figure 9.2 Based on material provided by EPA staff but modified by the author.

Figure 9.3 Based on U.S. Nuclear Regulatory Commission, *Information Digest: 2001 Edition* NUREG-1350, vol. 13 (Washington, DC: U.S. Government Printing Office, 2001), fig. 36, p. 76.

Figure 9.4 Adapted from U.S. Nuclear Regulatory Commission, *Recommendations to the NRC for Review Criteria for Alternative Methods for Low-Level Radioactive Waste Disposal,* NUREG/CR-5041, vol. 1 (Washington, DC: U.S. Government Printing Office, 1987), fig. 2.8.1, p. 2.8.2.

Figure 10.1 "Publications for Rat Control and Prevention Programs," *Public Health Reports* 80, no. 1 (January 1970), p. 40.

Figure 11.1 Adapted from Insurance Institute for Highway Safety, *Twenty Years of Accomplishment by the Insurance Institute for Highway Safety* (Arlington, VA: Insurance Institute for Highway Safety, 1989), p. 2.

Figure 11.2 Based on National Safety Council, *Injury Facts, 2001 Edition* (Itasca, IL: National Safety Council, 2001), unnumbered table, "Motor Vehicle Deaths and Rates, United States, 1913–2000," pp. 108–109.

Figure 11.3 Insurance Institute for Highway Safety, *IIHS Status Report* 28, no. 9 (24 July 1993), p. 3.

Figure 11.4 Courtney Humphries, "Injury Control: Child Firearm Deaths Tied to Gun Availability," *Focus: News of Harvard Medical, Dental, and Public Health Schools,* Boston, MA (8 March 2002), 1, 3, 5.

Figure 12.1 Prepared by author.

Figure 12.2 Prepared by author.

Figure 12.3 NCRP, *Screening of Humans for Security Purposes Using Ionizing Radiation Scanning Systems,* NCRP Commentary no. 16 (Bethesda, MD: National Council on Radiation Protection and Measurements, 2003), fig. 5.1, p. 18.

Figure 12.4 Prepared by author. Based on data in Donald T. Oakley, *Natural Radiation Exposure in the United States,* Report no. ORP/SID–72–1, U.S. Environmental Protection Agency (Washington, DC: U.S. Government Printing Office, 1972).

Figure 12.5 Prepared by author.

Figure 12.6 U.S. Nuclear Regulatory Commission, *Information Digest: 2001 Edition,* Report NUREG-1350, volume 13 (Washington, DC: Government Printing Office, 2001), fig. 17, "U.S. Commercial Nuclear Power Reactors," p. 40.

Figure 12.7 National Council on Radiation Protection and Measurements, *Ionizing Radiation Exposure of the Population of the United States,* Report no. 93 (Bethesda, MD: 1987), fig. 8.2, p. 55.

Figure 13.1 From Matthew Edel, *Economies and the Environment* 1973), chapter 5, "Environmental Fine Tuning," fig. 5.1, p. 87, "Total, marginal, and average costs and benefits of combustion." © reprinted by permission of Pearson Education, Inc., Upper Saddle River, NJ.

Figure 13.2 A. Myrick Freeman III, "Water Pollution Policy," in Paul R. Portney, and Robert N. Stavins, ed., *Public Policies for Environmental Protection,* 2nd ed. (Washington, DC: Resources for the Future, 2000), chapter 6, fig. 6.1, "Tasks Involved in Estimating the Benefits of Water Pollution-Control Policies," p. 192.

Figure 14.1 Drawn by the author. Adapted from J. H. Ausubel and H. E. Sladovich, eds., *Technology and Environment,* National Academy of Engineering (Washington, DC: National Academy Press, 1989), fig. 2, p. 101.

Figure 15.1 Prepared by the author. Based on data in Lauriston S. Taylor, "History of The International Commission on Radiological Protection (ICRP)," *Health Physics* 1, no. 2 (1959), 97–104; and ICRP, 1990 *Recommendations of the International Commission on Radiological Protection,* Publication 60, Annals of the ICRP, Vol. 21, no. 1–3 (New York: Pergamon Press, 1991).

Figure 15.2 Based on data in D. C. Kocher, "Review of Radiation Protection and Environmental Radiation Standards for the Public," *Nuclear Safety* 29, no. 4 (October–December 1988), 463–475.

Figure 16.1 D. W. Moeller, J. M. Moeller, D. A. Waite, and J. P. Corley, "Environmental Surveillance for Nuclear Facilities," *Nuclear Safety* 19, no. 1 (January–February, 1978), 66–79.

Figure 16.2 International Commission on Radiological Protection, *Principles of Monitoring for the Radiation Protection of the Population,* Publication 43, Annals of the ICRP, vol. 15, no. 1 (New York: Pergamon Press, 1985), fig. 3, p. 16.

Figure 16.3 Based on World Health Organization, "Guidelines on Studies in Environmental Epidemiology," *Environmental Health Criteria* 27 (Geneva, 1983), fig. 3.1, p. 109.

Figure 16.4 International Commission on Radiological Protection, *Principles of Monitoring for the Radiation Protection of the Population,* Publication 43, Annals of the ICRP, vol. 15, no. 1 (New York: Pergamon Press, 1985), fig. 2, p. 8.

Figure 16.5 Drawn by author.

Figure 17.1 U.S. Environmental Protection Agency, *Risk Assessment for Toxic Air Pollutants—A Citizen's Guide,* Report 450/3-90-024 (Research Triangle Park, NC, 1991), unnumbered figures, p. 8.

Figure 17.2 Richard Wilson and Edmund A. C. Crouch, *Risk-Benefit Analysis,* 2nd ed. (Cambridge, MA: Harvard Center for Risk Analysis, 2001), chapter 6, "Risk Management: Managing and Reducing Risks," fig. 6-1, p. 151.

Figure 18.1 Drawn by author, based on Marianne Lavelle, "Living without Oil," *U.S. News and World Report* 134, no. 5 (17 February 2003), 32–39.

Figure 18.2 Prepared by author.

Figure 18.3 Prepared by author. Based on data provided by the Council on Energy Awareness, Washington, DC, and on data in Council on Environmental Quality, *Environmental Trends* (Washington, DC: Executive Office of the President, 1989), p. 15.

Figure 18.4 Council on Energy Awareness, Washington, DC.

Figure 18.5 From USNRC, *Information Digest: 2001 Edition,* Report NUREG-1350, volume 13 (Washington, DC: U.S. Government Printing Office, 2001), fig. 7, "U.S. Electric Net Generation by Energy Source, 1999," p. 21.

Figure 19.1 D. W. Moeller, "A Review of Countermeasures for Radionuclide Releases," *Radiation Protection Management* 3, no. 5 (October 1986), fig. 1, p. 72.

Figure 19.2 Update of figure published in Atomic Energy Commission, *Radioactive Iodine in the Problem of Radiation Safety* (translated from USSR report) (Washington, DC: 1972). The revision was based on information in National Council on

Radiation Protection and Measurements, *Management of Terrorist Events Involving Radioactive Material,* Report no. 138 (Bethesda, MD: NCRP, 2001), section 4.4.4, p. 48.

Figure 19.3 Federal Radiation Council, *Background Material for the Development of Radiation Protection Standards: Protective Action Guides for Strontium-89, Strontium-90, and Cesium-137,* Report no. 7 (Washington, DC: U.S. Government Printing Office, 1965), table 2, p. 5.

INDEX